$$1 \text{ gallon} = 0.1337 ft^3$$
$$7.479 \text{ gall} = 1 ft^3$$

$$Hp = \frac{Q\gamma H}{550}$$

APPLIED HYDRAULICS IN ENGINEERING

HENRY M. MORRIS

JAMES M. WIGGERT

VIRGINIA POLYTECHNIC INSTITUTE
AND STATE UNIVERSITY

SECOND EDITION

JOHN WILEY & SONS
New York • Chichester • Brisbane • Toronto • Singapore

ISBN 0 471 06669-9

Library of Congress Catalog Card Number: 77–163950

PRINTED IN THE UNITED STATES OF AMERICA

20 19 18 17 16 15 14 13 12 11

PREFACE

The engineering development and conservation of the earth's water resources constitutes a field of great challenge and fascinating interest. The past decade has witnessed many advances in this field, as well as a significant upsurge of public awareness and concern.

The most essential tool for effective work in this discipline is a broad understanding of the principles of engineering hydraulics and hydrology, as applied in the design of structures and systems for hydraulic developments. To provide an introduction to this important and comprehensive subject is the purpose of this textbook. Since the number of interesting and worthwhile topics in hydraulics is almost inexhaustible, careful selection is required. Those included are thought to be of interest to engineers, geologists, regional planners, and others concerned with water planning, control, and utilization.

In this Second Edition, two new chapters have been added, one on engineering hydrology and one on groundwater hydraulics. In addition, new material has been incorporated into each of the other chapters. The book is suitable for both an introductory course in hydraulic engineering for juniors or seniors, and also for a sequence of one or more elective or graduate courses in special topics in applied hydraulics. It may also serve as a reference for self-study purposes for practicing engineers.

We express our thanks to the many users of the book, both in college and in practice, who have made suggestions for this Second Edition. We have tried to incorporate as many of these as feasible. The revisions and additions to previous chapters were typed by Mrs. I. H. Morris, and the two new chapters by Mrs. Bonnie N. Gore. Mrs. Naomi Wiggert lettered the Instructor's Supplement.

It is our hope that students will find the book adequately rigorous and yet clearly understandable and practical. We also hope that many will find engineering hydraulics to be such a highly interesting and rewarding study that they will be challenged to participate in the meeting of these vital human needs through a career in this field.

HENRY M. MORRIS
JAMES M. WIGGERT

Blacksburg, Virginia
October , 1971

CONTENTS

APPLIED HYDRAULICS
IN ENGINEERING

1

INTRODUCTION

1-1. Definition and Scope. The term *hydraulics* (from Greek *hydor*, "water") is the study of the mechanical behavior of water in physical systems and processes. By extension it has come to include in some contexts the study of other fluids, but the term *fluid mechanics* is now more commonly used for this purpose.

Hydraulics includes *hydrostatics*, the science of fluids at rest, and *hydrodynamics*, the science of moving fluids. The latter term is often limited to the study of an "ideal" fluid, with no viscosity, elasticity, or surface tension, although its root meaning (from *hydor* and *dynamis*—Greek for "power") is the study of water power. In any case the discipline of ideal-fluid hydrodynamics is now usually considered to have been united with engineering hydraulics in the broad field of fluid mechanics. The latter includes the study of fluids other than water, of course, especially air and the petrochemical fluids.

Nevertheless, hydraulics, as defined above, remains a highly important field in its own right. Water is still by far the most important of all fluids in earth processes and human life. The design of structures and other systems and processes for its conservation and use is the field of *hydraulic engineering*, or *applied hydraulics*.

Since such systems are normally large and intricate engineering works, the field of hydraulic engineering is normally considered as one of the major branches of civil engineering. Other engineers and scientists—especially agricultural engineers and geologists—are also associated with many hydraulic engineering projects.

At least 20 per cent of all civil engineers are working primarily in the field of hydraulic engineering, and almost all civil engineers are involved therein to some extent. The extent of research activity in the field is indicated by the fact that, typically, about 35 per cent of the articles published each year in the *A.S.C.E. Transactions* relate in some definite way to basic or applied hydraulics.

The field of hydraulic engineering may be further subdivided, more or less arbitrarily, into several divisional specialities. Some of these are: *applied hydrology*, the engineering application of the principles of hydrology, in-

3

cluding hydrometeorology, groundwater development, river forecasting, and urban hydrology; *irrigation and drainage engineering*, the planning and implementation of structures and systems for irrigation developments and surface and subsurface drainage; *waterway engineering*, the design of port and harbor facilities, and facilities for development and maintenance of inland waterways; *hydropower engineering*, the development of hydro-electric power through reservoir systems, turbines, and appurtenant facilities; *flood and sediment control*, the planning and construction of facilities for prevention of excessive damage by water in flood stages; *pipe-line engineering*, the transportation of water, oil, gas, and other fluids and mixtures over long distances through pressure conduits; *coastal hydraulics*, the design of coastal structures for harbor development, prevention of beach erosion, off-shore facilities, and similar uses; *water-resources engineering*, the basin-wide planning of reservoir systems and other facilities for optimum utilization of the over-all water resources of a region; *water supply and wastewater engineering*, that component of sanitary engineering concerned with collection and distribution systems for water for various uses, and systems for treatment and dispersion of waste waters after use. In addition, many hydraulic engineers are primarily involved in fundamental or applied research on hydraulic processes for one or more of these fields, and many are engaged in teaching hydraulics and fluid mechanics.

These and other fields involve a tremendous variety of hydraulic phenomena and hydraulic structures, machines, and systems. However, the same underlying principles of flow behavior can be applied to all of them. Thus the study of engineering hydraulics is not merely a collection of specialized design methods and details for a great number of particular applications, but rather is a study of general principles which, when supported by pertinent experimental data, can be applied to specific applications as needed.

However, these basic hydraulic phenomena are themselves so complex that completely rational solutions to specific problems are usually impossible. Therefore, hydraulic engineering necessarily involves much empiricism, but it is a scientifically controlled and rational empiricism which has proved very effective and powerful in the development of adequate and reliable data for engineering design and operation.

I-2. History of Hydraulic Engineering. Since water in its various aspects is of prime importance to man, the development of measures for its control and use was undoubtedly among the earliest human arts. In this sense, hydraulic engineering has been practiced since the beginning of recorded history. Large irrigation and drainage works, with dams, canals, aqueducts, sewers, etc., were built by the ancient Egyptians and Babylonians about 2500 B.C. or earlier. Not only were the early hydraulic engineers concerned with water supply and flood control, but with navigation as well. The

Egyptians, for example, constructed a canal 100 miles long connecting the Mediterranean and the Red Sea, which was in use for over a thousand years.

The great Marib masonry dam in the kingdom of Sheba (now Yemen), in Southern Arabia, supported an advanced irrigation economy for 1400 years until the failure of the dam in about A.D. 550. In the desert regions of the Negev, in Israel, in spite of an average rainfall of about 4 inches per year, a complex network of dams and canals provided water for a flourishing population for many hundreds of years. In one area here, of about 50 sq mi, at least 17,000 ancient dams have been discovered, most of them very small but nonetheless effective.

Even earlier than these—in fact before recorded history—such items as weapons (arrows, spears, stones, etc., with their implied experimental solution to problems of air resistance) and boats (with their problems of stability, streamlining, and wind propulsion) must have been developed. Sailboats, for example, are pictured in the earliest Egyptian monuments. The prototype of all sea-going vessels must have been the ark of Noah. So numerous are the prehistoric legends of the great Flood and of the ark, from nations and tribes all over the world, and so persistent the tales that the ark is still preserved in the perennial ice cover now topping Mount Ararat in Armenia, that it seems certain that this great prehistoric vessel must actually have existed.

Evidences of prehistoric navigation waterways and irrigation works are found in all ancient civilizations, especially Egypt and Babylonia. As a matter of fact, the very development of any type of civilization almost certainly presupposes the prior development of shipping and a reliable water supply for both agricultural and domestic uses. Although the engineering calculations—designing these ancient hydraulic works—have been lost, the works themselves testify to a high degree of at least empirical knowledge of hydraulics among prehistoric communities.

According to tradition, Nimrod, mentioned in the Bible as the founder of Nineveh and first king of Babylon, as well as the first important rebel against divine authority and revelation after the Deluge, built a massive earth dam across the Tigris River, diverting the flow into an irrigation network. The greatest of the canals was the Nahrwan Canal, still identifiable today, stretching for some 250 miles. Its first 9 miles was a steep chute, 50 ft deep and 65 ft wide. Then it broadened out into a 400-ft-wide channel, 15 ft deep.

Similarly, in Egypt the great Menes, founder of the first dynasty, is said to have built a masonry dam across the Nile at Memphis, together with canals and other works. These provided the foundation of the irrigation economy which sustained Egypt as one of the world's great empires for three thousand years.

There are also evidences of prehistoric irrigation systems in India, Iran, China, and other ancient nations. Not only dams and canals, but wells (some of them up to a thousand feet deep), various ingenious pumping and control devices, paddlewheels, windmills, and aqueducts were constructed in remote antiquity.

Once into the historic period, the investigator encounters an abundance of hydraulic structures and systems, all over the world. Extensive tunneling and drainage projects were undertaken by the Greeks. The Persians, about 1000 B.C., developed an extensive system of "qanats," or underground channels, which carried water from aquifers in the highlands by gravity flow to surface canals in the lowlands. These remarkable channels, which are still in common use after 3000 years, were copied by the later Assyrians, Romans, and others. About 750 B.C. King Hezekiah built a dam and conduit to bring water into Jerusalem; these too are still in existence.

In Iraq an ancient canal system has recently been unearthed which is believed to date from the earliest years of the period of the Sumerians, the first civilization in Babylonia. This was a very complex network, roughly paralleling the Euphrates River, and was used both for irrigation and navigation.

Similarly the ancient Egyptians, at about the same period, had built a dual-purpose navigation and irrigation canal from the Mediterranean to the Red Sea. About 1400 B.C. a similar canal was built from the Nile River to the Red Sea. These and other Egyptian canals apparently went through several cycles of silting-up and reconstruction before the present Suez Canal was built in the period 1859–69.

Canals of course require dams and various control facilities. Both masonry and earth dams have been used from the earliest historical periods in both Egypt and Babylonia. Some of the remains of a rubble masonry dam erected about 2500 B.C. or earlier can still be seen in the Wadi el-Garawi, about 18 miles south of Cairo, Egypt.

In the great Nahrwan canal system of Mesopotamia, traditionally attributed to Nimrod, traces of the masonry headworks and weirs, as well as the main canal itself, can still be seen. The Indians, Persians, Chinese, and others also constructed dams and canals in these early periods.

One of the most interesting ancient hydraulic projects was the Bhojpur Reservoir in north-central India, built around A.D. 1100, covering about 250 sq mi of surface area. Two large dams were necessary, both built with a core of earth and masonry faces. A protective spillway for the dams was cut out of solid rock two miles away from the larger dam. Water marks on the spillway crest indicate that the reservoir level came within 6 ft of the dam crests, but the spillway adequately prevented overtopping for 500 years, at which time the dams were torn down in order to open up the inundated valley for agriculture.

Conveyance of water in pipe lines is also an ancient art. The Chinese are said to have piped water through bamboo pipes as early as 2500 B.C. In 900 B.C. they used pipe lines to convey natural gas to brine evaporators. By about 200 B.C. the Romans had developed a water supply piping system that handled over 300 million gallons per day, using lead and bronze, as well as masonry for their pipes.

The Roman hydraulic works were justifiably famous and contributed substantially to the long supremacy of the Roman Empire. The great aqueducts were among the wonders of the world, some of them still in use today. Pipe and channel sizes seem to have been designed rather economically, in accordance with modern practice.

The treatises of Vitruvius and Frontinus, two outstanding Roman engineers, have come down to give something of an insight into the technology of the day. The designs, not only of the large tunnels and flumes, but also of fountains, distribution tanks, pumps, sand traps, and other devices, are discussed.

The largest tunneling undertaking of ancient times was the work of the emperor Claudius in draining Lake Fucinus. The tunnel was $3\frac{1}{2}$ miles long. Rome's harbor at Ostia was another great hydraulic engineering accomplishment.

How much of these hydraulic projects of antiquity was designed on the basis of rational understanding of the phenomena and how much on the basis of intuition and experience is uncertain. At any rate they were undoubtedly useful and effective for their purposes.

Perhaps the earliest known truly scientific contributions in the field of hydraulics, at least as known to us today, were those of the Greek Archimedes (287–212 B.C.). In approximately 250 B.C., Archimedes published a remarkable two-volume work on hydrostatics and flotation, in which he set forth in truly scientific fashion the laws of buoyancy and floating objects. Archimedes also developed the screw pump and other devices, but is justifiably given greater recognition as the father of hydrostatics.

Hero of Alexandria, sometime after 150 B.C., designed fountains and pumps of various types, many of which continued to be used in modern times. He was probably the first to recognize the relation between area, velocity, and quantity of flow, although later hydraulicians seem to have forgotten this discovery.

After the Romans, the Western world suffered a long period of intellectual and technological darkness. The Islamic nations were progressing, however, and the Moors in southern Spain introduced dam-building and other hydraulic arts into southern Europe during this period. As a result, the Spaniards were long superior to their European contemporaries in these fields, and carried much of their art to their colonies in the New World, where their early dams and irrigation systems continue to be used in many places.

Despite the remarkable hydraulic engineering accomplishments of the ancients, however, they apparently had very little understanding of the basic fluid mechanics involved. The Romans, for example, were unfamiliar even with such a basic concept as that of velocity. In fact, the whole science of kinematics and dynamics could hardly be developed until more accurate means of measuring short time periods became available.

However, the fall of Rome did stimulate the widespread development and use of waterwheels and windmills as a source of power, since the system of slave labor was no longer available. Some further developments could be made in hydrostatics. Stevin (1548–1620) codified these developments and seems to have been the first to recognize the basic hydrostatic equation, publishing an extensive work on hydrostatics in 1586.

Leonardo da Vinci (1452–1519) introduced the Renaissance period probably more influentially than any other single individual. His activities and inventions were wide-ranging, of course, but in hydraulics he made important observations on open-channel flow, on relative motion of fluids and immersed objects, on vortex motion, on wave phenomena, on the hydraulic pump, and varied other phenomena. He was evidently the first man, since Hero, to begin to understand the principle of continuity. This principle was finally clearly stated and elaborated by Benedetto Castelli (1577–1643), a pupil of Galileo's.

Sir Isaac Newton (1642–1727) made many studies in hydraulics and, in particular, formulated the law of viscous flow which still bears his name. His most important contributions, however, were his development of the calculus, formulation of the laws of motion, which later led to the recognition of the fundamental conservation principles, and establishment of the gravitation concept, all of which were essential for the future development of hydraulics.

With the Renaissance and the rapid advances in mechanics of that period, the *science* of hydraulics began to develop. Galileo (1564–1642) discovered the law of falling bodies, and his pupil, Torricelli (1608–1648), about 1640 applied it to a liquid, developing his famous theorem on flow through orifices and leading to the very important concept of velocity *head*. In 1732, Henri Pitot (1695–1771) invented his *Pitot tube* for measuring this head. In 1738, Daniel Bernoulli published the tremendously significant equation bearing his name.

The advent of Bernoulli brings us to the period of classical hydrodynamics. Before this, however, the French scientist, Blaise Pascal (1623–1662) brought to essential completion the science of hydrostatics. He is especially recognized for the principle that bears his name, to the effect that a pressure applied at any point in a static fluid will increase the pressure by that amount at all other points in the fluid.

Pascal's law served as the basis for the practical development of the hydraulic press by Joseph Bramah (1748–1814) in 1795. This foreshadowed the extensive applications today of hydraulic systems of many kinds for the transmission of power (hydraulic jacks, hydraulic brakes, hydraulic servomechanisms, etc.).

From this point on until the twentieth century, further development diverged in two independent directions, one mathematical and one empirical. The Swiss mathematician, Leonhard Euler, about 1755, founded the study of the mathematics of flow and force systems in an ideal, frictionless fluid, a study now known as *classical hydrodynamics*. This discipline of mathematics has been highly developed since that time, especially by such men as Helmholtz, Kelvin, Lamb, and Rayleigh. The practical flow problems encountered in engineering, however, were not amenable to this kind of analysis, since the neglected effects of friction and turbulence are actually of primary importance in real fluids.

Accordingly, a parallel development rapidly took place in the synthetic science of hydraulics. Large testing programs were set up, extensive field measurements made, and empirical formulas adapted to fit the actual data.

As far as the theoretical developments in mathematical hydrodynamics are concerned, opinion is somewhat divided as to whether Bernoulli or Euler should be recognized as the "father of hydrodynamics." Each made outstanding contributions, of fundamental significance in both applied hydraulics and theoretical hydrodynamics.

Daniel Bernoulli (1700–1782) was the most noteworthy member of an outstanding family, several of whom made significant studies in hydraulics. His father John Bernoulli (1667–1748) was, in fact, actually a rival of Daniel for recognition as the leading hydraulician of his day, and each published noteworthy volumes and numerous original contributions. It is generally recognized, however, that Daniel deserves credit for the famous Bernoulli theorem. Both of the Bernoullis shared in the development of the correct concept of fluid pressure.

Jean d'Alembert (1717–1783) was apparently the first man to develop potential theory and the use of complex variables in hydrodynamic analysis. He wrote extensively on many aspects of mechanics and the hydrodynamics of an ideal fluid.

It remained, however, for Leonhard Euler (1707–1783) to synthesize the contributions of Bernoulli, d'Alembert and other hydrodynamicists, as well as the hydrostatics of Pascal, along with his own vast contributions, into an elegant treatise published in 1755. This publication crystallized the discipline of classical hydrodynamics in essentially its permanent basic form.

Euler was followed by Joseph Louis LaGrange (1736–1813), who also made significant discoveries in the dynamics of shallow water waves, as did Pierre LaPlace (1749–1827). Franz von Gerstner (1756–1832) contributed to the theory of deep-water waves.

During the nineteenth century, a long succession of notable hydrodynamicists appeared, too many for individual recognition here. Like Euler, Bernoulli, and their predecessors, many of these men also made important contributions to fields other than hydrodynamics. A few representative names in this succession are those of George Green (1793–1841); Louis Navier (1785–1836), responsible with Sir George Stokes (1819–1903) for the standard equations of motion for a viscous fluid; Sir George Airy (1801–1892), who formulated the most important treatment of wave motion of the century; Hermann von Helmholtz (1821–1894), whose studies on vortex motion, free streamlines, dimensional analysis and other topics were of real importance; Lord Kelvin (1824–1907), who greatly advanced hydrodynamic theory in many fields, but whose greatest contribution perhaps was the codification and clarification of the first and second laws of thermodynamics; Lord Rayleigh (1842–1919), who first popularized the principles of similitude and dimensional analysis; Nikolai Joukowsky (1847–1921), who made foundational studies in aerodynamics and in the study of water hammer; and Sir Horace Lamb (1849–1934), whose famous textbook *Hydrodynamics* has served as the outstanding reference on this subject for the past century, down to the present day.

The developments in classical hydrodynamics took place more or less independently of the studies in experimental hydraulics, which also were increasing rapidly during the eighteenth and nineteenth centuries. The stimulus to hydraulics imparted by da Vinci, Galileo, and their followers maintained for Italy a lead in this field until the advent of a long line of notable French hydraulicians. One of the first of these was Henri Pitot, already mentioned. Another was Antoine Chezy (1718–1798), whose studies on hydraulic resistance led to the famous Chezy formula for flow in open channels, still definitive today although its importance was not adequately recognized until long after his death.

Jean Borda (1733–1799) clarified the problem of efflux through orifices and tubes. He was apparently the first to use the factor "$2g$" explicitly in hydraulic formulas.

A man of great influence was Pierre Louis Georges Du Buat (1734–1809), who attempted to assimilate all previous data on water flow, together with many experimental data of his own, into a systematic science and textbook of hydraulics. His work was of tremendous influence on the practice and further development of applied hydraulics throughout the nineteenth century. Other famous French hydraulicians included: Baron Riche de Prony (1755–1839); Jean Baptiste Belanger (1789–1874), best known for his analysis of backwater; Benoit Fourneyron (1802–1867), who developed the first practical hydraulic turbine; Gaspard de Coriolis (1792–1843), recognized for his studies on velocity distributions and the so-called Coriolis force in rotating systems; Jean Louis Poiseuille (1799–1869), a physician who shared in the development of the equation for laminar flow which bears his name; Jean de Saint-Venant (1797–1886) who wrote extensively in many fields, and made important contributions to open channel studies; Arsene Dupuit (1804–1866), best known for his formulas in well hydraulics; Emmanuel Boudin (1820–1893), who developed the present-day classification system of open-channel flow profiles; Antoine Charles Bresse (1822–1883), who made important studies on the calculation of flow profiles; and Henri Darcy (1803–1858).

Darcy's contributions were especially important in two fields. His law of pipe resistance, derived experimentally by him through extensive studies on pipes, is recognized today as essentially a basic dimensional equation applicable to any flow process. His studies on filter beds led to the Darcy law for flow through porous media.

Paul du Boys (1847–1924) made studies on bed movement in canals and rivers which are still definitive. Darcy's assistant, Henri Emile Bazin (1829–1917), contributed highly important data on flow through channels and weirs.

Although the hydraulic engineers of France dominated the scene during most of the eighteenth and nineteenth centuries, a number of men in other countries made noteworthy contributions. In England, John Smeaton (1724–1792), widely known as the first man to call himself a "civil engineer," was extremely active in the design of hydraulic works of many kinds. He invented the hydraulic ram and made hydraulic model experiments of water wheels and windmills of various designs.

The Italian physicist, Giovanni Venturi (1746–1822) studied the effect of changes in pipe and channel cross-sections on pressures and flow profiles. The Venturi meter, however, was first developed by an American engineer, Clemens Herschel (1842–1930) a century later. Giorgio Bidone (1781–1839) was the first to make a thorough analysis of the hydraulic jump and the bore in open channels. A Scottish engineer, John Scott Russell (1808–1882) pioneered in the study of channel waves and in the wave resistance of ships.

A German hydraulic engineer of this period, G. H. L. Hagen (1797–1884), made extensive tests on flow resistance in pipes and first noted the basic differences in laminar and turbulent flow friction relationships. Another German engineer, Julius Weisbach (1806–1871) was of great influence in codifying hydraulic theory and practice in an outstanding textbook which set the pattern for most hydraulics texts well into the twentieth century. He first wrote the Darcy head-loss equation and the Bernoulli equation in the form in which they are commonly used today by hydraulic engineers.

In America, significant contributions were made by Clemens Herschel, already mentioned, and by J. B. Francis (1815–1892). The latter's name is especially associated today with the Francis turbine and the Francis weir formula, but he made many other important hydraulic studies in his laboratories at Lowell, Massachusetts.

The widely-used Ganguillet-Kutter formula for uniform flow in open channels was developed by two Swiss engineers, Emile Ganguillet (1818–1894) and Wilhelm Kutter (1818–1888), largely on the basis of measurements made on the Mississippi River by U.S. Army Engineers. Their formula, though still used, has now been widely replaced by a formula attributed to an Irish engineer, Robert Manning (1816–1897). Apparently several others, including Hagen and P. G. Gauckler (1826–1905), a French engineer, had proposed the same formula as that of Manning somewhat earlier.

The last half of the nineteenth century saw the critical contributions of Osborne Reynolds (1842–1912) and William Froude (1810–1879), both in England. Reynolds developed the techniques of movable-bed model testing and the modern understanding of cavitation, as well as the important dimensionless number which now bears his name, with his definitive studies on laminar and turbulent flow and the critical conditions for onset of turbulence. Froude developed the towing-tank technique of ship testing and was responsible for the adoption, although he himself did not discover it, of the Froude Law of model testing. The techniques of similitude developed by Reynolds and Froude were in considerable measure a stimulus for the reunion of theoretical hydrodynamics and practical hydraulics which has occurred in the present century.

One important nineteenth century proponent of such a fusion was Joseph Boussinesq (1842–1929), who was a theoretician concerned with practical problems. Among other things, Boussinesq authored an extensive reference treatise on hydraulic theory and design.

At the turn of the century, leadership in hydraulic developments and in the emerging science of fluid mechanics largely passed from France to Germany. Extensive studies on model and prototype hydraulic structures and river hydraulics were carried out by Hubert Engels (1854–1945) and Theodor Rehbock (1864–1950). The outstanding hydraulic engineering compendium of this period was written by Philipp Forchheimer (1852–1933). Perhaps the outstanding English-language book was that of A. H. Gibson (1878–1959), who took over Osborne Reynolds' laboratory in Manchester, England.

In America, classic studies on sediment transportation were conducted at the University of California by Grove Karl Gilbert (1843–1918). Sherman Woodward (1871–1953) designed the influential laboratory and testing program of the Miami Conservancy District, in Ohio, and laid the groundwork for the founding of the Iowa Institute of Hydraulic Research, which has exerted significant leadership in American hydraulic practice for over half a century.

The most important developments of this period, however, were taking place in Germany. The founder of modern fluid mechanics is generally considered to be Ludwig Prandtl (1875–1953), who introduced the concept of the boundary layer in 1901. Prandtl's laboratory in Göttingen contributed enormously to the reunion of hydrodynamic theory and experimental hydraulics in the modern discipline of fluid mechanics. Many of his students made notable contributions, including Paul Heinrich Blasius (1883–   ), noted for his studies on smooth pipes and his recognition of the relationship between friction factor and Reynolds Number, and Johann Nikuradse (1894–   ), who made the classic studies on rough pipes which led to the adoption of the equivalent sand-grain diameter as a standard measure of surface roughness.

The most notable product of the Göttingen school, however, was Theodor von Kármán (1881–1963), who later became director of the aerodynamic laboratory at California Institute of Technology. Dr. von Kármán had to his credit a large array of original contributions in aerodynamics, hydrodynamics and elasticity. His studies of fluid turbulence and the laws of surface resistance are of special importance to the hydraulician. Other notable German engineers of this era included R. von Mises (1883–1952), L. Schiller (1882–   ), W. Spannhake (1881–1950) and Herman Schlichting (1907–   ).

Another fundamental development in fluid mechanics was the introduction of formalized techniques in dimensional analysis and similitude. Edgar Buckingham (1867–1940), at the National Bureau of Standards, contributed significantly to this with what he called the π-theorem.

Of unique importance as a catalyst in many of these developments were the efforts of John R. Freeman (1855–1932), who not only made numerous experimental studies of his own on the hydraulic resistance of pipes and fittings but, more importantly, acquainted American engineers with the developments and techniques in the German laboratories. Somewhat later B. A. Bakhmeteff (1880–1951) exerted a similar influence by incorporating European studies in turbulence and open-channel hydraulics into American practice.

Emphasis in the foregoing discussion has been placed on the history of the development of the basic principles of applied hydraulics. Of comparable interest, of course, is the actual utilization of these principles in the design and operation of hydraulic engineering systems and structures. In fact, most of the important laboratory studies in hydraulics during the nineteenth and twentieth centuries were made with specific design needs in mind.

Much of the early development of the United States, for example, centered around the growth of waterway transportation and water power. George Washington (1732–1799) actually planned and made the hydrographic surveys for one of the first American canals, connecting the Ohio River to the Potomac, built in 1784. The Erie Canal, joining the Hudson River above Albany to Lake Erie, has been called the "first American engineering school." It was built in the period 1817–1825, and extended for 363 miles through the wilderness, with a total of 83 locks. In all, over 4000 miles of canals were built in the United States in the next half-century. Although railroads eventually displaced most of the canals in the transportation economy, their development exerted a profound influence on the nation in its early years and, in particular, provided a great stimulus to the development of the civil engineering profession.

Waterway transportation in the system of navigable rivers and coastal canals is still, of course, a highly important activity. Such notable nine-teenth century engineers as James B. Eads (1820–1887) contributed leadership to the development of works for maintaining the navigability of the Mississippi and other rivers. It is not generally known that Robert E. Lee (1807–1870), at that time an officer in the Army Engineers, designed and built regulatory works which preserved St. Louis as a port city. Lee also was in charge of work in the New York harbor at a later period.

One of the earliest water power developments in this country was built in 1634 on the Neponset River near Dorchester, Massachusetts. This was a dam used to provide head for a grist mill, saw mill, and powder mill, the power being transmitted by waterwheels. Many such dams and mills were operated until the middle of the nineteenth century, when the water wheels began to be replaced by the new hydraulic turbines.

During this period, developing industrial cities often were planned around a water power plant, the first notable example being Lowell, Massachusetts, on the Merrimack River. One of the largest of such plants was on the Connecticut River, at Holyoke, Massachusetts.

The introduction of the steam engine, of course, radically changed this pattern, and hydropower fell into limited use until the invention of electric generation and transmission equipment near the end of the nineteenth century. The first important hydroelectric plant was installed at Niagara Falls in 1895. Although use of steam power continued to increase, hydro-electric power has also continued to grow markedly since that time.

With the growth of large cities, more and more ambitious water supply systems began to be designed, with complexes of reservoirs, tunnels, canals, pipe networks, and other hydraulic structures. Perhaps the outstanding of these was the system supplying water to Southern California from the Colorado River, centered in the famed 240-mile-long Colorado River Aqueduct built in 1932–1939.

The settling of America's western regions, most of which have an arid or semi-arid climate, was followed by expanding demands for water, especially for irrigation. The reclamation era began with the Mormon pioneers in Utah, about 1847, who diverted the waters of a small creek to irrigate their potato crops. Later, large numbers of small irrigation districts were formed all through the West, to build diversion works and ditches. These first centered around local communities, then around regional districts, and finally around large cooperative efforts of states and the federal government associated with the water users.

The Federal Reclamation Act of 1902 established the Bureau of Reclamation, which has taken the lead in the conservation and utilization of the water resources of the western states. Some of the most notable hydraulic engineering projects of all time, such as Hoover Dam (1931–36) and Grand Coulee Dam (1932–39) have been planned and built by the U.S. Bureau of

Reclamation. The multi-purpose, multi-reservoir, basin-wide project concept was developed in large measure by Bureau engineers. Although irrigation has been of basic importance; projects in these systems now commonly include power, recreation, industrial use, and other components in their operations. One of the nation's outstanding hydraulic laboratories and design centers is operated by the Bureau in Denver.

A hydraulic engineering organization of comparable stature is that of the U.S. Army Corps of Engineers, organized in 1802 and long charged with the improvement of our harbors and navigable waterways. An outstanding early accomplishment was the extensive hydrometric survey of the Mississippi River conducted for the Corps by A. A. Humphreys (1810–1883) and H. L. Abbot (1831–1927). Their 1861 report incorporating these studies was the greatest treatise ever written up to that date on river hydraulics.

More recently the Corps has been charged with the planning of many major works for flood control and drainage, in addition to provisions for navigation. The Fort Peck Dam, on the Missouri River in Montana, was built by the Corps of Engineers in 1935–40, the largest earth dam in the world.

The Mississippi River basin has been extensively developed by engineers of the Corps, in conjunction with the Mississippi River Commission, first established in 1879. A complex system of levees, revetments, wasteways, retention reservoirs, cut-offs, and other structures have transformed and largely protected the vast Mississippi delta region.

One of the world's finest hydraulic laboratories is operated by the Corps of Engineers at the U.S. Waterways Experiment Station in Vicksburg. The Corps is also responsible for the Coastal Engineering Research Board (known from its inception in 1930 until 1963 as the Beach Erosion Board) and its fine laboratories near Washington, D.C., for the study of coastal hydraulics and beach erosion.

The foregoing has, of course, not been in any sense a comprehensive history of hydraulics and hydraulic engineering. Instead, when dealing with the records of such a fundamental human need as water supply and control, any history can hardly hope to be much more than a random sample of the relevant information. The interested reader is urged to study some of the references listed at the end of the book for a greater appreciation of this heritage.

1–3. The Future of Hydraulic Engineering. World political conditions permitting, there can be no question that the well-trained hydraulic engineer is going to be in ever-increasing demand, for all the foreseeable future. This truism arises from the obvious fact of ever-increasing populations and ever-increasing demands for water for all sorts of uses, in contrast to the basic stability of the earth's water resources. Engineering skills of the highest order are going to be required to maintain supplies equal to needs, through increasingly efficient utilization of these resources.

The hydraulic engineer of the future must be a truly professional man in every sense of the word. Not only must he have a thorough knowledge of the continually developing sciences of fluid mechanics and hydrology, but he must maintain a good working facility in structural analysis, soil mechanics, geology, construction methods, chemistry, and other subjects ordinarily treated in his undergraduate curriculum. In addition, he needs to understand at least something of many other fields—agriculture, law, economics, statistics, sociology, nuclear energy, automation, meteorology, oceanography, sanitation, electronics, and others—all of which play vital roles in various types of hydraulic engineering projects. Most of this knowledge must be acquired through experience and personal reading—some perhaps through graduate study, which is becoming more important in all branches of engineering, including hydraulic engineering.

The use of electronic computers will contribute tremendously to hydraulic analysis and design, particularly in problems of flow-routing, multiple reservoir operation, economic studies, pipe and channel networks, flood-forecasting, flow profile analyses, and so forth. Electronic instrumentation and controls are becoming commonplace on hydraulic projects.

Reclamation of potentially arable lands through drainage and irrigation projects is less than half completed even in this country, and is essentially just getting under way in most other countries. Economical conversion of salt water to fresh, control of precipitation by cloud-seeding or other techniques, recharge of depleted ground water by controlled flooding, and other important innovations may soon revolutionize our concepts of water supply, with tremendous engineering programs resulting.

The expanding power needs of the world must be met in large part by hydroelectric projects. Less than 20 per cent of the potential hydroelectric power of even our own country has yet been harnessed. These projects may soon include even the power potential of the tides in the large estuaries.

River control problems—bank stabilization, harbor sedimentation, saline intrusions, and the like—are increasingly critical. Flood control facilities are being demanded wherever flooding problems exist. Soil conservation structures and measures will be augmented as the value of irreplaceable topsoils becomes more critical.

Modern industrial processes often require large amounts of water, so that availability of hydraulic facilities is more and more a key factor in the location of industries. Greater populations of course also mean greater provisions for municipal water supply and disposal. Problems of stream pollution, recreation and wildlife resources, navigation facilities, beach erosion, and drainage of transportation arteries and facilities are all primarily hydraulic problems and are becoming increasingly important.

Thus, all present indications point to an unprecedented expansion of hydraulic engineering projects continuing more or less indefinitely. At the same time, solutions for the problems encountered require a continually higher order of training and research in hydraulics and related fields.

The further development of knowledge and techniques in applied hydraulics will undoubtedly continue to center in the great hydraulic laboratories of the government agencies and the universities. Not only in this country, but also in Germany, France, Holland, and many other countries, the leading universities now have well-equipped hydraulic laboratories and are vigorously pursuing their research programs. All now have access to modern electronic computing facilities, which make many problems amenable to solution which formerly were considered hopelessly complex.

With the explosive increases in population and industrialization which are anticipated throughout the world during the remaining years of the twentieth century, and with man's water resources already strained and deficient in numerous regions, it is quite evident that water conservation and transmission programs will have to be undertaken on a scale far greater than any yet in use. This will call for a greater number of hydraulic engineers, and better qualified, than we can now visualize, all around the world.

Authorities generally agree that there is an adequate supply of fresh water in the world to meet all foreseeable needs for at least the next half century. The problem is one of location and quality, rather than total quantity. Not only arid regions, but also many highly urbanized and industralized areas in humid regions, already have grave shortages of usable water. It is obvious that the remedy is a combination of pollution control and purification, together with collection, storage and transmission of water where and when available, to times and places of water need. The role of the hydraulic engineer obviously looms large in the planning and implementation of all such future water resource projects.

2

BASIC PRINCIPLES OF HYDRAULICS

2-1. General. Although there are an infinite number of possible hydraulic systems, each with its distinctive boundary geometry and corresponding descriptive equation, it is helpful to note at the outset that certain basic principles are applicable to all such systems and processes. These principles, which are actually common to all physical phenomena, can usually be expressed in the form of equations which materially aid in the understanding of each particular phenomenon. The first step in the analysis or design of a particular hydraulic process or system should therefore normally be to write these equations as appropriate for that system.

The purpose of this chapter is to discuss these fundamental principles and some of their important applications in hydraulics. Most of this material is already familiar to the student who has taken an introductory course in fluid mechanics, and this chapter will serve therefore partially as a review. However, the integrated and generalized approach emphasized here should also provide for him a broader and more coherent perspective on the nature of hydraulic processes than is possible in such an introductory course.

The basic principles to which we have been alluding can be summarized in the following outline:

  I. External constraints on the process
    a. Conservation of mass
    b. Conservation of momentum
      1. Linear momentum
      2. Angular momentum
    c. Conservation of energy
 II. Internal mechanics of the process
    a. Increase of entropy
    b. Space-mass-time dimensional framework

There are other basic conservation principles in physics (e.g., conservation of electric charge) but the three listed above are those that are of particular application in hydraulics. They are strictly empirical laws, in the sense that everything in science is discovered and confirmed by observation and experimentation, but they have by now been so overwhelmingly verified that there is no doubt of their universal validity in hydraulic processes, as well as in practically all other processes (mass-energy conversions constituting the only known exception, in which case the mass and energy principles can be combined into a broader law of mass-energy conservation). These conservation laws, therefore, constitute the external constraints to which all processes must conform, regardless of what takes place internally in the process.

With respect to the internal mechanics of the process, it is a remarkable empirical fact of universal observation that all processes function in such a way that the "entropy" (i.e., the probability, or randomness, or disorder) of the system tends to increase. This means that the system will eventually wear out or run down unless continually augmented by an external source of ordering energy. In thermodynamics the term "entropy" is usually defined quantitatively in terms of the irreversible component of heat flow, stating that the entropy of an isolated system always increases in a real process. In hydraulics the increase in entropy is usually manifested in the irreversible conversion of flow energy into heat through friction as work is done on the fluid to maintain flow against the frictional forces resisting flow.

Thus every flow process (like all other physical processes) is basically a conservative process, in accordance with the energy conservation principle (first law of thermodynamics), and a decay process, in accordance with the entropy-increase principle (second law of thermodynamics). Although energy must be conserved in the process, some of the energy available for maintaining the flow must, in the process, be deployed into non-recoverable heat energy, which is eventually conducted through the fluid and its bounding walls, if any, into the atmosphere or other external medium and eventually dissipated through space.

The amount of energy so deployed depends upon the various types of force or energy affecting the system and on the boundary geometry. The quantitative formulation of this relationship may be called the *process equation*, and it must normally be determined experimentally, usually on a model of the system.

The necessary experimentation can be expeditiously organized in terms of a dimensional analysis, using the fundamental fact that all processes, no matter how complex, function in a universe which is a continuum of space and mass and time. Therefore the process equation must be expressible in terms of dimensions of space and mass and time (e.g., the foot-pound-second system, or something corresponding) and nothing more. This fact enables us to use the powerful tools of similitude and dimensional analysis as the means to an effective and efficient empirical determination of the particular process equation for any given system.

Finally, judicious use of the conservation equations and the process equation will permit the analysis or design of the system. These principles will now be discussed in more detail and their applications illustrated.

2–2. The Conservation Principles. In thermodynamics a *closed system* is defined as a system across the bounds of which no mass can be transferred. An *isolated system* is closed to both mass and energy transfer. In hydraulics it is often convenient also to use the concept of a *control volume*, which is a reference volume in the body of fluid, through the boundaries of which can pass both mass and energy. The system may be assumed either fixed or moving with the fluid. The control volume is generally assumed as fixed, and may be described as an *open system*.

In writing the conservation equations for hydraulic systems, it is convenient to combine the two concepts in the entity known as a *stream tube*. This is a control volume which is an isolated system in two dimensions but open in the third, so that mass, momentum, and energy may enter at one end and leave through the other, but may not pass through the sides. Thus the flow is *one-dimensional*. See Fig. 2–1.

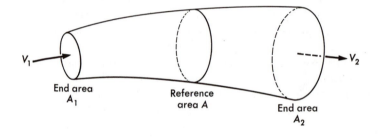

Fig. 2–1. Streamtube.

The streamtube may also be thought of as an aggregation of streamlines, a streamline being a line everywhere tangent to the velocity vector; in steady flow this coincides with the path followed by a particle of fluid as it traverses the streamtube. Each element of the streamtube surface is thus also a streamline.

Fluid is entering the end area A_1 of the streamtube and leaving at A_2, carrying mass, momentum, and kinetic energy with it. Consider a reference cross-section A in the tube, which is crossed by fluid particles having various point velocities v, each corresponding to an elemental cross-section dA.

The mass of fluid passing through an element dA at a velocity v in a time dt is $m = \rho(v\,dt)(dA)$, where ρ is the mass density, or mass per unit volume, and $v\,dt$ is the distance traversed by the particle in time dt.

Similarly, the momentum through dA in time dt is equal to $mv = \rho v^2(dt)(dA)$ and the kinetic energy is $\frac{1}{2}mv^2 = \frac{1}{2}\rho v^3(dt)(dA)$. Integrating over the cross-section, and dividing by dt, we obtain the expressions for *flux* of mass, momentum, and kinetic energy, respectively, as follows:

$$\text{Mass/Time} = \rho \int_A v\, dA = \rho K A V = \rho K Q \tag{2-1}$$

$$\text{Momentum/Time} = \rho \int_A v^2\, dA = \rho\beta A V^2 = \rho\beta Q V \tag{2-2}$$

and

$$\text{Kinetic energy/Time} = \tfrac{1}{2}\rho \int_A v^3\, dA = \tfrac{1}{2}\rho\alpha A V^3 = \tfrac{1}{2}\rho\alpha Q V^2 \tag{2-3}$$

In these equations V is defined as the average velocity over the cross-section, K is the mass distribution factor, β is the momentum distribution factor and α is the kinetic energy distribution factor, By definition:

$$V = \frac{1}{A} \int_A v\, dA \tag{2-4}$$

The volume per unit time is defined as the *quantity of flow*,

$$Q = \int_A v\, dA = AV \tag{2-5}$$

The various distribution factors may be evaluated as follows, substituting from above:

$$K = \frac{1}{A} \int \left(\frac{v}{V}\right) dA = \frac{1}{Q} \int v\, dA = 1 \tag{2-6}$$

$$\beta = \frac{1}{A} \int \left(\frac{v}{V}\right)^2 dA = \frac{A}{Q^2} \int v^2\, dA \tag{2-7}$$

and

$$\alpha = \frac{1}{A} \int \left(\frac{v}{V}\right)^3 dA = \frac{A^2}{Q^3} \int v^3\, dA \tag{2-8}$$

It may also be noted from these equations that

$$\alpha > \beta > K = 1 \tag{2-9}$$

The units of these quantities, in the English system, are as follows:

V, in ft per sec
Q, in ft^3 per sec
ρ, in lb-sec^2 per ft^4, or slugs per ft^3
Mass/Time, in lb-sec per ft
Momentum/Time, in lb
Kinetic energy/Time, in ft-lb per sec.

It should be noted that momentum flux has the units of force and kinetic energy flux units of power.

By the conservation principles, and by the definition of a streamtube, neither mass, momentum, nor energy can be created or destroyed within its walls. Therefore the mass flux entering must be the same as that leaving, and the same is true for momentum flux. Similarly the total energy of the flow entering (not kinetic energy only) must be the same as that of the fluid leaving plus the energy converted internally into heat energy. These equations are now considered individually in the following articles.

2–3. Mass Conservation. The Equation of Continuity. The law of conservation of matter states that (barring mass-energy interchange, as in nuclear fission) matter can change its state and can become more, or less, dense, but can be neither created nor destroyed. Therefore, for the stream-tube of Fig. 2–1, and from Eq. (2–1), it follows that:

$$\rho_1 A_1 V_1 = \rho_2 A_2 V_2 \tag{2-10}$$

It is often more convenient to express this equation, which is also known as the *continuity equation*, in terms of the weight flux (pounds per second) instead of the mass flux. In this case:

$$\gamma_1 A_1 V_1 = \gamma_2 A_2 V_2 \tag{2-11}$$

in which γ ($= \rho g$) is the fluid *specific weight*, in lbs/ft³, and g is the acceleration of gravity, 32.2 ft/sec/sec. For water, γ is approximately 62.4 lbs/ft³ and ρ is 1.94 slugs/ft³.

In the special case of *incompressible flow*, which is usually applicable in hydraulics, the density and specific weight are constants. Then:

$$A_1 V_1 = A_2 V_2 \tag{2-12}$$

It has also been assumed above that the flow is *steady flow*, in which the patterns and magnitudes of the velocities at each section remain constant with time. In the event of *unsteady flow*, combined with *compressible flow*, it is possible for the mass within the streamtube to increase or decrease with time. In such a case, the mass conservation equation becomes:

$$\rho_1 \int_{A_1} v_1 \, dA - \rho_2 \int_{A_2} v_2 \, dA = \left(\begin{matrix} \text{Volume in} \\ \text{streamtube} \end{matrix}\right) \frac{\partial \rho}{\partial t} \, dt \tag{2-13}$$

where dt is the total time for flow to move from A_1 to A_2 and $\dfrac{\partial \rho}{\partial t}$ is the rate at which the density increases with time within the streamtube. In most hydraulics problems, however, the steady flow assumption is sufficiently accurate.

Equation (2–12), for steady, incompressible flow through a streamtube, is equivalent to saying that the quantity of flow, or discharge, Q_2 is constant at all sections along the streamtube. It also states that the average velocity at a cross-section, for a given discharge, is inversely proportional to the area of the cross-section.

Since it is equivalent to work, the product of two vector quantities (force times distance), energy is a scalar entity. Therefore the total energy of a fluid mass is simply the sum of the different kinds of energy contained in it. Thus the total head is

$$H = \frac{p}{\gamma} + h + \alpha \frac{V^2}{2g} \tag{2-21}$$

Consider a body of fluid in the process of flowing from one position to another along a streamtube. Provided it can be considered essentially as an isolated energy system, its total energy must remain unchanged; that is,

$$\frac{p_1}{\gamma} + h_1 + \alpha_1 \frac{V_1^2}{2g} = \frac{p_2}{\gamma} + h_2 + \alpha_2 \frac{V_2^2}{2g} + h_f \tag{2-22}$$

Equation (2-22) is one form of the familiar *energy equation* of fluid mechanics, generally called the Bernoulli equation after the Swiss scientist who first propounded it in the late eighteenth century. In this form, it is assumed that the flow is one-dimensional, steady, and incompressible, assumptions which are adequately valid in most hydraulic engineering problems. Steadiness of flow implies that flow patterns and magnitudes do not change rapidly with time; incompressibility means that the fluid density remains essentially constant as the pressure varies.

This equation incorporates both the law of energy conservation (first law of thermodynamics) and of energy degradation (second law of thermodynamics).* It is undoubtedly the most important equation of hydraulics and fluid mechanics. In the above form, each term is a "head," or energy in foot-pounds per pound of the fluid. If use of the energy flux is more convenient, the energy and continuity equations can be combined to yield the *power equation*. Since power is defined as energy per unit time, the power represented in a flowing mass of fluid is the product of the weight rate of flow and the energy per pound. Thus

---

* The universality of the two laws of thermodynamics should be stressed. They have been substantiated empirically wherever it has been possible to test them, but have been widely accepted as foundational in science only since the work of Clausius, Kelvin, and others in the latter part of the nineteenth century. The reason for their universal scope cannot be determined by science but is clarified by theology. Thus, the first law enunciates the constancy of the totality of matter and energy in the universe, the reason being that the primeval processes of creation were terminated at the end of the six days of creation (see Gen. 2: 1-3; Heb. 4: 3-10). The second law in its broadest form states that there is a continual tendency toward disorder, decay, and death in the universe. This is best explained in terms of the curse pronounced by the Creator on the entire earth as a result of the introduction of moral rebellion into the world (see Gen. 1: 31; 3: 17-19; Rom. 8: 20-22). The obvious conflict of these scientific laws with the previously popularized philosophy of universal evolutionary progress has not yet been adequately recognized.

2-4. The Momentum Conservation Equations. By a similar analysis to that for mass conservation, the equation for *conservation of linear momentum* along a streamtube is obtained from Eq. (2-2). Thus:

$$\rho_1 \beta_1 A_1 V_1^2 = \rho_2 \beta_2 A_2 V_2^2 \tag{2-14}$$

From Eq. (2-10) this reduces to

$$\beta_1 V_1 = \beta_2 V_2$$

and since β_1 is nearly always very close to β_2 and to unity, this means that $V_1 = V_2$. Thus, linear momentum is conserved only if the velocity remains constant through the tube.

In general, this will not be true but the conservation principle remains valid since any change in velocity must be caused by an external force on the system. Such a force must be exactly equal to the corresponding change in momentum flux. Therefore:

$$F = \rho_2 \beta_2 A_2 V_2^2 - \rho_1 \beta_1 A_1 V_1^2$$

Since β_1 and β_2 can usually be assumed as unity, and $\rho_1 = \rho_2$, therefore:

$$F = \rho Q (V_2 - V_1) \tag{2-15}$$

This is the most common form of the linear momentum equation as used in hydraulics. The momentum flux, $\rho Q V$, has the dimensions of a force. The equation therefore is vector, rather than scalar, in application, with the velocity vector difference, $V_2 - V_1$, along the same line of action as the resultant of the external forces on the system, F. The equation may thus also be expressed in terms of x and y components.

$$F_x = \rho Q (V_{2_x} - V_{1_x}) \tag{2-16}$$

and

$$F_y = \rho Q (V_{2_y} - V_{1_y}) \tag{2-17}$$

If the fluid motion is curvilinear, rather than linear, the "moment of momentum," or *angular momentum*, is measured by the product of the momentum flux and the radius of curvature. The tangential component of the velocity is used, since the radial component produces no torque. Equation (2-15) can be modified to:

$$\sum [(F_t)r] = T = \rho Q (V_{2_t} r_2 - V_{1_t} r_1) \tag{2-18}$$

The external torque on the rotational motion is T. This equation describes what is known as the *forced vortex* and is used in the study of such processes as flow through a pump. If there is no external torque, it reduces to:

$$V_{2_t} r_2 = V_{1_t} r_1 \tag{2-19}$$

This is the equation of the *free vortex* (for example, a whirlpool on a water surface), in which angular momentum is truly conserved.

2–5. Energy Conservation: The Bernoulli Equation.

The most important of all physical laws, and undoubtedly the one supported by the greatest body of scientific verification, is that of energy conservation. In any isolated system, the energy contained therein may change its form but not its totality. Energy, defined as the capacity for doing work, is measured in the same units as work, i.e., foot-pounds; in hydraulics, it is usually convenient to think of energy in unit quantities, expressing it as *head*, the amount of energy per pound of fluid, thus in units of foot-pounds per pound, or feet.

Energy can exist in many forms (heat, light, sound, chemical, electrical, magnetic, atomic, etc.), and every process in nature, whether natural or artificial, is essentially a process of conversion of energy from one kind into another, often with work in the form of mechanical energy being accomplished in the process. Mechanical energy can be either kinetic energy (the energy of motion) or potential energy (energy due to position or configuration).

In a hydraulic process, the important forms of energy are: kinetic energy, the flow of which has been described in Eq. (2–3); potential energy in the form of the position of the fluid above the earth's surface or some other arbitrary datum; potential energy in the form of fluid pressure; and heat energy, resulting from the frictional forces generated by the fluid motion.

The kinetic energy flux can be converted to *velocity head*, by dividing by the weight flux. Thus, from Eqs. (2–1) and (2–3)

$$h_v = \frac{\frac{1}{2}\rho\alpha AV^3}{g(\rho AV)} = \alpha\frac{V^2}{2g} \qquad (2\text{–}20)$$

The kinetic energy distribution factor is usually quite close to unity and can thus normally (except for work of high accuracy or extremely non-uniform velocity distributions) be neglected.

The potential energy which the fluid contains by virtue of its elevation above an arbitrary datum is the work, per pound, required to raise it thereto. That is, if the elevation of a given differential volume, dV, of fluid is h, its *position energy* per pound is

$$\frac{\gamma(dV)(h)}{\gamma(dV)} = \frac{(\text{weight})(\text{distance raised})}{\text{weight}} = h$$

This distance h is known as the *position head*.

The pressure head may be considered as the potential energy contained in each pound of fluid in the form of pressure. That is, the compressed fluid, if released, could accomplish p/γ foot-pounds of mechanical work per pound. This is evident from consideration of the work done by the fluid on, say, a

piston in a cylinder, causing it to move a distance L (see Fig. 2–2):

$$\text{Work done per pound of fluid} = \frac{(\text{force})(\text{distance moved})}{(\text{pounds of fluid})} = \frac{pAL}{\gamma AL} = \frac{p}{\gamma}$$

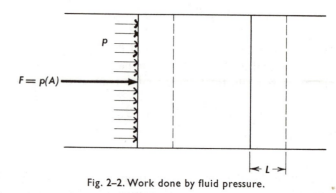

Fig. 2–2. Work done by fluid pressure.

The *friction head* is denoted by h_f, the amount of flow energy converted into heat energy in the flow process. This conversion is brought about through the shearing stresses resisting flow, the overcoming of which requires work to be done. This work generates heat, at the expense of some of the flow energy. In a truly isolated system, this heat energy would be retained by the fluid, manifested as increased temperature; practically, however, it is conducted through the fluid and its boundary channel walls, and dissipated into the environment. Thus, it is lost energy, as far as the flow process is concerned.

This phenomenon is an example of the universal scientific law known as the second law of thermodynamics. In any isolated physical system in which transformations of energy are taking place, a portion of the energy must be used to overcome frictional resistances and is therefore transformed into non-recoverable heat energy. *Entropy* is the term used as a measure of the energy so deployed. That is, the entropy continually increases in an isolated system, measuring the degree of unavailability of the energy of the system for performing useful work.

In nearly all hydraulic engineering problems, the determination of the friction head is basic to the understanding and effective functional design of the hydraulic structure or process. However, by virtue of its inherently complex nature, it is usually impossible to determine other than empirically, although great advances have been made toward more nearly rational methods of analysis. The process equation, as discussed in Art. 2–8, essentially represents the internal mechanics of the energy conversion process resulting in the friction loss. For the present, we shall simply use the friction head term, h_f, to reflect this process as it affects the over-all statement of energy conservation.

$$P = (\gamma Q)(H) = \gamma Q \left(\frac{p}{\gamma} + h + \alpha \frac{V^2}{2g} \right) \tag{2-23}$$

Since $Q_1 = Q_2$ by continuity, then for steady, incompressible, one-dimensional flow,

$$\gamma Q \left(\frac{p_1}{\gamma} + h_1 + \alpha_1 \frac{V_1{}^2}{2g} \right) = \gamma Q \left(\frac{p_2}{\gamma} + h_2 + \alpha_2 \frac{V_2{}^2}{2g} + h_f \right) \tag{2-24}$$

Each term dimensionally is now of the form γQE, or ft-lb/sec, and, if desired, may be converted to units of horsepower by dividing by the constant, 550.

It is also frequently convenient to write Eq. (2–22) as a group of pressure terms. Multiplying through by γ, and neglecting α:

$$p_1 + \gamma h_1 + \tfrac{1}{2}\rho V_1{}^2 = p_2 + \gamma h_2 + \tfrac{1}{2}\rho V_2{}^2 + \gamma h_f \tag{2-25}$$

In the special case of irrotational flow, the head loss term h_f will be zero, and it can be shown that under these conditions the remaining Bernoulli constant $(p + \gamma h + \tfrac{1}{2}\rho V^2)$ is the same at every point throughout the field of flow.

2–6. Application of Conservation Equations to Hydraulic Problems. It should be emphasized that all the above conservation principles must be satisfied in every hydraulic system or process. However, the structure of the particular system will determine which of the equations will give useful information.

For example, if the flow is *uniform* (the velocity distribution the same at all sections in the given reach) the continuity equation does not give any information that is not already known. However, in every case of *non-uniform*, or *varied* flow, it will specify the relationship of velocity and cross-section at all sections in the reach.

The momentum equations essentially constitute the equations of equilibrium for the force system external to the body of fluid within the system. A net external force produces a corresponding change in linear momentum and a net external torque a corresponding change in angular momentum. The momentum equations are directly useful when the external force system can be specified and when mechanisms for changing pressures and velocities internally are not of primary concern. Since force and momentum are vector quantities, it is important to keep the directions and algebraic signs correct when applying the equations.

The terms in the energy equation, on the other hand, are scalar quantities, and are therefore simply additive terms. The energy equation is the most universally applicable of all the equations of hydraulics, relating as it does the changes in mechanical energy to the internal process in the system causing friction loss.

Two examples are given below to illustrate application of the equations.

Example 2-1

A plate is placed across the lower half of a horizontal pipe, blocking half the pipe cross-section as shown. The lip of the plate is rounded so that no additional

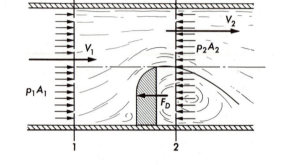

flow contraction takes place beyond the height of the lip. Assume no head loss takes place as the flow contracts. Derive a formula for the drag force on the plate, in terms of only the pipe area A, flow velocity V, and fluid density ρ.

Solution. Consider the free body of fluid between Arts. 1–1 and 2–2. The forces on this free body are as shown. The fluid pressures p_1 and p_2 both act over the entire area, since the pressure in the separation zone behind the plate is the same as in the *vena contracta* of the jet. The drag force on the plate is equal and opposite to the retarding force of the plate on the flow.

By the momentum equation:

$$p_1 A_1 - p_2 A_1 - F_D = \rho Q (V_2 - V_1) \tag{1}$$

and then, by the continuity equation:

$$Q = A_1 V_1 = A_2 V_2 = \tfrac{1}{2} A_1 V_2$$

from which,

$$V_2 = 2 V_1 \tag{2}$$

Substituting (2) in (1):

$$F_D = (p_1 - p_2) A_1 - \rho A_1 (V_1{}^2) \tag{3}$$

We can obtain the pressure difference from the energy equation:

$$\frac{p_1}{\gamma} + \frac{V_1{}^2}{2g} = \frac{p_2}{\gamma} + \frac{V_2{}^2}{2g} = \frac{p_2}{\gamma} + \frac{4 V_1{}^2}{2g}$$

Thus,

$$(p_1 - p_2) = \tfrac{3}{2} \rho V_1{}^2 \tag{4}$$

Now, substituting (4) in (3), and dropping subscripts:

$$F_D = \tfrac{1}{2} \rho A V^2 \qquad\qquad\qquad \textit{Answer.}$$

The effect of velocity distribution was neglected and so was friction loss, but these assumptions are normally adequate in a short contracting flow situation.

Example 2–2

A 90° horizontal reducing elbow carries a flow of 12 cfs (i.e., cubic feet per second). The initial diameter is 24 in. and the final diameter 12 in. The corresponding fluid pressures are 60 psi and 58 psi (pounds per square inch, respectively). Calculate the head loss around the elbow and the hydrodynamic force on the pipe.

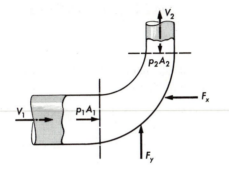

Solution. The forces on the body of fluid in the elbow are as shown on the sketch.

By the continuity equation:

$$V_1 = \frac{Q}{A_1} = \frac{12}{\pi(1)^2} = 3.82 \text{ fps (feet per second)}$$

and

$$V_2 = \frac{A_1}{A_2} V_1 = \left(\frac{D_1}{D_2}\right)^2 V_1 = 4V_1 = 15.28 \text{ fps.}$$

By the energy equation:

$$\frac{p_1}{\gamma} + \frac{V_1^2}{2g} = \frac{p_2}{\gamma} + \frac{V_2^2}{2g} + H_L$$

from which the head loss, H_L, is:

$$H_L = \frac{p_1 - p_2}{\gamma} - \frac{V_2^2 - V_1^2}{2g} = \frac{2(144)}{62.4} - \frac{15(3.82)^2}{64.4}$$

$$= 4.62 - 3.40 = 1.22 \text{ ft of water.} \qquad \textit{Answer.}$$

The momentum equation involves vector quantities and it is convenient to express it in terms of x and y components. Thus:

$$p_1A_1 - F_x = \rho Q(V_{2_x} - V_{1_x}) = -\rho Q V_1$$

and

$$F_y - p_2A_2 = \rho Q(V_{2_y} - V_{1_y}) = \rho Q V_2$$

From these equations:

$$F_x = p_1A_1 + \rho Q V_1 = 60(144)(\pi) + 1.94(12)(3.82) = 27{,}240 \text{ lb}$$

and

$$F_y = p_2A_2 + \rho Q V_2 = 58(144)\left(\frac{\pi}{4}\right) + 1.94(12)(15.28) = 6{,}910 \text{ lb}$$

Combining for the resultant, the resulting force on the elbow (equal and opposite to the above forces of the elbow on the water) is:

$$F = \sqrt{(27,240)^2 + (6910)^2} = 28,100 \text{ lbs.} \qquad \textit{Answer.}$$

acting at an angle θ with the horizontal equal to

$$\tan^{-1} \frac{F_y}{F_x} = \tan^{-1} \left(\frac{691}{2724} \right) = 14.2° \longrightarrow \qquad . \textit{Answer.}$$

2–7. Hydrostatics. The phenomena of hydrostatics can be considered as a special case in hydraulics in which all velocities become zero (or equal to the constant velocity of the boundaries. The conservation equations still apply but in this case simplify to:

$$\textit{Momentum Eqs.} \qquad \sum F = 0 \qquad \text{and} \qquad \sum T = 0$$

$$\textit{Energy Eq.} \qquad \frac{p}{\gamma} + h = \text{Constant}$$

The momentum equations thus become simply the equations of static equilibrium, for the force system on any free body in the hydrostatic system. The energy equation reduces to the familiar hydrostatic equation

$$\frac{p_1}{\gamma} + h_1 = \frac{p_2}{\gamma} + h_2 \qquad (2\text{–}26)$$

This equation states simply that the sum of the pressure and position heads is the same for every point in the static fluid. If point "1" is at the water surface, then the water pressure at that point is equal to atmospheric pressure, p_{atm}, which is approximately 14.7 psi or 2117 psf above absolute zero. Then the pressure at any point below the surface is

$$p_2 = p_{atm} + \gamma(h_1 - h_2) \qquad (2\text{–}27)$$

It is common in hydraulics to express pressures in terms of *gage pressure*, rather than *absolute pressure*, measuring all pressures above and below atmospheric. In this case,

$$p = \gamma D \qquad (2\text{–}28)$$

where p is the gage pressure at a depth D ($= h_1 - h_2$) below the liquid surface. In this text it will always be assumed that pressure means gage pressure, unless otherwise noted.

Pressure in a static fluid always acts in a direction perpendicular to the surface upon which it is acting. Otherwise the reacting pressure would have a component tangent to the surface and would initiate flow. Since such a surface may have any orientation, and since the intensity of pressure is determined solely by elevation, it also follows that the intensity of pressure at a point in a fluid at rest is equal in all directions.

The principle of Pascal states that an increment of pressure applied at any point in a static fluid is transmitted undiminished to every other point of the fluid. Otherwise, flow must occur. However, the pressure increment is not transmitted instantaneously but rather as a wave of pressure spreading out from the point of application, at a speed equal to that of sound in the fluid medium.

The hydrostatic equation is the basis for the use of piezometers and manometers to measure fluid pressure, both in static fluids and flowing fluids. In the latter case, as long as no flow enters or leaves the measuring tube, and as long as the streamlines just outside the tube have no acceleration component in the direction of the tube, the pressure in the static fluid just inside the measuring tube is equal to that in the moving fluid just outside. Note Fig. 2–3.

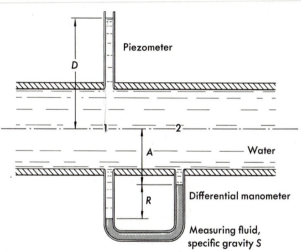

Fig. 2–3. Measurement of pressure.

The piezometer tube measures the "piezometric head," which is simply the sum of the pressure and position heads. The liquid in the pipe rises in the tube until it reaches a static position. Assuming parallel rectilinear streamlines in the pipe, the hydrostatic equation indicates that the height D is equal to the pressure head at point 1. The rate of change of the piezometric head in the direction of flow, $d\left(\dfrac{p}{\gamma} + h\right)\Big/dL$, is the *hydraulic gradient*.

If this height is too large for convenient measurement (or too small), then a manometer can be used with a measuring fluid of some different specific gravity than that in the pipe. The reading on the manometer is then converted into pressure in the pipe by step-by-step summation of pressure head terms from level to level in the hydrostatic equation. This procedure can also be illustrated on the differential manometer, which is used to measure pressure differences rather than actual point pressures.

Referring to Fig. 2–3, proceed stepwise around the manometer, from point 1 to point 2, adding pressure heads when descending and subtracting them when ascending, expressing them in feet of the fluid in the pipe. Thus:

$$\frac{p_1}{\gamma} + A + R - \frac{S\gamma}{\gamma} R - A = \frac{p_2}{\gamma}$$

from which:

$$p_1 - p_2 = \gamma(S - 1)(R) \tag{2-29}$$

where R is the differential manometer reading and S is the specific gravity of the measuring fluid, (assuming water is the pipe fluid; otherwise, S is the ratio of the specific gravities of the measuring fluid and the pipe fluid).

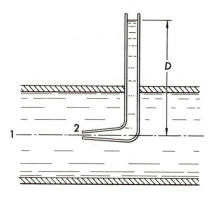

Fig. 2–4. Measurement of total head.

A similar tube, known as the Pitot tube, shown in Fig. 2–4, is a basic device used to measure the velocity at a point in a flowing fluid. The hydrostatic equation applies within the tube and the energy equation along the streamline outside the tube. The hydrostatic equation indicates that the pressure head at the nose of the tube, just as with the piezometer tube, is equal to D, after a static condition has been attained. Again assuming parallel streamlines, with no components of acceleration normal to the nose of the Pitot tube as the flow passes it, the streamline from 1 to 2 will have a "stagnation point" just outside the entrance. The energy equation along this streamline yields:

$$\frac{p_1}{\gamma} + \frac{V_1^2}{2g} = \frac{p_2}{\gamma} = D \tag{2-30}$$

which indicates that the Pitot tube measures the *total head* (piezometric head plus velocity head). The rate of decrease of the total head in the direction of flow,

$$d\left(\frac{p}{\gamma} + h + \frac{V^2}{2g}\right)\bigg/ dL$$

is the *energy gradient*.

If the velocity head is desired, it is equal to the difference in levels between the Pitot tube and the piezometer tube at the given point. There are many elaborations and modifications that are possible in manometric and Pitot devices, but all are based on the hydrostatic equation.

2–8. Hydrostatic Forces. It is often necessary to calculate total hydrostatic forces, either on a submerged structure or on a portion of fluid isolated as a free body. Consider the problem of determining the force exerted by the fluid on the irregular body "A" in Fig. 2–5. On each elemental surface area dA of the body will be exerted a normal hydrostatic force $\gamma h(dA)$, h being the depth of that element below the surface. Each such elemental force can be resolved into vertical and horizontal components. The total of such components acting on the body would then of course be the components of the resultant hydrostatic force. Consider the force system on one such element (Fig. 2–6), which can be assumed because of its differential size to be essentially plane. It is obvious that the vertical component is

$$(dF)_v = (dF)(\cos\theta) = \gamma h(dA)(\cos\theta) = \gamma h(dA)_h$$

$$= \text{weight of column of water vertically above } dA$$

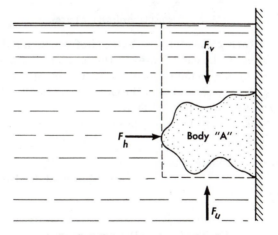

Fig. 2–5. Forces on submerged body.

Similarly, the horizontal component is

$$(dF)_h = (dF)(\sin\theta) = \gamma h(dA\sin\theta) = \gamma h(dA)_v$$

$$= \text{pressure force on vertical projection of area.}$$

Thus, by adding the elemental components,

F_v = weight of body of fluid vertically above the object, acting
through the centroid of the fluid body   (2–31)

F_h = hydrostatic force as calculated on the vertical projection
of the surface area of the object, acting through the centroid of the prism of horizontal pressure forces thus defined   (2–32)

The uplift force acting on the underside of the object, F_u, is evaluated in
the same manner as F_v, except that the height of the fluid column producing

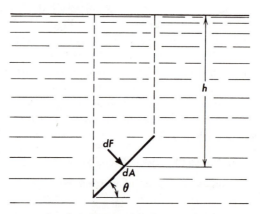

Fig. 2–6. Elemental hydrostatic forces.

the pressure on each elemental area extends of course all the way down
through the object to the surface area of the bottom. Consequently, F_u is
greater than F_v by an amount equal to the weight of the body of fluid
"displaced" by the object, and acts through the centroid of the combined
volume of the actual plus displaced fluid.

It should also be noted that this excess $(F_u - F_v)$ is known as the force
of buoyancy. Thus,

Buoyant force on submerged object = $F_b = F_u - F_v$
= weight of fluid displaced by object
(2–33)

Often the immersed object is composed of plane areas. Although the
relations of Eqs. (2–31), (2–32), (2–33) are quite general, the unidirectional
nature of the pressures acting on plane areas makes it possible to calculate
the resultant force directly, rather than through its components. By considering the pressure diagram on a plane submerged area, as in Fig. 2–7, with
its plane making an angle θ with the free surface, it is obvious that the pressure increases linearly with depth and therefore that the pressure prism, in
section, is trapezoidal in shape. Its resultant is the resultant hydrostatic

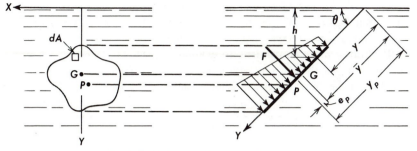

Fig. 2–7. Hydrostatic forces on plane area.

force on the plane area and, as with any parallel force system, it can be easily evaluated by the laws of static equilibrium. Each elemental force in the force system is

$$dF = \gamma h \, dA = (\gamma y \sin \theta) \, dA$$

and

$$F = \gamma \sin \theta \int y \, dA = \gamma \sin \theta \, \bar{y} A$$

Therefore

$$F = \gamma \bar{h} A \qquad (2\text{–}34)$$

where $\gamma \bar{h}$ is the pressure at the centroid of the area A. To determine the line of action of F, reference must be made to a set of coordinate axes; the X-axis is most conveniently chosen as the intersection of the plane of the area with the free surface, and the Y-axis may be any line in the plane of the area normal to the X-axis.

Summing moments about these axes gives

$$\int y \, dF = \gamma \sin \theta \int y^2 \, dA = y_p(F) = y_p \gamma \sin \theta \int y \, dA$$

and

$$\int x \, dF = \gamma \sin \theta \int xy \, dA = x_p(F) = x_p \gamma \sin \theta \int y \, dA$$

Thus the coordinates x_p, y_p of the *center of pressure*, the intersection of the line of action of the resulting force with the plane of the area, are

$$x_p = \frac{\int xy \, dA}{\int y \, dA} = \frac{H_{xy}}{Q_x} \qquad (2\text{–}35)$$

and

$$y_p = \frac{\int y^2 \, dA}{\int y \, dA} = \frac{I_x}{Q_x} \qquad (2\text{–}36)$$

where H_{xy} is the product of inertia of the area about the X and Y axis system, I_x is the moment of inertia of the area about the X-axis, and Q_x is the moment of the area about the X-axis.

If the Y-axis can be chosen as an axis of symmetry, as is frequently the case, then $H_{xy} = 0$. The center of pressure thus lies on any axis of symmetry which the area may have in the Y-direction.

It is often more convenient to evaluate e_p than y_p, where $e_p = y_p - \bar{y}$, representing the distance by which the center of pressure is located below the centroid of the area as measured in the plane of the area.

$$e_p = y_p - \bar{y} = \frac{I_x}{Q_x} - \bar{y} = \frac{I_g + A\bar{y}^2}{A\bar{y}} - \bar{y} = \frac{I_g}{A\bar{y}}$$

Therefore

$$e_p = \frac{I_g}{Q_x} \tag{2-37}$$

with I_g representing the moment of inertia of the area about its horizontal centroidal axis.

A case of such frequent occurrence as to warrant particular mention is that of the hydrostatic force acting on a vertical rectangular wall. For a one-foot strip of the wall, the total force is

$$F = \gamma \bar{h} A = \gamma \frac{h}{2}(h)(1) = \gamma \frac{h^2}{2} \tag{2-38}$$

Its line of action is located below the surface a distance

$$y_p = \frac{I_x}{Q_x} = \frac{(1/12)(1)(h)^3 + (1)(h)(h/2)^2}{(h/2)(1)(h)} = \frac{2}{3}h \tag{2-39}$$

These results also follow directly from the fact that, in this case, the pressure prism is triangular in cross-section, of altitude h and base γh.

2-9. The Process Equation. In Art. 2-1, it was noted that the conservation equations can be understood as the *external constraints* to which any hydraulic system or process must conform. The *internal mechanics* of the system can then be formulated empirically as the process equation for the system.

The above statements are true for processes and systems in all fields, not merely hydraulics. In hydraulics, the process equation will generally take the form of something equivalent to a *head loss function*. It seeks to express the relationship between the various types of energy acting within the system boundaries on the fluid mass and, in particular, the conversion of some of this energy into non-recoverable heat energy.

In general the interaction between the different types of energy or forces involved is so complex that the process equation can only be determined empirically. Some processes are sufficiently simple, and the necessary empirical support so firm and well known, that we can almost regard the resulting equation as completely rational (e.g., laminar flow in a circular tube). In most cases, however, both the mechanics of the process and the boundaries of the flow are so complex that experimental measurements are prerequisite to determination of the function.

However, it is not necessary to embark on a large program of more or less arbitrary experimentation for every hydraulic system that may become of interest. The empirical hydraulics of the eighteenth and nineteenth centuries often involved such an approach and, although it served the purpose at the time, accumulated much more data than were necessary and these data were often incomplete and inaccurate.

The modern tools of dimensional analysis and similitude, however, have made possible a scientific empiricism which can lead most directly and effectively to the desired process equations. Furthermore they also enable the writing of such equations in the most useful and widely applicable forms.

2–10. Dimensional Analysis of the General Flow Function.

In any problem with such a large number of independent variables that a fully rational analysis becomes impossible, the tool of dimensional analysis will be found very helpful. This tool recognizes the fact that the physical factors influencing a physical phenomenon should be related in an equation which is dimensionally homogeneous, and has been found especially useful in hydraulic analysis.

Consider a fluid flowing within or around some kind of boundary surface under the impetus of a differential of potential energy and resisted by shear stresses at the wall and in the fluid itself. It is obvious that these resistances will be affected by the roughness of the boundary surface and by the viscosity of the fluid, as well as by the nature of the fluid (i.e., its density) and by the geometry of the boundary. If the conduit is open, surface phenomena may be affected by gravity and possibly by surface tension. Under certain conditions, compressibility of the fluid may affect the flow.

Assume that all these factors can be related in an equation which will accurately describe the entire flow phenomenon. This equation can be written in the general functional form:

$$f(V, D, \rho, L, \Delta(p + \gamma h), \mu, \gamma, \sigma, E, h, \lambda, s, r, \ldots, A, B, C, \ldots) = 0 \quad (2\text{--}40)$$

in which

V = average velocity of flow (or some other representative velocity), ft/sec

D = diameter of cross-section (or *equivalent diameter* or some other representative transverse size-dimension of the boundary geometry), ft

ρ = fluid mass density, lb-sec^2/ft^4

L = representative longitudinal dimension of boundary geometry, ft

$\Delta(p + \gamma h)$ = differential of the unit potential energy (ft-lb/ft^3) occurring over the length L

μ = fluid coefficient of viscosity (as defined in the Newtonian equation, $\tau = \mu \, dv/dy$, τ being the fluid shear stress between fluid layers at a point where the transverse velocity gradient is dv/dy), lb-sec/ft^2

γ = specific weight of fluid, lb/ft^3

σ = unit surface tension, ft-lb/ft^2

E = fluid bulk modulus of elasticity, lb/ft^2

$h, \lambda, s, r, \ldots$ = dimensions of surface roughness elements, such as radial height, longitudinal spacing, peripheral spacing, radius of rounding, etc., as many being included as necessary to describe these elements geometrically, each measured in ft

$A, B, C, \ldots$ = other dimensions as may be required to define the boundary geometry, such as radius of bends, height of weir in channel, etc.

It should be noted that there are three basic kinds of parameters included in the general flow function, or general flow process equation, as expressed in this form. These are as follows:

1. *Geometric terms*, describing the flow system boundaries.
   a. A representative dimension of the main flow cross-section, in this case D.
   b. A representative dimension in the direction of flow, in this case L.
   c. Other dimensions as needed, sufficient to describe any aspect of the boundary geometry which could affect the fluid motion.
2. *Kinematic terms*, defining the flow movement.
   a. A representative velocity of the flow field, in this case V. It is not necessary to include other velocity terms, since they are dependent on V and the boundary dimensions, as required by mass conservation. If preferred, V could be replaced by the time, T, required for the flow to move, say, the distance L, since $T = L/V$. Other kinematic terms such as the discharge Q need not be included, since they also are dependent on variables already listed (thus, $Q = KD^2V$).
3. *Dynamic terms*, representing the types of energy (or force) that may have an internal effect on the process.
   a. *Kinetic energy* (or inertial force). This is always present in a moving fluid, and is specified by the mass density, ρ, of the fluid. Mass can be considered as that form of energy locked up in the matter of the fluid itself, and it is characterized by the property of inertia. The inertial force is, by Newton's law, $F = Ma$, and is the force required to produce the acceleration a on the mass M. The inertial force per unit cross-sectional area is, thus,

$$\frac{M(a)}{A} \propto \rho \frac{L^3}{L^2} \left(\frac{V}{2T} \right) = \tfrac{1}{2}\rho V^2$$

   where T is the time required for the force to accelerate the flow from rest to V. The term $\tfrac{1}{2}\rho V^2$ is the kinetic energy per unit volume. All of these relationships are implicit when we include the density, ρ, in the function.
   b. *Potential energy.* In like manner the potential energy per unit volume is equal to the force per unit area resulting from the fluid's pressure and position. Of course, if the total potential energy is the same throughout the length L, then the fluid is, by the hydrostatic equation, at rest. There must be therefore, in a moving fluid, always a *differential* of

potential energy in the direction of flow. This is expressed in the function by the term $\Delta(p + \gamma h)$.

c. *Viscous energy* (or shear, or friction). Flow is always resisted by internal shearing stresses between fluid particles and at the flow boundaries. These frictional effects must be overcome by work done by the fluid's mechanical energy, and this results in the conversion of a part of this energy into heat energy, which is then conducted through the boundary walls into the environment. The inability to reclaim this heat energy for reconversion to mechanical energy in the fluid results from the second law of thermodynamics and it is therefore treated as a "friction loss." The rate of this energy decay depends on the molecular structure of the particular fluid and is specified in terms of the viscosity, μ (also called the dynamic viscosity, or coefficient of viscosity).

d. *Gravitational energy* (or weight). All masses respond to a gravitational field in such a way that the force of gravity, or weight, $= Mg$, where g is the acceleration of gravity. This effect can be specified in the function by including the fluid *specific weight*, or weight per unit volume, γ.

e. *Surface energy* (or surface tension). This can be understood as the work required to form a free surface (or, better, interface between two fluids) against the tendency for mixing and dispersal. The surface energy per unit area of the interface is equivalent to the surface tensile force per unit length along any line in the interface. This also is a molecular phenomenon and is specified in terms of the unit surface tension, σ.

f. *Elastic energy* (or compression). Fluids will not resist tension, but they do respond to compressive stresses (i.e., their own fluid pressures) by a reduction in volume and a corresponding storage of elastic energy, which can of course accomplish work when released. This effect is measured by E, the bulk modulus of elasticity of the fluid, defined as $\dfrac{dp}{d\rho/\rho}$, the ratio of an increment of pressure to the corresponding unit change in density.

The several fluid properties associated with the different kinds of energy (ρ, μ, γ, E, and σ) all depend on the particular fluid. In addition, since they involve activity at the molecular level, they also vary to some extent with the "state" of the fluid, especially its temperature.

Tables of experimental values of these properties are given in the Appendix. If other types of energy may affect the flow process (thermal, electric, etc.) then corresponding terms should be incorporated in the equation. However, the ones shown normally suffice for hydraulic processes.

We now wish to examine the general flow function and to organize its terms in such a way as to give the best order for experimental determination of its exact form for any particular process.

Since there are in general three independent dimensional units (force, length, and time units), the number of variables in Eq. (2–40) can be reduced by 3, by the device of grouping the variables into a number of dimensionless groups or parameters.[1] Each parameter will consist of three of the variables, with exponents chosen to satisfy dimensional requirements, and one other variable whose exponent can be chosen arbitrarily. The three basic variables are chosen most conveniently as a representative velocity, a representative length, and a representative energy term or fluid characteristic, usually density. Letting the arbitrary exponent in each case be -1, Eq. (2–40) becomes

$$f\left[\frac{V^{x_1}D^{y_1}\rho^{z_1}}{L}, \frac{V^{x_2}D^{y_2}\rho^{z_2}}{\Delta(p+\gamma h)}, \frac{V^{x_3}D^{y_3}\rho^{z_3}}{\mu}, \cdots\right] = 0 \qquad (2\text{--}41)$$

Each of the above parameters is to be made dimensionless. Since the denominator in the first term is a length, the numerator must also be a length. Thus x_1 and z_1 must be zero and y_1 unity. The first term is therefore simply D/L.

Exponents x_2, y_2, and z_2 can be determined by three simultaneous equations expressing requirements for the parameter to be dimensionless. Writing the parameter in terms of its units, it becomes

$$\left(\frac{L}{T}\right)^{x_2}(L)^{y_2}\left(\frac{FT^2}{L^4}\right)^{z_2}\left(\frac{F}{L^2}\right)^{-1}$$

For the parameter to be dimensionless, the exponents of all length terms must total zero; similarly, the exponents of the force and time terms respectively must add to zero. This gives three simultaneous equations in the three unknown exponents:

$$x_2 + y_2 - 4z_2 + 2 = 0$$
$$z_2 - 1 = 0$$
$$-x_2 + 2z_2 = 0$$

Solution of these equations yields $z_2 = 1$, $x_2 = 2$, $y_2 = 0$. The dimensionless parameter resulting is therefore $\rho V^2/\Delta(p+\gamma h)$.

A similar analysis of the exponents of the third term yields the equations

$$x_3 + y_3 - 4z_3 + 2 = 0$$
$$z_3 - 1 = 0$$
$$-x_3 + 2z_3 - 1 = 0$$

These equations yield $z_3 = 1$, $x_3 = 1$, $y_3 = 1$. The third parameter then becomes $\dfrac{\rho V D}{\mu}$. Each of the other parameters can be obtained similarly.

---

[1] If preferred, the dimensional analysis may employ (mass, length, time) units or even (energy, length, time) units instead of the (force, length, time) units commonly used by engineers.

Thus:

$$f\left[\frac{D}{L}, \frac{\rho V^2}{\Delta(p + \gamma h)}, \frac{VD\rho}{\mu}, \frac{\rho V^2}{\gamma D}, \frac{\rho DV^2}{\sigma}, \frac{\rho V^2}{E}, \frac{D}{h}, \frac{D}{\lambda}, \frac{D}{s}, \frac{D}{r}, \frac{D}{A}, \frac{D}{B}, \frac{D}{C} \cdots \right] = 0$$

(2-42)

Since each of the original variables still appears in the function, nothing has been lost, even though the number of parameters is now smaller by 3. In addition, each parameter is dimensionless, a fact which will be seen to be of high utility in experimental investigation of the function. The geometric ratios in the function imply that it is not the actual geometric size that determines flow patterns, but rather relative size and geometric similarity. This immediately suggests the use of hydraulic models.

Similarly, the other terms, each of which involves some sort of force, suggests that relative force magnitudes, rather than actual force values, control flow phenomena. This can be better appreciated if these terms are written

$$\frac{\rho V^2}{\Delta(p + \gamma h)}, \quad \frac{\rho V^2}{\mu V/D}, \quad \frac{\rho V^2}{\gamma D}, \quad \frac{\rho V^2}{\sigma/D}, \quad \frac{\rho V^2}{E}$$

In this form, the numerator and denominator of each term appear dimensionally as a force per unit area, or stress. The numerator of each term, ρV^2, may be regarded as the unit inertial force inherent in the fluid's motion. The denominators of the respective terms may be recognized, dimensionally, as representative of the unit forces due to potential head, viscosity, gravity, surface tension, and elasticity, in order.

Each parameter therefore represents the ratio of the inertial forces, which are *always* present in a flowing fluid, to the other forces which *may* be present and affecting the patterns of flow. In like manner as the dimensionless geometric parameters imply that only the *relative* sizes affect flow patterns, so these dimensionless force ratios imply that the *relative* force magnitudes affect those patterns.

This fact leads to the recognition of the importance of geometric and dynamic similarity in the understanding of hydraulic phenomena. In other words, in two situations of similar character hydraulically (that is, in which all homologous dimensions and all homologous forces are in the same ratios to one another), either can be considered as a scale model of the other, and completely similar patterns of flow, pressure, etc., will exist. This is the basis of the widespread use of hydraulic models as a design tool for complex hydraulic structures and flow phenomena, as well as the scientific empiricism that has contributed so remarkably in recent decades to the advances in fluid mechanics and hydraulic engineering.

The term ρV^2 is also recognized as proportional to the kinetic energy per unit volume. Similarly, $\Delta(p + \gamma h)$ measures the potential energy per unit volume, E the elastic energy per unit volume, and so on. Thus each of the terms may also be thought of as energy ratios.

Each of the five dynamic parameters may also be modified to kinematic parameters (ratios of two velocities) by dividing numerator and denominator by the density ρ and then taking the square root. Obviously these operations do not change the dimensionless or "relative" character of the parameters, so that the numbers still represent essentially the same influence on the flow structure. In this form they become

$$\frac{V}{\sqrt{\Delta(p+\gamma h)/\rho}}, \quad \frac{V}{\sqrt{\mu V/\rho D}}, \quad \frac{V}{\sqrt{\gamma D/\rho}}, \quad \frac{V}{\sqrt{\sigma/\rho D}}, \quad \frac{V}{\sqrt{E/\rho}}$$

The denominator of each term now represents a velocity, each of which could be shown to be a significant velocity measure or index relating to the type of force associated with it. These, in turn, turn out to be proportional to—

a. the *efflux velocity* from, say, a pressurized chamber outlet, whereby the entire potential energy in the escaping liquid is converted into kinetic energy in the jet;

b. the *shear velocity,* $v^* = \sqrt{\tau_0/\rho}$, associated with boundary shearing stresses;

c. $V_w = \sqrt{gD}$, the velocity of a gravity wave on a free surface when it has a shallow depth of D;

d. the velocity of a capillary wave $\sqrt{\sigma/\rho D}$, where the dimension D can be shown to be proportional to the wave length λ; and

e. the velocity of a pressure wave (e.g., sound) in the fluid, known as c, the celerity, equal to $\sqrt{E/\rho}$.

Particular forms of these dimensionless parameters have come to be known by the name of the investigator who first recognized their respective importance in fluid studies. These names are as follows:

$$\text{Euler number, } N_E = \frac{1}{2}\left[\frac{V}{\sqrt{\dfrac{\Delta(p+\gamma h)}{\rho}}}\right]^{-2} = \frac{\Delta(p+\gamma h)}{\frac{1}{2}\rho V^2} \quad (2\text{-}43)$$

$$\text{Reynolds number, } N_R = \left[\frac{V}{\sqrt{\mu V/\rho D}}\right]^2 = \frac{DV\rho}{\mu} = \frac{DV}{\nu}, \quad (2\text{-}44)$$

where ν is the kinematic viscosity, μ/ρ.

$$\text{Froude number, } N_F = \frac{V}{\sqrt{\gamma D/\rho}} = \frac{V}{\sqrt{gD}} \quad (2\text{-}45)$$

(also frequently used as V^2/gD).

$$\text{Weber number, } N_W = \left[\frac{V}{\sqrt{\sigma/\rho D}}\right]^2 = \frac{\rho DV^2}{\sigma} \quad (2\text{-}46)$$

$$\text{Mach number, } N_M = \frac{V}{\sqrt{E/\rho}} = \frac{V}{c} \qquad (2\text{-}47)$$

(Sometimes $\rho V^2/E$ is called the Cauchy number.)

Equation (2–41) may thus finally be written

$$f\left(N_E,\ N_R,\ N_F,\ N_W,\ N_M,\ \frac{L}{D},\ \frac{h}{D},\ \frac{\lambda}{D},\ \frac{s}{D},\ \frac{r}{D},\ \frac{A}{D},\ \frac{B}{D},\ \frac{C}{D}, \ldots\right) = 0$$
$$(2\text{-}48)$$

Of the five dynamic parameters, the Euler number is the one of most ubiquitous concern, since it is essentially the ratio of kinetic to potential energy difference in the flow system. The presence of the potential energy differential may result from the activity of any or all of the other forms of energy on the system. That is, the fluid pressure which controls the flow may be present because of the effect of gravity, elasticity, surface tension, viscosity, or some combination. Stating it another way, the force due to pressure and/or elevation is determined by the other forces present, in a statement of force equilibrium.

It is nearly always convenient, therefore, to take the Euler number out of the flow function and regard it as dependent on the other parameters in the function. The resulting equation can be regarded as the general process equation. Thus:

$$\frac{\Delta(p + \gamma h)}{(\tfrac{1}{2})\rho V^2} = f\left(N_R,\ N_F,\ N_W,\ N_M,\ \frac{L}{D} \cdots\right) \qquad (2\text{-}49)$$

In the special, though common, case in which the flow is uniform (i.e., velocity patterns constant from section to section), the functional relationship of $\Delta(p + \gamma h)$ to L/D is obviously linear, and this may be written

$$\Delta(p + \gamma h) = \tfrac{1}{2}\rho V^2 \frac{L}{D} f\left(N_R,\ N_F,\ N_W,\ N_M,\ \frac{h}{D},\ \frac{\lambda}{D},\ \frac{s}{D},\ \frac{r}{D},\ \frac{A}{D},\ \frac{B}{D},\ \frac{C}{D}, \cdots\right)$$
$$(2\text{-}50)$$

or, in terms of pressure and velocity heads,

$$\Delta\left(\frac{p}{\gamma} + h\right) = \frac{V^2}{2g}\left(\frac{L}{D}\right)f \qquad (2\text{-}51)$$

Here, f is a dimensionless coefficient, which is seen in turn to be a function of the several dimensionless parameters pertinent to the particular flow situation.

These equations are fundamental and can be applied to most hydraulic phenomena. The experimental phase of hydraulics comes in the determination of the particular functional relationships for f.

The principles of dimensional representation are the basis of hydraulic design by models. As noted, two similar physical situations can be considered as scaled replicas of each other if all pertinent length ratios are the same (geometric similarity) and all pertinent force ratios are the same (dynamic similarity).

It is usually feasible to achieve geometric similarity in the laboratory by means of a scale model of the structure or field phenomenon being studied. Dynamic similarity is much more difficult to obtain, however, since the various types of forces acting would usually bear different ratios to the ever-present inertial forces. If all the pertinent force ratios (or all the pertinent dimensionless numbers of Eqs. (2–43) through (2–48)) could be made the same in model and prototype, then the factor f in Eq. (2–51) would become a constant applicable to both. It could be determined by one set of measurements on the model, of corresponding values of $\Delta(p + \gamma h)$ and V, and the resulting equation then used for the prototype.

Generally, this is impossible, and an attempt must be made to determine the function f in terms of the particular parameters which relate to forces that affect it. It may be possible in many instances, by a planned sequence of laboratory experiments, to determine f first as a function of N_R, holding other parameters constant, then of N_F, and so on, in order eventually to develop an explicit equation for the function. This is feasible mainly if the phenomenon or structure is of wide interest or application. It would probably not be feasible for use on only one particular structure.

In the latter case, the approach is usually to consider only the design values of velocity, discharge, etc., and to attempt to minimize (or simply to neglect) the effect of all except the one predominating type of force. Then, only one of the dimensionless force ratios must be made equal in model and prototype and this is usually quite possible.

Thus, in a phenomenon where viscosity predominates (such as flow around immersed objects or through closed boundaries), only the Reynolds number need be made equal in model and prototype. In situations where gravity is most important (such as flow in steep open channels, gravity waves, flow over weirs and through orifices, flow in spillways and stilling basins, etc.), the Froude number controls. In either case, model results can be transferred directly to the prototype by using the appropriate dimensionless number (Reynolds, Froude, etc.) to determine the basic relations between quantities in the model and the prototype. This is discussed more fully in Chapter 8.

2–11. Application of the Flow Function. We have seen that most hydraulic processes can be expressed in the form of the general flow function, Eq. (2–50). This has been derived by dimensional analysis (using the technique of Buckingham's π-theorem) and is quite general, including as it does provision for all the geometric, kinematic, and dynamic parameters which could normally be applicable in the system.

Therefore Eq. (2–50) can usually be considered as the starting point in developing a non-dimensional equation for a particular process. In some cases it may be preferable to go back to Eq. (2–48) and take one of the other parameters (instead of the Euler number) out of the function as the dependent variable.

In either case the equation may be considerably simplified by removing all the terms involving lengths or forces which have no immediate physical relevance to the particular process. The resulting equation will give the general form of the desired process equation but will still include a dimensionless coefficient which must be determined experimentally, as a function of the one or more dimensionless parameters remaining in the functional expression. The procedure can be illustrated in the following examples.

Example 2–3

A Venturi meter in a smooth circular pipe of diameter D causes a contraction to a throat diameter d. The pipe velocity is V. The liquid in the pipe has a density ρ, specific weight γ, surface tension σ, modulus of elasticity E, and viscosity μ. The length of the contracting section is L_1 and of the expanding section L_2. By dimensional analysis derive a formula for the head loss in the pipe caused by the meter.

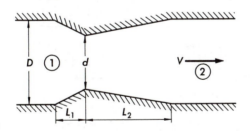

Solution. The liquid has no free surface, so that σ can be neglected, and also it can be assumed incompressible for ordinary flows, so that E can be neglected. Although gravity of course is acting, it acts on all fluid particles equally so that it has no relative effect on the flow structure. Since the pipe is smooth, the geometry of the boundary roughness elements need not be considered. If the pressure is constant across each cross-section, the general flow function becomes:

$$f\left(N_E, N_R, \frac{d}{D}, \frac{L_1}{D}, \frac{L_2}{D}\right) = 0$$

or

$$\Delta(p + \gamma h) = \tfrac{1}{2}\rho V^2 f\left(N_R, \frac{d}{D}, \frac{L_1}{D}, \frac{L_2}{D}\right) \tag{1}$$

By the energy equation, between points (1) and (2)

$$\frac{p_1}{\gamma} + h_1 + \frac{V_1^{\,2}}{2g} = \frac{p_2}{\gamma} + h_2 + \frac{V_2^{\,2}}{2g} + H_L$$

Since $V_1 = V_2$,

$$H_L = \left(\frac{p_1}{\gamma} + h_1\right) - \left(\frac{p_2}{\gamma} + h_2\right) = \Delta\left(\frac{p}{\gamma} + h\right) \tag{2}$$

Substituting equation (1) in equation (2)

$$H_L = \frac{V^2}{2g} f\left(N_R, \frac{d}{D}, \frac{L_1}{D}, \frac{L_2}{D}\right) \tag{3}$$

The function f in equation (3) may be considered as the dimensionless loss coefficient for the meter. It could then be evaluated experimentally on a laboratory model, in terms of the dimensionless parameters in the function.

Example 2–4

Derive by dimensional analysis an expression for the force resisting the motion of a surface ship through the water. In addition to the force, the pertinent variables include: the ship speed V, its hull length L, its width B, the hull surface roughness (assumed as expressed in terms of some roughness dimension ϵ), fluid density ρ, specific weight γ, viscosity μ, the ship draft h, bottom depth D, and a dimensionless shape factor α.

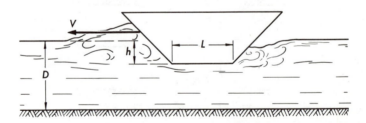

Solution. Consideration of the physical aspects of the problem seem to justify the assumption that the drag force on the ship will depend both on shear on its sides and the build-up of water ahead of its prow combined with a wake in its lee. The latter is expected to be more important and can be evaluated in terms of the resulting difference in hydrostatic pressures on the bow and stern. The repeating variables are chosen as h, V, and ρ. Then

$$f\left(N_E, N_R, N_F, \frac{B}{h}, \frac{L}{h}, \frac{\epsilon}{h}, \frac{D}{h}, \alpha\right) = 0$$

$$\Delta p = \tfrac{1}{2}\rho V^2 f\left(N_R, N_F, \frac{B}{h}, \frac{L}{h}, \frac{\epsilon}{h}, \frac{D}{h}, \alpha\right)$$

The pressure difference is acting effectively on the area $B(h)$ to cause the drag force due to wave pile-up. Thus:

$$F = \tfrac{1}{2}\rho V^2 (Bh)(C_D)$$

where the drag coefficient $C_D = f\left(N_R, N_F, \dfrac{B}{h}, \dfrac{L}{h}, \dfrac{\epsilon}{h}, \dfrac{D}{h}, \alpha\right)$ and would have to be determined experimentally.

PROBLEMS

2–1. A standpipe 12 ft high and 2 ft in diameter is filled with water.

    a. What is the total head at the top?

    b. What is the total head at the base?

    c. What is the total potential energy represented in the water in the entire tank, in foot-pounds?

2–2. A static pressure tube is used to measure the pressure in water flowing in a pipe line. It is connected to a differential manometer, the other leg of which is connected to a reference point in the pipe where the water pressure is known to be 30 psi. The manometer contains carbon tetrachloride (specific gravity 1.6) and gives a reading of 3 in., the lower level being in the leg adjacent to the reference section. The latter is 18 in. higher than the section being measured. What is the static pressure at the latter, in psi? *Ans.:* 30.6 psi.

2–3. A 6-in.-diameter vertical pipe is contracted to a 4-in.-diameter section to form a venturi meter. A differential manometer containing mercury (specific gravity 13.6) as the measuring fluid is attached to the full and throat sections. The throat section is 12 in. above the full section, with water flowing upward. Neglect friction losses. The manometer reading is 8 in.

    a. What is the difference in pressure heads, in feet of water?

    b. What is the pipe discharge, in cfs?

2–4. A horizontal pipe line 12 in. in diameter and 2000 ft long connects two reservoirs, with a 40-ft difference in surface elevation. Assuming the friction loss in the pipe line to be 20 ft, what is the rate of flow in the pipe, in cfs? *Ans.:* 28.2 cfs.

2–5. A 4-in. vertical pipe discharges vertically downward into the atmosphere. When the pipe discharges at a rate of 1.0 cfs, what is the diameter of the jet 10 ft below the pipe exit?

2–6. A hydraulic jump below a spillway forms in a channel 20 ft wide, jumping from a depth of 1.0 ft to a depth of 3.0 ft. What is the discharge in the channel? *Ans.:* 278 cfs.

2–7. A pipe 1 ft in diameter carries oil of specific gravity 0.90 at a mean velocity of 20 ft/sec. What is the rate of flow:

    a. In pounds per second?

    b. In slugs per second?

    c. In gallons per minute?

2–8. A waterfall discharging 200 cfs drops 1000 ft. What is the available horsepower in the falls? *Ans.:* 22,700 hp.

2–9. Water flows from a reservoir through a penstock to a turbine, discharging through a 1-in.-diameter nozzle. The total drop is 800 ft and the friction loss in the line is 200 ft. If the turbine efficiency is 90 per cent, what shaft horsepower can be developed on it?

2–10. An orifice plate with a rounded orifice 4 in. in diameter is placed at the discharge end of an 8-in. pipe. When the discharge is 4 cfs, what is the force exerted on the plate?

2–11. A horizontal bend in a 10-ft penstock goes through a change in direction of 60°. Neglecting friction, and assuming a pipe-line pressure of 12 psi, what is the total hydraulic force on the bend, when the discharge is 1000 cfs? *Ans.:* 160,500 lb.

2–12. A free vortex or "whirlpool" is formed above an open drain in a shallow tank. The tangential velocity is 0.20 ft/sec at a distance of 1.0 ft from the axis. What is the total drop in surface elevation at a distance of 1.0 in. from the axis? *Ans.:* 1.07 in.

2–13. A flow of 4 cfs of water is carried in a 6-in.-diameter horizontal pipe. A valve in the pipe line has a head loss coefficient of 0.4.
   a. If pitot tubes were inserted in the flow upstream and downstream from the valve, what would be the difference in water levels in the two tubes?
   b. If piezometer tubes were used instead of pitot tubes, what would be the difference in levels?
   c. Approximately what hydrodynamic force, in pounds, is exerted on the valve?

2–14. Water flows under a sluice gate in a horizontal open channel 10 ft wide. The flow depth upstream from the sluice gate is 8 ft and downstream is 2 ft. Calculate: (a) Discharge, cfs.; (b) force on sluicegate, pounds. (Assume no energy loss.)

2–15. Flow on a horizontal apron at the base of a spillway goes through a hydraulic jump, from a depth of 2 ft to a depth of 4 ft. The discharge is q cfs per ft of width.
   a. Write the continuity equation between the two sections, in terms of the velocities and the two known depths.
   b. Write the momentum equation between the two sections, in terms of q and the two known depths, *only.*
   c. Write the energy equation between the two sections, in terms of q, the two depths, and the head loss H_L, only.
   d. Determine the values of q and H_L.

2–16. An orifice plate with a 2-in.-diameter orifice is inserted in a 4-in.-diameter horizontal water pipe. The orifice has a contraction coefficient of 0.60. A differential manometer is attached to the pipe, with one leg just upstream from the plate and one just opposite the "vena contracta." When the manometer reads 2 in. of mercury, compute:
   a. pipe discharge (neglect energy losses); *Ans.:* 0.154 cfs.
   b. force on orifice plate (neglect shear forces). *Ans.:* 8.44 lb.

2–17. A Venturi meter in a horizontal pipe line carrying water contracts from a normal pipe section of 12-in.-diameter to a throat section of 4-in.-diameter. At a certain time, the velocity and pressure at the throat section are 10 ft/sec and 2 lb/in.², respectively. Neglect energy losses.

   a. What is the reading on a differential manometer attached to the meter? The indicating fluid is mercury (specific gravity 13.6).

   b. What is the hydrodynamic force, in pounds, on the contracting section?

2–18. A liquid of specific weight γ flowing in a horizontal open channel emerges under a sluice gate as shown, with upstream and downstream depths H and D. The unit discharge is q cfs/ft of channel width. Write the following equations for this system, in simplest form. Neglect energy losses.

   a. Conservation of energy, in terms of H, D, and q only.

   b. Hydrodynamic force on the sluice gate, in terms of H, D, and γ only.

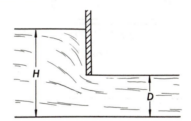

Prob. 2–18.

2–19. A Venturi meter in a horizontal pipe line carrying water contracts from a normal pipe section of 6 in. diameter to a throat section of 2 in. diameter. A differential manometer containing mercury (specific gravity 1.36) attached to the meter gives a reading of 1.47 in. and a piezometer tube attached to the upstream full section of the meter has a water level 4.62 ft above the pipe center line. Calculate (neglecting head losses):

   a. Pipe discharge, cfs.

   b. Hydrodynamic force on the meter, lb.

2–20. A sloping pipe line is carrying a discharge of 2 cfs. Between points A and B the diameter increases from 6 in. to 12 in. A carbon tetrachloride (specific gravity = 1.59) manometer attached to these two points gives a reading of 4 in., as shown. Calculate the head lost in the pipe between points A and B, in ft. *Ans.:* **1.29 ft.**

2–21. A semi-circular plate is placed in a 6-ft diameter pipe, as shown, blocking half the pipe cross-section. Assume the lip of the plate is well-rounded, so that no additional contraction, beyond the height of the lip, is caused. If the pressure upstream from the wall is 5 psi and the discharge is 100 cfs, calculate:

   a. The reading on a differential mercury manometer, with legs attached immediately upstream and downstream from the plate, in inches (specific gravity $= 13.6$)

   b. Energy loss caused by the plate, ft-lb/lb.

   c. Drag force on the plate, in lb.

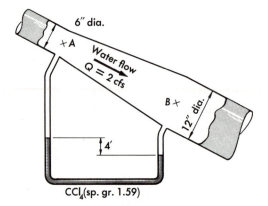

Prob. 2–20.

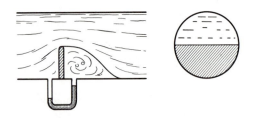

Prob. 2–21.

2–22. A masonry dam constructed of 150 pcf concrete is triangular in cross-section, with a vertical upstream face and a 45° slope on the downstream face. The dam is 50 ft high and retains water in a reservoir 42 ft deep. Compute:

   a. The force tending to cause sliding across the base. *Ans.:* 55,000 lb.

   b. The resultant foundation reaction, assuming an impermeable base. *Ans.:* 195,000 lb.

2–23. The top edge of a 3-ft-square gate on the upstream face of a dam lies 8 ft below the water surface. The face makes an angle of 60° with the horizontal. What is the load on the gate and where is it applied?

2–24. If the reservoir surface of Problem 2–9 rises 2 ft, what vertical force would have to be applied at the lower edge of the gate, which is hinged about its upper edge, to open it? *Ans.:* 6560 lb.

2–25. A closed cylindrical tank 2 ft high and 3 ft in diameter is filled with water under pressure. The pressure is measured by a piezometer tube, in which water rises 10 ft above the top of the tank. Draw to scale the diagram of vertical forces acting internally on the tank.

2-26. A sheet piling wall separates the sea water (specific weight 64 lb/cu ft) in an estuary from the fresh water upstream. The upstream depth is 25 ft. At what downstream depth will there be:

a. Zero resultant force on the wall? *Ans.:* 24.7 ft.

b. Zero resultant moment on the wall? *Ans.:* 24.8 ft.

2-27. A vertical sluice gate 4 ft high and 8 ft long weighs 1000 lb. It is raised in slots, the coefficient of friction between the gate and slots being 0.25. If its top edge is 1 ft below the water surface on one side, what vertical force is required to raise it?

2-28. An elevated water tank is in the form of a vertical cylinder with a hemispherical bottom. The tank is 20 ft in diameter. What is the resultant hydrostatic force on one quadrant of the hemispherical bottom when the water level is 50 ft above the bottom? *Ans.:* 386,000 lb.

2-29. A 6-ft-diameter circular gate is placed in the vertical wall of a water reservoir, pivoted about a horizontal axis 4 in. below its center. A sill at the bottom prevents it from opening, but there is no sill at the top. To what depth above the top of the gate can the water rise in the reservoir before the gate will open?

2-30. A 30-ft length of pipe, 4 ft in diameter, with ends sealed, is floated across a river. It weighs 4500 lb. How deep will it submerge?

2-31. A scow 12 ft wide, 24 ft long, and 4 ft deep is made of wood weighing 45 lb/cu ft. The sides and bottom are 10 in. thick. How many cubic yards of gravel, at 110 lb/cu ft, can it carry, with a freeboard of 1 ft?

2-32. A masonry dam has an upstream surface that makes an angle of 60° with the horizontal. The dam is 80 ft high. Calculate the magnitude, direction, and center of pressure of the hydrostatic force on the bottom half of this surface, when the reservoir is full. Also calculate the frictional force that would have to be provided at the base of the dam to prevent the dam from sliding across the foundation.

2-33. The upstream face of a dam follows the parabolic equation $y = 6x^{1/2}$, as shown. The dam is 60 ft high. When the reservoir is full, compute the magnitude of the resultant hydrostatic force on the upstream face of the dam.

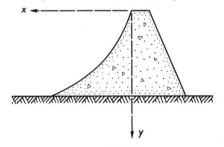

Prob. 2-33.

2-34. A dam has an upstream face with a slope of 4 vertical to 3 horizontal. What is the magnitude of the hydrostatic force on the dam when the center of pressure is 10 ft vertically below the water surface?

2-35. A 10 ft by 20 ft rectangular gate hinged at its lower edge closes the entrance to a tunnel as shown. The distance from the water surface to the centroid of the gate is 30 ft. The gate makes an angle of 60° with the horizontal. Compute the force on the stop along the upper edge of the gate.

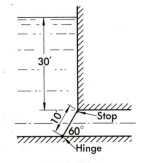

Prob. 2–35.

2-36. Compute the magnitude and direction of the resultant hydrostatic force and the location of the center of pressure on the tank wall $ABCD$ as shown.

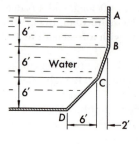

Prob. 2–36.

2-37. Compute the magnitude, direction, and line of action of the resultant hydrostatic force acting on the wall shown.

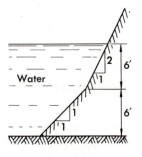

Prob. 2–37.

2–38. A paved levee embankment has a 45° slope. A rectangular gate, 6 ft square, in the face of the embankment has its center 20 ft above the bed.

   a. When the river water surface is 40 ft above the river bed, what is the total hydrostatic force on the gate and where is it applied?

   b. Calculate the height to which the water surface would have to rise in order for the resultant hydrostatic force on the gate to pass through the midpoint of the gate.

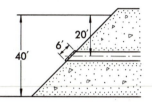

Prob. 2–38.

2–39. Water stands behind a spillway to a total depth of 100 ft above the river bed. After flowing down the spillway and becoming horizontal again, it has a depth of 10 ft as shown. Assuming continuous flow, streamlined transitions, and no air entrainment or energy loss, what will be the hydrodynamic force on the spillway? The crest length is 400 ft.

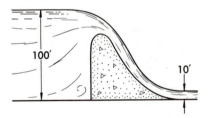

Prob. 2–39.

2–40. The spillway of Problem 2–39 is studied in the laboratory by means of a 25:1 scale model, using a model discharge of 0.8 cfs/ft. The head over the spillway crest is measured to be 0.5 ft and the velocity at the toe of the spillway to be 8 ft/sec. Calculate:

   a. Prototype discharge;

   b. Prototype crest head;

   c. Prototype toe velocity;

   d. Horsepower per foot in flow at toe of prototype spillway. *Ans.:* a. 40,000 cfs; b. 12.5 ft; c. 40 fps; d. 319 hp/ft.

2–41. Derive, from dimensional considerations, an equation for flow over a weir, assuming the rate of flow to depend on the length and height of the weir, head on the weir, density, specific weight, and surface tension of the fluid.

2–42. Derive by dimensional analysis an equation for the hydrodynamic force of wind blowing against a billboard, assuming it to be a function of the size and shape of the billboard and the velocity, viscosity, compressibility, and density of the moving air.

2–43. A submerged submarine is to cruise at 15 mph. At how many ft/sec must a 1:25 scale model be towed in a towing basin, to produce dynamic similarity? Assume sea water is used in the towing basin.

2–44. Water at 60 F flows in a tunnel of 10-ft hydraulic radius, and is to be studied by means of air flowing in a 1:20 scale model. Assuming geometric similarity includes full similarity of roughness patterns, what must be the ratios of velocity and discharge in the model and prototype to achieve dynamic similarity?

2–45. A spillway model is constructed to a scale of $1:L$ of its prototype. Determine for dynamic similarity the corresponding ratios of velocity, discharge, forces, and flow horsepower in the model and prototype, each in terms of L.

2–46. A disturbance on the surface of a shallow body of still water causes surface waves to radiate out from the source of the disturbance. Derive, by dimensional analysis, a formula for the velocity of the waves. Assume the velocity depends on the water density, specific weight, surface tension, and depth.
a. Assume effect of gravity more important than surface tension.
b. Assume effect of surface tension more important than gravity.

2–47. Derive, by dimensional analysis and the conservation equations, a general formula for the head loss at an orifice plate in a pipe line. Assume the variables that may affect the flow process include: (1) pipe diameter ; (2) orifice diameter; (3) radius of rounding of orifice opening; (4) density of fluid ; (5) pressure difference on the two sides of the plate; (6) flow velocity; (7) fluid viscosity; (8) approach length of straight pipe; (9) exit length of straight pipe.

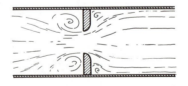

Prob. 2–47.

2–48. The velocity V with which an air bubble of diameter D will rise in a static liquid is believed to be dependent on:

$$\rho = \text{density of liquid}$$
$$\rho' = \text{density of air}$$

σ = surface tension of liquid-air
interface

E = compressibility of air

μ = viscosity of liquid

Δp = pressure difference across
bubble

Derive by dimensional analysis an expression for V.

3

HYDRAULICS OF FLOW IN CLOSED CONDUITS

3–1. Introduction. Water and other liquids can flow around or within an infinite variety of boundary geometrical configurations, and each one would in general be definable by a particular process equation. However, it is desirable to generalize these equations as much as possible, so as to reduce their number and enable each to apply to as broad a range of boundary geometries and flow structures as possible.

The two broadest categories are those of enclosed flow and open flow. In the former the fluid channels are completely enclosed by solid boundaries and the fluid pressure is normally something other than atmospheric. In the latter, the liquid has a "free surface," exposed to the atmosphere, and thus the fluid pressure at this surface is zero.

In a sense, these two categories could be held to include all hydraulic processes, but for practical reasons special treatment is required for numerous particular geometries and flow situations. In any case, the study of flow in ordinary closed conduits (or pipes) and flow in open conduits (or open channels) is basic in applied hydraulics in the broadest sense.

Since the only differences between flow in closed and open conduits are caused by the one free surface boundary in the latter, there are many ways in which the two processes are similar and many situations in which the same process equations can be used for both. The discussion in this chapter deals mainly with closed conduits and that in the next with free-surface conduits, but it will be noted that much of the treatment in this chapter can also be applied, with slight modifications, to open channels.

3.2. Uniform Flow. If a fluid is moving under either gravity or pressure in a straight, uniform conduit of any cross-sectional shape, the effects represented by the Weber number and by changes in section geometry need not be considered. Only such geometric terms are used as are necessary to describe the boundary geometry. Flow at constant velocity throughout a given reach is called *uniform flow*, and will obviously occur in this case. The general flow equation, Eq. (2–49), then becomes

$$\Delta(p + \gamma h) = \frac{\rho V^2 L}{2D} f\left(N_R, N_F, N_M, \frac{\lambda}{D}, \frac{h}{D}, \frac{s}{D}, \dots\right) \qquad (3\text{–}1)$$

Here it is assumed that the only dimension required to specify the cross-section size and shape is the *equivalent diameter* of a circular pipe having the same piezometric gradient. If, however, the section is of large aspect ratio (i.e., ratio of width to depth) then the reference dimension should be taken as the depth of section, with aspect ratio included as one of the dimensionless geometric parameters.

For most problems in hydraulic engineering, the aspect ratio is reasonably near unity and also the effect of compressibility is negligible. In open channels of mild slope, as well as in pipes, the relative effect of gravity may also be neglected. It is also usually more convenient to express the pressure drop in terms of pressure head. Equation (3–1) then becomes

$$\Delta\left(\frac{p}{\gamma} + h\right) = f\left(\frac{L}{D}\right)\left(\frac{V^2}{2g}\right) \tag{3–2}$$

where f is now primarily a dimensionless *friction factor*, since it is dependent on only the Reynolds number and the boundary roughness elements, both of which factors involve frictional resistance to flow. That is:

$$f = f\left(N_R, \frac{\lambda}{D}, \frac{h}{D}, \cdots\right)$$

It is also instructive to examine Eq. (3–2) from the point of view of the momentum and energy relationships. Consider a uniform length of pipe of any cross-section, as in Fig. 3–1. The force system acting on the body of fluid in the reach consists of its weight, the hydrostatic pressure forces on its ends (the radial pressures at the walls have no component in the flow direction), and the shearing force on its periphery. The latter is expressed in terms of a wall shearing stress τ_0. (Actually this is not a definite localized friction as in the case of a solid cylinder moving inside another solid,

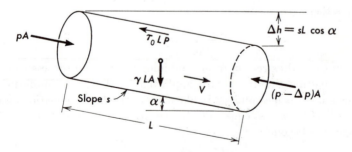

Fig. 3–1. Forces on flowing fluid.

hollowed cylinder; however, the aggregate effect of all internal resistances to flow can be treated for purposes of analysis as peripheral shearing stresses.)

Summing forces in the direction of flow (since there is no change in momentum in the reach), the momentum equation gives:

$$pA + \gamma LAS \, (\cos \alpha) = (p - \Delta p)A + \tau_0 PL$$

which simplifies to

$$SL(\cos \alpha) = -\frac{\Delta p}{\gamma} + \left(\frac{\tau_0}{\gamma}\right)\left(\frac{P}{A}\right)L$$

The ratio A/P is a length which reflects the size and shape of the cross-section and is known as the *hydraulic radius*. For a circular pipe it has the value $D/4$, and therefore can be replaced by the factor $\frac{1}{4}D$, where D represents the equivalent diameter. S is the slope of conduit, the ratio of the drop or rise, ΔH, to the horizontal length L ($\cos \alpha$). The equation can then be written

$$\Delta h + \frac{\Delta p}{\gamma} = \left(\frac{4\tau_0}{\gamma}\right)\left(\frac{L}{D}\right)$$

Since Δh and $\Delta p/\gamma$ both represent changes in potential energy, they can be combined as follows:

$$\Delta\left(\frac{p}{\gamma} + h\right) = \left(\frac{8\tau_0}{2g\rho}\right)\left(\frac{L}{D}\right) \tag{3-3}$$

Equations (3–2) and (3–3) may now be set equal to each other, yielding

$$\left(\frac{8\tau_0}{2g\rho}\right)\left(\frac{L}{D}\right) = f\left(\frac{L}{D}\right)\left(\frac{V^2}{2g}\right) \tag{3-4}$$

from which is obtained

$$fV^2 = \frac{8\tau_0}{\rho} \tag{3-5}$$

It now is evident that the wall shearing stress τ_0 and the friction factor f are directly related. It is convenient to regard the combination $\sqrt{\tau_0/\rho}$ as a velocity related to the wall velocity (although not numerically equal to it). It is called the *shear velocity* or *friction velocity*, v^*. In dimensionless form, Eq. (3–5) is then

$$\frac{V}{v^*} = \sqrt{\frac{8}{f}} \tag{3-6}$$

Equation (3–6) thus results from combining the general process equation and the momentum equation and is applicable to all cases of uniform flow in either closed or open conduits. In like manner the energy equation may be combined with the process equation. In the special case of uniform flow, the energy equation becomes:

$$\Delta\left(\frac{p}{\gamma} + h\right) = H_f \tag{3-7}$$

where H_f is the head loss in the reach due to normal boundary friction. Combining this with Eq. (3–2), the process equation becomes:

$$H_f = f\left(\frac{L}{D}\right)\left(\frac{V^2}{2g}\right) \tag{3-8}$$

In this form, the equation is known usually as the Darcy equation, after the French engineer who derived it empirically in 1859.

This equation gives a dimensionally correct expression, for evaluating energy dissipation in a uniform conduit, quite rational as far as it goes. However, its solution requires a knowledge of the friction factor f, which has been seen to depend on the Reynolds number for the flow and on the geometry of the wall roughness elements. In most cases, this function must be determined, at least in part, empirically.

Finally the momentum and energy equations, Eqs. (3–3) and (3–7), may be combined to yield an equation relating the head loss and shear stress.

$$H_f = \frac{4\tau_0}{\gamma}\left(\frac{L}{D}\right) \tag{3-9}$$

Equations (3–2), (3–3), (3–6), (3–7), (3–8), and (3–9) are thus all important general equations applicable to uniform flow in either open or closed conduits.

3–3. Shear Stress and Velocity Distribution in Circular Pipes.
The friction factor must be determined if a flow analysis is to be made. It reflects the irrecoverable conversion of flow energy into heat energy. This conversion process is accomplished in one of the following three ways:

1. Shearing stresses between adjacent laminae of fluid sliding over each other. Flow in which this phenomenon occurs is called laminar flow, and is characterized by a small value of Reynolds number (for circular pipes, less than about 2000).
2. Generation of turbulent vortices, the rotational energy of which is derived from the flow energy and is subsequently dissipated by viscous attrition with adjacent fluid particles. The vorticity generation results from contact of fast-moving fluid with slow or stagnant fluid in the laminar boundary film or at zones of separation and is itself the cause of most of the friction head loss in this type of flow, known as turbulent flow.
3. Combination of the laminar and turbulent mechanisms of energy dissipation, in so-called transitional flow. This zone is unstable and is limited to a small, but undefined, range of values of Reynolds number just above the upper limit for assured laminar flow.

Obviously, the factor of velocity distribution is basic to the dissipation process. If it were possible to have uniform velocity distribution in a flow, there could be no shearing stress set up between adjacent particles, and therefore no energy loss. However, the fluid particles immediately adjacent to the flow boundary will adhere to the boundary. This results in a differential velocity between it and the next layer, and so on through the entire fluid.

These velocity differentials create shearing stresses and dissipate energy through sliding friction or the generation of turbulence.

The velocity distribution for both laminar and turbulent flow in straight closed conduits, sufficiently far from the entrance for full development, is therefore directly related to the shear stress variation. In laminar flow, the shear stress follows the Newton equation:

$$\tau = \mu \frac{dv}{dy} = -\mu \frac{dv}{dr} \tag{3-10}$$

where v is the point velocity of flow at distance y from the wall (or r from the flow center line).

The apparent shear stress in turbulent flow has been derived by von Kármán, on the basis of similitude and statistical considerations relating to the turbulence mechanisms, to accord with the following equation:

$$\tau = \rho k^2 \frac{(dv/dy)^4}{(d^2v/dy^2)^2} \tag{3-11}$$

where k is a *universal constant* characteristic of all normal turbulent motion, experimentally about 0.38 and usually taken as 0.40.

The momentum equation, which for uniform flow simply states that the summation of forces on a reach in the direction of flow is zero, might be written in differential form by considering the forces on a small element of the fluid along the pipe axis, as shown in Fig. 3-1, assuming now that the filament has a circular cross-section of radius r. Thus, comparing with Eq. (3-3),

$$d\left(\frac{p}{\gamma} + h\right) = \frac{2\tau}{\gamma r} dL$$

where τ is the shear stress at radius r from the pipe center line. Therefore:

$$\tau = \frac{\gamma r}{2} \frac{d\left(\frac{p}{\gamma} + h\right)}{dL} \tag{3-12}$$

Since $\tau = \tau_0$ at $r = R$, the full pipe radius:

$$\tau = \tau_0 \frac{r}{R} = \tau_0\left(1 - \frac{y}{R}\right) \tag{3-13}$$

It should also be noted that $v = 0$ at $r = 0$ and $v = v_{\max}$ at $r = R$.

3-4. Laminar Flow.

For the process of laminar flow in a circular pipe, Eqs. (3-10) and (3-12) may be combined to yield:

$$dv = -\frac{d(p/\gamma + h)}{dL}\left(\frac{\gamma}{2\mu}\right)r \, dr$$

and therefore, by direct integration, since $v = 0$ at $r = R$,

$$v = \left(\frac{\gamma}{4\mu}\right) \frac{d(p/\gamma + h)}{dL} (R^2 - r^2) \tag{3-14}$$

which is the equation of a paraboloid of boundary radius R. Since the maximum velocity is at the center line, where $r = 0$, it follows that the hydraulic gradient can be expressed in terms of either the center-line velocity v_{max} or the average velocity V:

$$\frac{d(p/\gamma + h)}{dL} = \frac{4\mu v_{max}}{\gamma R^2} = \frac{32\mu V}{\gamma D^2} \tag{3-15}$$

since, for a paraboloid, the mean height is half the maximum. This is the equation of the *piezometric gradient*, or *hydraulic gradient*, for laminar flow in a straight circular pipe. Since the flow is uniform, the *energy gradient* (slope of the energy grade line, or the rate of decrease of total energy per unit length of pipe) is numerically equal to the hydraulic gradient.

The velocity distribution equation can be expressed in various alternate forms as convenient. Substituting (3–15) in (3–14):

$$v = 2V\left[1 - \left(\frac{r}{R}\right)^2\right] \tag{3-16}$$

or, substituting (3–12) and (3–13) in (3–14) and dividing by v^*,

$$\frac{v}{v^*} = \frac{\gamma}{4\mu} \frac{2\tau_0\left(\dfrac{r}{R}\right)}{\gamma r v^*} (R^2 - r^2) = \frac{v^*}{2\nu R}[R^2 - (R - y)^2]$$

and finally:

$$\frac{v}{v^*} = \frac{v^* y}{\nu}\left(1 - \frac{y}{2R}\right) \tag{3-17}$$

In this form, the dimensionless velocity v/v^* appears as a function of a "wall Reynolds number," $v^* y/\nu$ and the dimensionless wall distance, y/R.

Equation (3–15) can also be combined with the energy equation, Eq. (3–7) to give an expression for head loss in laminar flow:

$$H_f = \frac{32\mu V L}{\gamma D^2} \tag{3-18}$$

This is the familiar Poiseuille equation. It may be recast in the form of the Darcy equation

$$H_f = \frac{L}{D} \frac{V^2}{2g} \frac{64\mu}{DV\rho} = \frac{64}{N_R} \frac{L}{D} \frac{V^2}{2g} \tag{3-19}$$

Thus the friction factor for laminar flow in a circular pipe is given by the very simple function

$$f = \frac{64}{N_R} \tag{3-20}$$

3-5. Turbulent Flow. A similar analysis for turbulent flow encounters the equation

$$\tau = \tau_0 \frac{r}{R} = \tau_0 \left(1 - \frac{y}{R}\right) = \rho k^2 \frac{(dv/dy)^4}{(d^2v/dy^2)^2} \tag{3-21}$$

This is a differential equation describing the variation of velocity with distance y from the pipe wall. It can be rearranged, by taking the square root of both sides and replacing $\sqrt{\tau_0/\rho}$ by v^*, to

$$\left(\frac{dv}{dy}\right)^2 - \left(\frac{v^*}{k}\right)\sqrt{1 - \frac{y}{R}}\left(\frac{d^2v}{dy^2}\right) = 0 \tag{3-22}$$

This differential equation is difficult to solve explicitly, as it stands, but with certain simplifications it can be shown that an approximate solution is given by a logarithmic relation:

$$\frac{v}{v^*} = \frac{1}{k}\log_e\frac{y}{R} + c \tag{3-23}$$

This equation has been found satisfactory experimentally except for the wall region and for the pipe centerline. At the center line the velocity gradient must obviously be zero, a result not obtained with Eq. (3–23); the discrepancy, however, is quantitatively unimportant.

Near the wall, the form of the velocity distribution is modified by the nature of the wall surface. If the wall is "smooth," a thin boundary layer of laminar flow will prevail adjacent to the wall, with turbulent flow outside. If the wall is "rough" (i.e., if roughness elements are of radial height equal to or greater than the laminar film thickness), there will exist an abnormal turbulence near the wall, with flow separation zones behind each roughness element.

The velocity gradient in the core region is

$$\frac{dv}{dy} = \frac{v^*}{ky} \tag{3-24}$$

and has of course its minimum value (though not zero) at the center. If the dimensionless velocity distribution v/v^* is considered as a function of $\log_e(y/R)$, then the reciprocal of the turbulence constant k is the slope of the function and c is its value at $y/R = 1$. The term c, therefore, is v_{max}/v^*. Inserting this value in Eq. (3–23) gives a turbulent velocity distribution equation for established flow in any circular pipe:

$$\frac{v_{max} - v}{v^*} = \frac{1}{k} \log_e \frac{R}{y} \tag{3–25}$$

This equation is sometimes called the *velocity deficiency equation*, since it expresses the velocity at any point in a circular pipe as a decrement from the center-line velocity, this "deficiency" being logarithmically related to the distance y from the wall.

3–6. Turbulent Flow in Smooth-Walled Pipes.

If the conduit wall is smooth, a film of laminar flow will occupy a zone immediately adjacent to the wall. The laminar flow relations are therefore applicable in the laminar boundary layer and the turbulent flow equations in the central zone outside that layer, with a combination of these relations presumably applicable in the more-or-less ill-defined transitional zone. The criterion as to whether a conduit wall is "smooth" or "rough" is based on the radial height of the wall roughness protuberances. The wall is "smooth" if the roughness element height is somewhat less than the thickness of the laminar boundary layer, and "rough" otherwise. Since the laminar film thickness can be shown to decrease with increasing Reynolds numbers, a given wall may turn out to be smooth at low values of N_R and rough at higher values.

There have been numerous empirical or semi-empirical equations proposed for the friction factor relation for turbulent flow in smooth pipes. One of the simplest and most successful was the Blasius equation:

$$f = \frac{0.316}{N_R{}^{0.25}} \tag{3–26}$$

This equation corresponds with a seventh-root velocity-distribution relation

$$\frac{v}{v^*} = 8.7 \left(\frac{v^* y}{\nu}\right)^{1/7} \tag{3–27}$$

from which it can be derived via Eq. (3–6). These equations are simple to use and sufficiently accurate for values of N_R up to 100,000.

For a more generally accurate equation, especially for higher Reynolds numbers, the logarithmic velocity law should be used.

Equation (3–23) can be rearranged as follows:

$$\frac{v}{v^*} = \frac{1}{k} \log_e \frac{yv^*}{\nu} + \left[c - \frac{1}{k} \log_e \frac{v^* R}{\nu} \right]$$

Since all terms in the bracket are independent of y, they can be grouped into one constant A. The equation is then

$$\frac{v}{v^*} = \frac{1}{k} \log_e \frac{yv^*}{\nu} + A$$

The term yv^*/ν is of Reynolds number form and could be expected to indicate satisfactorily the dependence of the velocity distribution in smooth-pipe flow on the wall shear stress and the distance from the wall. It has been verified experimentally, with the boundary constant A found to have a value of 5.5. Thus the velocity distribution equation for smooth-pipe turbulent flow (k assumed to be 0.40) is

$$\frac{v}{v^*} = 2.5 \log_e \frac{yv^*}{\nu} + 5.5$$

$$= 5.75 \log_{10} \frac{yv^*}{\nu} + 5.5 \tag{3-28}$$

This equation may be set equal to Eq. (3–17), at the value $y = \delta$, where δ is the thickness of the laminar film. Thus:

$$\frac{v^*\delta}{\nu}\left[1 - \frac{\delta}{2R}\right] = 5.75 \log_{10} \frac{\delta v^*}{\nu} + 5.5$$

Solving this by trial, after assuming $\delta/2R$ is negligible:

$$\frac{\delta v^*}{\nu} = 11.6$$

which provides an expression for determining the thickness of the laminar boundary layer in smooth turbulent flow.

Therefore

$$\delta = \frac{11.6\nu}{v^*} \tag{3-29}$$

This is the nominal thickness of the boundary layer, although there is actually a gradual transition between the laminar and turbulent velocity distributions. This thickness decreases as the Reynolds number increases. Thus, combining Eqs. (3–6) and (3–29) gives

$$\frac{\delta}{D} = \frac{11.6\nu}{V\sqrt{f/8}D} = \frac{11.6\sqrt{8}}{N_R\sqrt{f}} \tag{3-30}$$

The velocity distribution equation can also be expressed in terms of the boundary layer thickness by combining Eqs. (3–28) and (3–30), resulting in the following:

$$\frac{v}{v^*} = 5.75 \log_{10} \frac{y}{\delta} + 11.6 \tag{3-31}$$

The velocity at the interface between the laminar and turbulent zones is called the *wall velocity*, v_w, and is found by inserting $y = \delta$ in Eq. (3–31). Then:

$$v_w = 11.6v^*$$ (3–32)

Combining this with Eq. (3–6), the following useful relation is obtained:

$$\left(\frac{v_w}{V}\right)^2 = 16.8f$$ (3–33)

The average velocity can be obtained by integrating over the cross-section A. Then the friction factor for the flow follows directly from Eq. (3–6):

$$f = 8\left(\frac{v^*}{V}\right)^2 = 8\left[\frac{Av^*}{\int v \, dA}\right]^2 = \frac{8A^2}{[\int (v/v^*) \, dA]^2}$$ (3–34)

Equation (3–34) is general and can be evaluated for any known velocity distribution, to give a resulting equation for friction factor.

For smooth turbulent flow in circular pipes, by neglecting the slight inaccuracies near the center and the wall, the factor becomes

$$f = \frac{8A^2}{\left\{\int_0^R [5.75 \log_{10} (yv^*/\nu) + 5.5]2\pi(R - y) \, dy\right\}^2}$$

Performing the integration, rearranging, and modifying slightly to conform to experimental determinations of friction factors in smooth pipes yields the following equation:

$$\frac{1}{\sqrt{f}} = 2 \log_{10} (N_R\sqrt{f}) - 0.80$$ (3–35)

This equation, known as the von Kármán–Nikuradse smooth-pipe equation, indicates the friction factor to be a decreasing function of Reynolds number. The latter enters the equation from the term yv^*/ν which in its limit is Rv^*/ν. Thus

$$\frac{Rv^*}{\nu} = \left(\frac{D}{2}\right)\left(\frac{V\sqrt{f}}{\nu\sqrt{8}}\right) = \frac{N_R\sqrt{f}}{4\sqrt{2}}$$ (3–36)

3–7. Normal Rough Turbulent Flow. For rough walls, it is reasonable that the roughness elements and their geometry would exert more influence on the apparent shearing stress and velocity distributions than would the viscous effects in the laminar film. If it can be assumed that one single roughness element dimension, ϵ, is indicative of the complete geometry of these elements (an assumption which is unrealistic but which, for simplicity, is often made), then Eq. (3–23) can be modified to include this roughness measure, in dimensionless form, as follows:

$$\frac{v}{v^*} = \frac{1}{k} \log_e \frac{y}{\epsilon} + \left[c + \log_e \frac{\epsilon}{R} \right]$$

$$= \frac{1}{k} \log_e \frac{y}{\epsilon} + B \tag{3-37}$$

The term B is independent of y and should reflect only the influence of the wall roughness elements, characterized by the length ϵ. However, it cannot be expected to be constant, as in the smooth-pipe equation, but will depend on the character of the roughness elements and on the chosen measure of roughness ϵ.

For pipes coated with densely packed and uniformly sized sand grains, the German investigator Nikuradse proved experimentally that, for sufficiently large values of Reynolds number, Eq. (3–37) was satisfied if ϵ was taken as the sand-grain diameter; the value of B then was found to be 8.48. Although it seems evident that this value should be applicable to no other type of roughness, the resulting equation has been very generally regarded as the *rough-pipe velocity distribution equation:*

$$\frac{v}{v^*} = \frac{1}{k} \log_e \frac{y}{\epsilon} + 8.48$$

$$= 5.75 \log_{10} \frac{y}{\epsilon} + 8.48 \tag{3-38}$$

Using the procedure of Eq. (3–34) for rough pipes, Eq. (3–38) yields

$$f = \frac{8A^2}{\left\{ \int_0^R [5.75 \log_{10} (y/\epsilon) + 8.48] 2\pi(R - y) \, dy \right\}^2}$$

With certain experimental adjusting, this leads to what is known as the Nikuradse rough-pipe equation:

$$\frac{1}{\sqrt{f}} - 2 \log_{10} \frac{R}{\epsilon} = 1.74 \tag{3-39}$$

Equation (3–39) indicates the friction factor to depend only on the *relative roughness* R/ϵ and not on N_R. If written in terms of the diameter, it becomes

$$\frac{1}{\sqrt{f}} - 2 \log_{10} \frac{D}{\epsilon} = 1.14 \tag{3-40}$$

The smooth-pipe equation, Eq. (3–35), may be recast for comparison with the rough-pipe equation, Eq. (3–39), by subtracting $2 \log_{10} \dfrac{R}{\epsilon}$ from both sides.

$$\frac{1}{\sqrt{f}} - 2 \log_{10} \frac{R}{\epsilon} = 2 \log_{10} \left(\frac{N_R \sqrt{f}}{R/\epsilon} \right) - 0.8 \tag{3-41}$$

3–8. The Colebrook Transition Function and the Moody Diagram.

Experiments indicate, however, that by far the majority of commercial pipe flows do not follow either the smooth-pipe or the rough-pipe law, but are intermediate between the two. Accordingly, Colebrook and White, two British engineers, suggested the use of an equation which would be asymptotic to both of them. Combining Eqs. (3–39) and (3–41):

$$\frac{1}{\sqrt{f}} - 2 \log_{10} \frac{R}{\epsilon} = 1.74 - 2 \log \left[1 + \frac{18.7}{N_R \sqrt{f}/(R/\epsilon)} \right] \tag{3-42}$$

Inspection of this equation reveals that it approaches the smooth-pipe equation at low values of N_R and the rough-pipe equation at high values of N_R.

The term $N_R \sqrt{f}/(R/\epsilon)$ will be recognized as the ratio of the two dimensionless parameters appearing in the smooth-pipe and rough-pipe equations. It also represents essentially the ratio of the roughness measure ϵ to the boundary layer thickness δ. It thus reflects the relative magnitude of the inertial and viscous effects at the wall. That it thus actually represents a *wall Reynolds number* is indicated by the following transformation:

$$\frac{N_R \sqrt{f}}{R/\epsilon} = \frac{DV \sqrt{f(2)}}{\nu(D/\epsilon)} = \frac{\epsilon v^* \sqrt{8(2)}}{\nu} = \sqrt{32} N_{Rw} \tag{3-43}$$

In this wall Reynolds number, N_{Rw}, the characteristic velocity is the shear velocity and the characteristic length is the roughness size. Physically, the Colebrook-White transition curve represents the change from boundary layer generation of turbulence to roughness separation zone generation of turbulence as the wall Reynolds number increases, or as the boundary film thickness decreases.

This may be further illustrated by introducing the laminar film thickness δ. From Eq. (3–43), and substituting $\dfrac{11.6}{\delta}$ for $\dfrac{v^*}{\nu}$,

$$\frac{N_R \sqrt{f}}{R/\epsilon} = 46.4 \sqrt{2} \, \frac{\epsilon}{\delta}$$

Substituting this in Eq. (3–42), the Colebrook-White equation can also be written:

$$\frac{1}{\sqrt{f}} - 2 \log_{10} \frac{R}{\epsilon} = 1.74 - 2 \log_{10} \left[1 + \frac{1}{3.5} \frac{\delta}{\epsilon} \right] \qquad (3\text{–}44)$$

In this form, the equation clearly indicates the importance of the ratio of the boundary layer thickness to the roughness size in controlling the type of flow. When δ/ϵ is large, smooth turbulent flow occurs; and when δ/ϵ is small, then normal rough-pipe turbulence prevails.

Many experimental data are available from commercial pipe tests which, in effect, give corresponding values of f and N_R (actually discharges and pressure drops were measured). These data permit Eq. (3–42) to be solved for the roughness measure ϵ.

It should be clearly understood that ϵ is not any actual roughness dimension, except in the case of uniform sand as in Nikuradse's tests. However, it has frequently been considered as a representative dimension identifying the actual roughness. It is often called the *equivalent sand-grain diameter*, since the roughness it represents results in the same friction loss as such an equivalent sand-grain roughness. With this limitation, Table 3–1 gives experimental values of ϵ for use in the Colebrook-White formula.

TABLE 3–I
Equivalent Sand Roughness

Pipe Material	ϵ (feet)
Riveted steel, few rivets	0.003
Riveted steel, many rivets	0.030
Concrete, finished surface	0.001
Concrete, rough surface	0.010
Wood-stave, smooth surface	0.0006
Wood-stave, rough surface	0.003
Cast iron, new	0.00085
Galvanized iron, new	0.00050
Asphalted cast iron, new	0.00040
Commercial steel, new	0.00015
Wrought iron, new	0.00015
Drawn tubing, new	0.000005
(glass, brass, copper, lead)	(essentially "smooth")

The character of the Colebrook-White equation is illustrated by the graph of Fig. 3–2. It is convenient to use a semilogarithmic plot, with the parameter $1/\sqrt{f} - 2 \log_{10} (R/\epsilon)$ (called the *generalized resistance function*) plotted on an arithmetical scale increasing downward, as the ordinate. The abscissa is the wall Reynolds number, $N_R \sqrt{f}/(R/\epsilon)$, plotted on a logarithmic scale.

The ordinate scale is arranged to increase downward in order to point up the similarity between the generalized resistance diagram of Fig. 3–2 and the more familiar friction-factor vs. Reynolds number curves which it summarizes. The Colebrook-White equation is inconvenient to solve algebraically, although not excessively so. It could also be solved graphically, using Fig. 3–2, but this is also inconvenient if the friction factor is the quantity sought.

Solution is considerably expedited by plotting the friction factor as a function of the Reynolds number, with relative roughness (R/ϵ, or D/ϵ, or ϵ/D, as preferred) parametric, yielding a family of curves rather than single curve as in Fig. 3–2. This has been done by Lewis Moody at Princeton, and

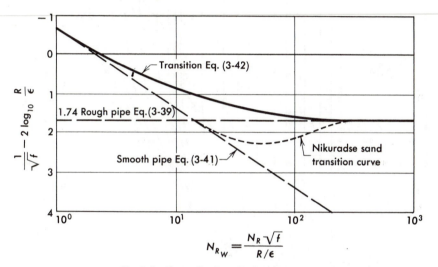

Fig. 3–2. Generalized resistance diagram.

his resulting *Moody curves* have now been quite widely adopted as a design and analysis tool in pipe flow problems. These curves are shown in Fig. 3–3.

In using the curves, the friction factor is normally needed in order to solve the Darcy equation:

$$H_f = f\left(\frac{L}{D}\right)\left(\frac{V^2}{2g}\right) = \frac{8fLQ^2}{\pi^2 g D^5} \qquad (3\text{-}45)$$

If the problem is one of determining the head loss for a given flow in a given pipe, the friction factor can be read directly from the curves for the known Reynolds number, by using the particular curve corresponding to the given relative roughness (interpolating if necessary).

If however, either the diameter or the discharge is to be determined for a given head loss, the problem must be solved indirectly, since the Reynolds

Moody Diagram

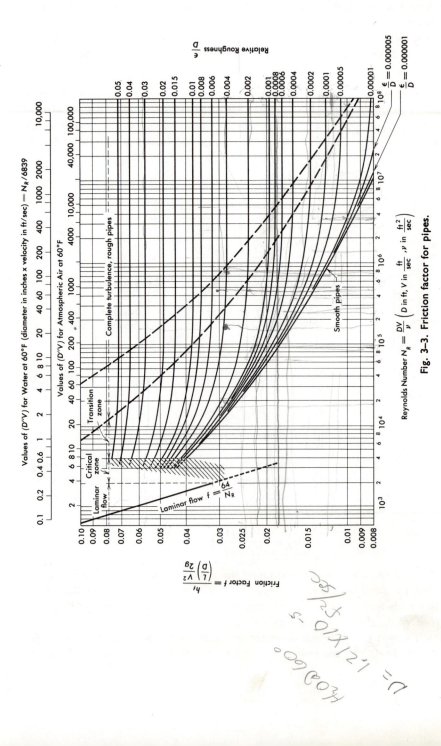

Fig. 3-3. Friction factor for pipes.

$D = 1.21 \times 10^{-5}$

$.000060$

number is dependent on both the diameter and the discharge. A reasonable friction factor can be assumed, and then Eq. (3–45) can be solved for the corresponding discharge (or diameter). The Reynolds number can then be computed and the assumed value of the friction factor checked from the curves. If agreement is not obtained, the process is repeated with the new friction factor until satisfactory convergence is attained.

In such a trial solution, the friction factor should always be assumed *first* as some reasonable value (*not* the discharge or the diameter), and then the Darcy equation solved for either Q or D as required. This is because Q depends on $f^{1/2}$ and D on $f^{1/5}$, and thus errors in the assumed value of f will yield a much smaller error in Q or D, and the solution will converge rapidly.

Example 3–1

Select a diameter of pipe to convey 100 cfs a distance of 5000 ft with a permitted head loss of 5 ft. The pipe is a smooth wood-stave pipe for which, from Table 3–1, ϵ is taken as 0.0006 ft. Assume water kinematic viscosity to be 1×10^{-5} ft²/sec, and neglect all losses except those due to normal boundary friction.

Solution. From Eq. (3–45),

$$D = \left(\frac{8fLQ^2}{\pi^2 g H_f}\right)^{1/5} = (252{,}000f)^{1/5} \tag{a}$$

The Reynolds number is

$$N_R = \frac{4Q}{\pi \nu D} = \frac{127}{D}(10)^5 \tag{b}$$

It is necessary to determine D by trial and error. First, assume some reasonable value of f, say 0.02. From Eqs. (a) and (b), the corresponding values of D and N_R are 5.5 ft and $2.31(10)^6$, respectively. The relative roughness ϵ/D is then 0.000109.

From the Moody diagram, for those values of N_R and ϵ/D, it is seen that f would be 0.0129. Repeating the solution with this value of f yields a D of 5.04 ft and N_R of 2.52×10^6. Further repetition does not produce any significant change in D. Probably a diameter of 5 ft would therefore be specified.

It may be helpful to note that three main types of problems are normally encountered. These are:

1. Given Q, D, ϵ, ν, L. Find H_f.
   Solve by finding f from Moody curves, after calculating ϵ/D and N_R. Then substitute directly in Eq. (3–45) to get H_f.
2. Given H_f, D, ϵ, ν, L. Find Q.
   Calculate ϵ/D to find which curve to use on Moody diagram.
   Assume f at some reasonable point along this curve. Substitute in Eq.

(3–45) and solve for Q. Then calculate N_R and check assumed f from Fig. 3–3. Repeat as necessary.

3. Given H_f, Q, ϵ, ν, L. Find D.

Assume f at some reasonable value. Then solve Eq. (4–35) for D. Calculate N_R and ϵ/D. Get f from Moody curves and check. Repeat as needed.

It should be noted that the values of ϵ given in Table 3–1 are for new pipes. It is certain, however, that many pipe materials deteriorate and become rougher with age. In effect, this means that an entirely different surface with a different $f - N_R$ curve must be postulated if flow problems are to be studied on an old pipe.

The magnitude of this effect is still quite uncertain, except that there is some evidence, for some types of pipe, that the increase in equivalent roughness is linearly proportional to age. The proportionality constant must be determined by field test, however.

3–9. Empirical Approximations to the Head Loss Function.

Prior to the developments in modern fluid mechanics and dimensional analysis which led to the currently recommended methods of pipe flow analysis, a number of empirical formulas had been developed for use under specified conditions. Many of these are still widely used and, if judiciously applied, are very satisfactory. Their scope and limitations can best be appreciated by an examination in the light of the Darcy equation. If the roughness can again be described in terms of one representative dimension, ϵ, this equation is

$$H_f = \left(\frac{L}{D}\right)\left(\frac{V^2}{2g}\right) f\left(N_R, \frac{D}{\epsilon}\right)$$

The friction factor function is actually rather complicated, as we have found. However, it is possible that it may be approximated, for a limited range, by some simpler exponential function. That is, assume

$$f = V^a D^b \nu^c \epsilon^d$$

since these are the variables that appear implicitly in the correct function. Then, the process equation becomes

$$H_f = \left(\frac{L}{D^{1-b}}\right)\left(\frac{V^{2+a}}{2g}\right)\nu^c\epsilon^d$$
$$= \left(\frac{L}{D^m}\right)\left(\frac{V^n}{2g}\right)\nu^c\epsilon^d$$

Experimental data will then yield values for m, n, c, and d. As a matter of fact, ϵ^d can be combined into a single roughness coefficient with ν^c, since water at ordinary temperatures is usually the fluid of interest in these empirical methods. The equation then assumes the form

$$H_f = K\left(\frac{L}{D^m}\right)\left(\frac{V^n}{2g}\right) \tag{3-46}$$

A large number of empirical formulas of this type have been published, usually in terms of the velocity, energy slope, and hydraulic radius. Thus, Eq. (3-46) can be rearranged as follows:

$$V = \left(\frac{2g(4)^m}{K}\right)^{1/n} R_h^{m/n}\left(\frac{H_f}{L}\right)^{1/n} \tag{3-47}$$
$$= CR_h^x S^y$$

where R_h is the hydraulic radius and S is the energy slope.

The most commonly used pipe-flow equations of this form are the Hazen-Williams, Manning, and Scobey formulas, although numerous others have been suggested. Each of them is applicable only to problems involving flow of water at normal temperatures and at a relatively high degree of turbulence, as well as to ordinary commercial pipes. They are obviously non-homogeneous dimensionally and thus can be applicable only within the range for which they were derived. Within these limitations, however, they are quite valid and, with the aid of nomographs, often easier to solve than the Darcy equation.

The Hazen-Williams equation, frequently used in sanitary engineering practice, is

$$V = 1.318 C_{hw} R_h^{0.63} S^{0.54} \tag{3-48}$$

where C_{hw} is the Hazen-Williams roughness coefficient.

The Manning equation, used widely in open channel problems and frequently also in pipe problems, is

$$V = \frac{1.49}{n} R_h^{2/3} S^{1/2} \tag{3-49}$$

where n is the Manning roughness coefficient.

There are three Scobey formulas, all used rather widely in irrigation and reclamation work:

$$\text{Concrete:} \qquad V = 356 C_s R_h^{5/8} S^{1/2} \tag{3-50}$$

$$\text{Steel:} \qquad V = \frac{85}{K_s^{0.526}} R_h^{0.58} S^{0.526} \tag{3-51}$$

$$\text{Wood-stave:} \quad V = 113 R_h^{0.65} S^{0.555} \tag{3-52}$$

Values of C_{hw}, n, C_s, and K_s are tabulated in Tables 3-2 through 3-5.

TABLE 3–2
Values of C in Hazen-Williams Formula *

Type of Pipe	Condition		C
	New	All Sizes	130
	5 years old	12″ and over	120
		8″	119
		4″	118
	10 years old	24″ and over	113
		12″	111
		4″	107
	20 years old	24″ and over	100
		12″	96
		4″	89
Cast iron	30 years old	30″ and over	90
		16″	87
		4″	75
	40 years old	30″ and over	83
		16″	80
		4″	64
	50 years old	40″ and over	77
		24″	74
		4″	55
Welded steel	Values of C the same as for cast-iron pipe, 5 years older		
Riveted steel	Values of C the same as for cast-iron pipe, 10 years older		
Wood stave	Average value, regardless of age		120
Concrete or concrete lined	Large sizes, good workmanship, steel forms		140
	Large sizes, good workmanship, wooden forms		120
	Centrifugally spun		135
Vitrified	In good condition		110

* Values are as given in *Handbook of Applied Hydraulics* (C. V. Davis, ed., New York, McGraw-Hill Book Co., Inc., 1969), p. 2–10.

TABLE 3-3
Values of Coefficient of Roughness n in Manning Formula *

Type of Pipe	Condition	n			
		Best	Good	Fair	Bad
Cast iron	Clean, uncoated	0.012	0.013	0.014	0.015
	Clean, coated	0.011	0.012	0.013	
	Dirty or tuberculated			0.015	0.035
Wrought iron	Commercial, black	0.012	0.013	0.014	0.015
	Commercial, galvanized	0.013	0.014	0.015	0.017
Lock bar or welded	Smooth and clean	0.010	0.011	0.013	
Brass or glass	Smooth	0.009	0.010	0.011	0.013
Riveted steel or spiral steel	Clean	0.013	0.015	0.017	
Vitrified sewer pipe		0.011	0.013	0.015	0.017
Common clay drainage tile		0.011	0.012	0.014	0.017
Concrete	Rough joints		0.016	0.017	
	Dry mix, rough forms		0.015	0.016	
	Wet mix, steel forms		0.012	0.014	
	Very smooth		0.011	0.012	
Wood stave		0.010	0.011	0.012	0.013

* Values are as given in *Handbook of Applied Hydraulics* (C. V. Davis, ed., New York, McGraw-Hill Book Co., Inc., 1969), p. 2–6.

TABLE 3-4
Values of C_s in Scobey's Formula for Concrete Pipes *

Class	Condition	C_s
1	Old California cement pipes; generous supply of mortar in joints, mortar squeeze not removed. Also; class 2 pipes conveying sewage	0.267
2	Dry-mix precast in short units, washed inside with cement mortar, moderate care; monolithic pipe over rough wood forms, cement gun finish, not troweled	0.310
3	Wet-mix precast in short units; dry-mix precast in long units; monolithic pipe on steel forms; small cement-lined iron pipe; concrete pipe made under pressure and mechanically troweled with neat cement. Also class 4 pipes conveying sewage or detritus-laden water	0.345
4	Glazed-interior pipes; large cement-lined iron pipes; monolithic pipes with joint scars or surface irregularities removed; highest quality precast-pipe, made against oiled steel form, with joints as smooth as remainder, untouched with brush or "wash" process	0.370

* Values are as given in *Handbook of Applied Hydraulics* (C. V. Davis, ed., New York, McGraw-Hill Book Co., Inc., 1969), p. 2–8.

TABLE 3–5
Values of K_s in Scobey's Formulas for Steel Pipes *

Class	Condition	Type	K_s
1	Full-riveted pipe, having both longitudinal and girth seams held by one or more lines of rivets with projecting heads from capacity standpoint; pipe with countersunk rivetheads on interior belongs in class 3	a. New sheet metal up to $\frac{3}{16}''$ thick	0.38
		b. New plate metal $\frac{3}{16}''$ to $\frac{7}{16}''$ thick, with either taper or cylinder joints	0.44
		c. New plate metal $\frac{1}{2}''$ up, with either taper or cylinder joints, and for plate $\frac{1}{4}''$ to $\frac{7}{16}''$ thick, when butt-jointed	0.48
		d. New butt-strap pipe of plate $\frac{1}{2}''$ up	0.52
2	Girth-riveted pipe, having no retarding rivetheads in the continuous-seamed longitudinal joints, but having the same girth seams as full-riveted pipe	New sheet- and plate-metal pipe, such as lock-bar and hammer-weld pipe with lap or flange-riveted field (girth) joints; electric weld, hammer-weld and drawn pipe with riveted bump joints; and all other types with surface continuous except for girth belt of rivetheads between field units	0.34
3	Continuous interior pipe, having interior surface unmarred by plate offsets or by projecting rivetheads in either longitudinal or girth seams. Not necessarily described as smooth	New sheet- and plate-metal pipe such as pipe with full-welded crimped slip joint, lock bar with welded flange or leaded sleeve connections, bell and spigot, bolted coupling pipes all belong to this class	0.32

* Values are as given in *Handbook of Applied Hydraulics* (C. V. Davis, ed., New York, McGraw-Hill Book Co., Inc., 1969), p. 2–8.

Example 3–2

Solve the problem illustrated in Example 3–1, using the Hazen-Williams formula.

Solution. From Table 3–2, the value of C_{hw} is 120. Substituting in Eq. (3–48) and replacing R_h by $D/4$ and S by H_f/L:

$$Q = AV = \left(\frac{\pi}{4}D^2\right)(1.318)(C_{hw})\left(\frac{D}{4}\right)^{0.63}\left(\frac{H_f}{L}\right)^{0.54}$$

$$D = \left[\left(\frac{4^{1.63}}{1.318\pi}\right)\left(\frac{Q}{C_{hw}}\right)\left(\frac{L}{H_f}\right)^{0.54}\right]^{1/2.63}$$

$$= 1.377\left(\frac{Q}{C_{hw}}\right)^{1/2.63}\left(\frac{L}{H_f}\right)^{1/4.87}$$

Inserting the given numerical values:

$$D = 1.377\left(\frac{100}{120}\right)^{1/2.63}\left(\frac{5000}{5}\right)^{1/4.87} = 5.30 \text{ ft.}$$

This compares well with the value of 5.04 ft obtained by the Moody diagram. It is obvious that in either case, however, the selected wall roughness may materially influence the result.

In all the above formulas it is important to use the dimensions corresponding to the constant terms used with the formula, since the formulas themselves are not dimensionally homogeneous. In the form of the equations as given above, R_h is in feet, S is dimensionless (i.e., feet per foot) and V is in feet per second.

Formulas of this type are widely used in engineering practice, and have the following obvious advantages:

1. They can be solved directly, rather than by successive approximations, as required for the solution of certain problems by the Colebrook-Moody procedure.
2. Although the fractional exponents are inconvenient algebraically, this difficulty is removed by the use of nomographs, charts, tables, or special slide rules, all of which are conveniently available from various sources.
3. They have been successfully used in engineering practice for many years, being based primarily on actual field tests of pipes installed and in service.

On the other hand, these formulas have serious limitations and disadvantages, as follows:

1. They are dimensionally non-homogeneous; essentially this means that the formulas fail to specify the principles of similitude for pipes geometrically similar. Therefore the formulas can legitimately be applied only within the limited range of diameter, velocity, and hydraulic slope covered by the experimental data on which they are based.
2. No account is taken of fluid viscosity. Consequently, they can be used only for the flow of water at ordinary temperatures. Even within the range of temperatures commonly encountered, however, the viscosity of water can change by a factor of 2 or more, which in effect changes the roughness coefficient by some unknown amount.
3. They are based on data obtained at fairly high Reynolds numbers, with therefore a high degree of turbulence. They cannot be used for very small velocities or diameters such that laminar flow is approached.
4. Because of the fractional exponents, they are inconvenient to use in connection with those pipes in which *minor losses* are an important factor. Such minor losses are usually expressed as proportional to the velocity head and must be added to the normal friction losses to account for the total energy expenditure. It is more convenient, therefore, for the normal losses also to be expressed as a factor multiplied by the velocity head, as in the Darcy equation.

5. The most serious defect of formulas of this type is probably the arbitrary and dimensionally meaningless method of including the effect of the boundary roughness. There is no actual relationship established between the flow and the roughness element dimensions which control its magnitude and character. In view of the continual development of new pipe and conduit materials and of new methods of forming pipes of the standard materials, as well as of new methods of laying pipe lines, the older coefficients are becoming obsolete. There is need therefore for quantitative relationships between the flow characteristics and the actual roughness geometry.

Of course, this latter defect is inherent also in the Colebrook-Moody procedure, with its physically unrealistic representation of the roughness elements by a hypothetical equivalent sand roughness. The method of the articles following, however, represents an attempt to overcome these deficiencies.

3–10. An Evaluation of the Colebrook-Moody Procedure. The Moody curves, which give the friction factor as a function of the Reynolds number and the ratio of equivalent sand roughness to pipe diameter, provide a compact design tool for use in flow problems. However, several logical fallacies are implicit in this procedure:

1. The Moody curves and the empirical values of equivalent sand roughness used with them are based on the Colebrook-White equation, which in turn describes an artificial curve asymptotic to the von Kármán–Nikuradse smooth and rough pipe curves. This is a transition function, showing the friction factor to be a decreasing function of the Reynolds number. However, the actual experimental transition function of Nikuradse, on whose results this procedure is based, is exactly opposite in character to the Colebrook function, showing the friction factor as primarily an *increasing* function of Reynolds number. If there were really anything "equivalent" about Nikuradse's sand roughness, it should show the same form of transitional relationship as obtained on the commercial pipes designed on the basis of it.

2. The Colebrook-Moody curves make no provision for conduit surfaces producing a rising friction factor–Reynolds number relation (such as corrugated metal pipes or flumes, sand-lined channels and conduits, etc.) or those with a fully horizontal characteristic (such as conduit surfaces of so-called "random roughness").

3. Even for surfaces which yield descending f-N_R curves, there are many data indicating that the Colebrook equation is not an adequate description of this relation. Some tests on large conduits show that friction factors tend to decrease up to much higher values of Reynolds number than predicted by the Colebrook relation. Certain tests have yielded systematic variations in empirical values of equivalent sand roughness for a given surface, proving it to be unreliable as a means of predicting friction factor unless these variations are taken into account.

4. The large number of roughness dimensions influencing the flow turbulence indicates the impossibility of representing the element by any single dimension —especially some non-existent "equivalent" dimension—and the several types of possible f-N_R relationships indicate the impossibility of using any single function to define this relationship.

These defects are aggravated by a rather widespread habit generated by the method, that of thinking of the equivalent sand-grain diameter in terms of an actual radial height of element from the wall. Actually, as seen before, the radial height is only one of the roughness dimensions influencing the turbulence, and not the most important one at that.

The actual mechanism of friction loss in rough-pipe flow is not one of longitudinally uniform generation of small-scale vorticity at a wall boundary layer, as in the case of smooth-pipe flow. Rather, at each roughness element, the flow separates from the wall at some point on the crest, leaving a "wake" of more or less dead fluid appended to the downstream side of the element. Contact of the moving fluid outside of the wake with the dead fluid results in the generation of a large vortex in the wake through friction. After attaining a certain size the vortex separates from the element and moves out into the flow, permitting a new vortex to be generated at the element. A continual stream of relatively large-scale vorticity is thus generated at each roughness element, each vortex requiring a substantial drain on the flow energy for its creation. Probably more energy is lost through friction in the generation process than the actual energy finally imparted to the vortex. The latter, also, is eventually dissipated through friction in the process of turbulent mixing.

Since each element is thus a vortex "mill," it is obvious that the total drain on the flow energy depends directly on the number of elements. This means that the *longitudinal spacing* of the elements must be the most important dimension of the roughness geometry. This fact has as yet received only limited recognition.

3–11. Regimes of Turbulent Flow. Several distinct regimes of turbulent flow are possible in conduits, each with its own peculiar characteristics. Turbulent flow is characterized by the continual generation and dissipation of vortices, with most of the energy loss occurring during the process of generation, the remainder in the process of viscous dissipation. In general, the generation of vorticity is a viscous phenomenon resulting from fluid moving at relatively high velocities in contact with fluid moving at relatively low velocities. The phenomenon typically occurs either at a laminar boundary layer or at a zone of separation behind a roughness element.

In either case, the vorticity so generated washes out into the main body of flow. The processes of mass and momentum transfer, as well as viscous attrition, combine to reduce the turbulence structure to a certain typical pattern in the interior region of the conduit, regardless of the method by which the original vorticity was generated.

This *normal turbulence* can be treated statistically, yielding the velocity distribution equation, Eq. (3–25):

$$\frac{v_{max} - v}{v^*} = \frac{1}{k} \log_e \frac{R}{y}$$

The constant k is the von Kármán universal turbulence constant characteristic of this normal turbulence. Equation (3–25), if plotted on semilogarithmic paper, yields a straight line, of slope $1/k$. It is assumed that the section under discussion is sufficiently removed from the inlet or other disturbance to permit full development of the turbulent boundary layer.

However, normal turbulence will usually prevail only in the central regions of the conduit. Near the wall, the velocity distribution will be materially affected by the nature of the vortex-generating mechanism. If the wall is smooth, with a laminar boundary layer, the very small-scale vorticity being generated, together with the transitional effects from laminar

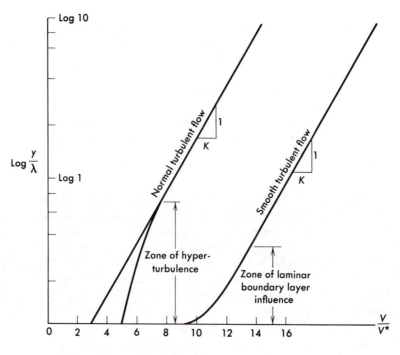

Fig. 3–4. Dimensionless velocity distributions for turbulent flow.

to turbulent flow, causes a steepening of the velocity gradient for some distance from the wall.

On the other hand, if the wall has roughness elements which pierce the laminar film, large-scale vortices are continually generated in the separation zones behind each element. This is manifested by a relative flattening of the velocity gradient for a region near the wall. These various velocity distribution phenomena are illustrated schematically in Fig. 3-4.

On a conduit surface with only occasional roughness elements, both these effects are combined. It becomes obvious, then, that the longitudinal frequency of vortex-generating roughness elements is of determinative significance with reference to the regime of turbulence that will prevail. With these preliminary considerations in mind, several regimes of turbulent flow in conduits may be considered.

3-12. Normal Turbulent Flow. Normal turbulence is defined as that characterized by the universal turbulence constant k. Normal turbulent flow would therefore involve a wall-to-wall homogeneity of the turbulence conforming to Eq. (3-25). There could be no wall zone of either subnormal or abnormal turbulence. Vortices generated at the wall are almost immediately broken up into patterns statistically equivalent to the normal turbulence of the central region. The term *normal* is thus a statistical term rather than a term meaning "usual," as defined here.

The velocity distribution equation for normal turbulent flow may be written

$$\frac{v}{v^*} = \frac{1}{k} \log_e \frac{y}{\lambda} + A \tag{3-53}$$

Here λ is the longitudinal spacing of the roughness elements, which is the roughness dimension characterizing the frequency of wall vorticity sources. (It may be noted also that λ is numerically equal to the sand-grain size in the special case of uniform sand roughness as used by Nikuradse.) In normal turbulent flow the other roughness dimensions are irrelevant, except that the crests of the elements determine the physical limits within which the velocity distribution equation is applicable. Whatever shape the elements may have has no effect on the turbulence pattern, which is fixed in terms of the constant k.

Experimental work on circular pipes has yielded a value of about 8.7 for the constant A, for various types of roughness elements. The value of about 7.8 seems somewhat better for two-dimensional closed flow or rectangular open channel flow. In either case the corresponding friction factor equation, as determined by use of Eq. (3-34), is found to be

$$\frac{1}{\sqrt{f}} = 2 \log_{10} \frac{R}{\lambda} + 1.75 \tag{3-54}$$

Here R is the pipe radius, or equivalent radius in the case of a non-circular pipe or open channel. In this type of flow, the friction factor is independent of both the Reynolds number and the type of roughness, depending solely on the relative roughness spacing. Note that the friction factor is defined in terms of the actual roughness dimension, not some "equivalent roughness."

3–13. Smooth Turbulent Flow. If the conduit wall is smooth, so that the turbulence-generating mechanism is solely one of minute vorticity shed from the laminar boundary layer, the velocity distribution near the wall is influenced by the *subnormal turbulence* in the region, and the velocity distribution and friction factor are independent of all wall roughness dimensions. They depend rather on the thickness of the boundary layer, which is a decreasing function of Reynolds number. The velocity distribution equation is Eq. (3–29):

$$\frac{v}{v^*} = \frac{1}{k} \log_e \frac{yv^*}{\nu} + 5.5$$

The corresponding friction equation, showing the friction factor to be a decreasing function of Reynolds number, is Eq. (3–35):

$$\frac{1}{\sqrt{f}} = 2 \log_{10} N_R \sqrt{f} - 0.8$$

This equation was derived for closed conduits. The corresponding smooth turbulence equation for open channels is

$$\frac{1}{\sqrt{f}} = 2.62 \log_{10} N_R \sqrt{f} - 3.16 \qquad\qquad (3\text{--}55)$$

However, Eq. (3–35) may be used for open channels without serious error, at ordinarily high Reynolds numbers.

It is interesting that some of the newer types of pipe (e.g., pipes lined with certain special coatings) have been found to give friction factors somewhat smaller than specified by Eq. (3–35), indicating that perhaps the "smooth" pipes of the earlier investigators may not have been truly smooth after all, possibly due to joint roughness. However the data on such ultrasmooth pipes are not yet adequately substantiated for design use and it is on the conservative side to continue to regard Eq. (3–35) as the smooth flow equation.

3–14. Semismooth Turbulent (Isolated-Roughness) Flow. If an ordinarily smooth conduit surface is interspersed with occasional isolated roughness elements, the over-all friction factor will be that due to the friction drag at the laminar boundary layer plus that due to the form drag forces on the roughness elements, a phenomenon illustrated in Fig. 3–5. Since both skin friction and form drag are fundamentally viscous phenomena, the same general function can be used to characterize both, the main difference lying

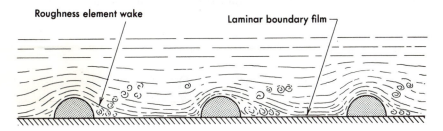

Fig. 3-5. Semismooth turbulent (isolated roughness) flow.

in the controlling boundary geometry. Starting from the general flow equation, and eliminating extraneous forces and dimensions, gives the function

$$H_f = (f_s + f_r)\left(\frac{L}{D}\right)\left(\frac{V^2}{2g}\right) = \left[f_s + \phi\left(N_R, \frac{\lambda}{D}, \frac{h}{D}, \frac{s}{D}, \cdots\right)\right]\left(\frac{L}{D}\right)\left(\frac{V^2}{2g}\right) \quad (3\text{-}56)$$

The function ϕ can be approximated by analogy with the corresponding equation for drag force on an immersed body. The latter is usually expressed in terms of the projected area of the body and a dimensionless drag coefficient. The drag coefficient is basically equivalent to a friction factor, but depends mainly on the shape and size of the wake behind the body, and therefore on the shape of the body and the Reynolds number. Thus,

$$F_r = \frac{\rho}{2} v_w^2 A_r C_D = (\Delta p_r)A \quad (3\text{-}57)$$

where F_r is the drag force on the roughness element, whose projected area is A_r, v_w is the wall velocity and the resulting pressure drop over the entire cross-section A due to the roughness element is Δp_r. Since A_r is equal to $h(p)$, where h is the roughness height and p is the part of the total wetted perimeter P that is occupied by roughness elements, this equation becomes:

$$\Delta(H_f)_r A = \frac{v_w^2}{2g} h(p)(C_D)$$

in which, also, $(\Delta p)_r$ has been replaced by $\gamma \, \Delta(H_f)_r$. Thus the head loss due to one peripheral group of roughness elements is

$$\Delta(H_f)_r = C_D \frac{v_w^2}{2g} \frac{ph}{A} \quad (3\text{-}58)$$

In a total longitudinal distance L, with a longitudinal spacing λ between successive peripheral element groups, the total head loss due to roughness elements is

$$(H_f)_r = C_D\left(\frac{v_w^2}{2g}\right)\left(\frac{ph}{R_hP}\right)\left(\frac{L}{\lambda}\right)$$

$$= \left[4C_D\left(\frac{h}{\lambda}\right)\left(\frac{v_w}{V}\right)^2\left(\frac{p}{P}\right)\right]\left(\frac{L}{D}\right)\left(\frac{V^2}{2g}\right) \tag{3-59}$$

in which A has been replaced by $R_hP = \frac{1}{4}DP$, D being the equivalent diameter.

The bracketed term is obviously the function ϕ. This is the portion of the friction factor due to the roughness elements

$$f_r = \frac{4C_D}{\lambda/h}\left(\frac{v_w}{V}\right)^2\left(\frac{p}{P}\right) \tag{3-60}$$

The drag coefficient C_D is left in this form since many data are already available, both analytical and experimental, for its value for various body shapes. The ratio h/λ may be called the *roughness index*. The *peripheral roughness ratio* p/P is simply the ratio of the total peripheral length occupied by roughness elements to the pipe perimeter. If the roughness element extends as a strip all around the circumference, then of course the peripheral roughness ratio is unity. The *wall velocity ratio* v_w/V is approximately $\frac{1}{2}$ to $\frac{2}{3}$, varying with the wall roughness and the Reynolds number. Since the velocity v_w is that impinging on the roughness elements, and since it is controlled by the smooth wall just upstream from each element, it is reasonable to insert Eq. (3-33) in Eq. (3-60). Finally, then, the resulting friction factor equation for semismooth turbulent (isolated-roughness) flow becomes

$$f = f_s[1 + 67.2(\textstyle\sum E)] \tag{3-61}$$

when f_s is the smooth turbulence friction factor at the given Reynolds number, and E is an element characteristic defined as follows:

$$E = C_D\left(\frac{p}{P}\right)\left(\frac{h}{R}\right)\left(\frac{R}{\lambda}\right) = C_D\left(\frac{p}{P}\right)\left(\frac{h}{\lambda}\right) \tag{3-62}$$

p/P is the peripheral roughness ratio (the ratio of total peripheral length filled with roughness elements to total conduit wetted perimeter). h/R is the relative roughness height, R/λ has already been termed the relative roughness spacing, and h/λ is the roughness index. The coefficient C_D is the drag coefficient for the particular element shape, determined as for an airfoil shape by the streamline configuration bounding the zone of separation.

A value of E can be computed for each repeating type of roughness element on the surface. The individual values are added to get the sum effect to use in the equation. Equation (3-61) indicates the friction factor relation for semismooth flow to be similar to that for smooth flow, with the friction factor computed as the smooth turbulence friction factor multiplied by a number greater than unity, the value of the number being $(1 + 67.2 \Sigma E)$. The friction factor for semi-smooth turbulent flow therefore normally decreases with increasing Reynolds number and with decreasing values of ΣE.

It would be slightly more accurate, in the above derivation, to assume f_s acting only over that part of the wall not occupied by roughness elements and their wakes if this were known. However, since f_s is usually small compared to Σf_r, and since it is on the safe side to do so, it has been assumed for simplicity that the laminar film in effect covers the entire wall, with the effect of the roughness elements superimposed thereon. Because of these combined effects, it is not feasible to specify a standard velocity-distribution equation for semismooth turbulent flow.

3-15. Hyperturbulent (Wake-Interference) Flow.

If the wall roughness elements are sufficiently close together, the wake behind each may extend to or nearly to the next element. There is then essentially no part of the wall over which a laminar boundary layer exists. Furthermore the vortex generation and dissipation phenomena associated with each wake will interfere with those at the adjacent elements, so that the individual effects are not additive as in the case of semismooth flow.

The over-all phenomenon of wake interference results in a zone near the wall of abnormally intense turbulence and mixing. The velocity distribution will be normal in the central regions, but the average slope near the wall will be somewhat flatter than normal, indicating a higher relative degree of turbulent mixing in this zone. This type of flow may be called *hyperturbulent flow*, and is illustrated in Fig. 3–6.

Fig. 3–6. Hyperturbulent (wake-interference) flow.

The velocity distribution in this type of flow is shown in Fig. 3–7.   Since obviously the dimension λ is the dimension most affecting the wake turbulence, the dimensionless length parameter is taken as R/λ, the relative roughness spacing. The radial distances y are measured from the crests of the roughness elements, near which points the separation zones originate.

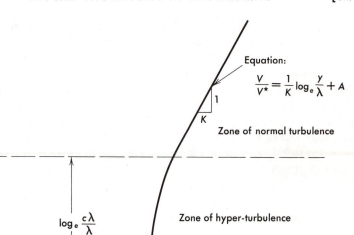

Fig. 3–7. Velocity distribution in hyperturbulent flow.

The velocity distribution equations are also shown on Fig. 3–7. The slope of the wall velocity distribution corresponds to $1/k$ in the central zone and would approach $1/k$ as the wall Reynolds number, defined as $N_R\sqrt{f}/(R/\lambda)$, increases. The boundary between the central and wall zones is at the distance $y = c\lambda$. This boundary is of course not sharply defined, but is where the velocity gradient first diverges appreciably from the value $1/k$. The coefficient c would evidently decrease with increasing N_{Rw}, and increases with increasing values of R/λ. The slope ψ of the wall distribution is taken as the *average* slope in this region.

It may be noted that the values of the boundary terms, A and A_w could be obtained from experimental velocity traverses as the values of v/v^* at $y = \lambda$, for the central and wall zone distribution curves. Since ψ is dependent on N_{Rw}, it would be expected that A_w also varies with N_{Rw}. However, the constant A is independent of the wall Reynolds number and relative roughness spacing and even of the form of elements. It is thus a truly *universal constant*. It is merely the value of v/v^* in the core distribution at a distance $y/\lambda = 1$. It is well established that v/v^* in normal turbulence for any y is independent of N_R and the roughness form. The velocity distribution in fully developed statistically normal turbulence, in other words, depends not on the geometry of the roughness elements themselves, but on the geometry of the wakes which they create, since these overlap each other and in effect obliterate the actual elements. Thus the longitudinal frequency of the crest separation points (i.e., λ), controls the velocity distribution and turbulence structure and the significant similitude parameters must be the dimensionless wall distance y/λ and the wall Reynolds number $N_R\sqrt{f}/(R/\lambda)$, or $(\sqrt{32})\lambda v^*/\gamma$.

The roughness element geometry has, however, two effects in this type of flow. First, the roughness height, h, controls the lateral extent of the velocity distributions. In effect, below this height, there is no flow, but only the wake regions. Thus the effective pipe diameter D is the internal crest-to-crest distance between roughness elements. Second, the form of the elements determines the longitudinal location of the actual separation points and thus the shape of the wake region. This may vary somewhat with the wall Reynolds number, especially if the element is streamlined.

An expression for the friction factor for hyper-turbulent flow in a circular pipe can now be derived from the velocity distribution equations, as follows:

$$\frac{1}{\sqrt{f}} = \frac{V}{\sqrt{8}v^*} = \frac{\dfrac{1}{A}\displaystyle\int_A v\,dA}{\sqrt{8}v^*} = \frac{1}{\pi R^2\sqrt{8}}\int_0^R \left(\frac{v}{v^*}\right)2\pi r\,dr$$

$$= \frac{1}{R^2\sqrt{2}}\left[\int_0^{c\lambda}(R-y)\left(A_w + \psi\log_e\frac{y}{\lambda}\right)dy + \int_{c\lambda}^R (R-y)\left(A + \frac{1}{K}\log_e\frac{y}{\lambda}\right)dy\right]$$

Omitting intermediate steps, this expression can be reduced to

$$\frac{1}{\sqrt{f}} - \frac{1}{k\sqrt{8}}\log_e\frac{R}{\lambda} = \frac{1}{\sqrt{8}}\left[\left(A - \frac{3}{2k}\right) + \left(\frac{1}{k} - \psi\right)\left\{2\left(\frac{c\lambda}{R}\right) - \frac{1}{2}\left(\frac{c\lambda}{R}\right)^2\right\}\right] \quad (3\text{-}63)$$

The term $c\lambda/R$ is the ratio of the boundary zone thickness to the total radius and, being dimensionless, should be independent of the relative roughness spacing R/λ. However, it will be influenced both by the form of the elements (which controls the separation point and therefore the wake boundary) and by the wall Reynolds number.

Inserting the value of 0.40 for k and 8.7 for A in Eq. (3-63) and converting to a decimal logarithmic base yields

$$\frac{1}{\sqrt{f}} - 2\log_{10}\frac{R}{\lambda} = 1.75 + \frac{\sqrt{2}}{4}(2.5 - \psi)\left[2\left(\frac{c\lambda}{R}\right) - \frac{1}{2}\left(\frac{c\lambda}{R}\right)^2\right] \quad (3\text{-}64)$$

The left-hand member is the generalized resistance function, defined in terms of the relative roughness spacing R/λ. The additive term in the right-hand member involves the terms ψ and $c\lambda/R$, which, as already pointed out, depend only on the wall Reynolds number and the geometric form of the roughness elements. It could be written in functional form since ψ and c would have to be determined experimentally anyhow, as:

$$\phi(N_{Rw}, \text{element shape}) = \frac{\sqrt{2}}{4}(2.5 - \psi)\left[2\left(\frac{c\lambda}{R}\right) - \frac{1}{2}\left(\frac{c\lambda}{R}\right)^2\right] \quad (3\text{-}65)$$

The resistance function equation thus becomes

$$\frac{1}{\sqrt{f}} - 2 \log_{10} \frac{R}{\lambda} = 1.75 + \phi\left(\frac{N_R \sqrt{f}}{R/\lambda}, \text{element shape}\right) \qquad (3\text{--}66)$$

This equation is the defining equation for this flow regime, which, because of the nature of its turbulence generating mechanism, may be called *hyperturbulent flow* or, if the boundary action is to be emphasized, *wake-interference flow*.

The function ϕ, for a given form of roughness, will decrease with increasing values of N_{R_w}. This follows primarily from the fact that the width $c\lambda$ of the zone of hyperturbulence becomes smaller. It is reasonable that, for sufficiently high values of N_{R_w}, regardless of the roughness form, the resistance function may approach a constant value of 1.75. Thus, hyperturbulent flow may become normal turbulent flow at high values of N_{R_w}. However, the hyper-turbulent zone may persist to indefinitely high N_{R_w} values for some types of roughness elements, in which case ϕ may approach some constant value other than zero. In any case, it must be determined by model testing for each form of elements.

However, this function is not applicable for the entire transition from smooth to normal turbulent flow. Equation (3–61) indicates the friction factor f to increase with increasing Reynolds number, whereas for both smooth and semismooth flow the friction factor decreases with Reynolds number. It seems likely that for surfaces which will yield hyperturbulent flow, and ultimately normal flow, there is a brief regime of smooth and then semismooth flow before hyperturbulent flow begins. This sequence is evidenced by the familiar dip-and-rise of the Nikuradse rough-pipe transition function.

Hyperturbulent flow is produced over such surfaces as corrugated metal and sand-coatings. The exact form of the transition function depends on the form of roughness elements, and must be determined experimentally. It is very significant, however, that for sufficiently high Reynolds numbers, each transition function approaches the normal flow equation, for which the friction factor depends solely on the relative roughness spacing.

3–16. Quasi-smooth Flow. Still another type of flow must occur over surfaces composed either of small depressions or of roughness elements spaced so closely as to form a more or less smooth pseudo-wall composed of the element crests and the enclosed pockets of dead fluid. Within these pockets or depressions will be stable vortices, unable to separate and commingle with the bulk flow because of the closeness of the downstream wall of the element. The depression vortices will be maintained through transmission of shear stress from the flowing fluid at their upper limbs. In addition, small-scale vorticity will be generated continuously along the pseudo-wall somewhat analogously to the process in smooth turbulent flow. Figure 3–8 illustrates the phenomenon, which is herein called *quasi-smooth flow* or *skimming flow*.

The energy expenditure is partially to generate the quasi-smooth vorticity at the pseudo-wall and partly to maintain the stable depression vortices.

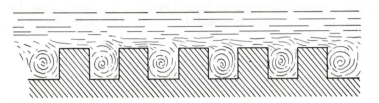

Fig. 3–8. Quasi-smooth (skimming) flow.

Such a vortex is shown schematically in Fig. 3–9. The width of the depression, and therefore the vortex radius, is j. Since it is stable, its energy can be evaluated in terms of the rotational energy of all its concentric cylindrical rings of fluid.

The energy of flow dE per unit time through any concentric cylindrical shell of the vortex is

$$dE = \tfrac{1}{2}v_r^2\,dm = \tfrac{1}{2}\rho S v_r^3\,dr = \tfrac{1}{2}\rho S(\omega r)^3\,dr$$

where S is the transverse length of the depression and ω is the angular velocity (assumed constant) of the vortex.

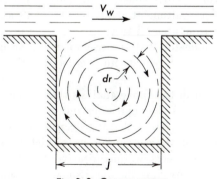

Fig. 3–9. Groove vortex.

At the upper limb of the vortex, its peripheral velocity is $\tfrac{1}{2}\omega j$. Since the vortex is kept spinning by means of shear between the conduit fluid, moving at wall velocity v_w, and the vortex, it may be assumed there is a definite relation between the wall and vortex velocities. That is, $\frac{1}{2}\omega j = c_w v_w$, where c_w is some fraction less than unity. The total vortex energy (in total foot-pounds per unit time) is then

$$E = 2 \int_0^{j/2} \frac{1}{2} \rho S \left(\frac{c_w v_w}{(1/2)j} \right)^3 r^3 \, dr = \frac{\rho S j (c_w v_w)^3}{8}$$

If the depressions are spaced longitudinally a distance λ apart, then the head that must be expended to maintain them in a length of pipe L is

$$H_v = \left[\frac{\rho S j (c_w v_w)^3}{8} \right] \left(\frac{L}{\lambda} \right) \left(\frac{1}{AV\gamma} \right) = \frac{S}{P} \left(\frac{j}{\lambda} \right) \left(\frac{c_w v_w}{V} \right)^3 \left(\frac{V^2}{2g} \right) \left(\frac{PL}{4A} \right)$$

$$= \left[\frac{S}{P} \frac{j}{\lambda} \left(\frac{c_w v_w}{V} \right)^3 \right] \left(\frac{L}{D} \right) \left(\frac{V^2}{2g} \right) \tag{3-67}$$

The term in brackets is obviously f_v, the part of the friction factor attributable to the maintenance of the depression vortex. Thus, the total friction factor for the quasi-smooth flow regime is

$$f_q = f_s + f_v = f_s + \left[\left(\frac{S}{P} \right) \left(\frac{j}{\lambda} \right) \left(\frac{c_w v_w}{V} \right)^3 \right] \tag{3-68}$$

In this relation, S/P is the ratio of depression length to total peripheral length, essentially identical with what was termed the peripheral roughness ratio in semismooth flow. Similarly the ratio λ/j is analogous to the roughness index. It is obvious that λ/j is greater than unity and that quasi-smooth flow cannot occur if j is significantly greater than the radial depth of the depression.

The ratio v_w/V will be fairly constant and such experimental data as are available indicate its magnitude at about $\frac{2}{3}$. The coefficient c_w is uncertain, but probably ranges from about $\frac{1}{2}$ to $\frac{3}{4}$.

Equation (3-68) indicates that the friction factor in quasi-smooth flow decreases with increasing Reynolds number, and with increasing relative spacing.

It has been assumed in the above analysis that no mass transfer of fluid takes place into and out of the depression and, also, that the depth of the groove is immaterial provided only that it is deep enough to maintain a stable vortex. Though neither of these assumptions is strictly correct, experiments have indicated that the effects are quantitatively negligible in comparison with the other factors controlling the quasi-smooth friction factor.

Since S is the transverse length of the depression, it may be slightly longer than the actual perimeter at the wall surface. For a circular pipe with a depression around the full perimeter, S/p becomes $\pi(D + h)/\pi D = (1 + h/D)$, where h is the depth of the depression. Thus, in Eqs. (3-67) and (3-68), the peripheral roughness ratio can be taken as $(p/P)(1 + h/D)$, where p is the peripheral length of depressions as measured around the actual pipe wall surface perimeter. Normally, h/D is so small that this effect can be ignored and the peripheral roughness ratio taken simply as p/P.

3-17. Design Curves for Flow Regimes.

In order to facilitate calculations as much as possible, design curves have been plotted for each of the various turbulent flow regimes. These are of the same general type as the Moody curves, giving friction factor as a function of Reynolds number, and are used in exactly the same way. The only difference is in the roughness parameters. As discussed previously, it is incorrect to attempt to lump all roughness effects together in terms of any kind of equivalent dimension. Each flow regime requires both its own roughness parameter and its own friction characteristic curve.

Figure 3–10 contains the curves for semismooth turbulent flow, the lowest curve being that for smooth flow. The curves are plots of Eq. (3–61) for parametric values of the roughness function ΣE defined in Eq. (3–62). Values of drag coefficient C_D (ordinarily assumed constant with sufficient accuracy for design purposes, at usual values of N_R) to use in evaluating E for each repeating roughness element are tabulated in Fig. 3–11. Use of these curves of course requires that the roughness element height h be great enough to pierce the laminar boundary film; otherwise, smooth flow will prevail. The thickness of this boundary layer is a decreasing function of Reynolds number, as given in Eq. (3–30). As long as the roughness height is significantly smaller than δ, it can be assumed that smooth flow will prevail. For simplicity, and for design purposes, no attempt has been made to delineate a transition function from smooth to semismooth flow.

The quasi-smooth function is shown in Fig. 3–12, again the lowest curve being that for smooth flow. These curves are plots of Eq. (3–68), with the roughness parameter $\Sigma\chi$ defined as $\Sigma[(p/P)(j/\lambda)(c_w v_w/V)^3]$. The ratio v_w/V may range from $\frac{3}{8}$ to $\frac{6}{8}$, perhaps averaging $\frac{2}{3}$. The coefficient c_w is yet unknown but would perhaps range from about $\frac{1}{2}$ to nearly unity. The term $(c_w v_w/V)^3$ would therefore lie within the range of, say, 0.01 to 0.40. In the absence of better information, a value of 0.10 would be probably reasonable and conservative. This corresponds to a c_w of $\frac{2}{3}$ and V_w/V of 0.7.

Since the friction factor in hyperturbulent flow depends on both the relative roughness spacing R/λ and the form of elements, a different set of curves would be required for each form of element.

On Fig. 3–13, curves have been plotted for a corrugated surface, the most common commercial type of rough surface producing this type of flow, as well as its limiting state of fully normal turbulent flow. The curves in the latter range, that of normal turbulent flow, will apply not only to corrugations but also to roughnesses of other forms producing this type of flow.

This set of curves is plotted with the relative roughness spacing R/λ as parameter. It may be noted that these curves, since they are experimental

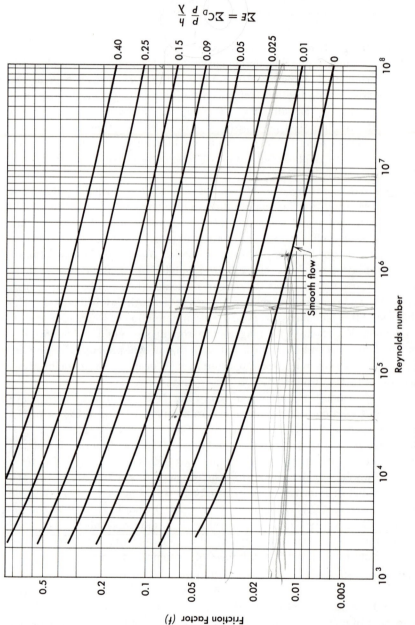

Fig. 3-10. Friction factor for semismooth turbulent flow (isolated roughness flow).

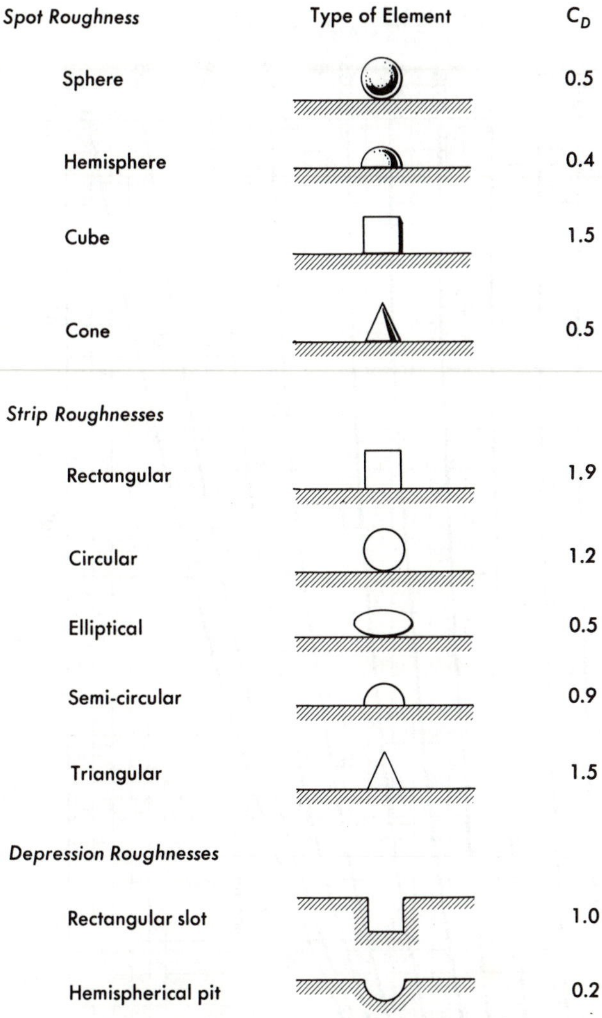

Spot Roughness	Type of Element	C_D
Sphere		0.5
Hemisphere		0.4
Cube		1.5
Cone		0.5
Strip Roughnesses		
Rectangular		1.9
Circular		1.2
Elliptical		0.5
Semi-circular		0.9
Triangular		1.5
Depression Roughnesses		
Rectangular slot		1.0
Hemispherical pit		0.2

Fig. 3–11. Drag coefficients for roughness elements.

curves, define the entire transition from smooth flow to normal turbulent flow for corrugated surfaces. As noted before, this includes a segment of semi-smooth flow as well as the hyperturbulent regime proper.

Similar families of experimental curves are plotted on Figs. 3–14 and 3–15, for sharp-edged strip roughnesses and uniform spot roughnesses, respectively. The latter are identical with the Nikuradse test curves on uniform sand-coated surfaces, since the sand-grain diameter and roughness spacing are identical in this case.

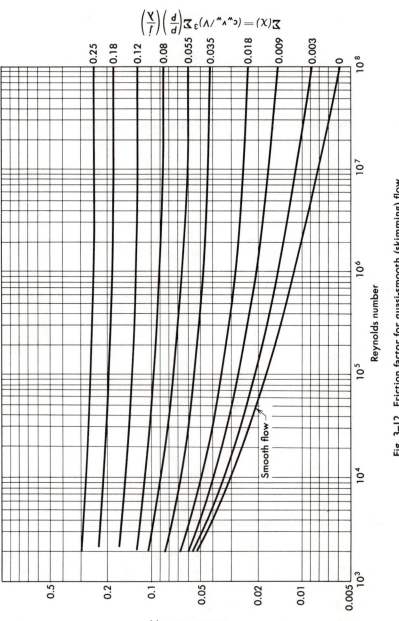

Fig. 3–12. Friction factor for quasi-smooth (skimming) flow.

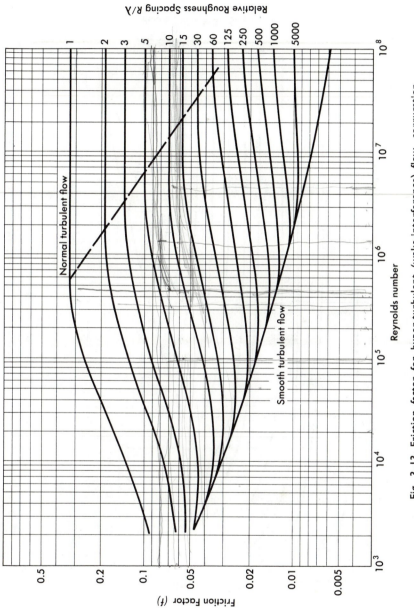

Fig. 3–13. Friction factor for hyper-turbulent (wake-interference) flow, corrugation strip roughness.

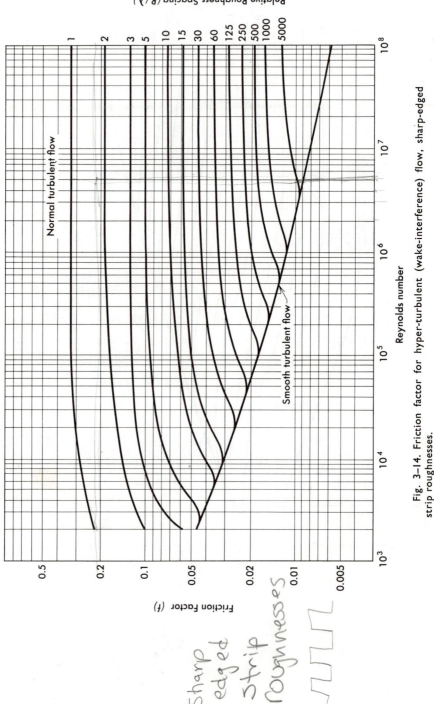

Fig. 3-14. Friction factor for hyper-turbulent (wake-interference) flow, sharp-edged strip roughnesses.

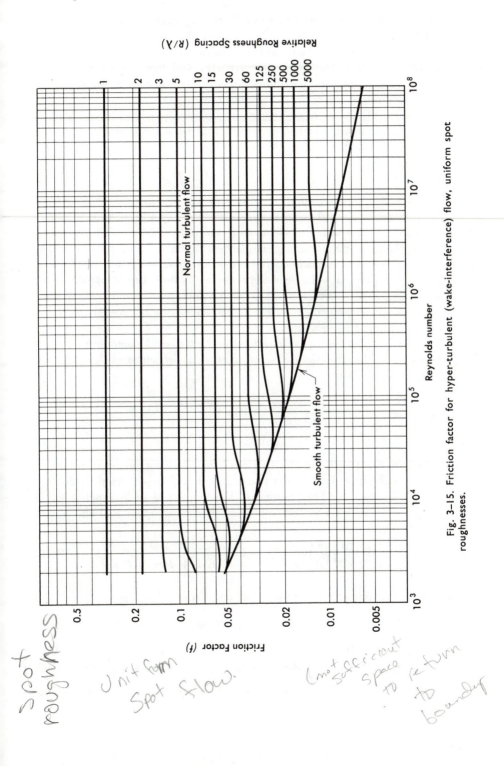

Fig. 3–15. Friction factor for hyper-turbulent (wake-interference) flow, uniform spot roughnesses.

spot roughness

Uniform spot flow.

(most sufficient space to return to boundry

3–18. Discrimination Between Flow Regimes.   The curves of Figs. 3–10 through 3–15 will permit the solution of most ordinary problems in conduit flow (in which elastic and gravitational effects can be neglected), provided only that the roughness geometry and flow regime are known. However, the problem of discriminating between semismooth, quasi-smooth, and hyperturbulent flow has not yet been discussed.

Quasi-smooth flow will occur over roughness depressions in which stable vortices can be maintained. This means that these depressions must be at least as deep as they are long. Otherwise they will develop either semismooth or hyperturbulent flow.

The problem of determining whether, for a given surface, semismooth or hyperturbulent flow will occur is more difficult. For a given type of roughness element, the spacing of elements is the critical factor. The boundary between the two regimes, for a given Reynolds number and conduit would be *that spacing for which the friction factor attains its maximum value.* For spacings exceeding this critical value, causing semismooth flow, Eq. (3–61) shows the friction factor to decrease as the spacing increases. For closer spacing, producing hyperturbulent flow, Eq. (3–66) shows the friction factor to decrease as the spacing decreases.

The critical spacing (or perhaps another factor) could therefore be determined by equating the friction factor expressions for semismooth and hyperturbulent flow. Thus,

$$\frac{1}{\sqrt{f_s(1 + 67.2E)}} - 2\log_{10}\left(\frac{R}{\lambda}\right) = 1.75 + \phi \qquad (3\text{–}69)$$

Solution of Eq. (3–69) for the spacing λ, or whatever factor is unknown, determines the boundary between semismooth and hyperturbulent flow for the given conditions. The presence of the function ϕ makes such solution quite difficult. As a reasonable approximation, assume $f_s = 0.01$ and $\phi = 0$. Then,

$$67.2\left(\frac{R}{\lambda}\right)\left(\frac{h}{R}\right)\left(\frac{p}{P}\right)C_D + 1 = \frac{100}{[1.75 + 2\log_{10}(R/\lambda)]^2} \qquad (3\text{–}70)$$

Equation (3–70) is plotted on Fig. 3–16, giving the critical relative roughness spacing $\frac{R}{\lambda}$ as a function of the relative roughness height, with the factor $(p/P)(C_D)$ parametric.

For a given roughness element geometry, the values of h, $\frac{p}{P}$, and C_D will be known. If the radius R (or equivalent radius, $2R_h$, if the pipe is non-

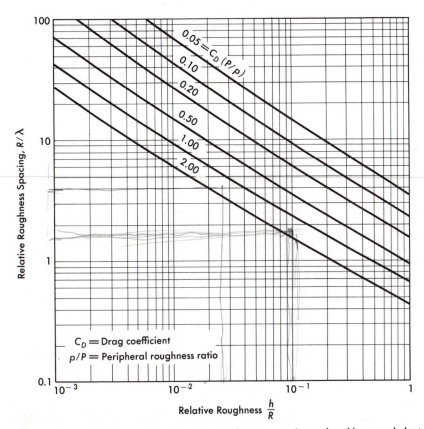

Fig. 3–16. Boundary function for discrimination between semismooth and hyperturbulent flow.

circular) is given or assumed, then one may read off the value $\dfrac{R}{\lambda}$ from the curves corresponding to the known values of $\left(C_D\,\dfrac{p}{P}\right)$ and $\dfrac{h}{R}$. This gives the "critical spacing," λ_c. If the actual element spacing is λ_a, then the flow is hyperturbulent if $\lambda_a < \lambda_c$ and semismooth if $\lambda_a > \lambda_c$.

The simplifying assumptions in the above analysis (namely that $\phi = 0$, $f_s = 0.01$, $C_D =$ constant, etc.) are not strictly valid, but will not seriously affect the value of the critical spacing. Furthermore, the values of the friction factor are approximately the same as computed from either the semismooth or hyperturbulent formulas, when λ is near the critical spacing.

3-19. Applications in Practical Hydraulic Analysis. The application of these concepts and design curves to practical flow problems requires knowledge of the geometry of the surface roughness elements. In the case of surfaces of regular roughness pattern, such as corrugated metal or uniform sand or gravel, this geometry is known or easily determined. Such surfaces, if the roughness elements are closely spaced, nearly always produce hyperturbulent flow and, at sufficiently high Reynolds numbers, normal turbulent flow.

Materials commonly used for water and sewer pipe, and for gas and oil transmission—concrete, cast iron, clay, welded steel, etc.,—are, at least when new, essentially smooth, except for isolated roughness elements and for the joints. If rivets, bolts, etc., are present, their effects are easily computed by the semismooth flow equation. The joints may cause either a protuberance or a depression, and the effects are calculated accordingly. Although the exact geometry of the elements may not be known, it is believed that the designer could estimate these factors with at least as much confidence and accuracy as a roughness coefficient or equivalent sand roughness. Materials of this type will usually produce either semismooth or quasi-smooth flow, and the appropriate curves can be used accordingly.

Surfaces of variable or irregular roughness are of course more difficult to evaluate. However, the principles of the various turbulence regimes must still be valid, so that friction factors for such surfaces could still be estimated by judicious application and extension of these principles.

If, for example, the surface consists of a combination of several types of isolated protuberances and depressions, the semismooth and quasi-smooth equations could be combined, as follows:

$$f = f_s\left[1 + \sum 67.2C_D\left(\frac{p}{P}\right)\left(\frac{h}{\lambda}\right)\right] + \sum\left(\frac{p}{P}\right)\left(\frac{j}{\lambda}\right)\left(\frac{c_w v_w}{V}\right)^3 \qquad (3\text{--}71)$$

$$= f_s[1 + \sum E] + \sum \chi$$

If the elements are close enough together to produce hyperturbulent or normal flow, the hyperturbulence equation must be modified to allow for elements of various forms and spacings. If only the spacing is variable, its average value may be used with sufficient accuracy, since it is basically the number of turbulence sources that determines the wall zone turbulence patterns. If both the spacing and the roughness form are variable, the average value of spacing may again be used in the hyperturbulent flow equation, but judgment must be exercised as to which form of transition function ϕ should be used, depending upon the prevalent form in the roughness pattern.

When occasional elements project much farther into the flow than others, the effect is that of superimposing isolated wakes and drag forces on the typical hyperturbulent flow near the smaller elements. It is reasonable in this situation to compute the friction factor due to the hyperturbulent flow near the ordinary elements, and then add increments of friction factor corresponding to the larger elements. The resulting relation is

$$f = \frac{1}{[2 \log_{10} (R/\lambda_1) + 1.75 + \phi]^2} + 2\Sigma \left(\frac{p}{P}\right) C_D \left(\frac{h}{\lambda_2}\right) \qquad (3\text{--}72)$$

The second term in Eq. (3–72) is derived from Eq. (3–60) and is an approximation based on the assumption that the velocity near the wall which impinges on the projecting element is about 0.7 of the average velocity in the cross-section. The element height h must be measured from the crests of the smaller adjacent elements.

In the case of a conduit with part of its perimeter very rough and part smooth (such as a corrugated pipe with paved invert, or a pipe with longitudinal rows of closely spaced rivet projections) it is obvious that the over-all friction factor will be intermediate between that for a smooth flow and that for a hyperturbulent flow corresponding to the rough section of the surface. Most of the energy expenditure due to friction in a flow occurs in the process of vortex generation at the conduit surface. Thus it is reasonable to take the proportion of rough surface area to smooth surface area as the basis for determining the bulk friction factor. The resulting equation is

$$f = \frac{p_s}{P} f_s + \left(\frac{p}{P}\right) \left\{ \frac{1}{[2 \log_{10} (R/\lambda) + 1.75 + \phi]^2} \right\} \qquad (3\text{--}73)$$

where p_s and p are the smooth and rough segments of the periphery, respectively.

In the relatively rare case of a surface composed of random roughness elements, of various shapes, sizes, and spacings, the tendency is for the friction factor–Reynolds number curve to be horizontal throughout the entire turbulent range. That is, such a statistically random roughness distribution produces a statistically random vorticity pattern near the wall, and therefore normal turbulent flow, even at low Reynolds numbers. The friction factor for such a surface can probably be estimated from the normal turbulent flow equation, by using the average spacing of the predominating larger elements on the surface.

Example 3–3

A 12-in. clear-diameter riveted steel pipe (measured between crests of rivets) is used to carry crude oil, at a temperature such that its kinematic viscosity is 0.0001 ft²/sec, a distance of 12 miles, with a permitted head loss of 50 ft. The longitudinal joints are double-riveted butt joints, with rivet heads 1 in. in diameter on a center-line pitch of 6 in. The girth joints are on a 6-ft spacing and of the same design except for being single-riveted butt joints, with a $\frac{1}{2}$-in. butt strap. Assuming that semismooth flow occurs, what discharge will the pipe carry?

Solution. Assuming semismooth flow, the roughness parameter $\Sigma E = \Sigma C_D(p/P) \times (h/\lambda)$ is determined as follows:

$$\text{Longitudinal rivets:} \quad E_1 = 0.4\left[\frac{4(1)}{12\pi}\right]\left(\frac{0.5}{6}\right) = 0.00354$$

$$\text{Girth rivets:} \quad E_2 = 0.4\left(\frac{1}{6}\right)\left(\frac{0.5}{36}\right) = 0.00093$$

$$\text{Butt strap:} \quad E_3 = (1.9)(1)\left(\frac{0.5}{72}\right) = 0.01320$$

$$\Sigma E = \overline{0.01767}$$

The Reynolds number, $N_R = 10,000\ V$, and the Darcy equation yields the relation $V = 0.2255/\sqrt{f}$. From the family of curves for semismooth flow, Fig. 3–10, using a curve corresponding to $\Sigma E = 0.01767$, trial solution of these equations yields values of $f = 0.075$, $V = 0.82$ ft/sec, and $N_R = 8240$. Then the discharge, $Q = (\pi/4)(1)^2(0.82) = 0.645$ cfs.

Example 3–4

A 36 in. corrugated metal pipe arch conduit has a cross-sectional area of 6.32 sq ft and a perimeter (all data based on inside clear dimensions) of 9.43 ft. The corrugations are $\frac{1}{2}$ in. deep and spaced at $2\frac{2}{3}$ in. What hydraulic gradient is necessary to cause the pipe to discharge 15 cfs, at a water temperature of 68 F?

Solution. Flow in standard corrugated pipes may usually be assumed hyperturbulent. The relative roughness spacing is $R/\lambda = 2A/\lambda P = [2/(8/36)] \times (6.32/9.43) = 6.04$. The Reynolds number is $4Q/\nu P = [4(15)/1.09(9.43)]10^5 = 584,000$. From the family of curves for hyperturbulent flow, Fig. 3–13, $f = 0.076$. Substituting in the Darcy equation yields $H_f/L = fQ^2P/8gA^3 = 0.00247$.

Example 3–5

What diameter of smooth concrete pipe would be required for the same discharge at the same hydraulic gradient? Assume the pipe functions as a smooth pipe except for its joints, which are spaced at 6-ft intervals and which form a peripheral groove $1\frac{1}{8}$ in. deep and 1 in. wide.

Solution. Assuming quasi-smooth flow, with a hydraulic gradient of 0.00247, as obtained above, the Darcy and Reynolds number relations are $D = \sqrt[5]{2290f}$

and $N_R = (1.75/D)10^6$. In absence of more precise information, it may reasonably be assumed that the factor $(C_w V_w/V)^3$ is no more than 0.1 (equivalent to assuming $C_w = 0.7$ and $V_w = \frac{2}{3}V$). Then the roughness parameter $\chi = 0.1(p/P)(j/\lambda) = 0.1(1)(1/72) = 0.00139$. From the quasi-smooth flow curves, Fig. 3–12, for a curve corresponding to $\chi - 0.00139$, trial solution of the Darcy and Reynolds number relations yields $f = 0.014$, $N_R = 875{,}000$, and $D = 2.0$ ft.

Example 3–6

A smooth circular pipe of 60 in. wall-to-wall diameter is strengthened by internal circumferential rings placed at intervals. The rings are 1 in. high and $\frac{1}{4}$ in. thick. What is the maximum and minimum longitudinal spacing of the rings that will produce hyperturbulent flow?

Solution. The boundary spacing between semi-smooth and hyperturbulent flow may be estimated by means of the discriminating function curves of Fig. 3–16. The parameter $C_D(p/P) = 1.9(1)$, and $h/R = 1/29 = 0.0345$. From the curves, $R/\lambda = 3.0$; therefore $\lambda = R/3 = 29/3 = 9.67$ in. The minimum spacing for hyperturbulent flow will be that for which $j = h = 1$ in. Thus, $\lambda = 1 + 1/4 = 1.25$ in. Hyperturbulent flow will occur if 1.25 in. $\leq \lambda \leq 9.67$ in. For smaller spacings, the flow will be quasi-smooth, and for larger spacings semismooth.

It will be noted that, in the above examples, all calculations are based on the *minimum internal diameter*, between crests of roughness elements, and not on the wall-to-wall diameter. This is because the roughness element wakes originate on these crests, so that the bulk flow and its distribution of velocities must pass through this minimum diameter. Even in the case of semismooth flow (in which case, of course, the laminar film component of the friction factor originates at the outer wall), use of the minimum diameter in the equations seems to give a better correlation with most of the available experimental data. This is presumably because the isolated-roughness elements usually produce a significantly larger part of the friction factor than does the smooth wall. If the isolated-roughness elements are pits or depressions in the wall surface, however, or if they are very widely separated (i.e., for ΣE less than about 0.01) then the wall-to-wall diameter should be used.

3–20. Selection of Design Method.

It is obvious from the foregoing discussion that there are various alternate methods available for analyzing or designing pressure conduits. Numerous other empirical equations and methods, for which space is not available here, have appeared in the technical literature from time to time. In spite of the fact that the problem of flow in a uniform pressure conduit is one of the most important and basic problems in fluid mechanics and hydraulics, much research is still needed before the

TABLE 3-6
Uniform Flow Regimes in Closed Conduits

Regime	Friction Factor Eq.	Variation of Friction Factor		Criterion
		As N_R Increases	With Roughness	
1. Laminar	$f = 64/N_R$	Decreases	No Effect	$N_R \leq 2000$
2. Smooth	$\dfrac{1}{\sqrt{f_s}} - 2\log N_R\sqrt{f_s} = -0.8$	Decreases	No Effect	$h \nless \dfrac{16\,D}{N_R\sqrt{f_s}}$
3. Quasi-Smooth	$f = f_s + \dfrac{p}{P}\dfrac{1}{\lambda}\left(\dfrac{C_w V_w}{V}\right)^3$	Decreases	Increases with $\dfrac{p}{P}, \dfrac{1}{\lambda}$	$j \leq h$
4. Semi-Smooth	$f = f_s\left[1 + 67.2\,C_D\,\dfrac{p}{P}\dfrac{h}{\lambda}\right]$	Decreases	Increases with $C_D, \dfrac{p}{P}, \dfrac{h}{\lambda}$	$\dfrac{\left[1 + 67.2\,C_D\,\dfrac{p}{P}\dfrac{h}{\lambda}\right]}{\left[\dfrac{10}{1.75 + 2\log\dfrac{R}{\lambda}}\right]^2} <$

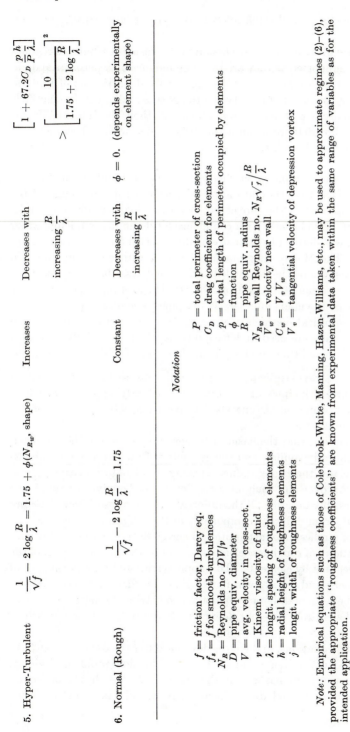

5. Hyper-Turbulent

$$\frac{1}{\sqrt{f}} - 2\log\frac{R}{\lambda} = 1.75 + \phi(N_{R_w},\ \text{shape})$$

Increases

Decreases with increasing $\frac{R}{\lambda}$

$$\left[1 + 67.2 C_D\,\frac{p}{P}\,\frac{h}{\lambda}\right] > \left[\frac{10}{1.75 + 2\log\frac{R}{\lambda}}\right]^2$$

6. Normal (Rough)

$$\frac{1}{\sqrt{f}} - 2\log\frac{R}{\lambda} = 1.75$$

Constant

Decreases with increasing $\frac{R}{\lambda}$

$\phi = 0.$ (depends experimentally on element shape)

Notation

f = friction factor, Darcy eq.
f_s = f for smooth-turbulences
N_R = Reynolds no. DV/ν
D = pipe equiv. diameter
V = avg. velocity in cross-sect.
ν = Kinem. viscosity of fluid
λ = longit. spacing of roughness elements
h = radial height of roughness elements
j = longit. width of roughness elements

P = total perimeter of cross-section
C_D = drag coefficient for elements
p = total length of perimeter occupied by elements
ϕ = function
R = pipe equiv. radius
N_{R_w} = wall Reynolds no. $N_R\sqrt{f}\,/\frac{R}{\lambda}$
V_w = velocity near wall
C_w = $V_v,\ V_w$
V_v = tangential velocity of depression vortex

Note: Empirical equations such as those of Colebrook-White, Manning, Hazen-Williams, etc., may be used to approximate regimes (2)–(6), provided the appropriate "roughness coefficients" are known from experimental data taken within the same range of variables as for the intended application.

process functions for even this relatively simple process are really known and understood.

It should be evident from the discussion of the various turbulence regimes as given above, supported as they are by an abundance of experimental data, that no single process equation can be applied to all situations of uniform turbulent flow in rough conduits. This is the greatest fallacy in the methods now in common use (e.g., the Colebrook-Moody method, the Hazen-Williams method, etc.), although there are numerous other deficiencies in these methods, as discussed previously in Arts. 3–9 and 3–10.

Nevertheless, these empirical methods are convenient and satisfactory for design use provided that they are applied strictly within the range of the empirical data upon which they are based. Thus the Hazen-Williams formula, for example, can often be used with confidence for designing municipal water lines because it was derived by measurements of actual flows and head losses in such pipes. The Moody diagram can be used with confidence only in design situations where the equivalent-sand-roughness diameter has been determined by experimental data from the same type of pipe operating over the same range of Reynolds numbers as the pipe to be designed.

In general, empirical methods such as these are not only valid but even preferable when they are used under conditions which essentially are interpolated from the conditions under which they were derived. However, it is never safe to extrapolate experimental results unless such extrapolation is based on sound principles of similitude consistently applied. And it is this requirement which unfortunately has been neglected in most of the methods in current use.

Similitude requires that the basic turbulence flow structures be geometrically and dynamically similar, and the geometry of the boundary roughness elements, especially their longitudinal spacing, is of determinative significance in this. An over-all "roughness coefficient" or "equivalent roughness size" cannot possibly describe all the different complexities of flow structure which can result from different components of the roughness geometry.

A number of studies have attempted to correlate the friction factor with the "concentration" of roughness elements—that is, the fractional part of the wall surface occupied by elements. This approach is valid for a given type of roughness element and for a given flow regime, but does not recognize the actual physical effects of different types and geometries of roughness element patterns and the entirely different flow regimes and process functions which they can cause.

The concepts, equations, and curves developed in Arts. 3–11 through 3–18 do, however, provide a design and analysis tool for flow problems which is substantially rational in nature, in the sense that it is based on sound principles of similitude and dimensional analysis and takes the necessary cognizance of all the different effects of specific boundary roughness geometries and the physical phenomena generated thereby.

These concepts and equations have been found to correlate data obtained from many different sources, on many types of roughnesses, not only in pipes but also for uniform tranquil flow in open channels. In recent years, the concept of the different flow regimes (i.e., isolated-roughness, wake-interference and skimming flows) has been widely adopted in the literature, but the specific design equations have been used only sparingly.

Nevertheless, the equations have been found to correlate practically all the available and relevant experimental data, within a degree of accuracy which is well within ordinary design requirements. Even these equations, of course, must rely on a certain amount of experimental data and have involved a number of simplifying assumptions, but they do provide a more rational approach and sounder base for extrapolation than any of those now in use. Although further study is needed, judicious use of the equations and methods of Arts. 3–11 through 3–18 will suffice to give design results of adequate reliability for situations where specific data for proper use of the older empirical methods may not be available.

Since the newer approach is necessarily somewhat more complex than the others, it is helpful to summarize here the various possible flow regimes and their characteristics. This is done in Table 3–6, page 104.

3-21. Energy Loss in Non-uniform Flow.

The preceding discussion of conduit friction loss applies only to straight, uniform reaches of conduit, in which the average velocity remains unchanged from one section to the next. The friction factors relate therefore to shearing stresses at the boundary and are dependent only on the conduit Reynolds number and the geometry of the boundary roughness elements.

If, however, the size, shape, or direction of the conduit changes, there must be corresponding changes in the magnitude or direction of the velocity, or both. Such changes can be accomplished only by the imposition of a force on the flow, opposite to the hydrodynamic force exerted by the changing flow on the pipe, as required by the momentum equation.

The component of this force which is normal to the direction of flow, if any, must be resisted structurally by the pipe assembly. The tangential component is a shearing force or *drag* force which sets up turbulence phenomena in the contiguous reach different from those in the uniform reaches. These abnormal shearing stresses and the turbulence generated thereby result in an abnormal drain on the flow energy.

The energy losses from causes of this nature are often called *minor losses* to distinguish them from the ordinary uniform losses due to boundary friction. Quantitatively, they sometimes are actually major losses, depending of course on the particular conduit and circumstances. The drag forces are called *form drag* forces in contradistinction to the *friction drag* forces on the walls in the uniform reaches.

The flow conditions through such a non-uniform reach may be described in terms of the general flow function, Eq. (2–48), modified as follows:

$$\Delta\left(\frac{p}{\gamma} + h\right) = \frac{V^2}{2g} f\left(N_R, \frac{D_a}{D}, \frac{r}{D}, \theta, \frac{B}{D}, \frac{B_a}{D}, \frac{L}{D}, \frac{h}{D}, \frac{\lambda}{D}, \dots\right)$$

(3–74)

In this equation, $\Delta(P/\gamma + h)$ represents the change in potential energy through the reach. The velocity of the downstream end of the reach is V. The equivalent diameter D is also measured at some reference section in the reach, commonly its downstream extremity. The other terms reflect the various geometrical factors that may influence the flow and turbulence structure; thus D_a is the equivalent diameter at the upstream extremity of the reach, r is the radius of pipe bend if the flow changes directions, θ is the central angle of bend, B is the transverse width of conduit if non-circular, B_a is the width at the upstream end of the reach, L is the length of reach, h, λ, etc., are wall-roughness element dimensions. As many similar terms must be included as may have an influence on the flow.

Usually the roughness terms and length L are relatively negligible when these form effects are being considered. Also, the aspect ratio B/D has little effect unless it is uncommonly large or small. If these terms are neglected, the head loss due to the form drag can be written

$$H_L = \Delta\left(\frac{p}{\gamma} + h + \frac{V^2}{2g}\right) = \frac{V^2}{2g} f\left(N_R, \frac{D_a}{D}, \frac{r}{D}, \theta, \dots\right) + \Delta\frac{V^2}{2g}$$

Since, by continuity, $\Delta V = V[(D/D_a)^2 - 1]$, this simplifies to

$$H_L = K_L \frac{V^2}{2g}$$

(3–75)

where K_L is a coefficient of loss given by

$$K_L = f\left(N_R, \frac{D_a}{D}, \frac{r}{D}, \theta, \dots\right)$$

(3–76)

The functional relationship will depend explicitly on the geometrical form of the varying reach. The most common cases are those involving pipe expansions, contractions, bends, or combinations thereof. Contracting flows often re-expand in part after the contraction; this is particularly true of contractions caused by gates or valves in the conduit partially obstructing the flow. Bends also frequently cause flow separation with subsequent re-expansion. Consequently, most of these form losses result at least in part from flow expansions.

3–22. Flow Expansions in Pipe Lines.   Consider a flow expansion in a pipe, as indicated in Fig. 3–17. As the flow expands, the average velocity decreases from V_a to V and therefore the pressure increases from p_a to p. The expansion cannot be accomplished in an infinitesimal length, and thus the flow separates from the wall in an expanding jet. The pressure throughout the separation zones remains more or less constant at p_a.

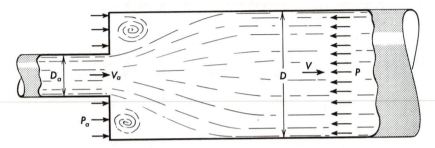

Fig. 3–17. Pipe expansion.

Within the separation zones, vortices are generated by shearing stresses at the jet boundaries. These grow, and then separate from the wall and wash out into the flow, to be followed by new vortices, and so on. The creation of these vortices requires a very substantial drain on the flow energy; their energy also is finally dissipated in the flow turbulence.

The form drag created by the expansion can be computed from the momentum equation, by considering the force system on the fluid mass from the section just downstream from the change in pipe cross-section to the section where the re-expansion is complete and uniform conditions have been established, with all the abnormal vorticity converted into normal turbulence. Thus

$$\gamma A \Delta L \sin \alpha + p_a A - p A = \rho Q (V - V_a)$$

or

$$\Delta h + \frac{p_a - p}{\gamma} = \frac{V}{g}(V - V_a) = \frac{V^2}{g}\left(1 - \frac{A}{A_a}\right)$$

The head loss caused by the expansion, as obtained from the energy equation, is then

$$
\begin{aligned}
H_{\text{exp}} &= \left(\frac{V_a^{\,2}}{2g} - \frac{V^2}{2g}\right) + \left(\frac{p_a}{\gamma} - \frac{p}{\gamma}\right) + \Delta h \\
&= \frac{V^2}{2g}\left[\left(\frac{A}{A_a}\right)^2 - 1\right] + \frac{V^2}{2g}2\left(1 - \frac{A}{A_a}\right) \\
&= \frac{V^2}{2g}\left[\frac{A}{A_a} - 1\right]^2 = \left[\left(\frac{D}{D_a}\right)^2 - 1\right]^2\left(\frac{V^2}{2g}\right) \qquad (3\text{–}77)
\end{aligned}
$$

In the above analysis, ΔL is the length of the expansion, α is the angle of pipe slope, and Δh is the drop in pipe elevation in length ΔL, equal to $\Delta L \sin \alpha$. The loss coefficient function for a sudden pipe expansion is therefore given by the simple expression,

$$K_{\exp} = \left[\left(\frac{D}{D_a} \right)^2 - 1 \right]^2 \tag{3-78}$$

and is essentially independent of the Reynolds number.

If desired, Eq. (3-77) can be written in terms of the approach velocity head, as follows:

$$H_{\exp} = \frac{V_a^2}{2g} \left(\frac{D_a}{D} \right)^4 \left[1 - \left(\frac{D}{D_a} \right)^2 \right]^2 = \left[1 - \left(\frac{D_a}{D} \right)^2 \right]^2 \frac{V_a^2}{2g} \tag{3-79}$$

From Eq. (3-77), the head loss expression can also be written as

$$H_{\exp} = \left(\frac{V_a}{V} - 1 \right)^2 \frac{V^2}{2g} = \frac{(V_a - V)^2}{2g} \tag{3-80}$$

In the extreme case of an infinite expansion ratio, this loss becomes equal to the approach velocity head. Practically, this means that the entire velocity head in a pipe discharging into a reservoir is lost.

Much of the separation zone vorticity can be eliminated by using a gradual expansion, materially reducing the energy loss. The loss coefficient will then depend on the expansion ratio D/D_a and the expansion slope L/D, and slightly on the Reynolds number. It has been determined experimentally for a considerable range of these parameters, and these data are available in most fluid mechanics and hydraulics textbooks. The exact function is very complicated, involving a combination of boundary friction, which changes as the diameter changes, and superimposed vorticity due to the adverse pressure gradient and partial separation.

The coefficient reaches an experimental minimum of about $\frac{1}{8}(A/A_a - 1)^2$ when the diffuser slope on each side is about 20:1. It then increases, as the slope increases, approximately linearly to the maximum value as given by Eq. (3-78) when the slope becomes equal to or less than about 3:1. Under some conditions it may be even somewhat greater than its theoretical maximum value. A diffuser is therefore of no value unless its sides diverge on a slope of about 3:1 or less (corresponding to a divergence central angle of about 40° or less).

As an approximation, therefore, the total head loss (expansion plus boundary friction) in a pipe expansion may be estimated as:

$$H_{\exp} = \left[\frac{1}{8} + \frac{(20 - S)(7)}{136} \right] \left(\frac{A}{A_a} - 1 \right)^2 \frac{V^2}{2g} \tag{3-81}$$

The above expression reduces to Eq. (3–77) when S decreases to 3. For values of the slope, S, less than 3, Eq. (3–77) should be used. On the other hand, when S becomes greater than 20, the expansion loss coefficient can be considered constant at

$$\frac{1}{8}\left(\frac{A}{A_a} - 1\right)^2$$

but a boundary friction loss should be added for that part of the diffuser length greater than $20R$, where R is the pipe radius.

3–23. Pipe Contractions.   The form loss due to a contraction is primarily caused by the re-expansion of the flow following contraction, and can therefore be calculated approximately by the expansion loss equation. This is illustrated in Fig. 3–18. By considering the flow expansion from the contracted section, of area A_c, to the normal section, of area A, the head loss as given by Eq. (3–77) is

$$H_{\text{cont}} = \left(\frac{A}{A_c} - 1\right)^2 \left(\frac{V^2}{2g}\right)$$

$$= \left(\frac{A}{C_c A} - 1\right)^2 \left(\frac{V^2}{2g}\right) = \left(\frac{1}{C_c} - 1\right)^2 \left(\frac{V^2}{2g}\right) \qquad (3\text{--}82)$$

The contraction coefficient, C_c, is equal to A_c/A, and depends on the area ratio A/A_a, the nature of the contraction geometry (whether sharp or rounded), and slightly on the Reynolds number.

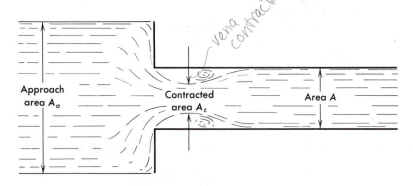

Fig. 3–18. Pipe contraction.

For a sharp-edged contraction, with a large approach area, permitting the boundary streamlines to approach the smaller pipe essentially at right angles to its axis, it can be shown that the coefficient of contraction is approximately $\pi/(\pi + 2) = 0.61$. For this case, therefore, which corresponds to the case of a flush entrance to a pipe from a reservoir, the loss is

$$H_{ent} = \left(\frac{1}{0.61} - 1\right)^2 \left(\frac{V^2}{2g}\right) = 0.41\frac{V^2}{2g} \cong \frac{1}{2}\left(\frac{V^2}{2g}\right) \qquad (3\text{--}83)$$

If the pipe entrance projects into the reservoir, the contraction is increased. In the extreme case, where the bounding streamlines must negotiate a turn of 180°, the contraction coefficient is $\frac{1}{2}$ and therefore the head loss is

$$H_{proj} = \left(\frac{1}{1/2} - 1\right)^2 \left(\frac{V^2}{2g}\right) = \frac{V^2}{2g} \qquad (3\text{--}84)$$

By rounding the contraction, the flow separation can be reduced or prevented, and therefore the re-expansion loss very much reduced. If the radius of rounding r is to be made such as to eliminate the contraction completely, the loss could theoretically be eliminated. To do this, the area at the beginning of the contraction must be such that the downstream area can be obtained by multiplying it by the theoretical coefficient of contraction, 0.61.

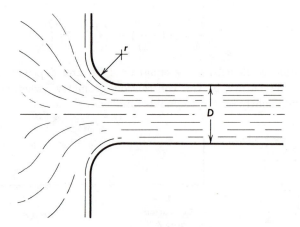

Fig. 3–19. Rounded entrance.

With reference to Fig. 3–19 this requirement is specified as follows:

$$\frac{\pi}{4}D^2 = 0.61\left(\frac{\pi}{4}\right)(D + 2r)^2$$

Therefore,

$$1 = 0.61\left(1 + 2\frac{r}{D}\right)^2$$

and

$$\frac{r}{D} = \frac{1}{2}\left(\frac{1}{\sqrt{0.61}} - 1\right) = 0.14 \cong \frac{1}{7}$$

A radius of rounding of one-seventh the diameter will therefore theoretically prevent the contraction loss. Actually boundary friction is present such that the minimum contraction loss will still be of the order of 5 per cent of the velocity head.

A radius of rounding less than this will partially reduce the contraction. Similarly, partial elimination of the contraction (therefore increase of the contraction coefficient and reduction of head loss) is accomplished as the area ratio A_a/A decreases.

3-24. Pipe Bends. The head loss caused by bends is a function of the bend radius and central angle, and of the pipe roughness and Reynolds number. The detailed mechanics are very complex, involving separation and re-expansion at the inner radius, "piling up" at the outer radius, a transverse gradient of pressure and flow superimposed on the axial flow, with a resulting spiraling component of flow, and other factors. A theoretical function has not yet been successfully devised. An empirical function that appears to satisfy many experimental data is the following:

$$H_{bend} = \frac{\theta°}{90°}\left[280 f_u{}^3\left(21.8 + \frac{43.7}{r/D}\right) + \frac{1}{5\sqrt{r/D}}\right]\left(\frac{V^2}{2g}\right) \qquad (3\text{-}85)$$

The head loss at a bend obviously decreases as r/D increases. However, increasing the bend radius means increasing the bend length and presumably its cost. It is generally agreed by designers that an r/D from 4 to 6 gives the optimum benefits, with the bend loss coefficient on the order of 0.15 for a 90° bend.

Numerous experiments have shown that a bank of guide vanes placed in the bend cross-section will materially reduce the bend loss. However, this is expensive and is usually not warranted except for extremely sharp bends. Equation (3-85) obviously is not valid for bends approaching zero radius. Experimentally, the limiting bend head loss coefficient for 90° miter bends ($r = 0$) is approximately 1.20. Equation (3-85) can be used up to such an r/D as to reach this value for K_b.

3-25. Losses at Fittings and Junctions. Losses of head for valves and other flow obstructions are essentially contraction head losses and can be determined approximately by the contraction loss equation, Eq. (3-82). Obviously they may have any value from zero to infinity, depending on the amount of flow constriction produced. They are commonly determined empirically, and tables of loss coefficients are available for all commercial types of fittings, valves, etc., from the manufacturers. Typical values for estimating purposes may be obtained from Table 3-7.

TABLE 3–7
Loss Coefficients for Common Fittings

Globe valve, fully open	10.0
Angle valve, fully open	5.0
Swing check valve, fully open	2.5
Gate valve, fully open	0.2
Gate valve, $\frac{3}{4}$ open	1.0
Gate valve, $\frac{1}{2}$ open	5.6
Gate valve, $\frac{1}{4}$ open	24.0
Short-radius elbow	0.9
Medium-radius elbow	0.8
Long-radius elbow	0.6
45° Elbow	0.4
Closed return bend	2.2
Tee, through side outlet	1.8
Tee, straight run	0.3
Coupling	0.3
45° Wye, through side outlet	0.8
45° Wye, straight run	0.3

The values in Table 3–7 should be understood as approximate only. More exact values, and values for other types of fittings, may be obtained from individual manufacturers or trade associations.

Frequently in practice fitting losses are expressed in terms of the *equivalent length* of straight pipe that would give the same friction loss. If the fitting loss coefficient is K_f, then equating the Darcy and minor loss equations yields the equivalent length as

$$L_{\text{equiv}} = \frac{K_f D}{f} \tag{3–86}$$

A common problem in many areas of hydraulic engineering practice is that of converging flows or pipe junctions, where flow from a lateral pipe enters a straight main pipe. By continuity the discharge in the downstream main is the sum of discharges in the upstream main and the lateral. The head loss in the main pipe caused by the mixing flows is expressed as a coefficient times the downstream velocity head. Extensive tests by Blaisdell and Manson at the St. Anthony Falls Hydraulic Laboratory have verified the following equation (based on application of the conservation equations to flow at pipe junctions) for the head loss coefficient in the main pipe caused by the junction:

$$K_j = 2\frac{Q_B}{Q_M} = \left(1 + 2\frac{A_M}{A_B}\cos\theta\right)\left(\frac{Q_B}{Q_M}\right)^2 \tag{3–87}$$

A_M and Q_M are the area and discharge in the downstream main pipe, respectively, and A_B and Q_B corresponding terms in the branch pipe. The angle θ is the convergence angle between the main and the branch.

In a conduit composed of reaches of various sizes of pipes, various types of fittings, etc., the total head consumed must be calculated by summation of all the uniform friction losses plus the various non-uniform losses. If H is the total head on the conduit producing flow, then

$$H = \Sigma f\left(\frac{L}{D}\right)\left(\frac{V^2}{2g}\right) + \Sigma K_L\left(\frac{V^2}{2g}\right) \tag{3-88}$$

3-26. Flow in Pipe Networks.   Although the calculations become tedious, the solution of flow problems in pipe networks is dependent on the same basic physical principles as for a single pipe or series of pipes. That is, the principles of energy conservation and continuity must be satisfied throughout the network and, for each pipe in the system, the friction loss can be calculated by an appropriate equation such as the Darcy equation; if necessary, the minor losses at each point of velocity change can be calculated also.

In general, a series of equations can be written for the network, as follows:

At each pipe juncture: $\Sigma Q = 0$

Around each closed circuit: $\Sigma H_f + \Sigma H_m = 0$

In these equations, Q is the rate of flow, in cfs, in any given pipe; H_f is the uniform flow boundary friction loss, and H_m represents the "minor losses" due to segments of non-uniform flow. Each Q term has the form $(A)(V)$, each H_f term has the form $f(L/D)(V^2/2g)$, and each H_m term has the form $K(V^2/2g)$. In each equation, careful attention must be paid to algebraic sign.

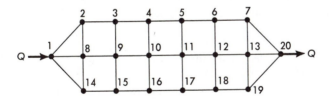

Fig. 3-20. Pipe network.

Depending on the unknowns in the problem, it should usually be possible to set up a sufficient number of independent equations of these two types to solve for the unknowns. A typical problem, for example, would be to determine the distribution of flows in a given pipe network when a given total discharge enters and leaves the network. The discharge equation is written for each juncture in the network, in terms of the given cross-sectional areas and the unknown velocities. The energy equation is written for enough circuits in the network to yield, when added to the number of discharge equations, a total number of equations equal to the number of unknown pipe velocities. These equations may then be solved simultaneously.

This procedure is quite suitable for small networks consisting of only a few pipes. For more complex networks, the number of equations becomes unwieldy. In the network sketched in Fig. 3–20, for example, there are 33 pipes, requiring 33 independent simultaneous equations. These could be obtained by writing the continuity equation for each of the 20 junctures and the energy equations for 13 independent circuits. In addition, if the friction factors and loss coefficients are also unknown, except in terms of the unknown velocities, this would mean in effect writing an additional equation for each of these unknowns also, each equation expressing the functional relation of f or K to V (actually to the Reynolds number) and boundary geometry. It is obvious that algebraic solution of the network equations is completely impracticable.

Nevertheless, solution of the simultaneous equations is the only really correct method of solving a pipe network problem. Solutions may be feasibly obtained by means of programing the equations on a high-speed digital computer, if this is available. The hydraulic analogy to electrical circuits may be utilized for solution on a form of analog computer known as an *electrical analyzer*.

In the absence of a computer of adequate capacity, some form of approximate solution must be utilized. Various methods are described in sanitary engineering textbooks. Of these the best in current practice is a relaxation technique developed by Hardy Cross.

In essence, the Cross method consists of the following steps, when adapted for the Darcy equation and minor loss terms:

1. Assume a reasonable distribution of flows throughout the network, selected such that $\Sigma Q' = 0$ at each juncture.

2. Assume appropriate friction factors and loss coefficients, for each pipe and juncture.

3. Compute head losses in each pipe and juncture for the assumed flows and coefficients.

4. If the correct flows Q have been assumed, then for each circuit in the network

$$\Sigma \frac{8(fL + KD)}{g\pi^2 D^5} Q^2 = \Sigma CQ^2 = 0 \qquad (3\text{–}89)$$

Instead, the assumed flows Q' may yield a summation for each circuit, $\Sigma CQ'^2 \neq 0$.

5. If ΔQ is the error in assumed flows in a given circuit, then

$$\Sigma C(Q' \pm \Delta Q)^2 = \Sigma CQ'^2 \pm 2\Sigma CQ'\Delta Q + \Sigma C(\Delta Q)^2 = 0$$

From this, the correction to apply to the assumed flows in the circuit is

$$\Delta Q = \pm \frac{\sum (CQ')}{\sum C} \left[1 - \sqrt{1 - \frac{[\sum (CQ'^2)](\sum C)}{(\sum CQ')^2}} \right] \tag{3-90}$$

This correction is calculated and applied to the assumed flows, using the plus or minus sign as necessary to make the losses balance.

6. The same procedure is followed for other circuits, adjustments in which may necessitate further modifications in the flows in circuits previously balanced. Eventually all flows should be such that $\sum Q = 0$ at each juncture and $\sum CQ^2 = 0$ around each circuit.

7. With the flows balanced, Reynolds numbers are computed for each pipe, and the assumed values of friction factors and loss coefficients checked and possibly revised.

8. The entire procedure is then repeated, if necessary.

In view of the repeated balancing of circuits that is necessary, it may be expeditious to use a simpler, though less accurate, formula for the correction ΔQ. Thus, neglecting the last term in the expansion of $\sum C(Q' \pm \Delta Q)^2$:

$$\Delta Q = \pm \frac{1}{2} \frac{\sum CQ'^2}{\sum CQ'} \tag{3-91}$$

In sanitary engineering practice, it is customary to ignore minor losses (or else to estimate them as equivalent lengths of straight pipe) and to use the Hazen-Williams formula rather than the Darcy formula to compute head losses. The basic procedure, however, is similar to that outlined above.

Pipe networks are now analyzed almost exclusively by digital computer programs. The technique outlined above can be expeditiously programmed for such solution and a number of such programs are available, of sufficient flexibility for adaptation to most individual situations. If necessary the Hardy Cross method can still be utilized for hand solution of networks, but in general this is no longer justified. Numerous textbooks are available, especially in sanitary engineering, in which examples of such hand calculation procedures are shown.

PROBLEMS

3-1. A straight pipe 8 in. in diameter slopes downward on a 20 per cent slope and is carrying a discharge of 2 cfs of water, at a temperature of 68°F. Pressure gages indicate the pressure gradient in the direction of flow to be a drop of 0.0010 psi per foot of horizontal length. Compute the following:

a. Wall shearing stress, in lb/sq ft. *Ans.:* 2.10 psf.

b. Shear velocity, v^*, in ft/sec.

c. Darcy friction factor, f. *Ans.:* 0.265.

d. Reynolds number for the flow, N_R.

3–2. If the pipe of Problem 3–1 is assumed to have a very smooth wall surface, compute the following:

a. Velocity at center of pipe, ft/sec.

b. Thickness of laminar boundary layer, ft.

c. Darcy friction factor, f. (Compare with measured friction factor, and thereby draw a conclusion as to the actual nature of the pipe surface.)

3–3. Assuming the pipe of Problem 3–1 to have a hydraulically rough surface, compute its equivalent sand-grain roughness ϵ, in feet:

a. Assuming turbulence to be "normal" throughout the pipe cross-section, i.e., friction factor independent of Reynolds number.

b. Assuming turbulence to be transitional between smooth turbulence and normal rough turbulence, and as based on the Colebrook transition function.

3–4. A smooth pipe 2 inches in diameter carries a fluid of kinematic viscosity 0.0001 ft²/sec, and specific gravity 0.9:

a. What is the maximum discharge that the pipe will carry at laminar flow? *Ans.:* 0.0262 cfs.

b. If the average velocity in the pipe becomes 4 ft/sec, what is the Darcy friction factor? *Ans.:* 0.035.

c. For this condition, how much power is dissipated per foot of line? *Ans.:* 0.26 ft-lb.

3–5. A smooth-walled circular pipe 4 in. in diameter carries a liquid of specific gravity 0.85 and viscosity 0.000105 lb-sec/sq ft. What is the thickness of the laminar boundary film when the hydraulic gradient is 0.003? *Ans.:* 0.0078 ft.

3–6. A 6-in.-diameter smooth-walled pipe carries a fluid of density 1.84 slugs/ft³ and dynamic viscosity 0.00092 lb-sec/ft². Calculate the discharge in cfs.

a. When the energy gradient is 0.0020. *Ans.:* 0.197 cfs.

b. When the energy gradient is 0.0200. *Ans.:* 0.79 cfs.

3–7. A smooth horizontal pipe is 24 inches in diameter. If the permitted power loss is 40 ft-lbs per second per foot of length of pipe, what maximum flow rate in cubic feet per second can the pipe transmit? The kinematic viscosity of the water is 0.000016 sq ft/sec. *Ans.:* 41.8 cfs.

3–8. What discharge can be transmitted in a pipe 4 feet in diameter under a hydraulic gradient of 1 foot per 10,000 ft if the pipe itself is on a slope of 1 foot per 8100 ft and is hydraulically a "smooth" pipe? The pipe is carrying water at kinematic viscosity of 1×10^{-5} ft²/sec. What is the pressure difference, in psi per 1000 ft.?

3–9. What diameter of smooth copper tubing is necessary to deliver a flow of 0.01 cfs of water with an available pressure differential of 1.0 psi per 100 ft? The pipe is straight and horizontal and the water kinematic viscosity is 1×10^{-5} ft²/sec.

3–10. A 4-inch-diameter smooth pipe carries a liquid with a specific gravity of 0.9 and a kinematic viscosity of 0.0001 sq ft/sec. The pipe is on a downward slope of 10 per cent, and the pressure drop permitted is 1.0 psi per 1000 ft of pipe. What discharge, in cfs, may the pipe carry? *Ans.:* 0.87 cfs.

3–11. Determine the lost head in 30,000 ft of a 36-in.-diameter horizontal riveted steel pipe line carrying 4 cfs of water at a temperature of 73°F. Assume the equivalent sand roughness for this material to be 0.01 ft. *Ans.:* 1.34 ft.

3–12. What diameter of new cast-iron pipe one mile long is required to discharge 4.4 cfs of water with a head loss not more than 50 ft? Water temperature is 50°F. *Ans.: D* = 1.0 ft.

3–13. What would be the discharge capacity of the pipe of Problem 3–5 after 10 years of service, assuming the pipe surface has roughened to the extent that, at the original design discharge. the friction factor would have doubled? *Ans.:* 3.14 cfs.

3–14. A pressure drop of 10 psi is available to carry 400 gallons per min of water at 60°F between two points 1000 ft apart on horizontal ground. What is the minimum size standard wrought iron pipe required, based on the Moody curves?

3–15. An oil has viscosity of 0.002 lb-sec/sq ft and a density of 2 slugs/cu ft. It flows with a velocity of 4.2 ft/sec in a 4-in. horizontal pipe line, 2400 ft long, discharging into the atmosphere.

a. Show whether the flow is laminar or turbulent.

b. Compute the pump horsepower necessary to maintain the flow. *Ans.:* 3.87 hp.

3–16. What size rectangular galvanized-iron conduit with width twice its height is needed to carry 3 cfs of water at 120°F with a hydraulic gradient of 0.006?

3–17. A pump delivers 1.0 horsepower to a water pipe line against a static head of 2.0 ft per 1000 ft of line. The pipe has an equivalent roughness of 0.01 ft and is of 12-in. diameter. The water is at 60°F. Determine the rate of flow, in gallons per minute.

3–18. A flow of 3.8 cfs of oil (μ = 0.00021 lb-sec/ft^2 and γ = 55 lb/ft^3) is pumped through an 18-in.-diameter horizontal riveted steel pipe (ϵ = 0.02 ft). Pumps used on the line can each increase the pressure by 125 psi. How far apart can they be placed along the pipeline?

3–19. Water at 120°F is carried from a reservoir on top of a building to a point in the basement 120 ft below the reservoir surfaces through 115 ft of 1.0-in.-diameter smooth pipe. At what velocity will the water leave the pipe when it is open at the lower end?

3–20. A flow of 80 cfs is being carried in a 48-in.-diameter water line. The pipe is made of new cast iron. Kinematic viscosity = 10^{-5} sq ft/sec.

a. How many ft-lb per sec are dissipated by friction in each foot of line? Use the Colebrook-Moody method. *Ans.:* 11.2 ft-lb per sec.

b. If the pipe were replaced by a wrought-iron pipe of the same size, what flow could be maintained with the same head loss as in part a? *Ans.:* 90.1 cfs.

3-21. Determine by the Colebrook-Moody method, the diameter of new, horizontal wood-stave pipe to transport 300 cfs of water under an energy gradient of 0.0001. Assume water temperature at 60°F (kinematic viscosity 0.00001217 sq ft per sec).

3-22. What diameter of new cast-iron pipe should be specified to transmit a flow of 200 cfs of water? Use the Moody curves and assume the kinematic viscosity is 0.0000125 ft²/sec. Flow is maintained by pumps spaced at intervals of 2000 ft, each transmitting a net of 300 horsepower to the flow. *Ans.:* 4.65 ft.

3-23. A flow of 3.5 cfs of oil ($\mu = 0.00021$ lb-sec/ft²) and ($\gamma = 55$ lb/ft³) is pumped through an 18-in.-diameter horizontal riveted steel pipe ($\epsilon = 0.02$ ft). Pumps used on the line can each increase the pressure by 100 psi.
 a. What horsepower must be supplied to each pump, assuming 75 per cent efficiency? *Ans.:* 122 hp.
 b. How far apart can the pumps be placed along the pipeline? *Ans.:* 27.8 miles.

3-24. A new cast-iron pipe is laid on a downward slope of 0.002 and has a diameter of 6 ft. Pumps of 75% efficiency are spaced every 24,000 ft along the line to maintain the flow of 300 cfs in the line. The water has a kinematic viscosity of 0.00001 sq ft/sec. Assume head losses are determined by the Colebrook-Moody method and that pipeline pressures are kept at the same level immediately downstream from every pump. What horsepower must be supplied to each pump? *Ans.:* 2270 hp.

3-25. A new cast-iron pipe carries water under a hydraulic gradient of 0.0064. The pipe is horizontal and 4 ft in diameter. Kinematic viscosity is 10^{-5} ft²/sec.
 a. Compute the pipe discharge by the Colebrook-Moody method. *Ans.:* 132 cfs.
 b. For these conditions, what percentage of the pipe friction factor is due to roughness elements? *Ans.:* 38.6%.

3-26. A pump delivers Q cfs of water through a 12-in.-diameter pipe line. The pipe is 1000 feet long and is horizontal. The pressure at the end of the 1000 feet is to be kept equal to that in the pipe on the suction side of the pump. The pipe is rough concrete with an equivalent sand-grain diameter of 0.01 ft. What is Q, in cfs, if the pump supplies 100 hp to the flow? Assume $v = 1.2 \times 10^{-5}$ sq ft/sec. *Ans.:* 9.75 cfs.

3-27. A 4-ft diameter steel pipe line carrying water slopes downward in the direction of flow at a slope of 4%. The pressure in the pipe decreases at a rate of 4.0 psi per 100 feet. The equivalent sand diameter for the pipe material is 0.0002 ft and the kinematic viscosity 10^{-5} sq ft per sec. What is the rate of flow in cu ft per sec? $S = 0.0107, \quad V = 56.4 \frac{ft}{sec}, \quad N_R = 22.6 \times 10^6$

3-28. A concrete-lined tunnel, 6 feet wide and 4 feet deep, carries water from a reservoir and discharges it into the atmosphere. The reservoir surface is 400 feet above the discharge point and the tunnel is one mile long. The concrete surfacing is rough, with an "equivalent sand diameter" of 0.01 feet. The water has a viscosity of 1.94×10^{-5} lb-sec/ft^2. Compute the flow carried through the tunnel, in cfs.

3–29. Determine the diameter of a concrete pipe to carry 40 cfs of water. The pipe slopes upward at a slope of 0.0005 and the permitted pressure drop is 1.5 ft of water per 1000 ft of pipe. Use the Colebrook-Moody method and assume the equivalent sand diameter for the pipe surface to be 0.01 ft. The water has a kinematic viscosity of 10^{-5} sq ft per second. *Ans.:* 4.0 ft.

3–30. Determine the diameter of a new asphalted cast-iron pipe to carry 80 cfs of water when the permitted pressure drop is 10 psi per 1000 ft and the pipe slopes upward at 10 ft per 1000 ft. Assume the kinematic viscosity is 0.00001 sq ft/sec. Use the Colebrook-Moody method.

3–31. Determine the discharge pressure in psi that must be developed by a pump in order to pump 4 cfs of water (temperature 73°F) through a 36-in.-diameter horizontal riveted steel pipe line 30,000 ft long with atmospheric pressure at the discharge end of the pipe. Compare results by the following methods:

    a. Colebrook-Moody, with $\epsilon = 0.01$ ft.
    b. Hazen-Williams, with $C = 110$.
    c. Manning, with $n = 0.015$.
    d. Scobey, with $K_s = 0.45$.

3–32. Compute the maximum permissible flow rate in cfs for a pump situated 8 ft above the lake from which it is pumping, if the suction pressure is not to drop below 12 psi. The suction pipe is 4 in. in diameter and 10 ft long; the discharge pipe is 3 in. in diameter and 100 ft long, and discharges into a water tank whose surface is 20 ft above the lake level. Use the Manning formula, with $n = 0.012$, and neglect minor losses. Also compute the head supplied to the flow by the pump.

3–33. A lined concrete tunnel of horseshoe cross-section has a cross-sectional area of 30 sq ft, and a perimeter of 20 ft, and is 2000 ft long. If its Hazen-Williams coefficient is 120, what total head loss is experienced when the discharge is 240 cfs? If the water temperature is assumed to be 75°F, what would be the value of the equivalent sand roughness ϵ in the Colebrook equation for these conditions?

3–34. If the Hazen-Williams coefficient for a 3-ft-diameter conduit is known to be 120, and the flow is 8 cfs for a given energy gradient, how much water at 80°F will a new wrought-iron pipe of the same diameter carry with the same gradient, as computed by the Colebrook-Moody method? *Ans.:* 10.2 cfs.

3–35. A 4 ft $\times$ 4 ft square pressure tunnel has a Manning roughness coefficient of 0.020 and is laid on a slope of 0.0009. What pump hp is necessary to convey 100 cfs a distance of 10,000 ft through this tunnel, with no loss in pressure *Ans.:* 685 hp.

3–36. A concrete box culvert flowing full is to be rectangular in cross-section, with the width twice the depth. The Manning roughness coefficient for the surface is 0.0149. The culvert slope is horizontal and is 400 feet long. A head loss of 20 feet can be permitted through the length of the culvert barrel. What should be the culvert dimensions for a design discharge of 200 cfs? *Ans.:* 2.3 ft × 4.6 ft.

3–37. What diameter concrete pipe should be specified for a pipe line which must carry 20 cfs if the permissible boundary friction loss is 1.0 ft per 1000 ft of pipe?

  a. Use the Darcy-Colebrook-Moody procedure, with an equivalent sand diameter of 0.002 ft. Assume kinematic viscosity of 0.00001 ft²/sec.

  b. Use the Scobey method, with an assumed Scobey coefficient of 0.35.

3–38. A cast-iron pipe is 4 feet in diameter and is on a downward slope of 0.0001. A series of pumping stations 5 miles apart maintain a steady flow of 100 cfs through the pipe. What horsepower must be supplied by each pump? Neglect minor losses. Assume $v = 0.00001$ ft²/sec.

  a. Use Colebrook-Moody method, with equivalent sand diameter 0.001 ft.

  b. Use Manning formula, with roughness coefficient 0.013.

3–39. A corrugated metal pipe drain is 3 ft in diameter (between crests of corrugations) and carries water (kinematic viscosity 1×10^{-5} ft²/sec) under an energy slope of 0.0001. The corrugations are $\frac{1}{2}$ in. high and spaced $2\frac{2}{3}$ in. on centers.

  a. What is the discharge, in cfs? Assume hyper-turbulent flow. *Ans.:* 4.08 cfs.

  b. Assuming normal turbulent flow for the computed discharge and given energy slope, what would be the equivalent sand grain diameter? *Ans.:* 1.08 in.

3–40. A pipe has a wall-to-wall diameter of 40 inches and carries a discharge of 120 cfs of water ($v = 0.00001$ ft²/sec). Sharp-edged steel strips two inches in height are attached to the pipe wall at intervals, each extending completely around the inside perimeter.

  a. At what longitudinal spacing, λ, will these strips cause the maximum head loss? *Ans.:* 10.3 inches.

  b. At this spacing, how much horsepower is lost to the flow in each cycle from one strip to the next? *Ans.:* 3.8 hp.

3–41. Compute the discharge in cfs that can be carried in a 5-foot diameter (minimum cross-section) corrugated metal pipe under a hydraulic gradient of 0.0036. The corrugations are 1 in. in height and on a 3-in. spacing. The water has a kinematic viscosity of 0.000015 sq ft/sec. *Ans.:* 83.5 cfs.

3–42. A 4-foot diameter pipe line carries 100 cfs of water, with a kinematic viscosity of 10^{-5} ft²/sec. The pipe is on a horizontal grade and is made of riveted steel. The steel surface is hydraulically "smooth" except for the rivets, which are placed in circumferential joints spaced at 4-foot intervals. Each joint consists of a single-riveted butt joint, with rivet heads 1 inch in diameter on a center-line pitch of 4 inches. The butt strap is $\frac{1}{4}$ inch in thickness. Calculate the hydraulic gradient for the flow, using each of the following methods:

  a. Hazen-Williams

  b. Colebrook-Moody

  c. Semi-smooth Flow Function

3–43. A discharge of 16 cfs is to be carried in a 25-in.-diameter horizontal pipe, wall-to-wall dimensions, the kinematic viscosity being 1×10^{-5} sq ft/sec. The pipe is smooth except for semicircular peripheral ridges $\frac{1}{2}$ in. high and spaced on 6 in. centers. What is the hydraulic gradient of flow? *Ans.:* 0.0121.

3–44. An irrigation subdrain is made of a standard 24-in.-diameter corrugated metal pipe, with corrugations $\frac{1}{2}$ in. high and spaced at $2\frac{2}{3}$ in. It is desired to have it carry 10 cfs of drain water (kinematic viscosity 1×10^{-5} sq ft/sec) at uniform flow at a depth equal to the pipe radius. On what slope should the pipe be laid? *Ans.:* 0.0314.

3–45. A new concrete pipe carrying 5 cfs of water is smooth except for its joints, which form a peripheral depression 1 in. deep and 1 in. wide. The pipe diameter is 18 in., and its joints are spaced at 36-in. intervals Assume that the wall velocity is $\frac{2}{3}$ of the average velocity in the cross-section and that the peripheral velocity of the groove vortex is $\frac{3}{4}$ of the wall velocity. Compute the "equivalent length" of unjointed pipe represented by each joint (i.e., the length of unjointed pipe that would cause the same friction loss as caused by the joint). Assume kinematic viscosity at 0.00001 ft²/sec. *Ans.:* 0.77 ft,

3–46. A 6-in.-diameter horizontal pipe has a surface which is hydraulically smooth and carries water at 75° F under an imposed hydraulic gradient of 0.02.
   a. What is the flow velocity in ft/sec? $S = 0.014, V = 6.78 \frac{ft}{sec}$
   b. If circumferential rings are inserted into the pipe at 12-in. intervals, each having a circular cross-section $\frac{1}{2}$ in. in diameter, what is the velocity? $\rightarrow S = 0.0379, V = 2.61 ft/sec$

3–47. A corrugated metal pipe is 36 in. in minimum internal diameter, and has standard corrugations $\frac{1}{2}$ in. deep with $2\frac{2}{3}$ in. spacing. Circumferential sharp–edged rings 2 in. high are placed in the pipe at intervals of 6 feet. Assuming a discharge of 4 cfs and a kinematic viscosity of 0.00001 ft²/sec, what is the head lost in a typical 6 ft length of pipe? *Ans.:* 0.00164 ft.

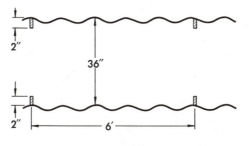

Prob. 3–47.

3–48. Water flows through an inside-threaded pipe with a mean velocity of 10 ft/sec. The diameter of the pipe, measured between crests of the threads, is 6 in. The threads are $\frac{1}{2}$ in. on centers, $\frac{1}{4}$ in. thick and $\frac{1}{4}$ in. deep. The velocity distribution is such the peripheral velocity of the stable vortex in the grooves is 0.5 of the mean velocity in the cross-section. Kinematic viscosity is 10^{-5} sq ft/ sec.

a. Calculate the head loss per ft of length, in ft.

b. Calculate the head loss per ft if the grooves are only $\frac{1}{8}$ in. deep (thread thickness and spacing remaining as before).

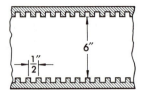

Prob. 3–48.

3–49. A pump of 80 per cent efficiency lifts water from one tank into another, the difference in surface levels being 20 ft. The pipe is "smooth," but contains three elbows, each with a loss coefficient of $\frac{2}{3}$. The pipe is 4 in. in diameter and 40 ft long. Kinematic viscosity of the water is 10^{-5} ft²/sec. What horsepower must be supplied to the pump in order to convey 2 cfs through the system? *Ans.:* 17.0 hp.

3–50. A standard corrugated metal culvert carries a spring flood discharge of 300 cfs, flowing full, with both inlet and outlet submerged. The pipe is 60 in. in diameter, 200 ft long, on a 5 per cent slope. The water temperature is 40°F. If the tailwater level is 1 ft above the crown of the culvert at its outlet, how high must the highway shoulder be above the culvert crown at the inlet to prevent overtopping the embankment? The culvert has a vertical, flush, sharp-edged headwall at both ends. *Ans.:* 5.9 ft.

3–51. A new 6-in.-diameter cast-iron pipe is to be used to convey hot water from an industrial plant to a stream flowing nearby. The water temperature will be about 143°F. The pipe will leave the bottom of the tank through a sharp flush inlet, will have one standard elbow ($K_L = 1.0$) and a control valve which, when fully open, has a loss coefficient of 5.5, and will discharge freely over the stream. The total head on the pipe (tank surface to pipe outlet) is 30 ft, and the total length is 100 ft. What is the peak flow that the pipe can discharge, in cfs? *Ans.:* 2.4 cfs.

3–52. A culvert through a highway embankment is constructed in two segments. The first is a concrete pipe, 3 ft in diameter, 200 ft long, on a 5 per cent slope. The second is a corrugated metal pipe, 4 ft in diameter, 100 ft long, laid horizontally. The inlet is set in a flush headwall and has a contraction coefficient of $\frac{3}{4}$. The concrete pipe is hydraulically smooth except for peripheral grooves at the joints, which are $\frac{1}{2}$ in. wide, $\frac{3}{4}$ in. deep, and spaced on 3-ft centers. The theoretical loss coefficient at the point of sudden diameter increase may be increased by 20 per cent by the simultaneous change in direction. The corrugated pipe has standard corrugations ($\frac{1}{2}$ in. deep and $2\frac{2}{3}$ in. on centers), and discharges directly into a tailwater pool, through a submerged outlet. Assume the water viscosity is 1.94×10^{-5} lb-sec/ft². At the design discharge of 120 cfs, what is the total difference in levels between the headwater and tailwater pools? *Ans.:* 8.7 ft.

3–53. If the culvert of Problem 3–52 had been designed as a straight culvert 300 ft long using the 3 ft concrete section throughout, on a 3 per cent slope, estimate the head difference by use of the Manning equation, using a roughness coefficient of 0.011. Include effects of any pertinent "minor losses." Compare the outlet velocities for the two designs.

3–54. A 5-ft-diameter smooth pipe carrying a flow of 80 cfs of water at 80°F goes through a 60° bend, on a center-line radius of 10 ft. What is the head loss due to the bend, in feet?

3–55. What horsepower must be supplied by a pump in a horizontal pipe line in order to maintain a uniform flow of water of 1.0 cfs? Pipe description as follows:

$$\begin{array}{lll} \text{First segment} & \text{4-in. dia., 100 ft long,} & f = 0.020 \\ \text{Second segment} & \text{8-in. dia., 400 ft long,} & f = 0.015 \\ \text{Third segment} & \text{4-in. dia., 100 ft long,} & f = 0.010 \end{array}$$

The transition from the first to the second pipe is made by a straight line transition 24 in. in length. The transition from the second to the third is made by a convergent section giving a contraction coefficient of 0.80. The pipe pressure at the end of the third segment is the same as that just upstream from the pump. *Ans.:* 2.30 hp.

3–56. A pipe segment is 4 ft in wall-to-wall diameter, 1000 ft long and horizontal. The segments upstream and downstream are both 2 ft in diameter and are joined to the central segment by sudden transitions. At the downstream transition the contraction coefficient is 0.635. The central segment has a smooth wall boundary except for sharp-edged peripheral strips 1 in. high, spaced on 6 in. centers. Kinematic viscosity is 10^{-5} sq ft/sec. What is the total head loss caused by this segment of pipe line, including the transitions, when the discharge is 10 cfs? *Ans.:* 0.48 ft.

3–57. A pipe line conveys 2 cfs of water ($\gamma = 10^{-5}$ ft²/sec) from one tank to a lower tank. The pipe inlet is submerged, with a sharp-edged, flush opening. The pipe leaving the first tank is a smooth pipe, 4 in. in diameter, extending horizontally for 20 ft. It then goes through a 90° bend and drops vertically for 10 ft; then it again proceeds horizontally for another 20 ft. Radius of bends is equal to diameter of pipe. At this point it expands suddenly to a 6-in.-diameter pipe. After another 40 ft of horizontal, smooth pipe, it then discharges freely into the atmosphere, falling 10 ft to the surface in the lower tank. What is the difference in surface levels between the two tanks? *Ans.:* 39.8 ft.

3–58. A corrugated metal culvert is 4 ft in inside diameter and has corrugations which are $2\frac{2}{3}$ in. long and $\frac{1}{2}$ in. high. It has an inlet with a contraction coefficient of $\frac{1}{2}$, is 100 ft long, horizontal, and with a straight submerged outlet. The discharge is 300 cfs and the kinematic viscosity of the water is 10^{-5} ft²/sec. What is the difference between headwater and tailwater levels? *Ans.:* 34.0 ft.

3–59. A pump conveys water between two reservoirs through the pipe line shown, with sharp, flush, inlet and outlet. The pipes are wrought iron, with equivalent sand roughness of 0.00015 feet. Calculate the net horsepower that must be supplied by the pump in order to pump a flow of 25 cfs. The kinematic viscosity of the water is 0.00001 ft²/sec. Contr. Coef. $= \frac{5}{8}$. *Ans.:* 239 hp.

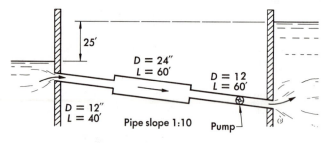

Prob. 3–59.

3–60. A corrugated metal pipe is 4 ft in inside diameter and has corrugations which are 3 in. long and $\frac{1}{2}$ in. high. It is used as an outlet pipe through a small dam. It has a sharp inlet (assume contraction coefficient of 0.75), is 100 ft long and horizontal, and has a straight outlet. If the pipe is submerged at both ends, how much discharge will it carry when the net head (difference in headwater and tailwater levels) is 25 feet? Assume kinematic viscosity $= 1 \times 10^{-5}$ ft²/sec. *Ans.:* 288 cfs.

3–61. A culvert through a highway embankment is constructed of a concrete pipe, 3 ft in diameter and 300 ft long, on a 3 per cent slope. The inlet is set in a flush headwall and has a contraction coefficient of $\frac{3}{4}$. The pipe discharges, at the design discharge of 120 cfs into a tailwater pool, through a submerged outlet. Assume the water viscosity is 1.94×10^{-5} lb-sec/ft². The pipe is hydraulically smooth except for peripheral grooves at the joints, which are $\frac{1}{2}$ in. wide, $\frac{3}{4}$ in. deep and spaced on 3-ft centers.

a. What is the total difference in levels between headwater and tailwater pools? Include effect of all "minor losses." *Ans.:* 9.92 ft.

b. What value of Manning's roughness coefficient n would have given the same result? *Ans.:* 0.0093.

3–62. What net head must be supplied by the pump in order to convey 2.0 cfs of water through the system shown? *Ans.:* 96.7 ft.

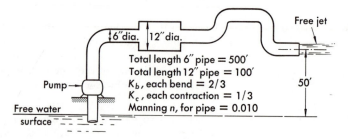

Prob. 3–62.

3–63. A corrugated metal culvert 6 ft in least diameter and 300 ft long is placed on a 1 per cent slope under a highway embankment. The design discharge is 200 cfs. The pipe flows full and discharges freely into the air above the outlet channel. The pipe has corrugations 4 in. center-to-center and 1 in. deep, over which it can be assumed the flow is hyper-turbulent. The pipe inlet loss coefficient is 0.5 and the kinematic viscosity of the water is 10^{-5} ft²/sec. How far is the headwater level above the culvert center line at the inlet? *Ans.:* 1.07 ft.

3–64. A culvert through a highway embankment is made of a concrete pipe 3 ft in diameter and 300 ft long, on a 3% slope. The inlet is set in a headwall and has a contraction coefficient of $\frac{3}{4}$. The pipe discharges 120 cfs into a tailwater pool through a submerged outlet. Assume the water kinematic viscosity to be 10^{-5} ft²/sec. The pipe is hydraulically smooth except for its joints, which are $\frac{1}{2}$ inches wide, $\frac{3}{4}$ inches deep and spaced on 3 ft centers. Assume that the average flow velocity in the pipe is equal to $(10)\frac{1}{3}$ times the peripheral velocity of the groove vortices at the joints. What is the total difference in levels between headwater and tailwater pools? *Ans.:* 9.45 ft.

3–65. A standard corrugated metal culvert is 100 ft long and on a slope of 5 per cent. Both inlet and outlet are submerged and the permissible headwater elevation is 16 ft above the tailwater elevation. For a discharge of 120 cfs., what diameter of culvert should be specified? The water viscosity may be taken as 1.94×10^{-5} lb sec/ft². *Ans.:* 3.12 ft.

3–66. A 6-ft inside-diameter culvert under a highway embankment is 200 ft long and laid on a slope of 0.015. The pipe is corrugated metal, with corrugations 1 in. deep, spaced 3 in. apart. However, the lower half of the pipe has been asphalted so that the surface has been made hydraulically "smooth." The pipe projects into the headwater and tailwater pools and the contraction coefficient at the inlet is 0.55. The design discharge of 200 cfs occurs in the early spring when the water is cold and has a kinematic viscosity of 1.6×10^{-5} ft²/sec. For this discharge the tailwater elevation is 5 ft above the outlet center-line. How high is the headwater above the inlet center-line? *Ans.:* 4.22 ft.

3–67. A pipe is 3000 ft long, in three segments of 1000 ft each, as follows:
First section, 2 ft diameter, new asphalted cast iron
Second section, 4 ft diameter, smooth except for sharp-edged peripheral strip 1 in. high and spaced on 6 in. centers.
Third section, same as first section.
The sections are joined together without any transition sections and the pipe is horizontal. The contraction head loss coefficient is $\frac{1}{3}$. Kinematic viscosity is 10^{-5} ft²/sec. A pump at the upstream end of the first length of pipe maintains a flow of 10 cfs of water through the system with no net loss in pressure between the point just upstream from the pump and the joint at the downstream end of the third length of pipe. What is the pump horsepower required if the pump is 90 per cent efficient?

4

FLOW IN OPEN CHANNELS

4–1. Characteristics of Open Flow. Most hydraulic engineering problems deal at least in part with the conveyance of water in natural or artificial open channels. In contrast to pressure conduits, which necessarily flow full and which may carry either liquids or gases, open channel flow must have a free surface[1] and therefore is concerned especially with liquids, usually water.

Despite this restriction, the mechanics of open channel flow is more complex than that of flow in closed conduits. All the forces that influence pressurized flow are present in open flow, and in addition the force of gravity and occasionally that of surface tension will affect the flow phenomena. Any of the flow regimes described for pipe flow may prevail as well in open channels. However, laminar flow is quite rare as an open channel phenomenon, one of the few examples in ordinary experience being so-called sheet flow, as rain water running off in a thin film over a uniform surface area such as a runway or highway.

Even smooth turbulent flow is quite rare in open channels, since most channel surfaces are relatively rough and the Reynolds number usually high. Consequently, most problems in hydraulic engineering deal with turbulent flow over surfaces which, at least in the hydraulic sense, are rough.

The same mechanisms of turbulence generation operate at the boundary surfaces in open channels as for pipes. Therefore the same rough-surface turbulent flow regimes can exist in open channels. However, there may be additional factors affecting the expenditure of energy. Because of the free surface, gravitational forces and surface tension forces will affect the velocity distribution and therefore the friction loss.

In addition, disturbances at the surface can be caused by changes in cross-section, changes in bed slope, wind stresses, objects dropped or floating on the surface, sediment transportation, and other factors. These may, in addition to affecting the internal turbulence, generate surface waves which may seriously complicate the energy picture.

---

[1] More correctly, the "free surface" is really an interface between two fluids of different densities; these may be two gases (e.g., flow of cold air under warm air), two liquids (e.g., density currents in reservoirs or estuaries), or a gas and a liquid, which is the commonest case.

Another important feature of open channel flow which increases the difficulties in its analysis is that it is seldom either *steady* or *uniform*. Steady flow has patterns and magnitudes which do not vary with time. Uniform flow implies a constant cross-section and velocity at every section in a given reach. Steady, uniform flow is quite common in pipe lines, but almost never occurs in open channels.

However, many open channel problems can be treated in terms of an approximate solution based on an assumption of steady flow, and the hydraulics of steady flow in open channels needs to be thoroughly understood as a prerequisite to successful consideration of unsteady flow problems. The latter deal mainly with wave phenomena and flow routing.

Uniform flow requires not only a longitudinally uniform cross-section, but also an equilibrium between gravitational and frictional forces. Since these forces act in opposite directions, one tending to accelerate and the other to decelerate the flow, and since for a given channel and discharge only one velocity will produce such an equilibrium condition, uniform flow can normally be attained only on a long uniform channel, and even that only after a long non-uniform transition. Therefore, practically all problems involve non-uniform (or varied) flow.

Nevertheless, as a starting point in the study of open-channel flow, a channel may be conceived in which, for a given reach, uniform flow prevails. Such a condition is sketched in Fig. 4–1.

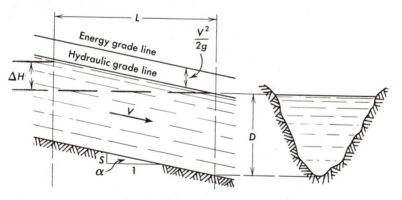

Fig. 4–1. Uniform flow in open channel.

It is assumed that the flow is rectilinear, so that surface tension forces may be neglected, and that the bed slope is relatively small, so that a hydrostatic pressure distribution exists. Assuming the hydrostatic forces on the two ends of the reach to be equal and opposite, the gravitational component in the direction of flow must balance the frictional resistance, since there is no change in momentum flux. Therefore:

$$(A \cos \alpha)(L \sec \alpha)(\gamma \sin \alpha) = \tau_0 (P \cos \alpha)(L \sec \alpha)$$

in which A and P, the area and wetted perimeter, respectively, are expressed in terms of the flow depth as measured vertically. Thus:

$$\gamma A \sin \alpha = \tau_0 P$$

It is usually more convenient to express this relation in terms of the bed slope, $S(= \Delta H/L = \tan \alpha)$. Therefore, the bed shear

$$\tau_0 = \gamma R_h S \cos \alpha \qquad (4\text{-}1)$$

where R_h is the hydraulic radius A/P. This is the momentum conservation equation for uniform flow in open channels. Usually α is small and τ_0 becomes simply $\gamma R_h S$. In the special case of a wide shallow channel, a condition approximately true for many natural streams, the hydraulic radius is

$$R_h = \frac{A}{P} = \frac{BD}{B+2D} = \frac{D}{1 + 2\dfrac{D}{B}} \cong D$$

In this case:

$$\tau_0 = \gamma D S \qquad (4\text{-}2)$$

The energy equation for uniform flow in an open channel is simply:

$$\Delta H = SL = H_f \qquad (4\text{-}3)$$

The pressure head is equal to the depth below the free surface. Both the velocity head and pressure head remain constant from section to section in uniform flow. Combining Eqs. (4-1) and (4-3):

$$H_f = \frac{\tau_0 L}{\gamma R_h \cos \alpha} \qquad (4\text{-}4)$$

which gives the head loss in terms of the bed shear.

The process equation for uniform open channel flow is:

$$\Delta H = \frac{fL}{4R_h \cos \alpha} \frac{V^2}{2g} \qquad (4\text{-}5)$$

with

$$f = f\left(N_R, N_F, N_W, \frac{h}{4R_h}, \frac{\lambda}{4R_h}, \frac{S}{4R_h}, \ldots\right) \qquad (4\text{-}6)$$

As written above, the equivalent diameter D, shown as $(4R_h)$, is the basic length parameter in the dimensional analysis. For ordinary flows in open channels on mild slopes, the effects of gravity and surface tension on the flow structure are negligible, so that N_F and N_W can be removed from the function. In this case the process equation becomes identical to that for uniform pipe flow, and the same equations can be used for both processes.

Combining Eqs. (4–3) and (4–5), the Darcy equation is again obtained, assuming $\cos \alpha \cong 1$:

$$H_f = f \frac{L}{4R_h} \frac{V^2}{2g} \qquad (4\text{–}7)$$

Similarly, Eqs. (4–1) and (4–5) combine to give:

$$\frac{V}{v^*} = \sqrt{\frac{8}{f}} \qquad (4\text{–}8)$$

As noted in Chapter 3, Eqs. (4–7) and (4–8) apply to uniform flow in both pipes and open channels.

Here, v^* is the shear velocity, $\sqrt{\dfrac{\tau_0}{\rho}}$. From Eq. (4–1) it may also be noted that:

$$v^* = \sqrt{gR_h S \cos \alpha} \qquad (4\text{–}9)$$

4–2. Empirical Open Channel Formulas. Most open channel problems involve flow of water over fairly rough surfaces at fairly high Reynolds numbers. For these conditions, it is often convenient to use an approximate formula, which is applicable only within this range of circumstances, and which may be easier to work with than the Darcy formula.

It has been noted that various formulas of this sort have been used for closed-conduit flow of water under these same conditions, especially those of Hazen-Williams, Manning, and Scobey. These may also be used for open channels, and there have been numerous other formulas of similar type suggested for open channels. The Kutter and Bazin formulas are still used occasionally in design practice, but the Manning formula has come into almost universal use in this country for open channel problems.

All such formulas are related in general form to the Darcy equation, which, when expressed in terms of the velocity, is known as the Chezy formula (named after the French engineer who derived it empirically in 1775, many years before Darcy presented his equation). Thus,

$$V = \sqrt{\frac{8g}{f}} \sqrt{R_h S} = C\sqrt{R_h S} \qquad (4\text{–}10)$$

In this equation, the coefficient C was originally thought to depend only on the wall roughness, but it is now obvious that it also depends on the Reynolds number and sometimes on the Froude and Weber numbers.

For the usual conditions, however, these factors may be insignificant. Even the Reynolds number will not have much effect at the high values normally encountered. The roughness effect, however, is not one of absolute roughness magnitude, but one of roughness relative to the size of channel. Consequently, the coefficient will be found to vary substantially with the hydraulic radius. (This relationship is also reflective of the remanent effect of the Reynolds number.)

The Bazin formula, widely used in Europe, has the following value for C:

$$C = \frac{157.6}{1 + m/R_h^{1/2}} \tag{4-11}$$

The Bazin roughness coefficient m is supposed to be an absolute measure of surface roughness, to be determined empirically.

The Kutter formula was formerly widely used both in England and America, although it is now substantially displaced by the Manning equation. Kutter's formula is still of historical interest, and is as follows:

$$C = \frac{1.811/n + [41.65 + (0.00281/S)]}{1 + (n/R_h^{1/2})[41.65 + (0.00281/S)]} \tag{4-12}$$

In this formula, n likewise is an empirically determined absolute measure of roughness.

It may be noted that when $R_h = 1$ m $= 3.2808$ ft, Kutter's formula reduces to

$$C = \frac{(1.811/n)[1 + (n/1.811)(41.65 + 0.00281/S)]}{[1 + (n/\sqrt{3.2808})(41.65 + 0.00281/S)]}$$

$$= \frac{1.811}{n}$$

since $\sqrt{3.2808} = 1.811$.

The Irish engineer, Manning, 1890, finding that the Kutter formula did not agree very well with many field measurements, made studies leading to the adoption of the more accurate and much simpler formula:

$$C = \frac{1.49}{n} R_h^{1/6} \tag{4-13}$$

It will be observed again, that for $R_h = 1$ meter, this reduces to $1.811/n$. Manning and others advocating this relationship attempted in this way to include in his formula the same roughness coefficient n as had been used in the Kutter formula. This can lead to errors, however, as the Kutter and Manning values of C diverge quite appreciably at other values of R_h.

If Eqs. (4–10) and (4–13) are combined, the familiar Manning equation[2] is obtained:

$$V = \frac{1.49}{n} R_h^{2/3} S^{1/2} \tag{4-14}$$

[2] The constant is often written as 1.486. It is obvious that this refinement is not warranted, and that 1.49, or even 1.5, is more realistic.

Values determined for the Manning roughness coefficient n are given in Table 4-1. Values of n originally derived by the Kutter formula are frequently used in the Manning equation at present. This approximation, however, is valid only for low values of n.

TABLE 4-1

Manning Roughness Coefficient, n

Channel Surface Type	Condition	n
Glazed coating or enamel	Perfect order	0.010
Timber	Planed boards, carefully laid	0.010
	Planed boards, old or inferior	0.012
	Unplaned boards, carefully laid	0.012
	Unplaned boards, old or inferior	0.014
Metal	Smooth	0.010
	Riveted steel	0.015–0.020
	Corrugated metal	0.022–0.026
	Cast iron, new	0.015
	Cast iron, tuberculated	0.020
Masonry	Neat cement plaster	0.010
	Sand and cement plaster	0.012
	Finished concrete	0.012
	Unfinished concrete	0.014
	Concrete in bad condition	0.020
	Brick, good condition	0.013
	Brick, rough	0.020
Stonework	Smooth, dressed ashlar	0.013
	Rubble set in cement	0.017
	Fine, well-packed gravel	0.020
	Dry rubble	0.025
Earth	Regular, good condition	0.020
	Regular, fair condition	0.023
	Regular, some stones and weeds	0.028
	Regular, poor condition	0.035
	Winding, irregular, but clear	0.035
	Obstructed with debris and weeds	0.050–0.150

The wide variation in n values, together with the very significant effect it has on flow calculations, places a high premium on judgment and experience in selecting proper design values. There is obviously need for more complete correlation with actual roughness dimensions. The methods of the previous chapter can be applied to open channels as well as pipes, but until they are adequately developed and recognized, the Manning coefficient will continue to be used, and the designer must exercise careful judgment in using and applying it.

A different type of variation of the roughness coefficient occurs in channels with movable beds, such as alluvial rivers. It is found that the patterns of ripples and dunes on the bed depend on the Froude number and the grain size; these ripples in turn serve as roughness elements and therefore must have an effect on the Manning coefficient, the friction factor, etc. As a result of these effects, the roughness coefficient for a channel with a movable sand bed and no vegetation may vary all the way from about 0.011 when no ripples are present to as much as 0.035 when strong dunes have been developed.

The Manning equation, as well as others based on the Chezy function, applies strictly only to uniform flow, in which the rate of head loss is equal to the bed slope. For a given discharge and channel, uniform flow will occur only at one particular depth. This depth is known as the *normal depth* for that channel (including a fixed shape, slope, and roughness) and discharge.

When the discharge and channel conditions are fixed, the Manning formula can be written

$$AR_h^{2/3} = \frac{nQ}{1.49S^{1/2}} = \text{constant} \qquad (4\text{--}15)$$

Since A and R_h are both functions of the depth, D, the normal depth can be obtained simply by solving this equation for D. The solution must often be obtained by trial and error, in view of the complicated nature of the function $AR_h^{2/3}$. If normal depths are needed in a given channel for several discharges, it may be easier to plot this function graphically, versus depth. Then, for the given discharge, roughness and slope, the value of $AR_h^{2/3}$ can be computed. The corresponding flow depth, obtained from the curve, is the normal depth for that discharge.

4–3. Friction Factors in Open Channel Flow.   The Manning equation is now used very widely for open channel design, especially in this country. However, it should not be forgotten that it is strictly an empirical formula, dimensionally non-homogeneous, with all the limitations implied by this fact. The roughness coefficient is physically meaningless and its selection must be based entirely on previous experience and judgment. Although the Manning formula has proved highly satisfactory for its purpose, it perhaps lends itself too readily to an indiscriminate handbook-type usage and can give dangerously erroneous results if injudiciously applied.

It is fundamentally more sound, of course, to use the dimensionally correct Darcy formula, with the friction factor computed from the correct functional expression relating it to the Reynolds number and the geometry of the boundary roughness elements. The Colebrook-White equation (Eq. 3–42) and the Moody diagram have been used with some success for open channels. Similarly, the friction factor equations for isolated-roughness flow, wake-interference flow, and skimming flow (Eqs. 3–61, 3–66, and 3–68), with their corresponding design diagrams, have been verified for open channels as well as pipes. However, the experimental data supporting all such relationships for open channel flow processes tend to show greater scatter than for pipe flow, primarily because of irregularities at boundary corners and on the free surface.

In general, the Manning equation is simple to use and is quite reliable, provided the roughness coefficient is known and the design situation is interpolated (not extrapolated) within the range of previous successful experience. It is occasionally desirable to convert Manning coefficients to Darcy friction factors and vice versa. Setting the slope terms ($S = H_f/L$) equal from the two equations:

$$\frac{n^2 V^2}{(1.49)^2 R_h^{4/3}} = \frac{f V^2}{2g(4R_h)}$$

and, simplifying:

$$n = \frac{f^{1/2} R_h^{1/6}}{10.8} \tag{4-16}$$

It is evident from Eq. (4–16) that n is not a constant for a given surface, as usually assumed. For surfaces with a falling $(f - N_R)$ characteristic, the effect of increasing the hydraulic radius is to some extent offset by the corresponding drop in friction factor, and n tends to remain more or less constant. For surfaces producing hyper-turbulent flow, however, such as corrugated metal, f will increase as R_h increases, and therefore n also will increase. This has been well verified by tests on both pipes and open channels with corrugated metal surfaces, which always show the Manning coefficient to increase with increasing Reynolds numbers.

4–4. Hydraulic Efficiency of Section. The normal depth in a channel depends not only on the discharge, slope, and roughness, but also on the shape of the cross-section, the design of which, at least within limits, is often adjustable. For this purpose, the Manning formula can be written in the form

$$Q = K \frac{A^{5/3}}{P^{2/3}} \tag{4-17}$$

where K is a channel constant. It is obvious that the discharge will be greatest for a given area when the wetted perimeter is a minimum. Or, for a given discharge, the required area will also be a minimum when the wetted perimeter is minimum.

There exists, of course, an infinite variety of possible cross-sectional shapes. The shape with the smallest possible perimeter for a given area is a semicircle, and this shape is occasionally used for metal flumes. However, a semicircular section is impracticable to construct or maintain for most materials.

The trapezoid (including the rectangle as a special case) is the most frequently used open channel section. In Fig. 4–2 is shown a trapezoidal section. The area and perimeter are given by

$$A = bD + zD^2 \tag{4-18}$$

and

$$P = b + 2D\sqrt{1 + z^2} \tag{4-19}$$

The relation between b, D, and z, to make P a minimum for a given A is defined by the simultaneous equations

$$\frac{\partial P}{\partial D} = 0 \quad \text{and} \quad \frac{\partial P}{\partial z} = 0$$

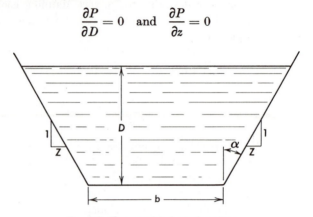

Fig. 4–2. Trapezoidal channel section.

The first of these equations yields

$$\frac{\partial P}{\partial D} = \frac{\partial}{\partial D}\left(\frac{A - zD^2}{D} + 2D\sqrt{1 + z^2}\right)$$

$$= -\frac{A}{D^2} - z + 2\sqrt{1 + z^2}$$

$$= -\frac{bD + zD^2}{D^2} - z + 2\sqrt{1 + z^2}$$

$$= -\frac{b}{D} - 2z + 2\sqrt{1 + z^2} = 0$$

Therefore

$$b = 2D(\sqrt{1 + z^2} - z) \tag{4–20}$$

If the side slope z has an arbitrary value, fixed perhaps by the angle of repose of the material, this equation gives the relation between b and D that must be provided if the section is to have maximum hydraulic efficiency. If z also may be varied, its most efficient value may be found from

$$\frac{\partial P}{\partial z} = \frac{\partial}{\partial z}\left(-\frac{A}{D} - zD + 2D\sqrt{1 + z^2}\right)$$

$$= -D + \frac{2zD}{\sqrt{1 + z^2}} = 0$$

[1] The same results can be obtained by writing and solving equations to make A a maximum for a given P.

from which

$$z = \frac{1}{\sqrt{3}} \qquad (4\text{-}21)$$

This side slope may also be too steep to be practicable, but is theoretically the best value. The corresponding slope angle α is $30°$. For this value, the relation between b and D becomes

$$b = \tfrac{2}{3}\sqrt{3}D$$

This is the same as each side slope, from the relation:

$$D\sqrt{1 + z^2} = \tfrac{2}{3}\sqrt{3}D$$

For this most efficient of all trapezoidal sections, therefore, the three sides are equal, with interior angles of $120°$. This means the section is a half-hexagon. It is significant that this section can be inscribed in a semicircle, which is the most efficient of all sections.

The hydraulic radius of a trapezoidal section of maximum efficiency, for any given side slope, is

$$R_h = \frac{bD + zD^2}{b + 2D\sqrt{1 + z^2}} = \frac{2D^2(\sqrt{1 + z^2} - z) + zD^2}{2D(\sqrt{1 + z^2} - z) + 2D\sqrt{1 + z^2}}$$

$$= \frac{D^2(2\sqrt{1 + z^2} - 2z + z)}{2D(\sqrt{1 + z^2} - z + \sqrt{1 + z^2})} = \frac{D}{2} \qquad (4\text{-}22)$$

The use of $D/2$ for the hydraulic radius, in problems involving maximum efficiency, materially simplifies solution of the Manning formula.

For the special case of a rectangular section, Eq. (4–20) becomes

$$b = 2D \qquad (4\text{-}23)$$

A rectangular channel is most efficient, therefore, when its depth is one-half its width.

It is frequently stated that, for the most efficient section of *any* shape, the hydraulic radius will be one-half the flow depth and the section can be inscribed in a semicircle. That the first statement is not necessarily true can be seen by determining the most efficient triangular section. For this case,

$$\frac{dP}{dz} = \frac{d(2D\sqrt{1 + z^2})}{dz} = \frac{d(2\sqrt{A/z}\sqrt{1 + z^2})}{dz}$$

$$= 2\sqrt{A}\left[\frac{1}{2\sqrt{(1 + z^2)/z}}\left(-\frac{1}{z^2} + 1\right)\right] = 0$$

from which, the best side slope, z is equal to unity.

The corresponding surface width is $2D$, so that the section *could* be inscribed in a semicircle. However, since the area A is equal to zD^2, and the perimeter P to $2D\sqrt{1 + z^2}$, the hydraulic radius is

$$R_h = \frac{A}{P} = \frac{zD^2}{2D\sqrt{1 + z^2}} = \frac{D}{2\sqrt{2}}$$

which is not equal to $D/2$.

The most efficient polygonal section of any specified number of sides can be found to be one which can be inscribed in a semicircle. However, it is obvious that this requirement could not be satisfied by the most efficient sections of other shapes (e.g., exponential, logarithmic, trigonometric, etc.).

The following tabulation compares perimeters required for a given area (as specified by a design discharge and permitted velocity, $A = Q/V$) for the most efficient sections of several shapes:

Shape of Section	A	$P = f(D)$	$P = f(A)$
Most efficient rectangle	$2D^2$	$4D$	$\sqrt{8A}$
Most efficient triangle $(z = 1)$	D^2	$2\sqrt{2}D$	$\sqrt{8A}$
Most efficient trapezoid $\left(z = \dfrac{1}{\sqrt{3}}\right)$	$\sqrt{3}D^2$	$2\sqrt{3}D$	$\sqrt{6.93A}$
Most efficient section (semicircle)	$\dfrac{\pi}{2}D^2$	πD	$\sqrt{6.28A}$

Example 4–1

A trapezoidal open channel has $45°$ side slopes, a Manning coefficient of 0.0149, and a discharge of 100 cfs. If the maximum velocity permitted in the channel is 2 feet per second and if the channel must be built on minimum possible slope, what base width should be specified?

Solution. The channel must have a minimum area, $A = Q/V = 100/2 = 50$ ft^2. Substituting in the Manning equation:

$$S = \left[\frac{nV}{1.49\left(\dfrac{A}{P}\right)^{2/3}}\right]^2 = \left[\frac{0.0149(2)(P)^{2.3}}{1.49(50)^{2/3}}\right]^2 = \frac{P}{449,000}$$

For the given area, the required slope will obviously be a minimum when P is minimum. The section must be that for maximum hydraulic efficiency for the given cross-section. Therefore:

$$b = 2D[\sqrt{1 + Z^2} - Z] = 2D[\sqrt{2} - 1] = 0.828D$$

Then,

$$A = 50 = bD + D^2 = \left(\frac{1}{0.828} + \frac{1}{(0.828)^2}\right)b^2 = 2.66b^2$$

Therefore, the required base width, $b = \left(\dfrac{50}{2.66}\right)^{1/2} = 4.34\ ft.\ Ans.$

The corresponding depth, $D = B/0.828 = 5.24$ ft and the slope,

$$S = \frac{1}{449,000}[B + 2\sqrt{2}D] = 4.6 \times 10^{-5}$$

4-5. Energy Regimes in Open Channel Flow. It has been pointed out that the force of gravity may play an important part in open channel phenomena. This is especially true when the bed slope is steep or when the flow depth changes rapidly along the channel reach. The relative magnitudes of the forces of inertia and gravity are indicated by the Froude number $V^2/gD = \rho V^2/\gamma D$. The numerator (except for a factor of $\frac{1}{2}$) represents the unit force due to inertia (dynamic pressure) and the denominator the unit force due to gravity (static pressure). The ratio might also be modified to the form $\frac{1}{2}(\rho V^2/\gamma)/(\gamma D/\gamma) = (V^2/2g)/D$, in which form it is seen to represent the ratio of kinetic head to potential head (as measured above the channel bed).

This ratio therefore can be expected to be very important in the understanding of open flow phenomena. It will be convenient to define the sum of the two terms, depth plus velocity head, as the specific head or specific energy at a section. These terms are illustrated below in Fig. 4–3 for a cross-section of any shape, with D the bottom depth and B the surface width.

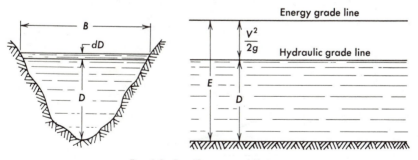

Fig. 4–3. Specific energy definition.

The water surface is identical with the hydraulic grade line, measuring as it does the total potential head of the water. The energy grade line, representing the total energy head, is above the hydraulic grade line a distance equal to the velocity head. It is assumed here that bed slopes are not excessively steep, flow is essentially rectilinear, and velocity distributions are fairly uniform. The specific energy E is shown as the total energy head measured above the channel bed. It is obvious that. E is the sum of the depth and velocity heads. The total discharge past the section, Q, is the product of the area and average velocity. Since the area is a function of depth, if the depth is changed, the discharge will change—unless the velocity is changed simultaneously just enough to keep the product of area and velocity constant. If this is done, however, it is clear that the same discharge could be obtained with àn infinite number of combinations of depth and velocity. These changes, however, would also result in a changing value of specific energy, which depends on both depth and velocity. For very high velocities, it is obvious that E would become very high and essentially equal to the velocity head. On the other hand, for very low velocities, D would have to be very large, and E would then become essentially equal to D.

These trends can be shown on a *specific-energy diagram*, with D plotted against E, as in Fig. 4–4. It is obvious from the diagram that there are two possible *alternate depths* giving the same specific energy, except for one *critical depth* for which the specific energy is a minimum.

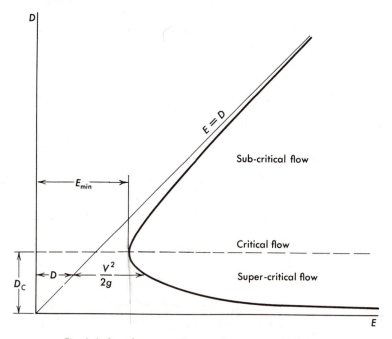

Fig. 4–4. Specific energy diagram, for constant discharge.

The magnitude of the critical depth may be determined by making the specific energy have a minimum value. Thus,

$$\frac{dE}{dD} = \frac{d}{dD}\left(D + \frac{V^2}{2g}\right) = \frac{d}{dD}\left[D + \left(\frac{Q^2}{2g}\right)\left(\frac{1}{A^2}\right)\right]$$

$$= 1 - \left(\frac{Q^2}{g}\right)\left(\frac{1}{A^3}\right)\left(\frac{dA}{dD}\right) = 0$$

critical Dep eqno.

It will be noted from Fig. 4–3 that $dA = BdD$. Therefore, the requirement for minimum specific energy is that the flow depth be such that

$$\frac{Q^2B}{gA^3} = 1 \tag{4–24}$$

Since A and B are functions of the depth, this relation defines a *critical depth* which is uniquely characteristic of the given discharge and channel shape. If D_m is defined as the mean depth of flow at the section, A/B, the equation may be written

$$\frac{V^2}{gD_m} = 1 = N_f^2 \tag{4–25}$$

Froude number

Thus, at critical depth, the Froude number (if expressed in terms of the mean depth) is unity. This criterion may also be expressed as the requirement that the velocity head equal one-half the mean depth of flow, for critical flow.

The corresponding *critical velocity* is given by

$$V_c = \sqrt{gD_m} \tag{4–26}$$

mean depth

If the velocity is less than critical, the flow is said to be subcritical, tranquil, or streaming. If greater than critical, the velocity is said to be supercritical, rapid, or shooting.

Critical flow is thus that regime of flow for which the specific energy is minimum for a given discharge. It may also be defined as the regime for which the maximum discharge is obtained with a given specific energy. This is illustrated in Fig. 4–5 and demonstrated by setting the first derivative of Q with respect to D equal to zero.

$$\frac{dQ}{dD} = \frac{d}{dD}\left[2gA^2(E - D)\right]^{1/2} = \sqrt{2g}\,\frac{d}{dD}\left(A\sqrt{E - D}\right)$$

$$= \sqrt{2g}\left[\sqrt{E - D}\,\frac{dA}{dD} + A\,\frac{-1}{2\sqrt{E - D}}\right]$$

$$= \sqrt{2g}\left[B\sqrt{\frac{V^2}{2g}} - \frac{A}{2\sqrt{V^2/2g}}\right] = 0$$

from which, as before,

$$\frac{V^2}{2g} = \frac{D_m}{2}$$

Thus, if

$$\frac{V^2}{2g} < \frac{D_m}{2} \qquad \text{flow is subcritical}$$

$$\frac{V^2}{2g} = \frac{D_m}{2} \qquad \text{flow is critical}$$

$$\frac{V^2}{2g} > \frac{D_m}{2} \qquad \text{flow is supercritical}$$

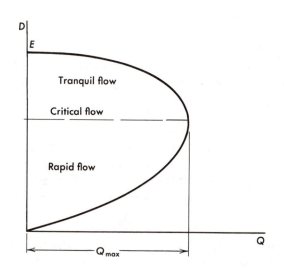

Fig. 4–5. Discharge function for constant specific energy.

The flow phenomena encountered in open channels will largely depend upon the flow regime. Consider for the moment a body of still water, on whose surface a disturbance is created, causing a local rise in water surface at some point. This rise will move radially outward under the action of gravity, in the form of a *gravity wave*. The velocity of the wave motion is called its celerity, and may be determined by simultaneous solution of the continuity and energy equations:

$$Ac = \text{constant } C_1 \tag{4-27}$$

and

$$\frac{c^2}{2g} + D = \text{constant } C_2 \tag{4-28}$$

As the wave moves, the average velocity of flow of water (relative to the wave as a moving frame of reference) is equal to c, except that in the wave itself the average velocity is $c - dc$, corresponding to the instantaneous surface elevation, $D + dD$. In differential form, therefore, Eqs. (4–27) and (4–28) become

$$A \, dc + c \, dA = 0$$

and

$$\frac{1}{g} c \, dc + dD = 0$$

Combining these equations yields

$$dD = \frac{1}{g} c \frac{c \, dA}{A} = \frac{2c^2 B \, dD}{2g A}$$

from which

$$\frac{c^2}{2g} = \frac{D_m}{2} \tag{4–29}$$

It is seen immediately that the celerity of a gravity wave (limited to ordinary surface undulations in relatively shallow channels) is the same as the critical velocity. This fact implies that disturbances on a *moving* surface will be transmitted upstream only if the flow velocity is less than the wave celerity—that is, if the flow is tranquil. If the flow is rapid, disturbance waves will be simply washed downstream and therefore can produce no effect on upstream conditions. The critical depth is thus very important as an indicator of flow characteristics, and it can be calculated, for a given discharge and channel shape, from Eq. (4–24) or Eq. (4–25).

In the special case of a rectangular channel, D_m is equal to D, and thus critical velocity head is one-half the critical depth, their sum being the minimum specific energy, equal to 1.5 times the critical depth. In terms of the discharge, critical depth is given by

$$D_c = \frac{V_c^{\,2}}{g} = \frac{Q^2}{B^2 D_c^{\,2} g} = \frac{q^2}{D_c^{\,2} g}$$

and therefore

$$D_c = \sqrt[3]{\frac{q^2}{g}} \tag{4–30}$$

where $q = Q/B$ is the discharge per foot of width.

If a channel bed slope is of such magnitude that the normal depth for a given discharge thereon is exactly equal to the critical depth, that slope is said to be the *critical slope*. Thus, if the Manning formula is used to define normal depth, the critical slope is given by

$$S_c = \left(\frac{nV_c}{1.49 R_h^{2/3}}\right)^2 = \left(\frac{nQ}{1.49 A_c R_c^{2/3}}\right)^2 \tag{4–31}$$

If the bed slope is greater than critical, it is said to be a *steep* or *supercritical* slope. A slope flatter than critical is a *mild* or *subcritical* slope. Critical slope obviously depends not only on critical depth and discharge, but also on the roughness of the channel.

In terms of the normal and critical depths, the following criteria are helpful:

If $D_n > D_c$ then $S < S_c$. Slope is mild.
If $D_n = D_c$ then $S = S_c$. Slope is critical.
If $D_n < D_c$ then $S > S_c$. Slope is steep.

Example 4–2   Mistake

An open channel carries a discharge of 625 cfs. The channel has a Manning roughness coefficient of 0.0298 and a slope of 0.000064. The channel is of triangular cross-section, with side slopes 4 horizontal to 1 vertical. Calculate: (a) Normal depth; (b) Critical depth.

Solution.

(a) By the Manning formula:

$$Q = \frac{1.49}{n} A R_h^{2/3} S^{1/2}$$

$$625 = (\cancel{52}\,50)(4D^2)\left(\frac{2}{\sqrt{17}} D\right)^{2/3}(0.008)$$

Solving, $D = D_n = 11.20\,\text{ft.}$ *Ans.*

(b) From the critical flow formula:

$$Q^2 B = gA^3$$

$$(625)^2(8D) = 32.2(4D^2)^3$$

Solving, $D = D_c = 4.34\,\text{ft.}$ *Ans.*

Since $D_n < D_c$, the channel is on a mild slope.

$D_n > D_c$

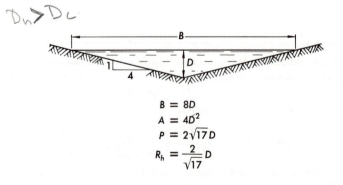

$$B = 8D$$
$$A = 4D^2$$
$$P = 2\sqrt{17}\,D$$
$$R_h = \frac{2}{\sqrt{17}} D$$

4-6. Flow Transitions. A mild slope will support a uniform sub-critical flow, and a steep slope will support a uniform supercritical flow. If the channel slope changes from mild to steep, there must therefore be a transition from subcritical to supercritical flow. Such a transition is sketched in Fig. 4-6. The normal depth for the mild slope is measured by D_m and that

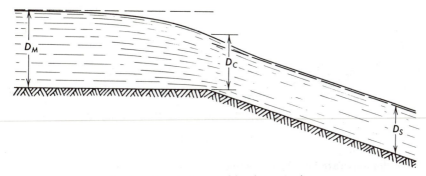

Fig. 4-6. Transition caused by change in slope.

for the steep slope by D_s. The critical depth D_c is intermediate between the two, and the transition must of course pass through D_c at some point, which point defines a *control section* for the profile. Since this section "controls" the flow profiles upstream and downstream, it is important to determine its location. Also, it can be seen from either Eq. (4-25) or Fig. 4-5 that discharge is a unique function of critical depth, and vice versa. That is, for a given channel shape, a given flow depth will be critical depth for only one discharge, and a given discharge can correspond to only one value of critical depth.

This location depends on the fact that the specific energy (distance from the channel bed to the energy grade line) is minimum at a control section. The slope of the energy grade line is, therefore, greater than the bed slope above the control section and milder than the bed slope below the control section. Since the energy slope measures the rate of friction loss, it is necessary that the friction loss be greater than normal above and less than normal below the control section. This means that the control section cannot be on the mild slope, since the energy loss rate after passing the control would still be greater than normal. Nor can it lie on the steep slope, since the loss rate would have to be less than normal before reaching the control. The only possible location of the control section, therefore, is exactly at the break in slope, which position satisfies all the above requirements.

A similar analysis could be made for other conditions producing a transition from subcritical to supercritical flow, such as a lateral constriction, a sudden bed rise or a sudden substantial decrease in channel roughness. In each of these cases, the control section will be located precisely at the station where the change occurs. This fact is used as a basis for measuring discharge in several types of open channel flow metering devices or techniques, as well as for providing a starting point for flow profile calculations.

If the slope changes suddenly from steep to mild, on the other hand, as in Fig. 4–7, it will be seen that no complete gradual transition from supercritical to subcritical can occur.

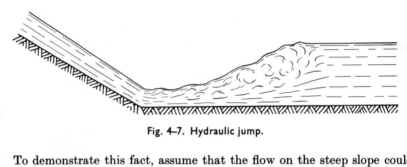

Fig. 4–7. Hydraulic jump.

To demonstrate this fact, assume that the flow on the steep slope could be somehow retarded on the upstream slope and caused to pass through critical depth, at which point the specific energy would be a minimum. The velocity and friction loss upstream from the control would have to become less than normal for the steep slope, but in order to approach critical depth, with E therefore decreasing, the energy slope would have to be greater than normal. These are two physically contradictory, and therefore impossible, requirements. The flow could thus not be retarded on the steep slope to approach critical depth, although further retardation after passing critical would be entirely possible.

On the mild slope, the flow could be retarded to approach critical depth but could not be further retarded after passing critical depth. This is true since energy loss requirements would then require the energy slope to exceed normal, but specific energy relationships would require the energy slope to be less than normal.

It is consequently impossible for the flow to be retarded *gradually* from supercritical to subcritical conditions. This is also intimated by the fact that the downstream subcritical profile could have no influence on the upstream profile, since such influences cannot be transmitted upstream in supercritical flow.

The conclusion is then that the transition from rapid to tranquil flow must be accomplished, by a sudden transition, or *hydraulic jump*. This is true whether the transition is caused by a decrease in slope, an increase in width, an increase in roughness, or other cause. The exact location of the jump cannot be predicted solely by energy considerations, but can be determined reasonably closely by momentum requirements, as shown in the next article.

If the change in slope, width, roughness, etc., is insufficient in magnitude to alter the flow regime, then the change is merely from one normal depth to another, for the same regime. Energy requirements for these situations will show that a subcritical transition must be accomplished upstream from the change section, and a supercritical transition downstream from the change section.

4-7. The Hydraulic Jump. An idealized sketch of a hydraulic jump is shown in Fig. 4-8. A relation is desired between the depths D_1 and D_2,

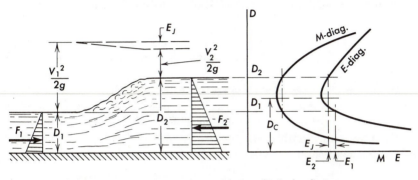

Fig. 4-8. Energy and momentum relations for hydraulic jump.

upstream and downstream from the jump, for a given discharge. This cannot be obtained from the energy equation, except in terms of the large, unknown energy loss E_j. However, momentum conservation requirements will yield the desired relationship.

If the body of water between sections 1 and 2 is considered as a free body, the only horizontal forces acting on it are the hydrostatic forces from the adjacent masses of water and a small shearing force along the perimeter. Neglecting the latter, since the length between sections is small, the momentum equation gives the following relationship:

$$F_1 - F_2 = \rho Q V_2 - \rho Q V_1 \qquad (4\text{-}32)$$

The hydrostatic forces are of the form $\gamma \bar{d} A$, where $\bar{d}$ represents the depth to the centroid of the cross-section. Equation (4-32) becomes

$$\bar{d}_1 A_1 + \frac{Q^2}{g A_1} = \bar{d}_2 A_2 + \frac{Q^2}{g A_2} \qquad (4\text{-}33)$$

A *pressure-momentum force* may be defined as the function $\gamma M = \gamma(\bar{d}A + Q^2/gA)$, and is obviously a function of the depth and of the cross-sectional shape, for a given discharge.

The function $M = \bar{d}A + Q^2/gA$ is also plotted in Fig. 4-8. It is seen to become infinite as D either approaches zero or becomes infinite. Each value of M corresponds to two values of D, known as *conjugate depths*.

Equation (4–33) states that the hydraulic jump depths are such that $M_1 = M_2$. That is, the depths upstream and downstream from the jump are conjugate depths. If one is known, the other could easily be determined graphically from the M-diagram for the given cross-section and discharge.

From requirements for jump formation, it is obvious that one of the conjugate depths must always be subcritical, the other supercritical. These are the *initial depth* and *sequent depth*, respectively. Consequently, the minimum value of M, as seen on the M-diagram, must occur at critical depth.

It is instructive to superimpose the specific energy diagram on the M-diagram, to the same depth scale, and this also has been done on Fig. 4–8. The head loss in the jump is immediately evident, since it is measured by $E_1 - E_2$.

In the special case of a rectangular section, which is by far the most frequently encountered in practice, it is possible to obtain convenient algebraic relationships for the solution of Eq.(4–33)and for determination of the head loss. In this case, since $d = D/2$ and $A = BD$, and letting $q = Q/B$, the discharge per foot of width, Eq. (4–33) becomes

$$\frac{D_1^2}{2} + \frac{q^2}{gD_1} = \frac{D_2^2}{2} + \frac{q^2}{gD_2} \qquad (4\text{--}34)$$

In this case, it may be convenient to define the function M as $D^2/2 + q^2/gD$; then the pressure-momentum force would be γBM.

Rearranged, Eq. (4–34) becomes

$$\frac{1}{2}(D_1^2 - D_2^2) = \left(\frac{q^2}{g}\right)\left(\frac{1}{D_2} - \frac{1}{D_1}\right) = \frac{q^2(D_1 - D_2)}{gD_1D_2}$$

Then

$$D_1 + D_2 = \frac{2q^2}{gD_1D_2} \qquad (4\text{--}35)$$

In quadratic form,

$$D_1^2 + D_2D_1 = \frac{2q^2}{gD_2}$$

Solving the quadratic yields

$$D_1 = \frac{D_2}{2}\left[-1 + \sqrt{1 + \frac{8q^2}{gD_2^3}}\right] \qquad (4\text{--}36)$$

Equation (4–36) can be used to give either jump depth if the other is known. Subscripts of course can be interchanged if D_2 is desired.

The energy lost in the jump, E_j, can now be computed by combining the energy equation with the jump equation. The energy equation is

$$E_j = \left(\frac{V_1^2}{2g} - \frac{V_2^2}{2g}\right) + (D_1 - D_2)$$

$$= \left(\frac{q^2}{2g}\right)\left[\frac{1}{D_1^2} - \frac{1}{D_2^2}\right] + (D_1 - D_2)$$

$$= (D_2 - D_1)\left[\frac{q^2(D_2 + D_1)}{2gD_1^2D_2^2} - 1\right]$$

Substituting $(D_1 + D_2)/2$ for q^2/gD_1D_2 from Eq. (4–35) gives

$$E_j = (D_2 - D_1)\left[\frac{(D_2 + D_1)^2}{4D_1D_2} - 1\right] = \frac{D_2 - D_1}{4D_1D_2}(D_2^2 + 2D_1D_2 + D_1^2 - 4D_1D_2)$$

Finally, the head loss in the jump is

$$E_j = \frac{(D_2 - D_1)^3}{4D_1D_2} \tag{4–37}$$

In a sense the jump is a control section and specifies a unique relation between the discharge and the two conjugate depths. That is, the discharge is fixed if the two depths are known. From Eq. (4–35):

$$q = \left[\frac{D_1 + D_2}{2}(gD_1D_2)\right]^{1/2} \tag{4–38}$$

Example 4–3

A rectangular open channel 20 ft wide carries a discharge of 400 cfs. The channel slope is 0.0025 and the Manning coefficient 0.0149. At a certain point in the channel, the velocity is 6 fps.

a. Is the flow at this point subcritical or supercritical?
b. If a hydraulic jump occurred at this point, what would be the conjugate depth at the jump?

Solution.

(a) $D = \dfrac{Q}{BV} = \dfrac{400}{20(6)} = \dfrac{10}{3}$ ft.

$$D_c = \sqrt[3]{\frac{Q^2}{gB^2}} = \sqrt[3]{\frac{(20)^2}{32.2}} = 2.32 \text{ ft.}$$

Since $D > D_c$, the flow is *subcritical. Ans.*

(b) The actual depth of 3.33 ft must be the sequent depth, since it is greater than the critical depth. The initial depth therefore is

$$D_1 = \frac{D_2}{2}\left[\sqrt{1 + \frac{8q^2}{gD_2{}^3}} = 1\right] = \frac{5}{3}\left[\sqrt{1 + \frac{8(400)(27)}{32.2(1000)}} - 1\right]$$
$$= 1.53\,ft.\;\;Ans.$$

4–8. Location of Hydraulic Jump. A very important problem in the design of spillway aprons or stilling basins is the determination of the location of the hydraulic jump. It may occur either on the steep slope or the mild slope, the only criterion being that M_1 must equal M_2 where it does occur. Since the profiles both upstream and downstream from the jump are in general non-uniform profiles, and since M is a function of depth, it is clear that the value of M continually changes in a longitudinal direction along the channel. Flow profiles must therefore be computed throughout the range of possible jump locations, for both the rapid and tranquil conditions, starting from an upstream control section (or point of known surface elevation) for the supercritical profile, and from a downstream control for the subcritical profile. For each flow profile, an M-line can be computed and plotted, showing how M varies along the reach of channel. The hydraulic jump must be located at the point of intersection of the M_1 and M_2 lines. This of course may not be exact, because of uncertainties in roughness coefficients and other factors, but will in general give reasonably reliable results. The jump does not take place in a vertical line, but experimentally seems to rise in an average slope of about one vertical to four or more horizontal. For further discussion of this subject, refer to the treatment of stilling basins in Art. 6–22.

4–9. Non-uniform Flow. Nearly all actual flows in open channels are non-uniform, so that the problem of determining the shape of flow profiles is of common occurrence. The fundamental approach to problems of this sort, as long as there is only a *gradual* variation in depth and velocity, is by means of the general energy equation. In the case of a *rapid* transition, such as in the hydraulic jump, the momentum equation is more effective, since it is difficult to estimate the rapid loss of energy otherwise.

Consider a very short reach of channel, dL in horizontal length, as shown in Fig. 4–9. Both the depth and the velocity head will be assumed to change by a differential amount in the differential length. By the energy equation (or simply by the geometry of the figure),

$$i_b\,dL + D + \frac{V^2}{2g} = i_e\,dL + D + dD + \frac{V^2}{2g} + d\left(\frac{V^2}{2g}\right) \qquad (4\text{–}39)$$

from which the following basic differential equation of gradually varied flow is obtained.

$$dL = \frac{d(D + V^2/2g)}{i_b - i_e} = \frac{dE}{i_b - i_e}$$

$$= \frac{d(D + Q^2/2gA^2)}{i_b - i_e} \qquad (4\text{--}40)$$

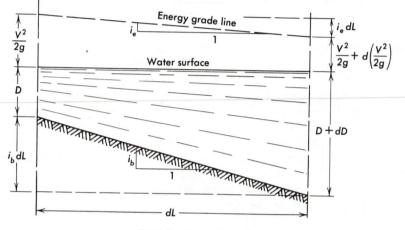

Fig. 4-9. Non-uniform flow.

The energy slope i_e depends on the rate of friction loss, and this is difficult to evaluate accurately for non-uniform flow. If it is approximated over a short reach by means of the Manning uniform flow formula, using mean values of area and hydraulic radius, it can be computed by

$$i_e = \left(\frac{nQ}{1.5A_m R_m^{2/3}}\right)^2 \qquad (4\text{--}41)$$

For a finite length ΔL, short enough so that such an average value of i_e can be used with reasonable justification, Eq. (4-40) becomes

$$L_2 - L_1 = \Delta L = \frac{(D_1 + V_1^2/2g) - (D_2 + V_2^2/2g)}{i_b - (nQ/1.5A_m R_m^{2/3})^2} \qquad (4\text{--}42)$$

where ΔL is the horizontal distance between sections 1 and 2 in the channel. This is the equation of the profile of the water surface, in terms of its co-ordinates L and D.

Computation of a flow profile must begin at some point of known co-ordinates, usually at a control section. Then, from Eq. (4-42) the distance ΔL to a point of assumed depth can be computed, thus giving the coordinates of another point. Every term in the right-hand member is a function of either the given depth or the assumed depth at the adjacent section, or both, so that ΔL is the only unknown in the equation, and it can therefore be computed by direct solution.

With the coordinates of the new point thus determined, another incremental length can be computed in the same manner to the next section of assumed depth, and so on until the entire profile is obtained. Computations proceed upstream in tranquil flow, downstream in rapid flow.

It should be noted that this procedure is applicable only to a channel of regular, known cross-sectional shape, for which the relation between depth and velocity is known at all points. A procedure for determining flow profiles in natural or irregular channels is discussed in the next article.

Because of the repetitive nature of the calculations, it is desirable to systematize them in tabular form as much as possible. A possible tabulation might be as indicated in Fig. 4–10. The tabulation is set up so that calculations may proceed vertically instead of horizontally, using appropriate increments of depth as the basis. In this way, all calculations of a given type may be made with one setting of the slide rule or calculating machine. The last two columns give the coordinates of the flow profile, corresponding to the station as identified in the first column.

The problem of algebraic signs is sometimes confusing. The exact arrangement depends on the particular type of flow profile involved and will be discussed in Chapter 5. However, for practical purposes in making the step calculations, it is sufficient to use the rule of thumb for calculating ΔL as follows:

1. In the numerator, use either $\left(\Delta D - \Delta \dfrac{V^2}{2g}\right)$ or $\left(\Delta \dfrac{V^2}{2g} - \Delta D\right)$, whichever is positive.

2. In the denominator, use either $(i_b - i_e)$ or $(i_e - i_b)$, whichever is positive.

Since $dA = B\,dD$, Eq. (4–40) may also be written in the following form:

$$L = \int_{D_1}^{D_2} \frac{1 - Q^2 B/gA^3}{i_b - (nQ/1.5A_m R_m^{2/3})^2}\,dD = \int_{D_1}^{D_2} f(D)\,dD \qquad (4\text{–}43)$$

Theoretically, the flow profile equation could be obtained in explicit form

Sta.	D_1	D_2	$\Delta D = D_1 - D_2$	$D_m = \dfrac{D_1 + D_2}{2}$	$V_1 = \dfrac{Q}{A_1}$	$V_2 = \dfrac{Q}{A_2}$	V_m	$\dfrac{V_1^2}{2g}$	$\dfrac{V_2^2}{2g}$	$\Delta\left(\dfrac{V^2}{2g}\right)$

Fig. 4–10. Headings for profile

by performing the indicated integration. Practically, however, the integration is impossible except for certain special cases and simplifying assumptions. Graphical integration may be employed, of course, but this amounts essentially to an incremental calculation of coordinates as described above.

Equation (4–43) does, however, indicate qualitatively the nature of the flow profile. The numerator vanishes at critical flow, for example, indicating a reversal of curvature at a control section. The denominator vanishes at uniform flow, corresponding to constant specific energy in the reach.

There have been numerous graphical aids and approximate methods published with the purpose of reducing the labor involved in these repetitive flow profile calculations. Some of these are discussed in Chapter 5. Similarly it is possible to program the computation process on a digital computer for specific types of profiles, and many such programs are available. However, it is essential that the designer first have a good understanding of the fundamental incremental method as outlined above in order to evaluate the reliability and applicability of other techniques.

Example 4–4

A channel is sufficiently wide that two-dimensional flow may be assumed (that is, the ratio of depth to width can be assumed negligibly small in calculating the peripheral shear). The slope is 0.0006 and the Manning coefficient 0.035 The discharge is 10 cfs/ft. At one point the depth of flow is 2.0 ft.

a. Determine the flow regime at this point.
b. If the flow is gradually varying in depth in the adjacent reach, how long is the reach between depths of 2.0 feet and 2.1 feet?

Solution.

a. The hydraulic radius, R_h, is equal to the depth in two-dimensional flow. In the Manning formula

$$q = \frac{Q}{B} = \frac{1.49}{n}(D)(D)^{2/3}(S)^{1/2} = \frac{1.49}{0.035}D^{5/3}(0.0006)^{1/2} = 10. \text{ cfs/ft.}$$

Solving for the normal depth, D_n.

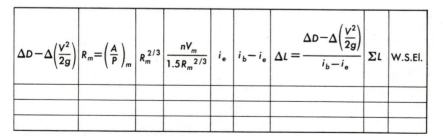

$\Delta D - \Delta\left(\dfrac{V^2}{2g}\right)$	$R_m = \left(\dfrac{A}{P}\right)_m$	$R_m^{2/3}$	$\dfrac{nV_m}{1.5R_m^{2/3}}$	i_e	$i_b - i_e$	$\Delta L = \dfrac{\Delta D - \Delta\left(\dfrac{V^2}{2g}\right)}{i_b - i_e}$	ΣL	W.S.El.

calculations in uniform channels.

$$D_n = \left[\frac{350}{1490(0.0006)^{1/2}}\right]^{3/5} = 3.89 \text{ ft.}$$

The critical depth, $D_c = \left(\frac{q^2}{g}\right)^{1/3} = \left(\frac{100}{32.2}\right)^{1/3} = 1.46 \text{ ft.}$

Since $D_n > D_c$, the channel slope is mild, and:

Since $D > D_c$, the flow regime is subcritical, and:

Since $D < D_n$, the profile is a *drawdown curve* (see Art. 5–1), with the flow depth gradually decreasing, away from D_n and approaching D_c, in the direction of flow.

b. The increment ΔL is to be calculated upstream from section 1 $(D_1 = 2.0$ ft) to section 2 $(D_2 = 2.1$ ft).

Assume a one-step calculation is adequate.

$$\Delta D = D_2 - D_1 = 0.1 \text{ ft}$$

$$\Delta \frac{V^2}{2g} = \frac{1}{2g}\left[\left(\frac{10}{2}\right)^2 - \left(\frac{10}{2.1}\right)^2\right] = \frac{1}{2g}(5^2 - 4.76^2) = 0.0365 \text{ ft}$$

The average energy slope in the reach can be approximated as:

$$(i_e)_m = \left(\frac{nq}{1.49 D_m^{5/3}}\right)^2 = \left(\frac{0.035(10)}{1.49(2.05)^{5/3}}\right)^2 = 0.00504$$

Then,
$$\Delta L = \frac{\Delta D - \Delta \dfrac{V^2}{2g}}{i_e - i_b}$$

$$= \frac{0.1000 - 0.0365}{0.00504 - 0.00060} = \frac{635}{44.4} = 14.3 \text{ ft.} \qquad Ans.$$

4–10. Computation of Water Surface Profile in an Irregular Channel.

The flow in a natural stream channel is usually in the tranquil regime, so that the shape of its profile will be dependent upon a downstream control section. Such a section may be created by a sharp channel constriction, a weir, rapids, or perhaps a controlling elevation at some base level into which the stream discharges. In any case, the computation of the profile must proceed *upstream* from some point of known elevation. Rapid flows in irregular channels are usually so complicated by wave phenomena that the uniform-flow friction assumption implicit in the use of the Manning formula is invalid. Principles of wave mechanics can often be employed in these cases, but accurate solutions are quite difficult.

The profile, for a given channel and discharge, is governed by the energy relationships, as expressed in the energy equation. Between two points in the channel, as illustrated in Fig. 4–11 the equation may be written

$$S_2 + \frac{Q^2}{2g A_2^{\,2}} = S_1 + \frac{Q^2}{2g A_1^{\,2}} + Li_e \qquad (4\text{–}44)$$

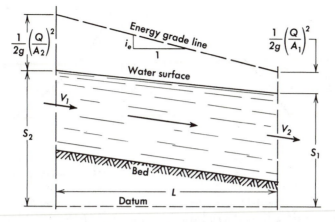

Fig. 4-11. Non-uniform flow in irregular channel.

S represents the water surface elevation and therefore the potential energy. The kinetic energy is $Q^2/2gA^2$. The energy loss per foot of length is indicated by the slope of the energy grade line i_e, which is determined approximately by the Manning formula, as in Eq. (4-41).

However, the slope i_e obviously changes as the depth changes, and consequently an average value must be used to solve Eq. (4-44). This requires that, for the equation to be reasonably realistic, the length of reach L must not be excessive and there must be no rapid changes of cross-section, bed slope, etc., within it. Furthermore, at each end of the section the relation between surface elevation, area, and hydraulic radius must be known. This normally requires that a cross-section profile be available.

To determine the flow profile in a channel for a given discharge, therefore, the channel must be divided into reaches, at each end of which the cross-section profile is available and has been plotted. For accuracy these sections should be spaced as closely as possible and so that no sharp or rapid changes occur between adjacent sections. Presumably the section farthest downstream is the one at which the controlling water elevation is known and at which the computations must therefore begin.

Since the profile must be computed in steps, from each section to the next adjacent upstream section, and since each step will involve a trial solution of Eq. (4-44), any systematization of the calculations will be helpful. It is suggested that, at each section, curves be plotted of A and i_e as functions of elevation, as in Fig. (4-12). This of course requires scaling the perimeter and area at each section for various elevations. The value of i_e computed at each depth would be the normal energy slope for that depth, that is, the slope that would be obtained in the case of uniform flow at that depth in a prismatic channel of the given cross-section.

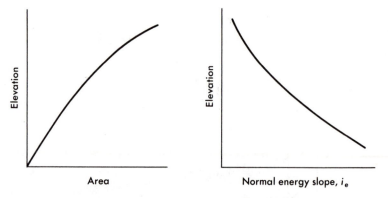

Fig. 4–12. Characteristics curves at channel cross-section.

At the first section the elevation and therefore the area and kinetic energy are known. But to determine the elevation S_2 it is also necessary to know A_2 and i_e, which also depend on S_2. Thus the equation must be solved by trial.

Probably the most expeditious way of making such a trial solution is to compute the slope i_e by two routes, the two values agreeing when the assumed value of S_2 is correct. Thus,

$$i_e = \frac{1}{L}\left[\left(S_2 + \frac{Q^2}{2gA_2{}^2}\right) - \left(S_1 + \frac{Q^2}{2gA_1{}^2}\right)\right] \qquad (4\text{--}45)$$

$$= \left(\frac{1}{L}\right)(E_{n+1} - E_n)$$

where E is the elevation of energy grade line. Also,

$$i_e = \frac{1}{2}(i_{e1} + i_{e2}) = \left(\frac{1}{A_1{}^2 R_1^{4/3}} + \frac{1}{A_2{}^2 R_2^{4/3}}\right)\left(\frac{1}{2}\right)\left(\frac{nQ}{1.5}\right)^2 \qquad (4\text{--}46)$$

where i_{e1} and i_{e2} can be read from the curves for the given value of S_1 and the assumed value of S_2.

The values usually converge rapidly, by at least the second or third trial. After the correct solution for S_2 is obtained, it then becomes S_1 for the next reach.

Sta.	Length	W.S.El.	Area	Vel. Hd.	E.G.L. El.	Normal slope
	L	S_n	A_n	$\dfrac{Q^2}{2gA_n{}^2}$	E_n	i_{e_n}

Fig. 4–13. Headings for profile

Because of the repetitive nature of the calculations, it is well to systematize them in tabular form. A suggested tabulation appears in Fig. 4–13.

4–11. Profile Calculation at Channel Constriction.

The previous discussion has assumed that flow is occurring within the main channel and that the changes in bed slope and cross-sectional form occur relatively gradually.

However, in the event of sudden changes, there will be introduced additional turbulent energy losses which are not accounted for in the Manning formula, on which the calculations have been based. This may be especially important in the case of sudden channel constrictions, such as those caused by bridge piers or other structures or islands in the river.

The restricted area at the contraction causes a conversion of some of the potential energy of the flow to increased kinetic energy in order for all the flow to pass through a reduced area. The velocity is increased and the depth decreased. After the constriction is passed, the reverse process takes place, except that a hydraulic expansion is less efficient than a contraction, with therefore a net loss of energy. This means the depth of flow below the constriction is less than that above it. Viewed in another way, the constriction causes a "backwater" above the normal flow depth.

The height of the backwater above the constriction depends primarily on the degree of constriction but also on many other factors, such as the shape of the constriction and its roughness, the velocity and depth of flow, etc. A completely rational analysis is probably impossible. There have been a large number of empirical investigations and methods derived therefrom. One of the simplest and most nearly rational is based on the momentum relationships.

If the depths upstream and downstream from the channel obstruction in Fig. 4–14 are D_1 and D_2, respectively, with D_0 the depth right at the obstruction, then the momentum equation between sections 1 and 2 is:

$$\frac{Q^2}{gA_1} + \frac{F_1}{\gamma} = \frac{Q^2}{gA_2} + \frac{F_2}{\gamma} + \frac{F_D}{\gamma} \qquad (4\text{–}47)$$

Trial W.S.El. S_{n+1}	Area A_{n+1}	Vel. Hd. $\dfrac{Q^2}{2g(A_{n+1})^2}$	E.G.L. El. E_{n+1}	Normal slope $i_{e_{n+1}}$	Av. normal slope i_e $\dfrac{1}{2}(i_{e_{n+1}} + i_{e_n})$	Actual slope $\dfrac{1}{L}(E_{n+1} - E_n)$

calculation in irregular channel.

In this equation F_1 is the hydrostatic force at section 1 ($= \gamma \bar{D}_1 A_1$) and F_2 the corresponding force at section 2. F_D is the drag force on the channel obstruction and can be written as:

$$F_D = \frac{\rho}{2} C_D A_0 V_1{}^2 = \frac{\gamma Q^2}{g A_1} \left(\frac{A_0}{A_1}\right) \left(\frac{C_D}{2}\right) \tag{4-48}$$

The drag coefficient, C_D, could be estimated for the particular shape of bridge pier or other obstruction. For a blunt-nosed pier, it would approach the value of 2; for stream-lined shapes it is smaller. Since other unknown losses may also be present it is reasonable and conservative to assume $C_D = 2$ unless more precise information is available. This approximation has been supported by numerous field tests on such constrictions. Equations (4–47) and (4–48) can then be combined to give

$$\left(\frac{Q^2}{g A_1}\right)\left(\frac{A_1 - A_0}{A_1}\right) + \frac{F_1}{\gamma} = \frac{Q^2}{g A_2} + \frac{F_2}{\gamma} \tag{4-49}$$

In the computation of a flow profile, the depth D_2 downstream from the constriction would be known and the problem is to determine the depth D_1 upstream. Thus the terms on the right-hand side of Eq. (4–49) can be determined and then those on the left computed from trial values of D_1 until the equation is satisfied. The area A_0 theoretically is based on the constriction

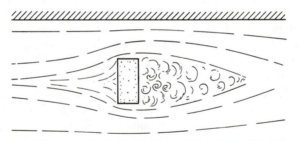

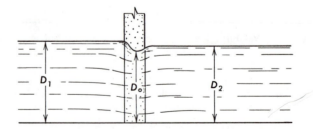

Fig. 4–14. Effect of channel constriction on flow profile.

depth D_0, but it is simpler, and on the safe side, to base it on D_1. The term F, the hydrostatic force on the water at the section is equal to $\gamma \bar{D} A$, where $\bar{D}$ is the depth to the centroid of the cross-section. For relatively shallow channels, this can be approximated as $D/2$. If the section is deeper and approximates a parabola in form, it may be assumed as $D/3$. Judgment may dictate an intermediate value and, if necessary, the centroid position could be computed exactly.

Equation (4-49) then becomes

$$\frac{Q^2}{gA_1^2}(A_1 - A_0) + (\bar{D}A)_1 = \frac{Q^2}{gA_2} + (\bar{D}A)_2 \qquad (4\text{-}50)$$

Equation (4-50) can be solved analytically, by trial, for D_1. If preferred, a graphical solution may be performed, as follows: Consider the expression in the left-hand member as a function of the depth of flow, $\phi(D)$:

$$\phi(D) = \frac{Q^2}{gA^2}(A - A_0) + \bar{D}A \qquad (4\text{-}51)$$

and the expression on the right as another function of depth, $\psi(D)$:

$$\psi(D) = \frac{Q^2}{gA} + \bar{D}A \qquad (4\text{-}52)$$

These functions can be computed for various depths and plotted on the same scale, with depth as ordinate and the functions ϕ and ψ, in cubic feet, as abscissa. The functions ϕ and ψ are equal when the proper depths upstream and downstream from the obstruction are inserted in Eq. (4-52). When one of these depths is known, the equation may be easily solved graphically by proceeding vertically from one curve to the other. Thus, if the downstream depth is given, the corresponding point on the ψ-curve is noted. The upstream depth is found from the position vertically above on the ϕ curve. This process is illustrated in Fig. 4-15.

4-12. Flow Profiles on Flood Plains.

Another type of special problem is encountered when flood flows are at such stages that a part of the flow moves along the banks and flood plains rather than in the channel, as in Fig. 4-16. This factor has two effects: (1) the roughness coefficient for the overbank flow is usually different from that of the river channel; (2) the channel length for the flow on the overbanks is shorter than in the main channel because of the meanderings of the latter, causing the hydraulic slope to be higher for the overbank flow.

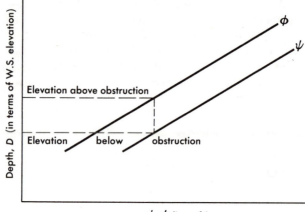

Fig. 4–15. Method for graphical determination of constriction effect on flow profile.

Those effects can be handled approximately by separating the two flows and adding them:

$$Q = Q_c + Q_o = \frac{1.5}{n_c} A_c R_c^{2/3} \left(\frac{E_2 - E_1}{L_c}\right)^{1/2} + \frac{1.5}{n_o} A_o R_o^{2/3} \left(\frac{E_2 - E_1}{L_o}\right)^{1/2}$$

$$= \left[\frac{1.5}{n_c} A_c R_c^{2/3} + \frac{1.5}{n_o(L_o/L_c)^{1/2}} A_o R_o^{2/3}\right]\left(\frac{E_2 - E_1}{L_c}\right)^{1/2} \qquad (4\text{–}53)$$

In Eq. (4–53) the subscripts c and o apply to main channel and overbank conditions, respectively. The term $E_2 - E_1$ is the difference between energy grade line elevations at the two ends of the reach under consideration, and is approximately the same for both divisions of the flow.

The calculations are carried out for the slope as based on the channel length, the same as for lower stages that are confined within its banks. However, the actual roughness coefficient for the overbank flow, n_o, is replaced by a smaller equivalent roughness coefficient, $n_o(L_o/L_c)^{1/2}$.

The normal energy slope for a given discharge would then be calculated by

$$i_e = \left\{\frac{Q}{(1.5/n_c)A_c R_c^{2/3} + [1.5/n_o(L_o/L_c)^{1/2}]A_o R_o^{2/3}}\right\}^2 \qquad (4\text{–}54)$$

This quantity can be calculated and plotted as a function of water-surface elevation (as in Fig. 4–12) for use in the detailed calculations. It may be noted that the wetted perimeter should be calculated only along the actual channel and overbank bed lines, not including the contact line between channel and overbank water areas, since the shear stress at the latter is much smaller than at the bed. (See Fig. 4–16.)

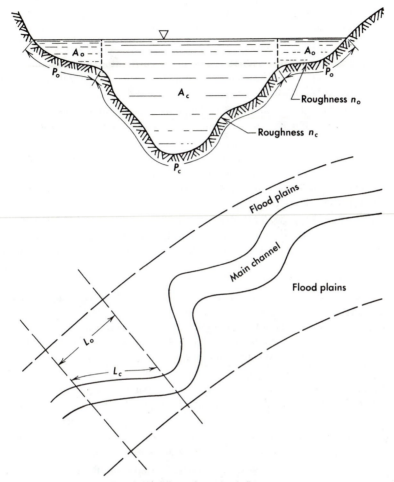

Fig. 4–16. Effect of overbank flow.

PROBLEMS

4-1. The Mississippi River at a gaging station near New Orleans was measured to have a discharge of 1,200,000 cfs, when the cross-sectional area was 202,000 sq ft, the wetted perimeter 2700 ft, and the slope 0.168 ft per mile. Assuming uniform flow, compute the Chezy C and the Manning n. *Ans.:* 121, 0.025.

4-2. A trapezoidal channel has a bottom width of 12 ft and a top width of 36 ft when flowing 6 ft deep. The slope is 0.0016 and the Manning n is 0.030. Compute the discharge. *Ans.:* 688 cfs.

4-3. An earth canal in good condition having a bottom width of 12 ft and side slopes of 2 horizontal to 1 vertical is designed to carry 200 cfs. If the slope of the canal is 2.1 ft per mile, determine the depth of water. Manning $n =$ 0.0225. *Ans.:* 4.0 ft.

4–4. A smooth-metal flume of semicircular cross-section has a diameter of 6 ft and a slope of 0.005. What diameter of corrugated metal flume will be required to have the same capacity? Roughness coefficients are 0.012 and 0.025 respectively. *Ans.:* 7.9 ft.

4–5. Calculate the normal depth of flow in a rectangular open channel 25 ft wide, with a slope of 0.0005 and a Manning coefficient of 0.025 when the discharge is 1000 cfs.

4–6. A corrugated metal culvert, 3 ft in diameter, carries uniform flow at a depth of 1.5 feet. The corrugations are standard ($\frac{1}{2}$ in. high, $2\frac{2}{3}$ in. spacing), and the culvert is on a 5 per cent slope. Assume water temperature at 75°F. What discharge is the culvert carrying?

4–7. A trapezoidal open channel is to be designed for a flow of 400 cfs, on a slope of 0.0016. The channel is unlined and, in order to prevent erosion, the maximum velocity allowed is 4.5 fps. The side slopes are 2 horizontal to 1 vertical and the Manning coefficient is 0.025. What flow depth and bottom width should be used? *Ans.:* 3.4 ft × 19.6 ft.

4–8. What should be the width and the depth of flow of a rectangular planed-timber flume of most efficient cross-section to carry 90 cfs, with a velocity of 5 ft/sec? What grade should the flume have? What percentage less flow would this flume carry if $d = 2$ ft and $b = 9$ ft? Assume Manning $n = 0.010$. *Ans.:* $D = 3$ ft, $b = 6$ ft, $S = 0.00066$, 5.35 per cent.

4–9. What should be the bottom width and the depth of flow in a concrete-lined canal of most efficient trapezoidal cross-section with side slopes $\frac{1}{2}$ horizontal to 1 vertical to carry 400 cfs on a grade of 3 ft per mile? The roughness coefficient is 0.010. *Ans.:* $D = 5.7$ ft, $b = 7.0$ ft.

4–10. A circular storm sewer has a diameter D and depth of flow d. Calculate the ratio of d/D for which the hydraulic radius is maximum.

4–11. For the circular sewer of Problem 4–10, calculate the maximum discharge that the sewer can carry without flowing under pressure, in terms of D, n, and S, and the value of d/D at which this discharge is obtained.

4–12. A highway ditch, triangular in cross-section, with side slopes of 4:1 on one side and 2:1 on the other side, is on a longitudinal slope of 0.0001, with a Manning coefficient of 0.020. The discharge is 50 cfs.

   a. What is the normal depth of flow?

   b. If it had been feasible to use the most efficient triangular section for the ditch, how much reduction in cross-sectional area of flow per lineal foot of ditch would have resulted? *Ans.:* 5.4 ft².

4–13. A rectangular flume carries a discharge of 306 cfs. The flow is uniform, the slope is 0.016, and the Manning coefficient is 0.018. What should be the width and depth of flow for maximum hydraulic efficiency?

4–14. An open channel carries 120 cfs with a bed slope of 0.0004 and a Manning coefficient of 0.0298. What should be the dimensions of a trapezoidal channel to carry this flow at normal depth and at maximum hydraulic efficiency?

Ans.: $b = 6.70$ ft; $D = 5.81$ ft; $z = \dfrac{1}{\sqrt{3}}$

4–15. A rectangular flume carries a flow of 100 cfs. The slope is 0.0009 and the Manning coefficient is 0.0149. What should be the width of flume if it is to carry this discharge at uniform flow with the smallest amount of material used to construct the flume?

4–16. On what slope should a trapezoidal metal flume be laid to assure the minimum amount of metal for the design discharge, for the following conditions?

$$Q = 10 \text{ cfs} \qquad n = 0.0149$$
$$z = 1:1 \qquad V = 2.5 \text{ fps}$$

For a 6-inch freeboard, what should be the dimensions of the flume? *Ans.:* 0.000935; b, $D = 1.23$ ft, 1.98 ft.

4–17. A rectangular channel 30 ft wide discharges water at a normal depth of 12 ft. The slope is 1:4000 and the Manning coefficient is 0.018. Determine:

a. The normal discharge, in cfs. *Ans.:* 1663 cfs.

b. The critical depth, in ft. *Ans.:* 4.59 ft.

4–18. Plot the specific energy and momentum diagrams for the channel and discharge of Problem 4–17.

4–19. Given a parabolic channel with central depth D and surface width B, derive formulas for:

a. Critical depth in the channel, in terms of discharge Q.

b. Critical velocity in terms of center depth.

4–20. Derive critical flow formulas, as in Problem 4–19, for:

a. Trapezoidal channel, with bottom width b and surface width B.

b. Triangular channel, with surface width B.

4–21. A trapezoidal channel has a bottom width of 5 ft and 1:1 side slopes. The hydraulic slope is 0.001, and n is 0.013. For discharges of 100, 250, and 500 cfs, respectively, determine both the normal and critical depths of flow. Hint: Graphical solution is easiest. Plot curves of "depth" vs. $A R_h^{2/3}$ and A^3/B, respectively, for the normal depth and critical depth functions.

4–22. In the upper Mississippi River, the St. Anthony Falls have been stabilized by construction of a steep apron to convey the water down the falls to the river below. At the head of the apron, the depth of flow is measured to be 3.2 ft. What is the river discharge per foot of width? *Ans.:* 32.5 cfs.

4–23. A concrete box culvert, 4 ft high and 6 ft wide, has a roughness coefficient of 0.015 and is on a slope of 1 per cent. It carries a discharge of 90 cfs, discharging freely at the outlet, and functioning as an open channel throughout its length. Is the flow tranquil or rapid?

4–24. What should be the width of flume, depth of flow, and bed slope, if a rectangular concrete flume is to be designed to convey water at uniform critical flow over an arroyo, at greatest hydraulic efficiency? The discharge is 200 cfs and the Manning coefficient 0.015.

4–25. A trapezoidal channel, with side slopes of 1:1 is lined with planed wood

(Manning $n = 0.012$). The design discharge is 300 cfs and bottom slope is 0.0001.

    a. Determine channel dimensions for the smallest amount of timber lining.

    b. Show whether flow is tranquil or rapid.

4-26. A rectangular wooden channel ($n = 0.010$) is 8 ft wide, on a 1 per cent slope, and carries a flow of 100 cfs. At a certain point the depth of flow is 2.5 ft.

    a. Show whether the flow at this point is subcritical or supercritical.

    b. Determine whether the channel slope is mild or steep.

4-27. The channel shown in the accompanying figure is carrying a discharge of 100 cfs. Determine whether the flow is streaming flow or shooting flow at the section shown.

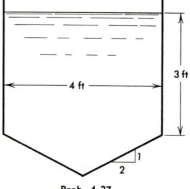

Prob. 4-27.

4-28. An open channel carries a discharge of 600 cfs. The channel has a Manning coefficient of 0.0298 and a slope of 0.000049. The channel is of triangular cross-section, with side slopes 3 vertical to 4 horizontal. Calculate: (a) normal depth; (b) critical depth.

4-29. An open channel carries 120 cfs, with a bed slope of 0.0004 and a Manning coefficient of 0.0298.

    a. What should be the dimensions of a rectangular channel to carry this discharge at uniform flow with a minimum of channel lining material? *Ans.:* 11.0 ft × 5.5 ft.

    b. Determine whether the indicated slope is a mild or steep slope. *Ans.:* Mild.

4-30. An open channel is of rectangular cross-section, 10 feet wide. The slope is 0.0016 and the Manning coefficient is 0.0149.

    a. If the section has been designed as one of maximum hydraulic efficiency, what is the design discharge? *Ans.:* 368 cfs.

    b. For this discharge, is the slope mild or steep? *Ans.:* Mild.

    c. If the discharge is doubled, in the same channel, will the flow be subcritical or supercritical at a point where the velocity is 15 feet per second? *Ans.:* Supercritical.

4–31. A rectangular wooden flume, 6 feet in width, is to carry a discharge of 80 cfs.

    a. What slope should the flume have in order to carry this flow with the minimum flume height? The Manning coefficient is 0.012.

    b. Show whether the resulting flow is subcritical or supercritical. *Ans.:* 0.000746; subcritical.

4–32. A triangular open channel has side slopes of 2 horizontal to 1 vertical. The Manning coefficient is 0.0298 and the slope 0.0004. For a discharge of 20 cfs at uniform flow, is the flow tranquil or rapid?

4–33. A rectangular, concrete-lined canal carries a discharge of 500 cfs. The channel is 25 feet wide, on a longitudinal slope of 0.0009 and has a Manning roughness coefficient of 0.0149. Calculate the following:

    a. Normal depth, ft.

    b. Critical depth, ft.

    c. Energy regime at a section in the channel where the velocity is 5.0 ft/sec.

4–34. A rectangular concrete lined canal carries a discharge of 500 cfs. The channel is 25 feet wide, on a longitudinal slope of 0.0009 and has a Manning n of 0.0149.

    a. What is the uniform flow depth? *Ans.:* 3.45 ft.

    b. Determine whether the channel slope is mild or steep. *Ans.:* Mild.

4–35. A triangular channel has an angle of 60° at the apex, as shown. For a discharge of 10 cfs, slope of 0.0004 and roughness coefficient of 0.0149, compute:

    a. Critical depth. *Ans.:* 1.80 ft.

    b. Normal depth. *Ans.:* 3.18 ft.

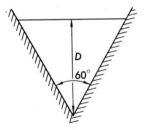

Prob. 4–35.

4–36. A trapezoidal channel of 45° side slopes, designed for maximum hydraulic efficiency at a discharge of 72 cfs, is on a slope of 0.000001 and has a Manning coefficient of 0.0298.

    a. Calculate the normal depth and base width. *Ans.:* 14.5 ft, 12.0 ft.

    b. Calculate the critical depth. *Ans.:* 1.01 ft.

4–37. A hydraulic jump forms on a horizontal apron at the foot of an overflow spillway. The design discharge is 30,000 cfs and the apron width is 150 ft. The jump occurs when the upstream depth reaches 8 ft. Compute:

    a. Downstream depth and velocity. *Ans.:* 14.0 ft, 14.3 fps.

b. Head loss in the jump. *Ans.:* 0.48 ft.

4-38. At the outlet from a chute spillway, it is observed that the water jumps from a depth of 3.0 ft to a depth of 6.0 ft. The section is rectangular in form. Compute:

a. Pressure-momentum force at the jump. *Ans.:* 1965 lb/ft.

b. Energy dissipation at the jump. *Ans.:* 0.375 ft.

c. Critical velocity for given discharge. *Ans.:* 11.8 fps.

4-39. A highway ditch is triangular in cross-section with side slopes of 4 horizontal to 1 vertical. During a rainstorm, water runs downhill in the ditch, entering a pool at the bottom of the hill leading to a culvert, by a hydraulic jump. The water depths upstream and downstream from the jump are, respectively, 0.8 ft and 2.4 ft. How much water is flowing in the ditch? How much energy is dissipated in the jump?

4-40. A hydraulic jump in a horizontal rectangular channel goes from a depth of 2 ft to a depth of 5 ft. Calculate:

a. Discharge, in cfs per ft.

b. Head loss, ft.

c. Pressure-momentum force at the jump, in lb per ft of width.

d. Minimum specific energy for the existing discharge, in ft.

Ans.: (a) 33.5 cfs/ft; (b) 0.675 ft; (c) 1217 lb/ft; (c) 4.90 ft.

4-41. A horizontal rectangular channel 30 ft wide at the base of a spillway experiences a hydraulic jump, in which the conjugate depths are 3.5 ft and 5.5 ft.

a. What is the discharge in cfs: *Ans.:* 1584 cfs.

b. What is the specific energy upstream and downstream from the jump? *Ans.:* 7.03 ft. 6.93 ft.

c. How much head is lost in the jump? *Ans.:* 0.104 ft.

4-42. A rectangular stilling basin at the base of a spillway is 30 ft wide and on a horizontal slope. At a point where the Froude number (expressed as $V_1/\sqrt{gD_1}$) for the upstream flow is 2.5, a hydraulic jump occurs. The discharge is 300 cfs. Calculate the energy that will be lost in the hydraulic jump, in feet. *Ans.:* 0.57 ft.

4-43. A flow of 100 cfs occurs in a horizontal rectangular channel 10 ft wide. A hydraulic jump occurs at a point where the flow depth has become 1.0 ft. Compute the following:

a. Critical depth. *Ans.:* 1.46 ft.

b. Alternate depth (same specific energy). *Ans.:* 2.25 ft.

c. Conjugate depth (same pressure-momentum force). *Ans.:* 2.05 ft.

d. Head loss in hydraulic jump. *Ans.:* 0.133 ft.

4-44. A discharge of 300 cfs is carried in a rectangular channel 15 ft wide. The channel slope is 1:9000 and the Manning coefficient 0.030. At a certain point in the channel the velocity is 5 ft per second.

a. Show whether the flow is tranquil or rapid.

b. Show whether the slope is mild or steep.

c. Compute the alternate pressure-momentum depth to the depth at the indicated section.

4–45. A horizontal rectangular channel 30 ft wide at the base of a spillway experiences a hydraulic jump. The depth upstream from the jump is 3.0 ft and the head loss in the jump is $\frac{3}{8}$ ft. Compute:

a. Discharge in the channel, cfs.

b. Normal depth for the given discharge in the channel, feet.

4–46. A horizontal rectangular channel at the base of a spillway experiences a hydraulic jump when the initial depth is 3.0 ft and the discharge is 51 cfs/ft. Compute the following:

a. Depth downstream from the jump, ft.

b. Subcritical depth with the same specific energy as the 3-ft depth.

c. Horsepower lost in the jump per ft of width.

4–47. A horizontal rectangular channel 30 ft wide at the base of a spillway experiences a hydraulic jump, in which the conjugate depths are 3.0 ft and 6.0 ft.

a. What is the discharge, in cfs? *Ans.:* 1530 cfs.

b. What is the critical depth, in ft? *Ans.:* 4.33 ft.

c. What is the horsepower dissipated in the jump? *Ans.:* 65.1 hp.

4–48. A hydraulic jump occurs in a rectangular channel 20 ft wide, with the depths above and below the jump being 2 ft and 8 ft, respectively. Compute the following:

a. Discharge in cfs.

b. Head loss, in ft.

c. Pressure-momentum force, in lbs.

d. Critical depth, in ft.

4–49. A rectangular open channel is on a slope of 0.0001 and has a Manning coefficient of 0.0298. The discharge per unit width is 10 cfs per ft. At a certain point in the channel, the flow depth is 2.0 ft.

a. If a hydraulic jump occurs at this point, how much energy will be lost in the jump? *Ans.:* 0.12 ft.

b. If no hydraulic jump occurs, how much energy will be lost as the flow depth gradually changes from 2.0 to 2.4 ft? *Ans.:* 0.29 ft.

4–50. A retarded flow of 100 cfs is carried in a rectangular wooden channel ($n = 0.010$), 8 ft wide, on a bed slope of 1.0 per cent. At a certain section the flow depth is 2.5 ft.

a. Show whether the slope is mild or steep.

b. Show whether the flow is tranquil or rapid.

c. How far upstream (based on a single-step computation) is the flow 2.35 ft deep? *Ans.:* 10.5 ft.

4–51. A rectangular open channel 20 ft wide carries a discharge of 400 cfs. The channel slope is 0.0025 and the Manning coefficient 0.0149. At a certain point in the channel the velocity is 6 ft/sec.

a. What would be the conjugate depth if a hydraulic jump occurred at this point? *Ans.:* 1.53 ft.

b. Assuming there is no hydraulic jump in the channel, how far from this section would the depth be 3.57 ft? Assume one step calculation is sufficient. *Ans.:* 108 ft.

4-52. A rectangular open channel is 10 ft wide, carries a discharge of 20 cfs, and has a Manning coefficient of 0.012. The bed slope is 0.0025. At a certain point the depth of flow is 1.0 ft.

    a. Determine whether the flow is tranquil or rapid at this point. $_\prime$ 5 $\mathcal{L}+$

    b. At what flow depth other than 1.0 ft would the pressure-momentum force be the same? $M_1 = M_2$, .227

    c. How far from this point will the flow depth be 2.0 ft? Assume one-step calculation is sufficient. $D_1 = 1.0$, $V_1 = 2 \mathcal{G}p_s$ $V_m = 1.55 p_s$

$\Delta L = \frac{}{clock}$ $D_2 = 2$ $V_2 = 1 \mathcal{F} p_s$

4-53. A rectangular open channel 20 ft wide carries a discharge of 400 cfs. The channel slope is 0.0025 and the Manning coefficient 0.0149. At a certain point in the channel, the velocity is 6 fps. Compute the water surface elevation (as measured from a datum through the bed of the channel at the given section) at another section 100 ft upstream from the given section. *Ans.: 3.80 ft.*

4-54. In a rectangular channel, the flow is accelerating between sections *A* and *B*. The channel width is 20 ft, the slope is horizontal, and the Manning coefficient is 0.0149. The discharge is 400 cfs. At section *B* the depth of flow is 4.00 ft. What is the flow depth at *A*, 100 ft upstream? Assume one-step computation is sufficient.

4-55. A rectangular channel 30 ft wide discharges water at a normal depth of 12 ft. The bed slope is 1:3000 and the Manning coefficient 0.018. A dam 16 ft high at the crest is placed across the stream, with an overflow spillway of slope 1:4. Using depth increments of 1.0 ft, calculate and plot the complete flow profile, including:

    a. Backwater curve above the dam, to a point where the depth is within 0.5 ft of normal depth.

    b. Dropdown curve on the spillway.

    c. Slowdown curve on the channel bed below the spillway, up to the point where critical depth would theoretically be reached.

    d. Location and dimensions of the hydraulic jump.

4-56. A backwater curve increment is being computed in a stream between two sections *A* and *B*, 460 ft apart. The hydraulic properties of the cross-sections are as follows:

Water Surface Elevation	Area (sq. ft)		Wetted Perimeter (ft)	
	Section A	Section B	Section A	Section B
1696.0	1333.5		151.0	
1697.0	1483.5	841.5	155.5	104.0
1698.0	1637.5	936.0	160.0	107.5
1699.0	1795.5	1062.0	166.8	111.0
1700.0	1957.5	1167.0	171.3	115.0
1701.0		1275.0		119.0

For a flow in the channel of 10,000 cfs and a Manning coefficient of 0.040, the water elevation at station *A* is 1697.5. Compute the surface elevation at station *B*, accurate to the nearest 0.1 ft.

4–57. Just upstream from station B (Problem 4–56), the flow is partially obstructed by a large bridge pier in the channel, presenting a surface 10 ft wide normal to the direction of flow. The channel can be considered roughly rectangular in cross-section, with bottom at station B at elevation 1690.0. Compute the water surface elevation immediately upstream from the pier.

4–58. Assume that the cross-sections at stations A and B (Problem 4–56) are valid up to elevation 1697 for station A and 1698 for station B, but that these elevations mark the flood-bank level. The latter extends horizontally for 200 ft on each side of the banks, then encountering slopes which rise essentially vertically. The overbank flow follows a direction such that the distance from station A to station B is only 300 ft, but the roughness coefficient on the flood plain is 0.045. For a discharge of 10,000 cfs and a water surface elevation at A of 1697.5, compute the water surface elevation at B.

<h1 style="text-align:center">5</h1>

OPEN CHANNEL TRANSITIONS
AND CONTROLS

5–1. General Analysis of Flow Profiles. An introductory discussion of the phenomena of non-uniform flow in open channels was given in the previous chapter. We shall now consider in somewhat more general fashion the characteristics of the various types of possible flow profiles.

The general differential equation of steady flow in an open channel (see Fig. 4–9) is:

$$dL = \frac{d(D + V^2/2g)}{i_b - i_e} \tag{5-1}$$

when $d(D + V^2/2g)$ is the differential change in specific energy in the differential channel length dL (horizontal). The slope of the bed is i_b and the slope of the energy grade line is i_e. If the latter is computed by the Darcy equation, then Eq. (5–1) can be modified in the following fashion:

$$dL = \frac{dD + (Q^2/2g)\, d(1/A^2)}{i_b - (f/8gR_h)(Q^2/A^2)} = \frac{dD - (Q^2/gA^3)(dA/dD)\, dD}{i_b - (Q^2/8g)(f/A^2R_h)}$$

$$= \frac{dD(1 - Q^2B/gA^3)}{i_b[1 - (Q^2/8g)(f/A^2R_h)/i_b]} \tag{5-2}$$

The function Q^2B/gA^3 is equal to unity at critical flow and the function $Q^2f/8gAR^2$ is equal to the bed slope for uniform, or normal, flow. Thus Eq. (5–2) can be written

$$\frac{dD}{dL} = \frac{i_b[1 - (A^2R_h/f)_n/(A^2R_h/f)]}{1 - (A^3/B)_c/(A^3/B)} \tag{5-3}$$

In this relation R_h is the hydraulic radius and the subscripts n and c denote normal and critical flow, respectively. This is another form of the general differential equation of steady flow in open channels. The equation indicates, of course, that $dD/dL = 0$ for uniform flow, or that $D =$ constant. It also indicates that, when a flow profile approaches critical depth, the water surface slope approaches vertical. The numerator is negative if the actual depth is less than normal depth; the denominator is negative if the depth is less than critical. In either case dD/dL must be negative; that is, the depth must be decreasing and the velocity is accelerating. However, if the depth is less than *both* normal and critical, the resulting fraction is positive and therefore the flow is decelerating. Similarly, the flow must be decelerating if the depth is *greater* than both normal and critical.

The above analysis leads to a four fold classification of varied flow profiles, as follows:

1. *Drawdown curve*, obtained on a mild slope when the flow depth is decreasing from normal to critical depth, as shown in Fig. 5-1.

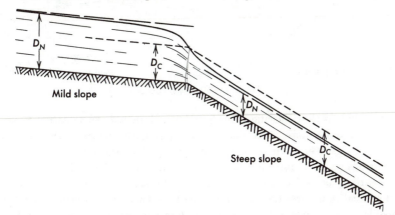

Fig. 5–1. Drawdown and dropdown curves.

2. *Dropdown curve*, obtained on a steep slope, when the flow depth is decreasing from critical depth to normal depth for the steep slope, as shown in Fig. 5–1.
3. *Backwater curve*, obtained when the flow depth is greater than both normal and critical depth and is being retarded, either by a downstream channel obstruction, if the slope is mild, or after a hydraulic jump, if the slope is steep. These conditions are sketched in Fig. 5–2.

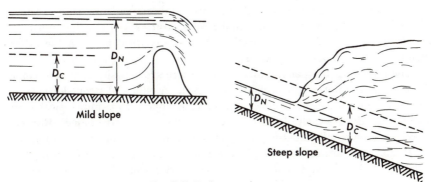

Fig. 5–2. Backwater curves.

4. *Slowdown curve*, obtained when the flow depth is less than both normal and critical and the flow is being retarded—toward normal if on a steep slope, and toward critical, and a hydraulic jump, if on a mild slope. These profiles are sketched in Fig. 5–3.

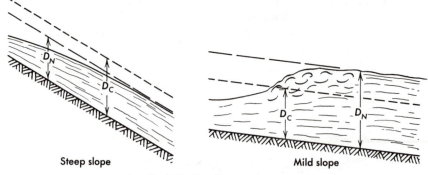

Fig. 5–3. Slowdown curves.

It is evident that classifications 3 and 4 each include two distinct types of profiles. Furthermore, this analysis has implicitly assumed that the bed slope, i_b is greater than zero, and is not equal to the critical slope. As a matter of fact, it is quite possible for the bed slope to be critical, horizontal, or negative, as well as mild or steep. If the slope is critical, then normal and critical flows are the same, and there are two non-uniform flow profiles possible—one a retarded flow with the depth increasing above normal, and one a retarded flow with the depth increasing toward normal.

In the case of either a horizontal or an adverse slope, there is of course *no* normal depth of flow, and therefore the energy equation in the form of Eq. (5–3) is not applicable. However, it can be rearranged as follows, from Eq. (5–2):

$$\frac{dD}{dL} = \frac{i_b - (Q^2/8g)(f/A^2R_h)}{1 - (A^3/B)_c/(A^3/B)} \qquad (5\text{–}4)$$

The numerator is negative for either a horizontal or an adverse slope. Thus, if the flow is subcritical, the denominator is positive and the flow must be accelerating, since dD/dL is negative. For supercritical flow, on the other hand, the flow would be retarded.

The foregoing discussion indicates that there are actually twelve distinct types of non-uniform flow profiles, three each on mild and steep slopes, and two each on critical, horizontal, and adverse slopes. It is impracticable to attempt to denote each of these by a particular descriptive name. Rather, a system of symbolic identification has been adopted, with the letters M, S, C, H, and A denoting the type of slope, and the numbers 1, 2, 3 denoting the relation of the actual flow depth to critical and normal depths. The region 1 is for actual depths exceeding both critical and normal, 3 is for depths less than both critical and normal, and 2 is for depths intermediate between critical and normal.

The general shape of any one of the twelve profiles can be determined by examination of Eq. (5–3), (or Eq. (5–4) for horizontal and adverse slopes).

Example 5–I

Determine the general shape of the M-1 and M-3 profiles.

Solution. The M-1 curve (the backwater curve proper) has depths exceeding both normal and critical, and therefore both numerator and denominator in Eq. (5–3) are positive, and the flow depth must be increasing. It is also evident that, as the depth increases, both numerator and denominator approach unity and thus dD/dL approaches i_b, which means that the water surface asymptotically approaches the horizontal. In the other direction, as the depth approaches normal depth, the numerator approaches zero, and therefore dD/dL approaches zero. The M-1 profile, therefore, is asymptotic to the uniform flow line upstream and the horizontal downstream.

The M-3 profile, which we have called the slowdown curve before the hydraulic jump, is obtained on a mild-slope apron. Equation (5–3) shows that the numerator and denominator are both negative, and therefore the flow must be retarded, approaching critical depth. Furthermore, dD/dL becomes infinite when the depth reaches critical (actually a hydraulic jump would intervene before this point), and so the water surface slope is concave upward near the jump. To determine the shape at the other end of the curve, rearrange Eq. (5–4) as follows:

$$\frac{dD}{dL} = \frac{(A^3/B)i_b - (A^3/B)(Q^2/8g)(f/A^2R_h)}{A^3/B - (A^3/B)_c} = \frac{A^3 i_b - (Q^2 f/8g)(A/R_h)}{A^3 - A_c^3(B/B_c)}$$

$$= \frac{A^3 i_b - (Q^2/8g)fP}{A^3 - A_c^3(B/B_c)} \tag{5–5}$$

where P is the wetted perimeter. As the depth approaches zero, A^3 approaches zero and P approaches B. Therefore dD/dL approaches $Q^2 f B_c/8g A_c^3 = f/8$, since $Q^2 B/g A^3 = 1$ at critical depth. This is a small positive number whose magnitude perhaps depends on the channel roughness, but which is still uncertain, corresponding as it does to an infinite Froude number. The same limiting value would obviously be obtained also on the S-3, H-3, C-3, and A-3 profiles. However, since the S-3 profile must approach normal depth asymptotically, rather than critical depth, its form will be concave downward.

The shape of each of the twelve basic flow profiles can be determined qualitatively, by examination of each in light of the basic differential equation, as illustrated in the foregoing example. Computation of actual coordinates on any given profile, however, would require step calculations in some such manner as described in Arts. 4–8 and 4–9. The twelve profiles are sketched in Fig. 5–4.

The discussion thus far has assumed that the Darcy friction factor f is to be used in the flow profile equations. If it is desired to use the Manning roughness coefficient n, then Eq. (5–3) becomes

$$\frac{dD}{dL} = \frac{i_b[1 - (A^2 R_h^{4/3})_n/A^2 R_h^{4/3}]}{1 - (A^3/B)_c/(A^3/B)} \tag{5–6}$$

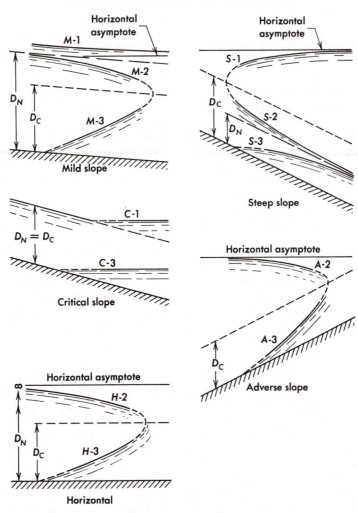

Fig. 5–4 Basic water surface profile. Flow is from left to right in every case.

This follows from equating the slope terms in the Manning and Darcy equations, which operation yields

$$f = 117 \frac{n^2}{R_h^{1/3}} \tag{5–7}$$

It is assumed also that Manning's n does not vary with flow depth, which is only approximately true.

In the special case of two-dimensional flow, then $A = BD$ and $P = B + 2D$, so that

$$R_h = \frac{BD}{B + 2D} \cong D \tag{5–8}$$

since D is small compared to B. Equation (5-3) is then

$$\frac{dD}{dL} = \frac{i_b[1 - (D^3/f)_n/(D^3/f)]}{1 - D_c^{\ 3}/D^3} \tag{5-9}$$

and Eq. (5-6) becomes

$$\frac{dD}{dL} = \frac{i_b(1 - D_n^{10/3}/D^{10/3})}{1 - D_c^{\ 3}/D^3} \tag{5-10}$$

5-2. Integration of the Varied Flow Equation. The general equation of the flow profile has thus far been considered only in differential form, and methods for numerical computation of profiles have similarly been based on incremental methods. Theoretically, however, the differential equation should be integrable, permitting evaluation of the flow profile equation in explicit form. Thus, from Eq. (5-2),

$$L = \int \frac{(1 - Q^2 B/gA^3)\, dD}{i_b - (Q^2/8g)(f/A^2 R_h)} \tag{5-11}$$

For a given discharge and channel, this equation is of the form

$$L = \int f(D)\, dD \tag{5-12}$$

and, if it could be solved, would give L directly in terms of D, so that the flow depth at any point in the channel would be known. As a practical matter, however, the function $f(D)$ is usually too complex to be integrated, except by means of graphical or numerical integration.

The equation can be integrated graphically by plotting, for the given cross-section, the function $f(D)$ vs. D. Then, of course, the length of reach between two sections of given or assumed flow depths is

$$(\Delta L)_{1-2} = (\text{area under } f(D) \text{ curve})_{1-2} \tag{5-13}$$

The procedure of numerical integration is simply the calculation of incremental lengths between successive sections of assumed depths as described previously.

It is possible under some conditions to simplify the equation sufficiently to permit explicit integration. A large number of special methods have been published with this purpose. For example, one of the most useful of these has been the backwater function and tables presented by the French hydraulician, Bresse, in 1860.

Bresse assumed the channel could be approximated by a rectangular section of sufficient width to permit the hydraulic radius R_h to be calculated as the depth of flow D. He also assumed that, for these conditions, the friction factor f was constant, for a given discharge. With these simplifications, Eq. (5-11) can be written

$$L = \int \frac{(1 - q^2/gD^3)\, dD}{i_b - q^2 f/8gD^3} = \int \frac{(1 - D_c^{\ 3}/D^3)\, dD}{i_b - i_b\, D_n^{\ 3}/D^3} \tag{5-14}$$

since, for the assumed conditions, $q^2/g = D_c^3$ and $i_b = q^2 f/8g D_n^3$. Rearranging gives

$$i_b L = \int \frac{D^3 - D_c^3}{D^3 - D_n^3} \, dD = \int \left[1 + \frac{D_n^3 - D_c^3}{D^3 - D_n^3} \right] dD$$

$$= \int dD + (D_n^3 - D_c^3) \int \frac{dD}{D^3 - D_n^3} \tag{5-15}$$

Integration yields

$$i_b L = D + \frac{(D_n^3 - D_c^3)}{3 D_n^2} \left[\frac{1}{2} \log_e \frac{(D - D_n)^2}{D^2 + D_n D + D_n^2} + \sqrt{3} \arctan \frac{2D + D_n}{-\sqrt{3} D_n} \right] + C$$

$$= D + D_n \left[\left(\frac{D_c}{D_n} \right)^3 - 1 \right]$$

$$\times \left[\frac{1}{6} \log_e \frac{(D/D_n)^2 + D/D_n + 1}{(D/D_n - 1)^2} + \frac{1}{\sqrt{3}} \operatorname{arccot} \frac{\sqrt{3}}{2D/D_n + 1} \right] + C'$$

Hence

$$\frac{i_b L}{D_n} = \frac{D}{D_n} + \left[\left(\frac{D_c}{D_n} \right)^3 - 1 \right]$$

$$\times \left[\frac{1}{6} \log_e \frac{(D/D_n)^2 + (D/D_n) + 1}{(D/D_n - 1)^2} - \frac{1}{\sqrt{3}} \left(\arctan \frac{\sqrt{3}}{2D/D_n + 1} + \frac{\pi}{2} \right) \right] + C' \tag{5-16}$$

$$= \frac{D}{D_n} + \left[\left(\frac{D_c}{D_n} \right)^3 - 1 \right] \left[\phi \left(\frac{D}{D_n} \right) - \frac{\pi}{2\sqrt{3}} \right] + C' \tag{5-17}$$

The constant of integration is determined by inserting the given depth at the point $L = 0$, whatever it may be. Or, usually more conveniently, the length from this initial point, with depth $= D_0$, to some other point may be computed as a definite integral, with limits from D_0 to D. Then,

$$\frac{i_b L}{D_n} = \frac{D - D_0}{D_n} + \left[\left(\frac{D_c}{D_n} \right)^3 - 1 \right] \left[\phi \left(\frac{D}{D_n} \right) - \phi \left(\frac{D_0}{D_n} \right) \right] \tag{5-18}$$

This equation, known as Bresse's equation, gives the flow profile in dimensionless form, for a channel approximating the assumed conditions. The function ϕ can be evaluated for various values of D/D_n and tabulated in this dimensionless form. Any desired number of points can be determined on a given flow profile by selecting and processing the corresponding numbers in the tabulation. This tabulation is given in Table 5-1, applicable to the M and S profiles.

Bresse's equation would also be applicable for a critical slope, but it reduces simply to

$$i_b L = D$$

or

$$\frac{D}{L} = i_b = \text{constant} \tag{5-19}$$

TABLE 5–I
Bresse's Backwater Function
Part I
For M-I, S-I, and S-2 Curves

D/D_n	$\phi(D/D_n)$	D/D_n	$\phi(D/D_n)$	D/D_n	$\phi(D/D_n)$	D/D_n	$\phi(D/D_n)$
1.000	∞	1.054	0.8714	1.29	0.3816	2.30	0.0978
1.001	2.1837	1.056	0.8599	1.30	0.3731	2.35	0.0935
1.002	1.9530	1.058	0.8489	1.31	0.3649	2.40	0.0894
1.003	1.8182	1.060	0.8382	1.32	0.3570	2.45	0.0857
1.004	1.7226	1.062	0.8279	1.33	0.3495	2.50	0.0821
1.005	1.6486	1.064	0.8180	1.34	0.3422	2.55	0.0788
1.006	1.5881	1.066	0.8084	1.35	0.3352	2.60	0.0757
1.007	1.5371	1.068	0.7990	1.36	0.3285	2.65	0.0728
1.008	1.4929	1.070	0.7900	1.37	0.3220	2.70	0.0700
1.009	1.4540	1.072	0.7813	1.38	0.3158	2.75	0.0674
1.010	1.4192	1.074	0.7728	1.39	0.3098	2.80	0.0650
1.011	1.3878	1.076	0.7645	1.40	0.3039	2.85	0.0626
1.012	1.3591	1.078	0.7565	1.41	0.2983	2.90	0.0604
1.013	1.3327	1.080	0.7487	1.42	0.2928	2.95	0.0584
1.014	1.3083	1.082	0.7411	1.43	0.2875	3.00	0.0564
1.015	1.2857	1.084	0.7337	1.44	0.2824	3.1	0.0527
1.016	1.2645	1.086	0.7265	1.45	0.2775	3.2	0.0494
1.017	1.2446	1.088	0.7194	1.46	0.2727	3.3	0.0464
1.018	1.2259	1.090	0.7126	1.47	0.2680	3.4	0.0437
1.019	1.2082	1.092	0.7059	1.48	0.2635	3.5	0.0412
1.020	1.1914	1.094	0.6993	1.49	0.2591	3.6	0.0389
1.021	1.1755	1.096	0.6929	1.50	0.2548	3.7	0.0368
1.022	1.1603	1.098	0.6867	1.52	0.2466	3.8	0.0349
1.023	1.1458	1.100	0.6806	1.54	0.2389	3.9	0.0331
1.024	1.1320	1.105	0.6659	1.56	0.2315	4.0	0.0315
1.025	1.1187	1.110	0.6519	1.58	0.2246	4.1	0.0299
1.026	1.1060	1.115	0.6387	1.60	0.2179	4.2	0.0285
1.027	1.0937	1.120	0.6260	1.62	0.2116	4.3	0.0272
1.028	1.0819	1.125	0.6139	1.64	0.2056	4.4	0.0259
1.029	1.0706	1.130	0.6025	1.66	0.1999	4.5	0.0248
1.030	1.0596	1.135	0.5913	1.68	0.1944	4.6	0.0237
1.031	1.0490	1.140	0.5808	1.70	0.1892	4.7	0.0227
1.032	1.0387	1.145	0.5707	1.72	0.1842	4.8	0.0218
1.033	1.0288	1.150	0.5608	1.74	0.1794	4.9	0.0209
1.034	1.0191	1.155	0.5514	1.76	0.1748	5.0	0.0201
1.035	1.0098	1.160	0.5423	1.78	0.1704	5.5	0.0166
1.036	1.0007	1.165	0.5335	1.80	0.1662	6.0	0.0139
1.037	0.9919	1.170	0.5251	1.82	0.1621	6.5	0.0118
1.038	0.9834	1.175	0.5169	1.84	0.1582	7.0	0.0102
1.039	0.9750	1.180	0.5090	1.86	0.1545	7.5	0.0089
1.040	0.9669	1.185	0.5014	1.88	0.1509	8.0	0.0077
1.041	0.9590	1.190	0.4939	1.90	0.1474	8.5	0.0069
1.042	0.9513	1.195	0.4868	1.92	0.1440	9.0	0.0062
1.043	0.9438	1.200	0.4798	1.94	0.1408	9.5	0.0055
1.044	0.9364	1.21	0.4664	1.96	0.1377	10.0	0.0050
1.045	0.9293	1.22	0.4538	1.98	0.1347	12.0	0.0035
1.046	0.9223	1.23	0.4419	2.00	0.1318	15.0	0.0022
1.047	0.9154	1.24	0.4306	2.05	0.1249	20.0	0.0013
1.048	0.9087	1.25	0.4198	2.10	0.1186	30.0	0.0006
1.049	0.9022	1.26	0.4096	2.15	0.1128	50.0	0.0002
1.050	0.8958	1.27	0.3998	2.20	0.1074	100.0	0.0001
1.052	0.8834	1.28	0.3905	2.25	0.1024	∞	0.0000

TABLE 5–1 (continued)

Part 2

For M-2, M-3, and S-3 Curves

D/D_n	$\phi(D/D_n)$	D/D_n	$\phi(D/D_n)$	D/D_n	$\phi(D/D_n)$	D/D_n	$\phi(D/D_n)$
0.00	0.0000	0.75	0.6856	0.940	1.4028	0.981	1.8001
0.10	0.1000	0.76	0.8742	0.945	1.4336	0.982	1.8185
0.20	0.2004	0.77	0.8923	0.950	1.4670	0.983	1.8379
0.25	0.2510	0.78	0.9110	0.952	1.4813	0.984	1.8584
0.30	0.3021	0.79	0.9304	0.954	1.4962	0.985	1.8803
0.35	0.3538	0.80	0.9505	0.956	1.5117	0.986	1.9036
0.40	0.4066	0.81	0.9714	0.958	1.5279	0.987	1.9287
0.45	0.4608	0.82	0.9932	0.960	1.5448	0.988	1.9557
0.50	0.5168	0.83	1.0160	0.962	1.5626	0.989	1.9850
0.52	0.5399	0.84	1.0399	0.964	1.5813	0.990	2.0171
0.54	0.5634	0.85	1.0651	0.966	1.6011	0.991	2.0526
0.56	0.5874	0.86	1.0918	0.968	1.6220	0.992	2.0922
0.58	0.6120	0.87	1.1202	0.970	1.6442	0.993	2.1370
0.60	0.6371	0.88	1.1505	0.971	1.6558	0.994	2.1887
0.62	0.6630	0.89	1.1831	0.972	1.6678	0.995	2.2498
0.64	0.6897	0.900	1.2184	0.973	1.6803	0.996	2.3246
0.66	0.7173	0.905	1.2373	0.974	1.6932	0.997	2.4208
0.68	0.7459	0.910	1.2571	0.975	1.7066	0.998	2.5563
0.70	0.7757	0.915	1.2779	0.976	1.7206	0.999	2.7877
0.71	0.7910	0.920	1.2999	0.977	1.7351	1.000	∞
0.72	0.8068	0.925	1.3232	0.978	1.7503		
0.73	0.8230	0.930	1.3479	0.979	1.7661		
0.74	0.8396	0.935	1.3744	0.980	1.7827		

In other words, the retarded-flow profile on a critical slope must be horizontal. (However, this conclusion is based on an assumed constant f; the curve is actually slightly concave downward.)

In the case of a horizontal or adverse slope, Bresse's equation cannot be used as it stands, since there is no applicable normal depth D_N. For a horizontal slope, however, the flow equation can be easily integrated directly. Thus, from Eq. (5–14), with $i_b = 0$,

$$L = \int \frac{(1 - D_c{}^3/D^3)}{-(f/8)(D_c{}^3/D^3)}\, dD = \int \left(\frac{8}{f} - \frac{8D^3}{fD_c{}^3} \right) dD$$

$$= \frac{8}{f} D - \frac{8}{f}\frac{1}{4D_c{}^3} D^4 + C$$

$$= \frac{8D}{f} \left[1 - \frac{1}{4}\left(\frac{D}{D_c} \right)^3 \right] + C \tag{5–20}$$

The constant of integration is avoided if the length is computed between two sections of given depths. Note that the profile reaches critical depth at a length $L = 6D_c/f + C$. If it is preferred to use the Manning roughness coefficient instead of the friction factor, the conversion equation

$$f = \frac{117n^2}{R^{1/3}} = \frac{117n^2}{D^{1/3}} \tag{5–21}$$

can be used. Then Eq. (5–20) becomes (after inserting in Eq. 5–14 and integrating)

$$L = \left(\frac{6D^{4/3}}{117n^2}\right)\left[1 - \left(\frac{4}{13}\right)\left(\frac{D}{D_c}\right)^3\right] + C \qquad (5\text{–}22)$$

or, in terms of q, since $D_c^3 = q^2/g$,

$$L = \left(\frac{6D^{4/3}}{117n^2}\right)\left(1 - \frac{4gD^3}{13q^2}\right) + C \qquad (5\text{–}23)$$

These equations could be applied to calculation of H-2 and H-3 profiles on wide horizontal aprons.

In the case of an adverse slope, Bresse's equation and tables can be used, except that the term D_n must be replaced by

$$D_n = \left(\frac{q^2 f}{8gi_b}\right)^{1/3} = \left(\frac{f}{8i_b}\right)^{1/3} D_c \qquad (5\text{–}24)$$

This same relation is used to compute the normal depth for use in the Bresse method for mild and steep slopes. The only difficulty is that the friction factor f varies somewhat as the flow depth changes, since it is a function of Reynolds number. However, it is usually found that an average value of f for the range of depth expected, taken as a constant, will give sufficiently accurate results for the purpose.

There are numerous other methods in the hydraulic literature that have been offered for the approximate solution of the varied flow equation. The only really accurate method is the basic step method, and that only if the steps are sufficiently short. The Bresse method is representative, and probably the most widely used, of the various approximate methods, and therefore is the only one discussed here. However, its accuracy depends on the extent to which its assumptions correspond with actual conditions. If the channel departs significantly from that required for two-dimensional flow, or if the friction factor changes significantly, then some other method should be used. Probably the most useful of these latter methods is that of Bakhmeteff, as modified and extended by Chow.

With the increasing use of digital computers, it is often expeditious to use the standard step method, programmed for computer solution, especially if the project is large and requires examination of several alternate flow profiles. Although the various approximate solutions of the varied flow differential equation could likewise be expedited by a computer program, there is little reason for using one of these when the more accurate standard step method can be programmed just as easily.

For rapid approximate profile estimation by hand methods, the Bresse method is probably the most useful and gives reasonably accurate results in most cases.

Example 5–2

A wide shallow channel has a slope of 0.0004, roughness coefficient of 0.015, and discharge of 10 cfs/ft. A low dam 4.0 ft high is placed across the channel. The depth of flow a certain distance upstream from the dam is 5.0 ft. Compute by the Bresse method the distance upstream to the point where the depth is 3.0 ft.

Solution. By the Manning equation, $q = D \dfrac{1.5}{n} (D)^{2/3} (S)^{1/2}$

$\therefore$ Normal depth, $D_n = \left[\dfrac{0.015(10)}{1.5\sqrt{0.0004}} \right]^{3/5} = (5)^{3/5} = 2.62$ ft

$\phi\left(\dfrac{D_0}{D_n}\right) = \phi\left(\dfrac{5}{2.62}\right) = \phi(1.91) = 0.146$ (from Table 5–1).

$\phi\left(\dfrac{D}{D_n}\right) = \phi\left(\dfrac{3}{2.62}\right) = \phi(1.15) = 0.561$

Critical depth, $D_c = \sqrt[3]{q^2/g} = \sqrt[3]{\dfrac{100}{32.2}} = 1.46$ ft

By Equation (5–18):

$$L = \frac{D - D_v}{i_b} + \frac{D_n}{i_b}\left[\left(\frac{D_c}{D_n}\right)^3 - 1 \right]\left[\phi\left(\frac{D}{D_n}\right) - \phi\left(\frac{D_0}{D_n}\right) \right]$$

$$= \frac{1}{0.0004}\left[-2 + \left\{ 2.62\left(\frac{1.46}{2.62}\right)^3 - 1 \right\}(0.561 - 0.146) \right]$$

$$= -7250 \text{ ft (i.e., 7250 ft upstream).} \qquad Ans.$$

5–3. Flow Profiles in Channel of Varying Bed Slope. A long open channel will usually contain two or more reaches of different bed slopes. The slope is usually forced to conform within certain limits to the requirements of the general topography or the dimensions of the various structures concerned, so that it is commonly not feasible for the channel to retain the same bed slope throughout its length. This, of course, is even more common in natural channels.

Since the type of flow profile depends in part upon bed slope, it is important to recognize the different types of profiles that may occur simultaneously in different reaches of the same channel. An initial qualitative study of the complete profile will indicate in what segments of the channel it will be necessary to compute detailed profiles.

Several basic requirements should be kept in mind, as discussed in Art. 4–6:

1. A change from subcritical flow on a mild slope to supercritical flow on a steep slope is such that critical flow is attained essentially at the break in grade.

2. A change from supercritical to subcritical flow is accomplished through a hydraulic jump, located at that point where the flows of momentum in the subcritical and supercritical flows are equal.

3. A change from one mild slope to another mild slope causes a flow profile transition curve that is completed entirely on the upstream slope.

4. A change from one steep slope to another steep slope causes a flow profile transition that is accomplished entirely on the downstream slope.

5. Flow profiles approaching, or diverging from, uniform flow for the given slope, do so asymptotically.

6. Flow profiles approaching, or diverging from, critical depth tend to do so at a vertical slope, although not actually becoming vertical.

7. Flow profiles should be computed (or sketched) upstream if the depth is greater than critical and downstream if the depth is less than critical.

If it is assumed, now, that the channel is prismatic, with uniform section throughout its length, there are two basic types of profile analysis that may have to be made:

1. For a given discharge, determine the flow profile throughout.

2. For a given water elevation at some point or points in the channel system (the most common being a given surface elevation in the reservoir supplying the channel), compute the discharge and the balance of the flow profile.

The more general problem of design of the channel itself would typically involve trial analyses of several possible channel designs, with comparative economic studies based thereon. We are concerned here only with the profile analysis for a given channel.

By considering first the case of known discharge, the normal depth for each slope segment, as well as the critical depth, can be computed. A sketch should be made to scale, with vertical scale highly exaggerated, showing bed line as a solid line, critical depth line as a dotted line, and the various normal depth lines as dashed lines. Possible control points for segments of the flow profile should be noted, these being of two basic types, critical depth control, where the flow goes through critical, and normal depth control, where the flow has become essentially uniform. Since both these depths are known numerically, if it is known where they may occur in the channel, they can be used as basic points on the flow profiles.

Example 5–3

Determine the general character of possible flow profiles for the channel of Fig. 5–5.

Solution. The critical depth and normal depth lines are shown by dotted and dashed lines, respectively. Possible control points on these lines are indicated by circled numerals. Sketching of profiles proceeds upstream for subcritical and downstream for supercritical flow.

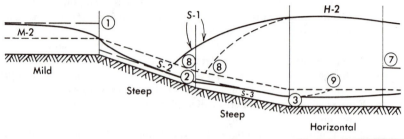

Fig. 5–5. Possible flow

An *M*-2 curve is sketched upstream and an *S*-2 curve downstream from the critical depth control at point 1. The latter is followed by an *S*-3 curve on the second steep slope. This will begin from control point 2 if the first steep slope is long enough to attain normal depth; otherwise it will begin at the actual water elevation at the breakpoint. Similarly, point 3 may serve as a normal depth control for the *H*-3 curve that develops on the horizontal slope.

The latter may intersect the critical depth line at point 9, or it may continue throughout the horizontal reach and be replaced by an *M*-3 curve on the next mild slope, depending upon the particular conditions. Similarly the *M*-3 curve may intersect the critical depth line at point 10, or it may continue and be succeeded by an *A*-3 curve on the adverse slope, and so on. Wherever the profile intersects the critical depth line, it may be stopped, since a hydraulic jump must occur somewhere before that point is reached.

To determine the subcritical profile to which the supercritical flow must jump, a downstream control point must be used as a starting point. Possible controls are located at points 4, 5, 6, and 7. Point 4, the farthest downstream, would probably be used unless it were reasonably certain that uniform flow could be assumed at point 5, 6, or 7. Point 4 is a critical depth control, and causes an *M*-2 profile upstream from it, which may or may not, attain uniform flow at point 5, or 6, or both.

On the adverse slope, an *A*-2 profile will follow, then an *M*-1 curve on the mild slope upstream from it. If the *A*-2 curve should have come out below the normal depth line on the latter slope (very unlikely, however), then an *M*-2 curve would have followed, approaching point 7. The horizontal slope causes an *H*-2 profile; then *S*-1 profiles must develop on the steep slopes. The latter will end at point 8, on the critical depth line, which may turn out to be on either of the two steep slopes.

Detailed profile calculations would be necessary to elucidate the exact forms and locations of the various profiles. The qualitative analysis does indicate, however, that point 7 will *not* be a control and that there will be an *M*-1 rather than an *M*-2 profile on the mild slope succeeding it. It is quite possible, also, depending mainly on the lengths of slopes that computations may begin at point 2 or 3 for the supercritical flow and at point 6 for the subcritical flow, with the antecedent portions neglected, since this reach will contain both the maximum flow depths and the hydraulic jump.

The jump, of course, must be located by determining a point where the

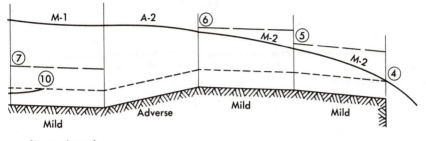

profiles in channel.

depths on the subcritical and supercritical flow profiles satisfy the requirement for conjugate depths in the jump equation. It will most likely occur on either the downstream steep slope or the horizontal slope, but it may possibly occur anywhere between the points (8 and 10) where the two profiles respectively reach the critical depth lines.

In the event that point 8 on the S-1 curve turns out to be above point 1 on the S-2 curve, it may be that the jump is "submerged" and cannot occur. The flow would follow the upper profile throughout, with either a truncated M-2 curve or an M-1 curve on the first mild slope, depending upon whether point 8 were below or above the normal depth line for that slope.

In the above example, it was assumed that the discharge was known. However, it may be that some particular surface elevation is fixed and the discharge will be limited thereby. The flow profile cannot be determined until this discharge is known.

This type of problem must be solved by trial. A discharge is assumed and the corresponding flow profile calculated. If this profile correlates with the controlling elevation, the assumed discharge is correct; if not, the process is repeated with another trial discharge.

For example, consider flow out of a lake or reservoir into a canal. The fixed lake elevation determines the discharge that can be supplied to the canal, and the flow profile in the canal must conform thereto. For simplicity, assume that the canal has a uniform slope throughout, discharging freely at its terminus, as shown in Fig. 5–6.

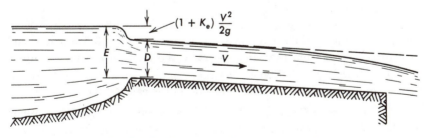

Fig. 5–6. Profile controlled by reservoir level.

At the canal inlet, the water surface must drop an amount equal to $(1 + K_e)V^2/2g$, where V is the unknown canal velocity and K_e is the entrance loss coefficient.

The discharge is unknown, but its upper limit would be determined by the available specific energy, E, measured by the difference between reservoir surface elevation and canal bed elevation at the inlet. The maximum discharge is determined by the critical flow equation, which is

$$Q = \sqrt{\frac{gA^3}{B}} = \sqrt{8gB^2(E - D)^3}$$
$$= \sqrt{gB^2(D_m)^3} \tag{5-25}$$

where B is the width and D the depth of flow. For a rectangular canal section, since the depth is $\frac{2}{3}$ the specific energy, this would simplify to

$$Q = B\sqrt{gD^3} = B\sqrt{\frac{8gE^3}{27}} \tag{5-26}$$

For this maximum discharge, the critical slope is then computed, and compared with the actual bed slope, to determine whether the latter is steep or mild for the given conditions.

If the slope is steep (or critical) it will be able to support the maximum discharge as computed, and this flow will then prevail throughout the entire channel. An S-2 curve will be formed in the canal after the flow passes the inlet section. If desired, the inlet loss can now be estimated from the critical velocity head, and applied as a correction to the assumed value of E, then the entire computation repeated.

If the slope for the computed maximum discharge turns out to be mild, then the canal velocity will be subcritical and the assumed discharge cannot be maintained. The true discharge must be determined by trial.

A discharge is assumed (somewhat less than the previous maximum discharge) and the flow profile computed therefrom. If the channel is long, it is probable that computation of the uniform flow depth will suffice. At the upstream end of the canal, the sum $D + (1 + K_e)V^2/2g$ is obtained and compared with E. If necessary, another trial discharge is used, until satisfactory agreement is obtained.

If the channel consists of a succession of reaches of different slopes, as in the earlier example, the same procedure is followed. The full profile would have to be traced from the canal outlet to the inlet, in order to compare $D + (1 + K_e)V^2/2g$ with E at the entrance, unless some other downstream control point is available from which to begin the calculations.

5–4. Changes in Cross-Section. When the cross-sectional dimensions of a channel change, there is a change in *both* normal and critical depth for a given discharge. However, the basic principles already discussed are still applicable, and the transition must take place in the form of one of the twelve fundamental flow profiles. Within the transition zone, the profile will correspond to *gradually changing* values of normal and critical depth. The latter in turn correspond to a gradually changing flow boundary, which may be caused directly by the gradual change of the channel cross-section as designed, or indirectly by flow separations in the case of sudden changes in channel dimensions.

Upstream or downstream from the point of cross-section change, the flow profile may be determined by methods already discussed. Similarly, if the changes occur so gradually that the controlling factor is channel roughness, rather than acceleration or deceleration, then the methods for calculating varied flow profiles in irregular channels may be used.

In this article we are concerned directly with the local flow profile within the transition zone of a material and a relatively rapid change in cross-section, for which accelerative forces become quantitatively more important than friction forces. Also, the present discussion will be limited to subcritical flow transitions, since supercritical transitions may involve special wave phenomena which significantly modify the effects.

In the usual problem of this sort, it is desired to design an economical and hydraulically efficient transition structure, to convey the flow between two reaches of channel of different cross-sections. Normally, the discharge is known, as well as the surface elevations at the beginning and end of the transition.

Efficiency of energy conversion requires that the flow profile be continuous and as smooth as possible. Since the beginning and end points of the profile are known, an arbitrary smooth transition profile is sketched in, as shown in Fig. 5–7.

The transition head loss, H_t, may be estimated on the basis of empirical data that are available for various types of transition sections, commonly given as a coefficient times the change in velocity head between the two ends of the transition. That is,

$$H_t = K_t\left(\frac{V_2^2}{2g} - \frac{V_1^2}{2g}\right) \tag{5–27}$$

Approximate experimental values of K_t are as shown in Table 5–2.

TABLE 5-2
Coefficient of Head Loss in Transitions, K_t

Type of Transition Section	Contracting Section	Expanding Section
Warped-surface transition	0.10	0.30
Wedge transition	0.20	0.50
Cylinder-quadrant transition	0.20	0.50
Sharp transition	0.40	0.75

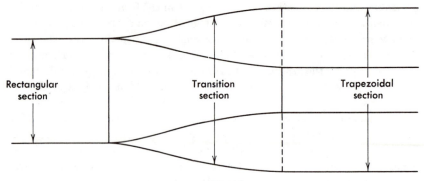

Plan View

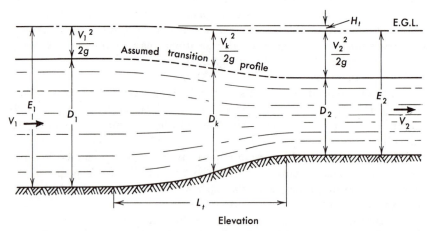

Elevation

Fig. 5–7. Subcritical channel transition.

The total head loss in the transition, H_t, may with sufficient accuracy be assumed to be distributed uniformly along the length of the transition. This then gives the elevation of the energy grade line at every point along the transition, as well as the elevation of the assumed water surface. At any point K in the transition, the velocity head $V_k^2/2g$ is scaled off the flow and energy line profiles, as shown.

The required cross-sectional area of flow at K is then computed from the given discharge:

$$A_k = \frac{Q}{V_k} = \frac{Q}{\sqrt{2g(V_k^2/2g)}} \qquad (5\text{–}28)$$

The area A_k is of course a function of the depth D_k and of the particular cross-sectional form. Usually one or more dimensions of the cross-section will be fixed by the geometric form of transition used, and the remaining dimensions can be determined as necessary to make the total area equal to A_k. For example, if a bed elevation is assumed, the channel width can be computed, and vice versa.

This process is necessarily arbitrary to the degree that sketching of the transition flow profile is arbitrary. The most efficient transition hydraulically will be one in which no flow separations occur at any part of the cross-section. Accordingly, not only the flow profile but also the bed profile and the channel width must change smoothly, gradually, and continuously.

However, the latter dimensions may not, upon first computation, exhibit such smooth form, but may have sharp curves or breaks. If so, the assumed flow profile should be modified as needed, and the calculations repeated. Finally, all dimensions of the transition, as well as the flow profile, should show the desired gradual, continuous change between end sections. This is especially important in expanding transitions, because of the inherent instability of expanding flows. For this reason, also, the expanding transition must be spread over a sufficiently great length to prevent flow separations. Experimental studies have indicated that the transition length, L_T, should be

$$L_T \geqq 2\tfrac{1}{4}(\Delta B) \tag{5-29}$$

where ΔB is the change in surface width through the transition. This can be reduced somewhat if splitter walls are used to guide the flow.

Although the warped-surface transition, designed as described above, is the best from a hydraulic standpoint, it may be too expensive to construct in a specific application. Various types of simpler transitions have been used to reduce construction costs, especially the wedge type and cylinder-quadrant type, as sketched in plan in Fig. 5–8.

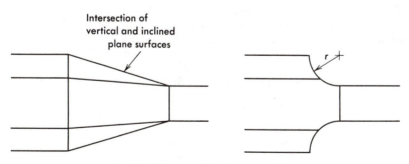

Fig. 5–8. Wedge and quadrant transitions.

If the cross-section dimensions are thus arbitrarily fixed, the energy loss will be somewhat increased, corresponding to discontinuities that will be induced in the computed theoretical profile. This additional turbulence will have the effect of smoothing out the actual flow profile.

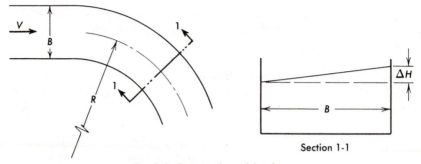

Fig. 5–9. Flow in channel bend.

5–5. Changes in Direction. When a channel changes direction, centrifugal effects cause a transverse hydraulic gradient, with the surface elevation higher on the outside of the curve. Assume a case of subcritical flow around a curve of local mean radius R; with surface width B, as in Fig. 5–9. If it is assumed that the velocity is uniform across the section, equilibrium of gravitational and centrifugal forces on a fluid particle at the outer edge of the bend yields the following relation (see Fig. 5–10):

$$\frac{MV^2}{R} \cong Mg\frac{\Delta H}{B}$$

or

$$\Delta H = \left(\frac{2B}{R}\right)\left(\frac{V^2}{2g}\right) = \frac{4(r_o - r_i)}{(r_o + r_i)}\frac{V^2}{2g} \tag{5–30}$$

This expression is only approximate, since the effect of non-uniform velocity and non-concentric streamlines has been neglected, both of which tend to increase the superelevation somewhat above the value given in Eq. (5–30). The difference is usually unimportant, however.

The greatest accuracy is apparently obtained if the flow in the bend is assumed to be a free vortex, for which the velocity distribution equation is

$$vr = C \tag{5–31}$$

where v is the velocity at the point of radius r. The average velocity is then

$$V = \frac{\displaystyle\int_{r_i}^{r_o} \frac{C}{r}\,dr}{r_o - r_i} = \frac{C}{r_o - r_i}\log_e\frac{r_o}{r_i} \tag{5–32}$$

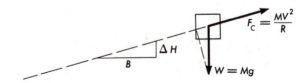

Fig. 5–10. Forces on particle in bend.

Similarly, the average depth of flow is given by

$$D_m = \frac{\int_{r_i}^{r_o} D\,dr}{r_o - r_i} = \frac{\int_{r_i}^{r_o} \left(E - \frac{C^2}{2gr^2}\right) dr}{r_o - r_i}$$

$$= E + \left[\frac{C^2}{2g(r_o - r_i)}\right]\left(\frac{1}{r_o} - \frac{1}{r_i}\right) = E - \frac{C^2}{2gr_or_i} \qquad (5\text{-}33)$$

in which E is the specific energy at the section. The discharge can then be expressed as

$$Q = VD_m(r_o - r_i) = C\left(E - \frac{C^2}{2gr_or_i}\right)\log_e\frac{r_o}{r_i} \qquad (5\text{-}34)$$

Since Q, E, r_o and r_i are presumably known for a given channel, the circulation constant C can be determined from Eq. (5–34). Then the superelevation is

$$\Delta H = D_o - D_i = \left(E - \frac{C^2}{2gr_o^2}\right) - \left(E - \frac{C^2}{2gr_i^2}\right)$$

$$= \frac{C^2}{2gr_o^2r_i^2}(r_o^2 - r_i^2) \qquad (5\text{-}35)$$

A major effect of channel bends is to produce asymmetric flow both upstream and downstream from the bend, a factor which may be important if the flow passes through some nearby transition structure. Also, a transverse flow is initiated along the bottom from the outer to the inner edge of the bend, because of the higher static head induced by the superelevation. This flow, superimposed on the forward component, produces a spiral secondary flow for some distance beyond the bend. The detailed hydraulics of flow in open channel bends is very complex, and little understood as yet. Model studies are commonly used in the design of important structures involving direction changes.

In general, the head loss induced in a bend appears from dimensional analysis to depend on certain parameters as follows:

$$H_b = K_b\frac{V^2}{2g} = \phi\left(N_R, \frac{R}{B}, \frac{D}{B}, \theta\right)\frac{V^2}{2g} \qquad (5\text{-}36)$$

θ being the total central angle of the bend. This head loss is understood to be over and above that from normal boundary friction along the length of the bend. The Froude number will have little effect as long as the flow is definitely subcritical.

The function of Eq. (5–36) is apparently quite complex and cannot as yet be expressed in terms of a single equation. The experimental curves obtained by A. Shukry represent the most reliable design data available at present. In general, K_b ranges from zero to slightly over unity. It increases with increasing values of θ, but decreases with increasing values of R/B and D/B. It decreases with increasing values of N_R, reaching a minimum at about $N_R = 160,000$, and then begins to increase with further increases in N_R, according to the results of Shukry.

5–6. Transitions in Supercritical Flow. The typical flow profiles in the supercritical regime as discussed previously are strictly valid only so long as the walls are parallel and rectilinear. Changes in wall direction (or any other disturbance in the channel) will set up standing waves which may materially affect the flow profile and the entire character of the flow. For this reason, the discussion of subcritical flow transitions in the foregoing article is also inapplicable to transitions in supercritical flow.

The chief physical difference between subcritical and supercritical flow is that the velocity of transmission of a gravity wave in an open channel is always smaller than supercritical velocities; thus any disturbances generated in such a channel will be washed downstream. This is illustrated by the ripple patterns that would be formed by a stone dropped into the surface of channels flowing at various velocities, as shown in Fig. 5–11.

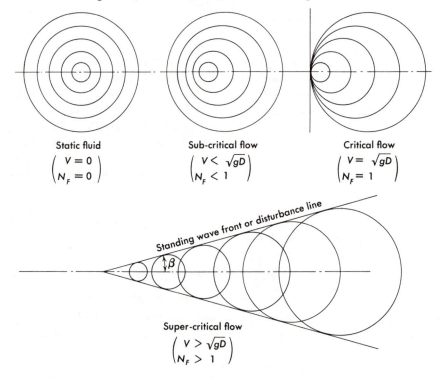

Fig. 5–11. Gravity wave patterns.

The relative velocity of a small gravity wave has previously been shown to be $\sqrt{gD}$, where D is the flow depth (valid only for small waves and relatively shallow channels, for which vertical acceleration components are unimportant). The absolute velocity of the wave is obtained by adding this vectorially to the flow velocity. The absolute wave velocity in the upstream direction is therefore

$$V_w = V - \sqrt{gD} \tag{5-37}$$

When V becomes equal to $\sqrt{gD}$, at critical flow, a standing wave front is thus produced, normal to the flow. When V exceeds the wave velocity, this standing wave front, also called a *disturbance line* or *shock wave* is washed downstream somewhat, the angle β obviously decreasing as the flow velocity increases. For a certain particle P on a ripple in supercritical flow, it is obvious that if P is located on the wave front, its absolute velocity is the vector sum of the radial wave velocity $\sqrt{gD}$ and the forward velocity V.

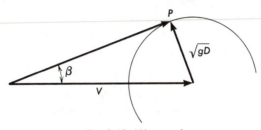

Fig. 5–12. Wave angle.

Thus, from the geometry of the velocity vector diagram (Fig. 5–12), it follows that the wave angle β is given by

$$\sin \beta = \frac{\sqrt{gD}}{V} = \frac{1}{N_F} \tag{5-38}$$

These relationships are all perfectly analogous to sound wave phenomena in gas flow problems. Thus the critical velocity or gravity wave velocity corresponds to the celerity of a sound or pressure wave. Subcritical and supercritical open channel flow correspond to subsonic and supersonic gas flow. The standing oblique wave in a high velocity water flow is mechanically similar to the shock wave at the nose of a supersonic missile. Disturbances associated with crossing the sonic barrier are analogous to the unstable conditions near critical flow, and so on. This "gas analogy" has proved very fruitful in the study of both open channel flow and high-speed gas flows (or high-speed motions of objects through gases).

However the analogy becomes somewhat distorted when the open channel waves become large. In such cases, the wave velocity becomes greater than $\sqrt{gD}$, and the wave angle somewhat larger than indicated in Eq. (5–38). Consider a standing wave generated by a channel wall deflection, as shown in Fig. 5–13. (Such a wave will be positive if the deflection is into the flow, negative if away from it.) As the flow traverses the wave front, the stream-

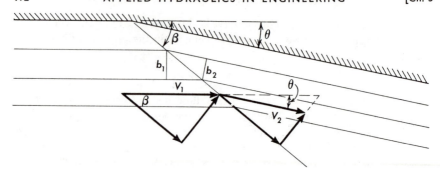

Fig. 5–13. Standing wave at wall deflection.

lines are deflected through the angle θ, the depth increasing from D_1 to D_2 and the velocity decreasing from V_1 to V_2. The components of V_1 and V_2 parallel to the wave front must be equal, however, since there is no shearing stress in that direction. That is,

$$V_1 \cos \beta = V_2 \cos (\beta - \theta) \tag{5–39}$$

The continuity and momentum equations can be written between sections upstream and downstream from the wave front, along a direction normal to the wave front (see Fig. 5–14).

The continuity equation is

$$D_1 V_1 \sin \beta = D_2 V_2 \sin (\beta - \theta) \tag{5–40}$$

The momentum equation is

$$\frac{\gamma}{2} D_1{}^2 + \frac{\gamma}{g} D_1 V_1{}^2 \sin^2 \beta = \frac{\gamma}{2} D_2{}^2 + \frac{\gamma}{g} D_2 V_2{}^2 \sin^2 (\beta - \theta) \tag{5–41}$$

Substituting $V_2 = V_1(D_1/D_2)[\sin \beta / \sin (\beta - \theta)]$ from Eq. (5–40) in Eq. (5–41) gives

$$\frac{V_1{}^2}{g} \left[D_1 \sin^2 \beta - D_2 \left(\frac{D_1}{D_2}\right)^2 \sin^2 \beta \right] = \frac{1}{2} (D_2{}^2 - D_1{}^2)$$

$$\frac{V_1{}^2 \sin^2 \beta}{g} (D_1) = \frac{D_2}{2} (D_2 + D_1)$$

The wave angle is thus given by

$$\sin \beta = \frac{\sqrt{g D_2}}{V_1} \sqrt{\left(\frac{1}{2}\right)\left(1 + \frac{D_2}{D_1}\right)} = \frac{\sqrt{g D_1}}{V_1} \sqrt{\left(\frac{1}{2}\right)\left(\frac{D_2}{D_1}\right)\left(1 + \frac{D_2}{D_1}\right)} \tag{5–42}$$

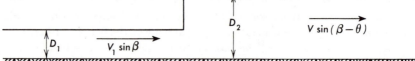

Fig. 5–14. Standing wave front.

or by

$$\sin \beta = \frac{1}{N_F} \sqrt{\left(\frac{1}{2}\right)\left(\frac{D_2}{D_1}\right)\left(1 + \frac{D_2}{D_1}\right)} \qquad (5\text{--}43)$$

The depth ratio D_2/D_1 is greater than unity for a positive wave. For small disturbances, the ratio is close to unity, and Eq. (5–43) simplifies to Eq. (5–38).

The normal component of approaching velocity is

$$V_1 \sin \beta = \sqrt{gD_1} \sqrt{\left(\frac{1}{2}\right)\left(\frac{D_2}{D_1}\right)\left(1 + \frac{D_2}{D_1}\right)} \qquad (5\text{--}44)$$

This must also be the wave celerity, since the standing wave is in equilibrium with this flow component. This expression simplifies to $\sqrt{gD_1}$ for small disturbances.

Equation (5–43) can be rearranged to yield the depth ratio D_2/D_1 if desired, as follows:

$$\frac{D_2}{D_1} = \left(\frac{1}{2}\right)\left[-1 + \sqrt{1 + (8 \sin^2 \beta)N_F{}^2}\right] \qquad (5\text{--}45)$$

Another expression for depth ratio is obtained by combining Eqs. (5–39) and (5–40):

$$\frac{D_2}{D_1} = \frac{V_1 \sin \beta}{V_2 \sin (\beta - \theta)} = \frac{V_1 \sin \beta}{[V_1 \cos \beta / \cos (\beta - \theta)] \sin (\beta - \theta)} = \frac{\tan \beta}{\tan (\beta - \theta)} \qquad (5\text{--}46)$$

Then, by combining Eqs. (5–45) and (5–46), an expression relating the wave angle to the approach flow Froude number and the channel deflection angle is obtained:

$$\frac{\tan \beta}{\tan (\beta - \theta)} = \left(\frac{1}{2}\right)\left[-1 + \sqrt{1 + (8 \sin^2 \beta)N_F{}^2}\right] \qquad (5\text{--}47)$$

This equation must be solved by trial to obtain the disturbance angle for given channel conditions. Then, after β is determined, the depth ratio is obtained from Eq. (5–45) and the velocity V_2 from the continuity equation, Eq. (5–40).

Consider now what happens when the shock wave strikes the opposite wall, assumed to be straight, as in Fig. 5–15. The wave is reflected back and forth between walls down the channel, somewhat as shown, in order A to B to C, etc. The initial wave angle β_1 is determined by the approach flow Froude number, N_{F_1}, and the deflection angle θ. When the wave strikes the opposite wall at B, another wave BC is set up; in this case the disturbance is generated at B by the wall which makes an angle θ with the direction of flow in the region ABC. Flow in this region has a Froude number N_{F_2}, which is less than N_{F_1} since the depth increased across wave front AB. The

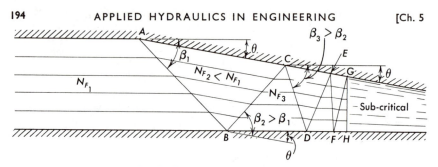

Fig. 5–15. Reflection of shock wave.

wave angle β_2 of the reflected wave BC is determined by N_{F_2} and θ, and will be larger than β_1 since N_{F_2} is less than N_{F_1}. A similar process is repeated at C, D, etc. In each case the angles β, which are measured between the wave front and the direction of approach flow, become larger as the Froude numbers decrease.

If the convergent channel is long enough, the wave fronts tend to come together finally at right angles to the flow, forming in essence a hydraulic jump to a subcritical flow condition downstream. If the convergence occurs on both sides of the channel at equal deflection angles, θ, the shock waves meet at the center line and each is reflected by the other in the same way as at a boundary. Thus, the wave pattern on *each* side of the center line is identical with that in Fig. 5–15, as sketched in Fig. 5–16. If the angles θ on the two walls are unequal, then the shock waves are of different intensities and an asymmetrical pattern will be produced which is more complicated to analyze, but which, if needed, could be worked out from the same basic principles.

The situation is not quite the same in a divergent channel, since a negative shock wave is not possible; that is, an expansion wave front cannot be assumed vertical, but must extend over a substantial distance (in like manner, a negative hydraulic jump is impossible). Individual streamlines (except the bounding streamline) will follow curvilinear paths, and equal depth lines will fan out from the boundary discontinuity. This is all sketched

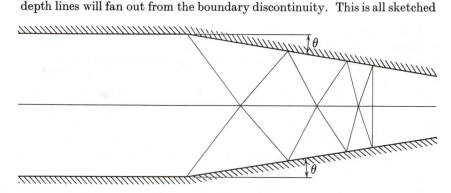

Fig. 5–16. Waves in symmetrically converging channel.

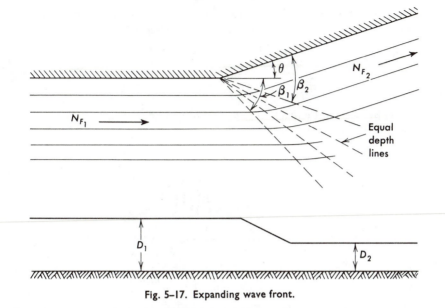

Fig. 5–17. Expanding wave front.

in Fig. 5–17. Since there is only a gradual change in depth, the initial and final wave angles may be computed by

$$\sin \beta_1 = \frac{1}{N_{F_1}} \quad \text{and} \quad \sin \beta_2 = \frac{1}{N_{F_2}} \tag{5–48}$$

Neglecting energy losses through the expansion, the sum of the depth and velocity heads at each end is the same. That is,

$$1 + \frac{V_1^2}{2gD_1} = \frac{D_2}{D_1} + \left(\frac{V_2^2}{2gD_2}\right)\left(\frac{D_2}{D_1}\right) = \left(\frac{D_2}{D_1}\right)\left(1 + \frac{V_2^2}{2gD_2}\right)$$

Rearranging gives

$$\frac{D_2}{D_1} = \frac{N_{F_1}^2 + 2}{N_{F_2}^2 + 2} \tag{5–49}$$

However, to obtain a relation between N_{F_1}, θ, and the wave dimensions, corresponding to Eqs. (5–45) and (5–47) for the positive wave, the effect of the *gradual* change in surface elevation and streamline direction must be considered. This specific problem can conveniently be approached in terms of the more general case of a gradually changing section, either converging or diverging, as distinct from the condition of a sudden directional change, as assumed in the analysis thus far.

Example 5–4

A high-velocity flow in an open channel encounters an abrupt positive wall deflection of 20°, which forms a transverse shock wave. The depths immediately upstream and downstream from the wave are measured to be 2.0 ft and 5.0 ft, respectively. What is the channel discharge, in cfs per ft of width?

Solution. From Equation (5–46) and Fig. 5–13:

$$\frac{D_2}{D_1} = \frac{5}{2} = \frac{\tan \beta}{\tan (\beta - \theta)} = \frac{\tan \beta}{\tan (\beta - 20°)}$$

Solving by trial (or by quadratic), $\beta = 36.5°$ or $73.5°$.
From Equation (5.44)

$$V_1 \sin \beta = \sqrt{gD_1} \sqrt{\frac{1}{2}\left(\frac{D_2}{D_1}\right)\left(1 + \frac{D_2}{D_1}\right)}$$

$$= \sqrt{32.2(2)\left(\frac{1}{2}\right)\left(\frac{5}{2}\right)\left(\frac{7}{2}\right)} = 16.8 \text{ fps}$$

$$V_1 = \frac{16.8}{\sin \beta} = \frac{16.8}{\sin 36.5°} = 28.3 \text{ fps}$$

or

$$V_1 = \frac{16.8}{\sin 73.5°} = 17.5 \text{ fps}$$

Thus, two different values of discharge could produce the depth ratio specified.

$$q = D_1 V_1 = 2V_1$$
$$= 56.6 \text{ cfs/ft or } 35.0 \text{ cfs/ft. } Ans.$$

5–7. Gradual Transitions in Supercritical Flow. If the change in cross-section is produced gradually, then it is possible to derive a general expression for the flow profile through the transition. It may be considered as made up of an accumulation of small surface displacements corresponding to differential directional changes.

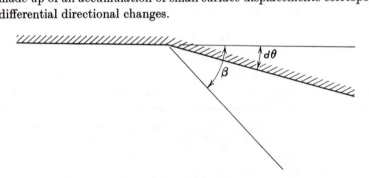

Fig. 5–18. Wave front at small direction change.

Consider the small wave created by a change in direction $d\theta$, as in Fig. 5–18. This will correspond to depth change dD and velocity change dV. Since the depth change is small, it is permissible to use the approximate equations:

$$\sin \beta = \frac{1}{N_F}$$

and

$$V \sin \beta = \sqrt{gD}$$

The momentum equation across the small wave front becomes

$$\frac{1}{2} D^2 + \frac{1}{g} DV^2 \sin^2 \beta = \frac{1}{2}(D + dD)^2 + \frac{1}{g}(D + dD)[V \sin \beta - d(V \sin \beta)]^2$$

Simplifying, and dropping differentials of higher order yields

$$D \, dD - \frac{D}{g}(2V \sin \beta) \, d(V \sin \beta) + \frac{1}{g}(V^2 \sin^2 \beta) \, dD = 0$$

Combining terms gives

$$dD \left[D + \frac{(V \sin \beta)^2}{g} \right] = \frac{2D(V \sin \beta)}{g} d(V \sin \beta)$$

Since $V \sin \beta$ is approximately $\sqrt{gD}$, this becomes

$$\frac{d(V \sin \beta)}{dD} = \frac{g}{(V \sin \beta)} \tag{5–50}$$

This is the differential equation of the flow profile in terms of the relation between depth, velocity, and wave angle. To relate these factors to the direction change θ, consider the velocity vectors on the two sides of the disturbance line in Fig. 5–19. As noted before, the tangential components of the two velocity vectors must be equal. The change in normal component $d(V \sin \beta)$ is thus determined by the geometry of the vector triangles. By the law of sines,

$$\frac{d(V \sin \beta)}{d\theta} = \frac{V}{\sin (\pi/2 + \beta - d\theta)} = \frac{V}{\cos (\beta - d\theta)} \cong \frac{V}{\cos \beta} \tag{5–51}$$

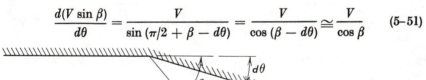

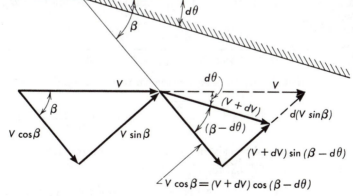

Fig. 5–19. Velocity relations at wavelet.

Combining Eqs. (5–50) and (5–51) gives

$$\frac{g}{V \sin \beta} \, dD = \frac{V}{\cos \beta} \, d\theta$$

The differential equation of the flow profile thus is

$$dD = \frac{V^2 \tan \beta}{g} \, d\theta \tag{5–52}$$

The term $\tan \beta$ can be expressed in terms of depth and velocity as follows

$$\tan \beta = \frac{\sin \beta}{\sqrt{1 - \sin^2 \beta}} = \frac{1/N_F}{\sqrt{1 - 1/N_F{}^2}} = \frac{1}{\sqrt{N_F{}^2 - 1}} = \frac{1}{\sqrt{V^2/gD - 1}} \tag{5–53}$$

Thus,

$$dD = \left(\frac{V^2}{g}\right)\left(\frac{\sqrt{gD}}{\sqrt{V^2 - gD}}\right) d\theta = \frac{V^2}{2g}\sqrt{\frac{2D}{V^2/2g - D/2}} \, d\theta \tag{5–54}$$

The velocity is a function of depth, depending in an unknown way upon channel friction. However, if this effect is neglected, and the specific head $D + V^2/2g$ therefore assumed constant, Eq. (5–54) can be put in integrable form. Let $E = D + V^2/2g$ be constant. Then,

$$dD = (E - D)\sqrt{\frac{2D}{E - (3/2)D}} \, d\theta$$

or

$$\theta = \int_{D_1}^{D} \frac{1}{E - D}\sqrt{\frac{E - (3/2)D}{2D}} \, dD \tag{5–55}$$

This equation has been integrated by von Kármán as follows:

$$\theta = \sqrt{3} \tan^{-1} \sqrt{\frac{D}{(2/3)E - D}} - \tan^{-1}\frac{1}{\sqrt{3}}\sqrt{\frac{D}{(2/3)E - D}} - \theta_1 \tag{5–56}$$

in which θ_1 is simply the value of the first two terms for $D = D_1$. Note also that the function

$$\frac{D}{(2/3)E - D} = \frac{D}{(2/3)(D + V^2/2g) - D} = \frac{1}{V^2/3gD - 1/3} = \frac{3}{N_F{}^2 - 1}$$

Equation (5–56) may thus be written, in terms of N_F,

$$\theta = \sqrt{3} \tan^{-1} \sqrt{\frac{3}{N_F{}^2 - 1}} - \tan^{-1} \sqrt{\frac{1}{N_F{}^2 - 1}} - \theta_1 \tag{5–57}$$

The term $\theta + \theta_1$ obviously ranges from $0°$ for an infinite Froude number to $(\sqrt{3} - 1)(90°) = 65° \, 53'$ for $N_F = 1$.

The function is plotted in Fig. 5–20. From the curve, values of N_F or D can be obtained for any value of θ, and vice versa. The angle θ_1 is first

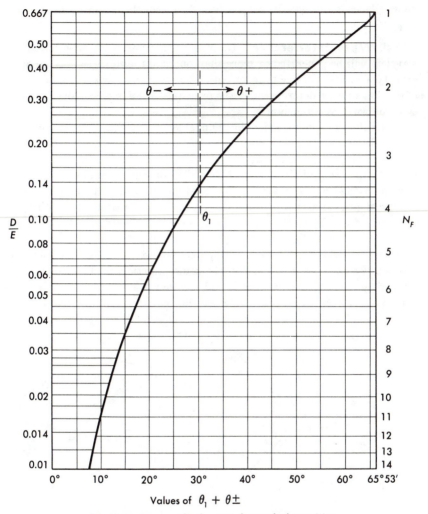

Fig. 5–20. Flow profile function for gradual transition.

obtained from the curve for the initial Froude number N_{F_1}. The angle θ is added to or subtracted from θ_1 for the abscissa, accordingly as the disturbance line is positive or negative.

Equation (5–57) may likewise be applied to the case of the expansion wave produced by the sharp corner illustrated in Fig. 5–17, and described by Eqs. (5–48) and (5–49), since the streamlines follow a gradual transition even though the boundary does not. Thus the downstream Froude number N_{F_2} can be determined from Eq. (5–57) for the given total deflection angle θ. The total angular region, measured at the corner in which the depth change takes place, is computed by $\beta_1 + \theta - \beta_2$, in which β_1 and β_2 are given by Eqs. (5–48).

5–8. Wave Interference. In most structures, of course, waves or wavelets will be generated at more than one point, and these may be of the same type or the opposite type, or both. The resulting surface characteristics of the flow may be determined by the principles of wave mechanics, since this is a true wave phenomenon.

As an illustration, consider the channel shown in Fig. 5–21, in which the flow is deflected through a small angle $d\theta$. On one side a positive wavelet is generated, on the other a negative. The disturbance angle β_1, as measured from the direction of the approach flow field of Froude number N_{F_1}, is the same on both sides; however, the flow field set up by the expansion wave is of smaller depth, and that by the positive wave of greater depth, than the approach flow. Thus these two fields may be characterized by Froude numbers $N_{F_2}{}^-$ and $N_{F_2}{}^+$ respectively.

The flow field $N_{F_2}{}^-$, encountering the positive wave CD, is deflected through the same angle $d\theta$ and in the same direction as the field N_{F_1}, encountering the positive wave BC, and the depth is increased by the same amount by which it had previously been decreased. Similarly, the flow field

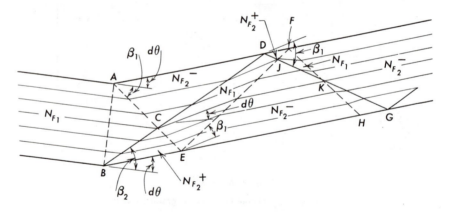

Fig. 5–21. Wave interference.

$N_{F_2}{}^+$, encountering the negative wave CE, is deflected through $d\theta$ and the depth is decreased to its original value. Consequently, downstream from CDE, the flow is restored to its original Froude number N_{F_1}, but is at an angle $d\theta$ with the boundaries. Note that positive fronts deflect the flow toward the front, and negative fronts away from the front.

When this flow field encounters the reflections of the waves CD and CE, it is again deflected through $d\theta$, thus becoming parallel to the wall. Flow fields corresponding to $N_{F_2}{}^+$ and $N_{F_2}{}^-$ are again established, but on the sides of the channel opposite their previous location. The angle β_1 for the reflections is the same as for the original disturbance lines, since it is determined by the magnitude of N_{F_1} as before, but it is measured with respect to the direction of the new approach flow. Thus EF tends to converge toward BD, and DG to diverge from AE.

A similar analysis could be made for the further reflections at F and G. As long as the positive and negative waves do not merge, there will be alternating regions of higher-than-normal and lower-than-normal depths established along both walls, with flow parallel to the walls. In the interior zones, such as $CDJE$, the flow will be at the original approach Froude number N_{F_1}, but its direction will be at $+d\theta$ or $-d\theta$ with the walls.

When positive and negative waves merge and cross, as at K, or if further boundary deflections are introduced, then the wave patterns become very complex. The principles already described are sufficient to determine directions of streamlines, wave lines, and flow depths, for any combination of waves, but the detailed analysis may become extremely tedious. A graphical method, known as the *method of characteristics* has been developed for computation of such complex wave patterns and may be employed when necessary.

However, the principle of wave superposition can be used in design to eliminate many of the downstream effects of the transition. This is illustrated in Fig. 5-22. The directional change at one side of the channel is offset so as to correspond with the reflection point for the wave coming from the opposite

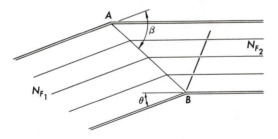

Fig. 5-22. Cancellation of wave front.

side. The waves cancel, so that there is no further wave disturbance downstream from the standing wave front AB. Of course, this requires that the approach Froude number N_{F_1} be known with sufficient accuracy for the calculation of β and the location of point B.

One other important aspect of wave combinations should be noted. Wavelets formed along a gradually curving wall will tend to diverge from one another if the boundary curves *away from* the flow, as in Fig. 5-23, because of both the decreasing depth (an increase in Froude number implies a decrease in the wave angle β) and the diverging boundary geometry.

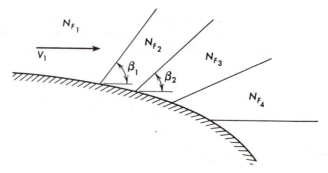

Fig. 5–23. Diverging wavelets.

On the other hand, a boundary gradually curving *into* the flow will cause increasing values of β and converging disturbance lines. Therefore, the small disturbance lines, at some distance from the boundary, will merge into a single large shock front, which will behave thereafter as a typical shock wave, reflected back and forth down the channel, essentially as though the total change in boundary direction had been accomplished instantaneously instead of gradually (see Fig. 5–24).

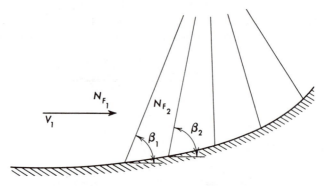

Fig. 5–24. Converging wavelets.

5–9. Design of Contractions in Supercritical Flow. Surprisingly, the most effective design of a contraction in supercritical flow is one with straight walls. This is because the shock wave produced by the convergence of disturbance lines from a curved contraction transition would have a depth magnitude dependent primarily on the total deflection angle. The minimum deflection angle for a transition of given length and reduction of width is the one for a straight-walled transition, as illustrated in Fig. 5–25.

The proper length for a straight-walled transition is that length which will just suffice for the positive waves to reach across the channel at the point where the contraction is completed. As shown in Fig. 5–26 the total

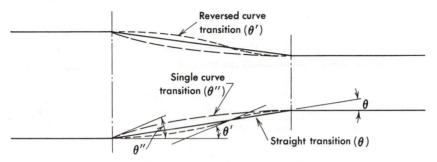

Fig. 5–25. Contraction design in rapid flow.

transition length is

$$L = \frac{B_1}{2 \tan \beta_1} + \frac{B_2}{2 \tan (\beta_2 - \theta)} \qquad (5\text{--}58)$$

The angle β_1 and the transition Froude number N_{F_2} are determined by the approach Froude number N_{F_1} and the angle θ. The angle β_2 in turn depends on N_{F_2} and θ. For an assumed value of θ, the length L can be computed from Eq. (5–58). If the value of θ has been chosen correctly, the waves will intersect the wall right at the boundary corner, in which case, the length is also given by

$$L = \frac{B_1 - B_2}{2 \tan \theta} \qquad (5\text{--}59)$$

The proper value of θ is thus determined by trial and error, until Eqs. (5–58) and (5–59) agree. For this condition, there is theoretically no further wave reflection down the channel.

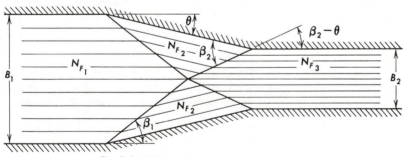

Fig. 5–26. Wave elimination in contraction.

5–10. Design of Expansions in Supercritical Flow. As shown previously, steep wave fronts will not exist in an expanding supercritical flow. However, when the expansion is terminated at the downstream end, the walls of the exit channel set up positive waves which may become a problem in that channel unless the expansion is properly designed.

By using the principles of wave mechanics to design the transition geometry precisely, it is theoretically possible to develop uniform rectilinear flow just at the outlet of the transition, thus eliminating all disturbance lines in the exit channel. However, this design turns out to require an impracticably long transition. Accordingly, in most actual structures, it is necessary to use some design other than the ideal, which would be economically practicable and yet not cause objectionable downstream disturbances.

The principles of similitude and dimensional analysis have been employed in model-testing programs to develop designs which would meet these criteria. The basic functional relationship, neglecting head losses in the expansion, is

$$\frac{D}{D_1} = f\left(\frac{x}{B_1}, \frac{y}{B_1}, N_{F_1}\right) \tag{5-60}$$

in which x and y are the horizontal coordinates of the water surface at any point at which the depth is D, measured from an origin at the center line of the beginning of the expansion. B_1, D_1, and N_{F_1} are the approach width, depth, and Froude number, respectively. The model studies were directed to the end of obtaining side-wall transition designs which would give flow surfaces free from gross disturbances and approximately rectilinear flow at the outlet.

The following tentative conclusions were drawn from results of these studies:

1. The approach channel should be at least $5D_1$ in length, in order to assure hydrostatic pressure distribution at the inlet to the transition.
2. If a straight flaring sidewall is used, it is impossible to eliminate standing waves in the downstream channel. However, these waves will not be of seriously objectionable height if the sidewall divergence angle is given by

$$\tan\theta \leqq \frac{1}{3\sqrt{N_{F_1}}} \tag{5-61}$$

3. For a gradually flaring sidewall, approximately 90 per cent of the theoretical flow (as based on an abrupt expansion, using the method of characteristics) will be contained in a boundary having the following equation:

$$\frac{B}{B_1} = 1 + \left(\frac{x}{B_1 N_{F_1}}\right)^{3/2} \tag{5-62}$$

where B is the expansion width at distance x from the beginning of the expansion.

4. If the enlargement is to be only to a definite width $B_2 \leqq 4B_1$, then the equation of the flare should be modified to

$$\frac{B}{B_1} = 1 + \frac{1}{4}\left(\frac{x}{B_1 N_{F_1}}\right)^{3/2} \tag{5-63}$$

The point at which the reverse curve should begin, as well as the form of the reverse curve, depends on the width ratio B_2/B_1. The experimental curves shown in Fig. 5–27 may be used to obtain suitable transition curves for the complete enlargement.

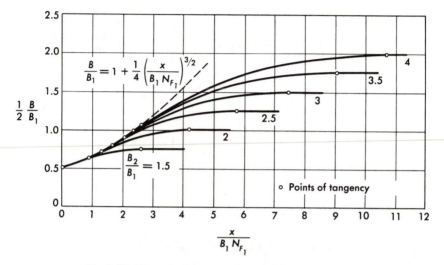

Fig. 5–27. Dimensionless curves for rapid-flow expansions.

5–11. High-Velocity Flow Around Channel Bends.

Consider a circular curve in a channel in which the flow is supercritical. At the beginning of the curve, a negative and a positive disturbance will be generated, as sketched in Fig. 5–28, each making an angle β_1 with the approaching flow, the value determined by the approach Froude number. On the outside of the curve, the surface will gradually rise, in accordance with Eq. (5–56) or Fig. 5–20. Similarly, the surface drops along the inside of the bend. When the negative and positive disturbances cross each other, their respective local deflections are added algebraically to give the actual deflection. When the first negative disturbance line reaches the outer wall and is reflected, the surface at the outer wall begins to drop. Similarly, when the positive wavelet generated at A reaches the inner wall at E, the surface along the inner wall, which had been falling, begins to rise. Thus the maximum surface depth is attained at D and the minimum at E.

The negative wave BD is a straight line until reaching the region of positive disturbance at C; thereafter it is deflected downstream somewhat. The exact location of D could be determined by wave mechanics; however, it is sufficiently accurate usually to approximate the distance AD by AD':

$$AD \cong AD' = \frac{B}{\tan \beta_1} \qquad (5\text{–}64)$$

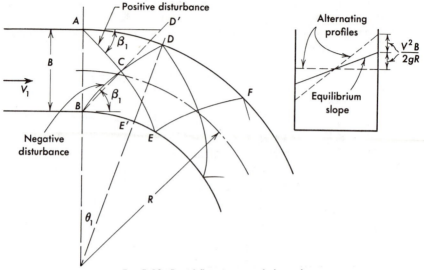

Fig. 5–28. Rapid flow in curved channel.

The corresponding central angle is given approximately by

$$\tan \theta_1 = \frac{B/\tan \beta_1}{R + B/2} \qquad (5\text{--}65)$$

This is also the deflection angle from the flow direction at A to that at D, and so can be used in Fig. 5–20 to obtain the maximum surface elevation along the wall at D.

It often is sufficiently accurate also to assume that the minimum depth at E is actually located at E', as defined by the same angle θ_1. If this is assumed, then it can be further inferred that the points F and G, where new reflection points occur, are defined by a central angle $2\theta_1$, the next pair by $3\theta_1$, etc.

The distance between successive points of maximum depth is therefore defined by an angle $2\theta_1$, and the same for successive points of minimum depth, points of maximum and minimum occurring alternately on opposite sides of the channel.

This wave pattern, in effect, is superimposed on the equilibrium super-elevation that the surface would assume because of centrifugal forces. This superelevation (excess of outer depth over inner depth) is given by Eq. (5–30) as $V^2 B/gR$. It is found experimentally that the actual maximum rise attained on the outer wall is about twice the equilibrium rise. Thus the disturbance pattern oscillating about the equilibrium depth has a wave length of $2\theta_1$ and amplitude of $V^2 B/2gR$ (see Fig. 5–28).

This disturbance pattern continues into the downstream tangent beyond the curve. The linear distance between successive maxima or minima in this reach is given by

$$L = \frac{B}{\tan \beta_1} \tag{5-66}$$

However, the amplitude of this disturbance may be as much as V^2B/gR, because the sudden elimination of the curvature introduces a new disturbance of equal, though opposite, magnitude to that at the entrance to the bend. This of course depends upon whether or not the new disturbance is in phase with the disturbance pattern in the bend.

Three different methods have been used to produce improved conditions over the conditions described above for supercritical flow in simple curves, and each is again based on the judicious application of the principles of wave interference.

The most effective method seems to be by the expedient of banking the channel bottom. The effect of this is to impart centripetal forces to the fluid particles opposing the centrifugal forces from the curvature. If these are made to balance, then there will be no change in the force of the side walls on the flow; consequently, no disturbance is caused in the flow by the walls. In an unbanked transition, the flow is "turned" by the centripetal force of the walls on the water, and this is the cause of the typical wave phenomena.

Complete equilibrium, of course, would require a cross slope equal to V^2/gR, which corresponds to the centrifugal rise in water surface as given by Eq. (5-30). The banking is usually most effectively accomplished by dropping the inside bottom elevation and leaving the outside at the same level. A spiral transition should be used before and after the main curve to accomplish the desired banking, in order to eliminate all potential sources of surface disturbance. The recommended length of transition is

$$L_t = 15 \frac{V^2B}{gR} \tag{5-67}$$

The equation of the transition may be taken as

$$rL = RL_t = \text{constant} \tag{5-68}$$

where L is the length along the transition to the point where the local radius is r. This equation implies a linear increase in the bed cross-slope with distance along the transition.

Another expedient, somewhat less costly in general, is to use a compound curve for the bend, as shown in Fig. 5-29. If the total direction change is θ, then

$$\theta = 2\theta_T + \theta_c \tag{5-69}$$

where θ_T is the angle subtending each transition curve and θ_c is the central angle for the main curve. Also, if the transition length is made to equal

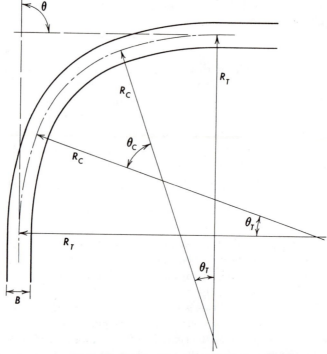

Fig. 5–29. Compound curve at bend.

one-half wave length, the following approximate equation applies:

$$\tan \theta_T = \frac{B/\tan \beta_1}{R_T + B/2} \tag{5-70}$$

where R_T is the radius of the transition curve. The transition radius is to be selected of such magnitude as to generate a disturbance one-half that caused by the main curve. Since the disturbance height is inversely proportional to the radius of curvature, as in Fig. 5–28, the transition radius should be

$$R_T = 2R_c \tag{5-71}$$

A transition curve of radius and central angle given by Eqs. (5-70) and (5-71) will therefore produce a disturbance of the same wave length as the main curve, but offset by one half-wave-length in phase. It thus attains its maximum depth just at the point where the disturbance due to the main curve would be generated, but this maximum depth is just the value required for centrifugal-centripetal equilibrium, so that no additional wall force is present and therefore no disturbance is generated. Or, in other terms, the negative disturbance from the beginning of the inner transition reaches the outer wall just in time to cancel the positive disturbance that would be generated there.

The cross-slope then remains at the equilibrium value through the main curve. An identical transition at the downstream end will similarly eliminate the disturbances that would otherwise be produced in the downstream tangent.

A third method consists of placing diagonal sills on the floor of the channel at the entrance to the curve, and again at the downstream end. The function of these sills is to produce an interference pattern which will mitigate the wave disturbance pattern produced by the circular curve. They impart a lateral velocity component to the lower layer of flow, which is spread throughout the entire body of flow by the process of momentum exchange.

Theoretical analyses, based on the momentum principle considering the hydrodynamic forces of the water on the sills, have yielded results which are only partially verified by experiment. Accordingly, model tests are usually desirable for an installation of this type.

At best, sills will not be as effective as transition curves or banking. Furthermore, they will tend to cause lodgment of debris if the stream contains such, and also they are subject to damage from local high velocities, which may cause erosion by sand near the bed, or even may cause cavitation. Their main application is as a corrective expedient in existing channel bends where wave disturbances have proved a serious problem.

5–12. Flow near Control Sections.   As defined here, a *control section* is a section in a channel at which the flow changes from subcritical to supercritical, passing through critical depth. At such a section, there exists a unique relation between depth and discharge, for a given cross-section shape. That is,

$$Q = \sqrt{g\,\frac{A^3}{B}} = f(D) \tag{5–72}$$

Inspection of Eq. (5–72) indicates immediately two main uses of the control section principle:

1. As a starting point in flow profile calculations, since the depth at this point can be computed for the given discharge.
2. As a means of measuring discharge, since the latter can be calculated directly from a simple depth measurement. This principle also can be applied in devices intended to control the flow to specified amounts.

However, a serious difficulty is encountered in the actual application of the control section theory. The derivation of Eq. (5–72) implicitly assumed that flow streamlines were all parallel and, therefore, that the pressure distribution was hydrostatic. Actually, the true specific energy of flow at a section is

$$E = \int_A dE = \frac{1}{Q\gamma} \int \left(\frac{p}{\gamma} + z + \frac{v^2}{2g}\right)\gamma v\, dA$$
$$= \frac{1}{Q} \int_A \left(\frac{p}{\gamma} + z + \frac{v^2}{2g}\right)v\, dA \tag{5–73}$$

wherein p, z, and v represent, respectively, the pressure, elevation above bed, and velocity of the water flowing through section dA. Obviously, in order to obtain E as a function of the depth D, the functional relation of p and v to depth must be known. In general, this information may be difficult or impossible to obtain analytically.

If the flow is curvilinear, the pressure will vary with radius of curvature, according to the requirement for equilibrium with centrifugal forces. For rectilinear flow, this factor is absent and the only internal forces are those due to fluid weight and pressure, a condition which immediately yields the hydrostatic equation, in which

$$\frac{p}{\gamma} + z = \text{constant} = D \tag{5-74}$$

Equation (5–73) then becomes

$$E = \frac{1}{Q} \int D(v\,dA) + \frac{1}{Q} \int \frac{v^3\,dA}{2g}$$

$$= D + \left(\frac{1}{2g(AV)}\right)\left(\frac{V^2}{V^2}\right) \int v^3\,dA = D + \left(\frac{V^2}{2g}\right)\left(\frac{1}{A}\right) \int \left(\frac{v}{V}\right)^3 dA$$

$$= D + \alpha \frac{V^2}{2g} \tag{5-75}$$

where α is the kinetic energy distribution factor. It is common, though not strictly correct, to assume this factor is unity, in which case Eq. (5–75) reduces to the simple definition of E upon which Eq. (5–72) was based. But in curvilinear flow, except for certain special cases, the true specific energy is difficult or impossible to calculate. Consequently the true critical depth, for which the true specific energy becomes a minimum, is also difficult or impossible to determine. And since, as we have seen in section 1 of this chapter, the water surface tends to assume an infinite slope at a control section, the actual flow profile near such a section is quite definitely curvilinear. Therefore, the actual depth at a control section is not really given by Eq. (5–72) as heretofore assumed, but rather by $dE/dD = 0$, where E is given by Eq. (5–73).

This difficulty is not ordinarily troublesome in flow profile calculations, since discrepancy between true and assumed critical depth is rapidly obliterated as the profile calculations proceed upstream or downstream from the control section. If, however, a control section is used as the basis of a flow meter (and most open-channel flow meters *are* based on this principle— weirs, Parshall flumes, overfall flumes, sluice gates, etc.), then obviously the meter must be calibrated. The process of calibration in effect determines empirically the relation between true critical depth and discharge for the particular meter. Actually, most such meters do not utilize the depth right at the control section but rather that at some convenient point on the flow profile upstream from the control section, the latter of course being "controlled" by the critical depth.

The flow profile in regions of rapidly accelerating flow, as near control sections, can be approximately determined graphically by use of a flow net, which consists of an orthogonal system of streamlines and equipotential lines, constructed by trial-and-error adjustment to form approximately "square" figures. To be truly valid, the flow net procedure requires that the flow be two-dimensional and irrotational, a condition which is seldom fulfilled. Nevertheless a rapidly accelerating flow in a rectangular open channel approximates the condition closely enough for the flow net to give a rather good picture of the stream dynamics in the region.

In irrotational flow, the Bernoulli equation applies at every point in the flow field. That is, if H is the total head available to maintain the flow, and if p, z, and v represent the pressure, elevation, and velocity at any arbitrary point in the flow, then

$$\frac{p}{\gamma} + z + \frac{v^2}{2g} = H = \text{constant} \tag{5-76}$$

Also, the continuity equation applies throughout the flow field. That is, for steady, incompressible, two-dimensional flow,

$$vs = \Delta Q = \text{constant} \tag{5-77}$$

where s is the spacing of adjacent streamlines in the flow net, in a region where the velocity is v. The flow net is so constructed that identical increments of flow ΔQ are conveyed between each pair of adjacent streamlines. If Q is the total discharge, then

$$\Delta Q = \frac{Q}{n_f} \tag{5-78}$$

where n_f is the number of flow channels in the flow net. Therefore, if the total head and discharge are given for a particular flow, and if the flow net can be constructed, then the velocity at any point in the flow field can be determined from Eq. (5–78) and the pressure from Eq. (5–76).

Construction of the flow net is fairly simple if the bounding streamlines are known. However, in an open channel flow, one and sometimes two of the bounding streamlines are free surfaces, the exact profile of which is unknown. Nevertheless, such surfaces can be determined from the requirement that the pressure at every point on a free surface must be zero.

Thus, the pressure distribution determined from the flow net as outlined above must yield zero pressures at all points on any free surface. The location of the free surface can therefore be determined by trial and error to conform to this restriction. That is, a trial profile is selected for the free surface and a flow net constructed to correspond to it. Then the pressure distributions are determined from the flow net for selected sections. If the free surface pressures so determined are not zero, then the profile is adjusted as needed and the process repeated.

Once the correct profile is determined, then also the true pressure distribution is known, and can be utilized as needed in structural analyses or energy computations. The true critical depth is the actual depth yielded by the flow net at any section where the flow must change from tranquil to rapid, as at a point of sudden increase in slope.

As an example, consider the flow over a sharp-crested rectangular weir, as sketched in Fig. 5–30, for which the final flow net also has been constructed. At any point in the flow field, the total head, assuming frictionless, irrotational flow, is

$$H = \frac{p}{\gamma} + z + \frac{v^2}{2g} \tag{5-76}$$

The free surfaces, which form flow boundaries, must be determined by trial, conforming both to flow net requirements and to the requirement that the pressure thereon everywhere be zero. That is, nappe surfaces are first assumed, then the flow net is sketched to conform. Velocities are determined at various points by

$$v = \frac{\Delta Q}{s} = \frac{Q}{n_f s} \tag{5-79}$$

Then the corresponding pressures are computed by Eq. (5–76). If the nappe surfaces have been chosen correctly, the surface pressures will turn out to be zero; if not, they can be adjusted and the process repeated.

Finally, both pressure and velocity distribution curves can be plotted where needed. If desired, the relation for E and D_c could then be worked out from Eq. (5–73).

Since the flow net is based on potential flow theory, other means of potential flow computations may be used when feasible. The theoretical equations may be used if the boundary equations are amenable to their

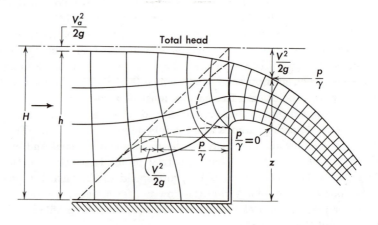

Fig. 5–30. Flow net for sharp-crested weir.

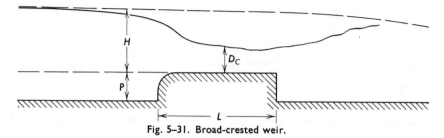

Fig. 5–31. Broad-crested weir.

solution. Otherwise, and more commonly, some numerical method of analysis, such as the relaxation method, can be applied. Electrical analogs may also be employed.

In the case of the sharp-crested weir, the rapidly contracting flow made it legitimate to ignore the effects of frictional resistance. On the other hand, if a transition from subcritical to supercritical flow is accomplished gradually, say by a small increase in bed slope, then it is sufficiently accurate to ignore the effects of flow curvilinearity and to base the profile calculations solely on energy and friction considerations.

In some situations, however, the effects of *both* friction and curvilinearity must be considered. The broad-crested weir, as sketched in Fig. 5–31, is a case in point.

If the tailwater level is sufficiently low, then critical depth will occur somewhere on the weir. The exact location depends on the weir profile, particularly on whether or not the upstream end is rounded, as well as on the slope and roughness of the crest, and it may well be in one of the zones of curvilinearity.

If it were assumed that D_c occurs at a section of hydrostatic pressure distribution, then the ordinary rectangular-channel critical flow formula could be used. That is,

$$Q = BD_c\sqrt{gD_c} = B\sqrt{g}\ D_c^{3/2}$$
$$= B\sqrt{g}\left(\frac{2}{3}H\right)^{3/2} = 3.09BH^{3/2} \qquad (5\text{–}80)$$

Actually the critical section may well be in a zone of strong curvilinearity; furthermore, boundary friction along the crest may substantially reduce the true specific energy from the assumed value of H by the time the flow reaches the critical section. Thus a more general formula would be

$$Q = CBH^{3/2} \qquad (5\text{–}81)$$

where the coefficient is dependent on various dimensionless parameters that may be evaluated empirically:

$$C = \phi\left(\frac{P}{H}, \frac{L}{H}, N_R, \text{shape, roughness}\right) \qquad (5\text{–}82)$$

The range of values of C obtained experimentally on actual broad-crested weirs is from about 2.67 to 3.05.

If the weir is long enough and the tailwater low enough, the flow becomes equivalent to that at a *free overfall*, which is another type of control section of frequent interest (see Fig. 5–32).

It is found experimentally that the true critical depth, which occurs at the brink, is somewhat less than the ordinary critical depth, D_c, obtained on the assumption of parallel flow, which is therefore located at some distance L_c upstream. Thus,

$$D_b = K_1 D_c \quad \text{and} \quad L_c = K_2 D_c \tag{5–83}$$

Both K_1 and K_2 vary somewhat with bed slope i_b, bed roughness, Reynolds number, etc. Various tests indicate that K_1 ranges from about 0.63 to about 0.72. A number of different theoretical approaches have been made to the problem, using analyses based on the momentum principle, potential theory, velocity distributions, etc., all leading to different results, but in fair agreement quantitatively. Rouse found that, for most cases, K_1 was fairly constant at about 0.715. If so, the free overfall can be used as a flow meter, of the equation

$$Q = B\sqrt{gD_c^{3/2}} = \left[\frac{\sqrt{g}}{(0.715)^{3/2}}\right] BD_b^{3/2}$$

$$= 1.65\sqrt{g}BD_b^{3/2} = 9.35BD_b^{3/2} \tag{5–84}$$

Rouse also found that K_2 varied between 3 and 4. The section at which D_c occurs, should be used as the starting point in flow profile computations. This is located, therefore, at about

$$L_c = 3.5\left(\frac{D_b}{0.715}\right) \cong 5D_b \tag{5–85}$$

Another form of control structure is the gate, several types of which are discussed in Art. 6–21. The sluice gate is another example. In most cases,

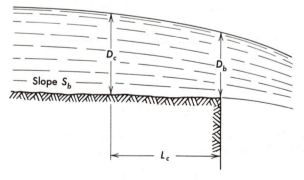

Fig. 5–32. Free overfall.

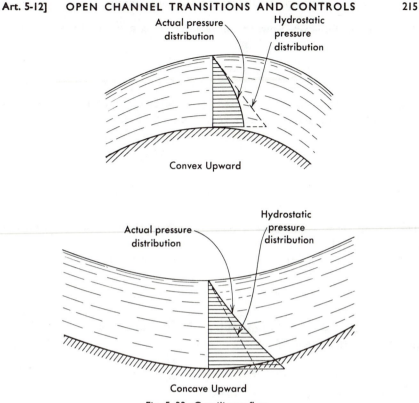

Fig. 5–33. Curvilinear flow.

the contraction coefficients and pressure distributions for gates may be determined most expeditiously by flow net analysis of the same sort as described above for the sharp-crested weir, or by analogous methods.

In many cases the total gate thrusts can be obtained by a simple momentum analysis, provided the contraction characteristics are known or predictable. However, this will not yield the distribution of pressure or the line of action of the resultant.

Finally, it remains true that the hydraulic characteristics of most gates must still be determined empirically, especially in those cases where the outlet conditions involve flow expansions and separations of uncertain location.

In some cases of curvilinear flow, whether or not at a control section, it may be possible to make an analysis in terms of the centrifugal forces on the individual streamlines, provided the local radii of curvature can be estimated. Figure 5–33 shows the effect of both concave and convex curvilinearity on the pressure distribution. As indicated, pressures are less than hydrostatic when the flow curvature is convex upward, and greater than hydrostatic when concave upward. Consider a mass of fluid in a curvilinear flow at the section where the streamlines become horizontal (see Fig. 5–34). Assuming the centrifugal force acts upward (curvature convex upward), the summation of forces yields

$$dp = \gamma \, dH - \rho \frac{v^2}{r} \, dH$$

where v is the streamline velocity and r is the streamline radius of curvature. Since both v and r may vary with H, this equation may be written

$$p_2 - p_1 = \gamma \int_{H_1}^{H_2} \left(1 - \frac{v^2}{gr}\right) dH$$

$$= \gamma(H_2 - H_1) - \frac{\gamma}{g} \int_{H_1}^{H_2} \frac{v^2 \, dH}{r} \qquad (5\text{–}86)$$

The second term would have a positive sign in the case of flow which is concave upward.

The pressure at depth H below a free surface can thus be written

$$p = \gamma H \pm \frac{\gamma}{g} \int_0^H \frac{v^2}{r} \, dH$$

In the special case of a constant radius of curvature and a uniform velocity distribution, this becomes

$$p = \gamma H \left(1 \pm \frac{v^2}{gR}\right) \qquad (5\text{–}87)$$

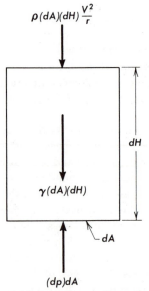

Fig. 5–34. Forces in curvilinear flow.

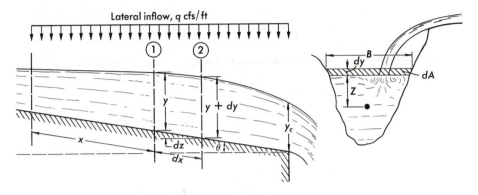

Fig. 5–35. Spatially varied flow.

5–13. Spatially Varied Flow. It has been assumed thus far that the open-channel discharge remains constant at all points through the channel reach under consideration. However, there are a number of practical situations in which inflow or outflow may occur laterally along the channel, causing the discharge to change continuously along the given reach. The conservation equations for this type of flow, known as *steady spatially varied flow*, are developed below.

Let Q be the discharge at a point x in the channel and q the lateral inflow or outflow (taken as positive or negative, respectively) per unit length along the channel, as shown in Fig. 5–35. The lateral inflow in distance dx is dQ, equal to $q\,dx$. At the point $(x + dx)$, the discharge is $Q + dQ = Q + q\,dx$.

The continuity equation is simply:

$$q = \frac{dQ}{dx} = V\frac{dA}{dx} + A\frac{dV}{dx} \qquad (5\text{–}88)$$

To obtain the momentum equation, the forces on the free body of fluid between secs. 1 and 2 must be equated to the change in flux of momentum between those sections. Assume streamlines parallel to the channel bed, hydrostatic pressure distribution, and uniform velocity distribution. In addition to the hydrostatic forces on the end areas, the component of fluid weight in the direction of flow and the frictional resistance to flow must be considered. It is possible that the lateral flow may enter (or leave) the channel at an angle ϕ with the channel velocity V, in which case the momentum component of the lateral flow must be included in the force balance.

Summing components in the direction of flow along the channel bed, the momentum equation states that:

$$\gamma\bar{z}A - \gamma(\bar{z} + dy)(A) + \gamma\sin\theta A\,dx - \tau_0 P\,dx$$
$$= \frac{\gamma}{g}(Q + dQ)(V + dV) - \frac{\gamma}{g}(QV) - \frac{\gamma}{g}q\,dx(U\cos\phi)$$

In the above equation, $\bar{z}$ is the distance from the water surface to the centroid of the area A, θ is the bed slope angle, P is the wetted perimeter, τ_0 is the peripheral shear stress, and U is the velocity of the lateral flow as it enters or leaves the channel. Dividing through by $\gamma A \, dx$

$$-\frac{dy}{dx} + \sin\theta - \frac{\tau_0 P}{\gamma A} = \frac{Q \, dV + V \, dQ}{gA \, dx} - \frac{qU \cos\phi}{gA}$$

Introducing the bed slope, i_b, defined as $\sin\theta = dz/dx$, the nominal friction slope, i_e, which is equivalent to $\tau_0 P/\gamma A$ and may be calculated approximately from a uniform flow equation such as the Manning formula, the above relationship becomes:

$$\frac{dy}{dx} = i_b - i_e + \frac{qU \cos\phi}{gA} - \frac{V}{g}\left(\frac{dV}{dx} + \frac{1}{A}\frac{dQ}{dx}\right)$$

Further manipulation of this equation, by expanding $\dfrac{dV}{dx}$ as $\dfrac{d}{dx}\left(\dfrac{Q}{A}\right)$, replacing dQ/dx by q, and substituting $B \, dy$ for dA, where B is the flow surface width, yields the following:

$$\frac{dy}{dx} = i_b - i_e + \frac{qUV \cos\phi}{gQ} - \frac{V}{g}\left(\frac{A \, dQ - Q \, dA}{A^2 \, dx} + \frac{A \, dQ}{A^2 \, dx}\right)$$

$$= i_b - i_e + \frac{qUV \cos\phi}{gQ} - \frac{Q}{gA^3}\left(2Aq - QB\frac{dy}{dx}\right)$$

Solving for $\dfrac{dy}{dx}$, and simplifying:

$$\frac{dy}{dx} = \frac{i_b - i_e + \dfrac{qUV \cos\phi}{gQ} - \dfrac{2Qq}{gA^2}}{1 - \dfrac{Q^2 B}{gA^3}} \tag{5-89}$$

Equation (5–89) may be considered as the general differential equation of the spatially varied steady flow profile, as derived from the momentum principle. When there is no lateral inflow or outflow, it simplifies to the standard equation for steady varied flow. At the downstream control, $y = y_c$, the slope dy/dx approaches the vertical, since the function $Q^2 B/gA^3$ approaches unity.

Specific evaluation of the flow profile for a given channel and lateral flow condition requires evaluation of Q, A, B, and i_e as functions of x. In principle the equation may then be integrated to give y explicitly in terms of x but the complexity of the function normally requires solution by numerical methods.

The energy equation may also be derived for the same conditions, as follows:

$$\gamma Q\left(dz + y + \frac{V^2}{2g}\right) + \gamma q\, dx\left(z_L + y_L + \frac{U^2}{2g}\right)$$

$$= \gamma (Q + q\, dx)\left(y + dy + \frac{(V + dV)^2}{2g}\right) + i_e\, dx(\gamma)\left(Q + \frac{dQ}{2}\right) + H_L \gamma q\, dx$$

In this equation, the term $(z_L + y_L)$ represents the piezometric head available initially in the lateral flow and $U^2/2g$ the velocity head. H_L is the head lost in the lateral flow as it enters or leaves the channel. Dropping higher-order differentials and simplifying:

$$dz + \frac{q\, dx}{Q}\left(z_L + y_L + \frac{U^2}{2g}\right) - i_e\, dx$$

$$= \frac{q\, dx}{Q}\left(y + \frac{V^2}{2g}\right) + d\left(y + \frac{V^2}{2g}\right) + H_L \frac{q\, dx}{Q}$$

Replacing dz by $i_b\, dx$, and dividing by dx

$$i_b - i_e = \frac{q}{Q}\left(y + \frac{V^2}{2g} - z_L - y_L - \frac{U^2}{2g} + H_L\right) + \frac{dy}{dx} + \frac{d}{dx}\left(\frac{Q^2}{2gA^2}\right)$$

The last term in the above equation is:

$$\frac{1}{2g}\frac{2Q}{A}\left(\frac{1}{A}\frac{dQ}{dx} - \frac{Q}{A^2}\frac{dA}{dx}\right) = \frac{Qq}{gA^2} - \frac{Q^2 B}{gA^3}\frac{dy}{dx}$$

Substituting in the above, and solving for dy/dx:

$$\frac{dy}{dx} = \frac{i_b - i_e - \dfrac{q}{Q}\left[\left(y + \dfrac{V^2}{2g}\right) - \left(z_L + y_L + \dfrac{U^2}{2g}\right) + H_L\right] - \dfrac{Qq}{gA^2}}{1 - \dfrac{Q^2 B}{gA^3}} \qquad (5\text{–}90)$$

Equation (5–90) is the flow profile equation for steady spatially varied flow as derived from the energy conservation principle. Like Eq. (5–89), it reduces to the standard varied flow equation when there is no lateral inflow or outflow.

Within the assumptions mentioned earlier, the two principles of momentum and energy conservation should both be satisfied in the actual flow process and thus should give equivalent expressions for the flow profile. Equations (5–89) and (5–90) may be modified, respectively, to the following:

$$\frac{dy}{dx} = \frac{(i_b - i_e) + \dfrac{q}{Q}\left(\dfrac{UV \cos \phi}{g} - 4\dfrac{V^2}{2g}\right)}{1 - \dfrac{Q^2 B}{gA^3}} \qquad (5\text{–}91)$$

and

$$\frac{dy}{dx} = \frac{(i_b - i_e) + \frac{q}{Q}\left(z_L + y_L + \frac{U^2}{2g} - H_L - y - 3\frac{V^2}{2g}\right)}{1 - \frac{Q^2 B}{g A^3}} \qquad (5\text{-}92)$$

Examination of Eqs. (5–91) and (5–92) indicates that, for them to be equivalent, the following relation must hold:

$$\frac{UV \cos \phi}{g} - 4\frac{V^2}{2g} = z_L + y_L + \frac{U^2}{2g} - H_L - y - 3\frac{V^2}{2g}$$

which expression yields an equation relating the flow profile to the head loss H_L.

$$y = \frac{V^2}{2g} - \frac{UV \cos \phi}{g} + \left(z_L + y_L + \frac{U^2}{2g}\right) - H_L \qquad (5\text{-}93)$$

Specific applications of these equations depend on the channel geometry. A discussion of spatially varied flow in side-channel spillways is given in Art. 6–17. Since the lateral flow head loss H_L is large and unknown in the case of lateral inflow, the momentum approach and Eq. (5–91) are normally used for this case. For lateral outflow, H_L is essentially zero and the energy method is frequently used, especially if the exit angle ϕ is uncertain.

PROBLEMS

5-1. Determine by study of the general differential equation for varied flow, the general shape of the M–2 profile, including upstream asymptotic slope, downstream limiting slope, and direction of curvature of surface.

5-2. Determine the asymptotic downstream slope, the limiting upstream slope, and the direction of curvature of the S–2 profile.

5-3. For a wide rectangular channel, assume the hydraulic radius R_h is equal to D, and the friction factor f is constant. Derive the differential equation of the water surface, from Eq. (5–3), in terms of D_n and D_c. Evaluate the C_1 and C_3 profiles in light of this equation. What difference would it make if f is proportional to $1/R_h^{1/3}$, as according to the Manning equation?

5-4. Sketch an actual open-channel condition which would produce each of the twelve profiles.

5-5. A flow of 100 cfs is carried in a rectangular wooden channel ($n = 0.010$), 8 ft wide, with a bed slope of 0.2 per cent. At a certain section the depth of flow is 2.5 ft. Which of the twelve basic flow profiles is represented in the reach containing this section? *Ans.*: S–1.

5-6. A wide, shallow open channel has a slope of 0.0025 in a certain reach, and a Manning coefficient of 0.015. The discharge is 10 cfs per foot of width. At a certain point, the velocity of flow is 2.5 ft/sec. Sketch and identify the type of flow profile prevailing at this point. *Ans.: M–1.*

5–7. A sluicegate admits water into a rectangular canal, 10 ft wide, which consists of a long straight channel of uniform mild slope 0.0004. The channel roughness coefficient is 0.010 and the discharge is 275 cfs. Assume that the water emerging beneath the sluicegate has an initial depth of 1.0 ft. How far downstream will there be a hydraulic jump?

5–8. Calculate the backwater curve of Problem 4–55 by the Bresse method, and compare with the profile as obtained by the step method.

5–9. A rectangular wooden channel ($n = 0.010$) is 8 ft wide, on a 1 per cent slope, and carries a flow of 100 cfs. At a certain point the depth of flow is 2.5 feet. Compute the depth of flow 10 ft upstream:

    a. By the step method.
    b. By the Bresse method.

5–10. A wide, shallow open channel has a slope of 0.0003. A low dam is placed across the channel. The depth of flow a certain distance upstream from the dam is 3.0 ft. The discharge is 10 cfs/ft and the roughness coefficient 0.015. Compute by the Bresse method the distance upstream to the point where the depth is 2.9 ft. What flow profile type is obtained?

5–11. Water emerges beneath a sluicegate onto a horizontal floor of roughness coefficient 0.015. The depth at maximum contraction is 2.0 ft and at 363 ft downstream it is 3.0 ft. Assuming supercritical flow, what is the discharge per foot of width? *Ans.: 66 cfs/ft.*

5–12. A channel consists of several segments with different slopes, in the following order:

    a. Mild   e. Mild
    b. Critical   f. Adverse
    c. Horizontal   g. Steep
    d. Steep

Sketch and identify the various profile types for the flow profile that will develop in this channel.

5–13. A long rectangular canal is 10 ft wide, and has a bed of slope 0.001 and roughness coefficient of 0.015. The canal draws water from a reservoir, the surface elevation of which is 8 ft above the channel invert elevation. The inlet loss coefficient is $\frac{1}{4}$. Compute the discharge in the canal.

5–14. A channel consists of the following segments, in order:

    a. Sluicegate, with rapid flow emerging   e. Critical
    b. Mild   f. Mild
    c. Steep   g. Adverse
    d. Steeper   h. Steep

Sketch and identify the various profile types that may develop in this channel.

5–15. A long canal draws water from a reservoir whose surface elevation is H ft above the canal bed at its inlet. The inlet is rounded so that the inlet loss is negligible. The canal has a uniform slope S and roughness coefficient n, and is triangular in cross-section, of side slope Z horizontal to 1 vertical. Derive expressions for the discharge, in terms of H, S, n, and Z only, for:

a. Slope S supercritical.
b. Slope S subcritical.

5-16. For a channel consisting of the following segments, sketch and identify the various profiles that may occur:

a. Reservoir
b. Steep
c. Steeper
d. Mild

e. Horizontal
f. Steep
g. Adverse
h. Vertical dropoff

5-17. A concrete-lined canal ($n = 0.013$) carrying a flow of 400 cfs on a slope of 0.0001 changes from a trapezoidal to a rectangular cross-section. The trapezoidal section has a base width of 20 ft, with 45° side slopes, and the rectangular section has a base width of 16 ft. Assume uniform flow in both sections. Design a transition for one section to the other, and compute the transition head loss for:

a. Warped-surface transition
b. Cylinder quadrant transition
c. Wedge transition
d. Sharp transition

5-18. A rectangular canal is 20 ft wide and carries a discharge of 640 cfs at a normal velocity of 8 ft/sec. The canal goes through a 45° bend, with a center-line radius of 60 ft. How much super elevation of the water surface will occur in the bend, assuming:

a. Uniform velocity distribution?
b. Free-vortex velocity distribution?

5-19. Derive an expression giving the downstream Froude number, N_{F_2}, in terms of the upstream Froude number, N_{F_1} and the depth ratio D_1/D_2, for an oblique positive standing wave, i.e., of the form, $N_{F_2} = f(N_{F_1}, D_1/D_2)$. *Hint:* Use the respective velocity vector relations on the two sides of the standing wave, together with the continuity and momentum equations.

5-20. Derive an expression giving the deflection angle θ for a standing wave of angle β_1, explicitly in terms of β_1 and the approach Froude number, N_{F_1}, i.e., of the form, $\theta = f(\beta_1, N_{F_1})$. Use Eq. (5–47) and solve explicitly for θ.

5-21. Derive an expression giving θ explicitly as a function of β_1 and D_2/D_1.

5-22. A supercritical flow encounters an abrupt positive wall deflection of 20°, which creates a transverse shock wave. The depth immediately upstream from the wave is measured to be 2.0 ft and the discharge in the channel is 50 cfs per foot of width. What is the wave angle and what is the downstream depth?

5-23. For an abrupt *negative* wall deflection of 20°, which causes a gradual drop in water surface from an initial depth of 5.0 ft to a final depth of 2.0 ft. What is the discharge? Compute the angle at the corner subtending the region through which this change takes place. Assume rapid flow throughout.

5-24. A rectangular channel 10 ft wide carries water flowing at a depth of 3 ft and a velocity of 15 ft sec. The channel is turned abruptly through an angle of

30°. In order to prevent wave disturbances in the downstream channel, how far should the inner corner be offset longitudinally from the outer corner, and what will be the width of the downstream channel:

   a. If the inner corner occurs first? *Ans.:* 11.5 ft, 14.4 ft.
   b. If the outer corner occurs first?

   5-25. A rapid flow of 100 cfs/ft encounters a gradually expanding wall, through a total deflection of 20°. The initial depth is 4.0 ft. Determine:

   a. Depth and velocity along wall at end of deflection. *Ans.:* 1.1 ft, 28.5 fps.
   b. Disturbance angle of initial and final wavelets, with initial and final wall directions, respectively. *Ans.:* 26.0, 11.1°.

   5-26. A rectangular channel 12 ft wide carrying a discharge of 600 cfs at a design depth of 2.0 ft must be contracted to a width of 8 ft. Design a transition section which will prevent formation of waves in the downstream channel.

   5-27. The downstream channel of Problem 5–26 must again be enlarged some distance downstream, but to a width of 20 ft. Design and draw a scale sketch in plan of an efficient expansion transition section.

   5-28. A rectangular channel 20 ft wide carrying a flow of 800 cfs at a velocity of 16 ft/sec must be turned through 90°. A central curve of 200-ft radius is selected for the bend. In order to prevent wave formation in the curve and in the tangent below the bend, how much should the bottom slope be banked in the bend? Design a spiral easement to attain this superelevation.

   5-29. Instead of using bottom superelevation, design a transition consisting of a circular transition curve at beginning and end of the main curve, for the bend of Problem 5–28.

   5-30. For the weir and flow net of Fig. 5–30, determine the discharge over the weir and the hydraulic force on the weir, in terms of H, P, and L, where H is the specific head, P is the weir height, and L is the weir crest length.

   5-31. The depth of flow at a control section in a 30-ft rectangular channel is measured to be 4.0 ft. Determine the approximate channel discharge if the control section is:

   a. Broad-crested weir with depth measured in center of weir in zone of parallel flow.
   b. Free overfall, with depth measured at the brink.
   c. Sharp-crested suppressed weir, with depth measured above level of weir crest, but at section of approximately horizontal flow upstream from the weir.
   d. Rounded-lip sluicegate, of opening D ft, with depth measured as difference between headwater and tailwater levels.

   5-32. A spillway flip bucket carrying a flow of 100 cfs/ft has a 60-ft radius of curvature. When flow over the bucket is at a velocity of 25 ft/sec, what is the pressure on the base of the bucket, in psf? Assume uniform velocity distribution and no air entrainment.

6

HYDRAULIC STRUCTURES

6–1. Classification and Functions of Hydraulic Structures. Because of the almost innumerable ways in which water and other liquids affect the activities of man, there exists a great variety of structures and other devices which serve to control and utilize these effects. Any classification of hydraulic structures will therefore be somewhat arbitrary and with ill-defined boundaries between the different categories. A classification based on use (i.e., structures for water supply, for irrigation, for drainage, etc.) is not satisfactory since many examples of virtually identical structures serving entirely different uses might be cited (for example, a highway culvert and a filling conduit for a lock chamber are very similar hydraulically and structurally).

A functional classification is suggested here, attempting to group structures of equivalent hydraulic behavior into each of the defined categories and subcategories.

- A. Storage structures—for storage of water or other fluids, under essentially hydrostatic conditions
  1. Dams and reservoirs
  2. Tanks, open or closed

- B. Conveyance structures—for guiding and controlling the flow of fluids from one location to another
  1. Closed or pressure conduits, including pipe lines and pipe systems, tunnels, penstocks, siphons, draft tubes, etc.
  2. Open conduits at ordinary velocities, including canals, flumes, sewers, transitions, aqueducts, etc.
  3. Open conduits at high velocity, including spillways, chutes, rapid flow transitions, etc.
  4. Short conduits, which may function either as open or closed conduits, such as culverts, reservoir outlet conduits, lock-chamber filling and emptying conduits, etc.

- C. Energy dissipation structures—for prevention of erosion or other structural damage due to high fluid energies
  1. Surge tanks and chambers, for dissipating elastic energy of pressure waves caused by gate or valve closures
  2. Stilling basins, for dissipating kinetic energy of shooting flows through a stabilized hydraulic jump

3. Spillway bucket end drop structures

4. Check dams, series of small dams or weirs in steep channel to reduce hydraulic slope, used also in fish ladders

D. Flow measurement or control structures—for apportioning or determining discharge through the conduit

1. Closed-conduit metering structures, especially orifices, nozzles, venturi meters, elbow meters, and pitot installations

2. Open-conduit metering structures, such as weirs (including spillways), venturi flumes, etc.

3. Gates and valves

4. Turnouts and wasteways, for diverting and wasting surplus flows in the conduit

E. Sediment and chemical control structures—for controlled removal or addition of non-hydraulic components in the flow

1. Sedimentation tanks and basins, for removal of silt or other components requiring slow settling

2. Traps, depressions in channel bed or other devices for trapping and removal of large particles in bed load

3. Racks and screens, at entrances to conduits to prevent admission of large particles and debris

4. Sluiceways, for flushing of entrapped sediments

5. Filtration beds, for removal of fine materials

6. Mixing basins, for apportioning and mixing chemical or other additives to flow

F. Collection or Diffusion structures

1. Intake structures, for controlled admission of water from reservoir to conduits

2. Infiltration galleries and drains, for collection and conveyance of ground water to discharge points, or for application of recharge water to depleted ground water

3. Wells, both pumped and artesian, and for extraction or augmentation of ground water

4. Surface drainage inlets, for collection of surface run off and transmission to storm sewers

5. Perforated pipe irrigation or drainage systems

G. Waterway stabilization structures—for protection of shore or bank developments and for maintenance of water transportation

1. Levees, to confine river flows within specified channel

2. Cutoffs, channels constructed across the narrow necks of tortuous bends in alluvial stream to shorten the flow path

3. Dikes, groins, jetties, and revetments, for training of flow in prescribed portion of channel and prevention of possible resulting erosion

4. Navigation locks and dams, maintaining navigable depths by reducing velocities, with provision for passing vessels around dams via lock chambers

    5. Breakwaters, for stabilizing harbors and other coastal areas
    6. Seawalls, for prevention of beach erosion and coastal flood damages

H. Hydrofoil structures, providing for hydrodynamically efficient relative
   motion of water to structure
    1. Stationary structures, including piers, current and sediment meters,
       cables, etc.
    2. Moving structures, i.e., ships, submarines, underwater missiles, barges,
       etc.

I. Energy conversion structures, i.e., hydraulic machines
    1. Pumps, for converting mechanical energy into hydraulic energy,
       including propulsive devices
    2. Rams, for converting hydrokinetic energy into elastic and pressure
       energies
    3. Turbines, for converting hydraulic energy into mechanical energy
    4. Hydrodynamic transmissions; various combinations of pump and
       turbine devices for transfer of mechanical energy through intermediate
       hydraulic energy stage, including various types of hydraulic servo-
       mechanisms and controls

The foregoing list is fairly comprehensive and representative of the various
types of hydraulic structures, though by no means exhaustive. It is immediately obvious that detailed consideration of all of them is out of the question
in the scope of this treatment. Only a few of the more commonly encountered
structures can be considered, and even these in only a cursory fashion.
General or basic principles will be stressed in so far as possible, with the
broadest possible range of application to specific structures. Certain types,
though common and important, are rather specialized in certain fields (such
as hydrofoil structures in marine engineering and some types of collection
and control structures in sanitary engineering) and so will not be treated
here.

6–2. The Role of Hydraulic Design. The design of a structure
involves many elements and considerations. Functional, esthetic, economic,
and similar factors must be considered thoroughly and often are determinative in the decision as to type of structure or even whether or not it can or
should be built at all. Correct decisions based on these factors involve a
higher level of engineering experience and ability than the details of structural
dimensioning and proportioning, important as these latter may be.

In the case of hydraulic structures, the structural design and detailing
must be preceded by a thorough hydraulic analysis and design. In the first
place, the efficient functioning of the hydraulic phenomena controlled by
the structure requires thorough knowledge of these phenomena and the
design of the structure to accommodate them most efficiently. Secondly, the
structural design depends basically upon a knowledge of the nature and
magnitudes of the loads to be resisted. Since the hydrostatic and hydro-

dynamic loads constitute an important part of the force system, a detailed hydraulic analysis is necessary to determine these loads. No matter how carefully and accurately a stress analysis is performed, its results are invalid if the assumed loading system is not realized in the actual structure.

Moving fluids are much more complex in their behavior than solids, and therefore inherently more difficult to analyze. A completely rational analysis of all the stresses, velocities, momenta, and other factors in a field of fluid moving within or around given boundaries will probably always be impossible, primarily because of the factors of turbulence and friction.

However, great advances (both theoretical and experimental) have been made in recent years in the understanding of these factors. The tools of statistical and dimensional analysis, together with the basic conservation laws, can be applied in a scientific empiricism which can give design information of a high order of reliability. The modern methods and data of hydraulic engineering can and should certainly be intelligently applied to the design of any hydraulic structure.

Thus the hydraulic and structural design of a hydraulic structure are not at all contradictory, but are complementary. The hydraulic analysis is prerequisite to the structural analysis, in order for the hydraulic loading to be both accurately known and as easy as possible to support structurally. This in turn will permit a structural design which is functionally more effective and can justify a smaller factor of safety.

The present treatment will deal primarily with the hydraulic aspects of the design, but some structural considerations will be introduced where appropriate, particularly in connection with massive structures, this type being limited almost exclusively to hydraulic structures (e.g., dams and levees). In general, however, the structural design of a hydraulic structure is the same in principle as for any other structure, once the loading system has been determined by the hydraulic analysis.

6-3. Dams and Reservoirs. The first type of hydraulic structure to be considered is the dam. There are many different kinds of dams and, particularly for large dams, the design is so complex and involves so many different components that a thorough study of any one of them is beyond the scope of this text. The reader is referred to the references on dams, at the end of this chapter, for works giving more detailed treatments.

A dam is simply a barrier placed across a watercourse to prevent or retard the normal flow of water therein. It is one of the most ancient of all structural types and scores of thousands of them have been built in the course of history. With greatly augmented modern demands for water conservation and control around the world, large numbers of them are under construction or being planned at present. There are over 11,000 medium and high dams in current use in the world, of which more than 3500 are in the United States.

Dams may be classified either as to their purpose or as to their structural type. Classifications on each of these bases are given in the following tabulations.

6-4. Classification of Dams According to Purpose

A. Stage control dams

    1. Diversion—to raise water level at entrance to canal or other diversion, to permit gravity or semigravity flow into diversion system

    2. Navigation—to maintain navigable depths of flow in stream reach above dam, with lock provided for shipping bypass

    3. Check—to retard velocity in steep channel, either for debris barrier or for erosion control

B. Storage dams

    1. Flood control—to store peak discharges behind dam, with releases controlled to minimize downstream flood damage

    2. Water supply—to store inflows until times of water demand, with releases controlled to make optimum use of available water in stream

      a. Irrigation supply

      b. Municipal water supply

      c. Industrial supply

      d. Stock pond

    3. Hydroelectric power —to raise stage for provision of head, as well as to provide storage of water, since amount of power available is proportional to product of head and discharge

    4. Sedimentation—to remove sediment from stream by retarding velocity sufficiently to induce settling

    5. Recreation—to provide water area for swimming, boating, fishing, hunting, etc

    6. Groundwater recharge—to detain flood flow long enough for infiltration to occur

    7. Pollution control—to provide clean water for discharge as needed for dilution purposes at low stages in downstream reaches

C. Multipurpose dams—to provide storage or stage control, or both, for two or more purposes harmonized or compromised as necessary for optimum use

D. Barrier dams

    1. Levees and dikes—to protect land areas from overbank flow

    2. Cofferdams—for temporary dewatering of construction sites

6-5. Classification of Dams According to Material

A. Masonry dams

    1. Concrete gravity dams—designed so that water and other loads are resisted by weight of dam

      a. Massive gravity dams

      b. Hollow gravity dams

      c. Flexible gravity dams

2. Concrete arch dams—loads resisted by arch action carried to abutments
   a. Constant-center arch dams
   b. Variable-center arch dams
   c. Double-curvature (shell) arch dams
3. Gravity-arch dams—loads resisted by combination of gravity and arch action
4. Buttress dams—loads resisted by slab or arch action between successive buttress supports
   a. Slab-and-buttress dams
   b. Multiple-arch dams
   c. Multiple-dome dams
   d. Massive-head dams
   e. Truss-buttress dams
   f. Columnar-buttress dams
5. Stone-masonry gravity dams
6. Stone-masonry arch dams

B. Earth-fill dams—designed to resist loads by gravity and by mechanics of embankment stability
   1. Homogeneous embankment dams
   2. Zoned-earth embankment dams—with internal core of relatively impermeable material
   3. Diaphragm-type embankment dams—with central core wall of concrete, steel or timber

C. Rock-fill dams—with action intermediate between earth-fill and masonry gravity dams

D. Steel dams
   1. Steel slab-buttress dams
   2. Sheet-steel cofferdams
   3. Cellular-steel cofferdams

E. Timber dams
   1. Timber slab-buttress dams
   2. Timber crib dams, with rock-filled cribs

F. Inflatable dams

Dams are also classified as low dams (under 50 ft in height) medium dams and high dams (over 300 ft high).

6–6. Factors Governing Selection of Type of Dam. In general, it is possible to design a satisfactory dam at a given site using any one of several possible structural types and materials. The decision to use a specific type must therefore be based largely on comparative economic studies. The following outline indicates important factors that may influence the cost and therefore the optimum type.

A. Site conditions
  1. Foundation materials (masonry dams, especially arch dams, require stronger foundations)
     a. Allowable foundation stresses
     b. Percolation values
     c. Excavation requirements
     d. Potential settlements
  2. Topography of site
     a. Arch dams usually limited to narrow canyons with strong walls
     b. Earth dams usually suitable for all conditions except very high or narrow sections
  3. Availability of materials
     a. Convenient source of earth or aggregate may indicate earth or masonry dam
     b. Buttress dam requires smallest quantity of materials

B. Hydraulic factors
  1. Spillway requirements
     a. Overflow spillways best for large capacities, most feasible on gravity or slab-buttress dams
     b. Side-channel and tunnel spillways adaptable to any type
  2. Diversion requirements
     a. Costs usually greater for earth dam, because of greater base thickness
  3. Outlet works and penstocks
     a. Arch dams not adapted to large or numerous openings

C. Climatic effects
  1. Spalling of concrete in cold climates disadvantage of thin arch and buttress dams

D. Traffic factors
  1. Crest highways costly for thin arch and buttress dams
  2. Navigation locks precluded for arch dams

E. Social factors
  1. Gravity dams provide greatest safety against sudden destruction due to earthquake, bombing, etc., with resultant damage to affected communities
  2. Benefits to be derived may control cost of dam; temporary dam may have to suffice if benefits are small or short-range
  3. Volume of employment, particularly of local labor
  4. Esthetic considerations

6-7. Forces on Gravity Dam. The simplest and most common dam is the concrete-masonry massive gravity dam. An outstanding example is Grand Coulee Dam (Fig. 6-1). In cross-section such a dam usually has either a vertical or slightly battered upstream face and a more strongly sloping, perhaps curvilinear, downstream face. The major forces acting on a gravity dam are indicated in Fig. 6–2, and defined in the ensuing discussion. Of course, most of these forces also act on any other type of dam. It is assumed that the cross-section shown is typical for the entire width of dam and the forces therefore are expressed in pounds per foot of width.

F_H = horizontal component of hydrostatic pressure, acting along a line $H/3$ feet above the base

   = $\frac{1}{2}\gamma H^2$, where γ = specific weight of water.

F_v = vertical component of hydrostatic pressure

   = weight of fluid mass vertically above the upstream face, acting through the centroid of that mass.

W = weight of dam = (area of cross-section of dam)$\times(S\gamma)$, where S = specific gravity of masonry, approximately 2.4 or 2.5, acting through centroid of cross-section.

F_u = uplift force on base of dam, as determined by foundation seepage analysis, and integration of point pressure intensities over base area; if foundation is homogeneous and permeable, pressure varies approximately linearly from full hydrostatic head at the heel to full tailwater head, and F_u is approximately $\frac{1}{2}\gamma HB$, acting at $B/3$ from the heel. This value is often multiplied by some fraction less than 1 if the foundation is relatively impermeable, but it is on the safe side to assume uplift over the entire base area.

F_s = additional hydrostatic force due to silt deposits near the heel, approximately $\frac{1}{2}\gamma(S_s - 1)(h_s^2) \cong 14h_s^2$, where h_s is the depth of silt and S_s is the specific gravity of the mixture of silt and water, about 1.45; it is sufficiently accurate normally to neglect any vertical component.

F_{Q_D} = earthquake force on dam due to quake acceleration,

   = $(W/g)a$, where the acceleration a is usually taken as $\frac{1}{10}$ g in the horizontal direction, and about $\frac{1}{12}$ g in the vertical direction, acting through centroid of dam.

F_{Q_W} = earthquake force due to acceleration of water behind the dam, equal approximately (by von Kármán's analysis) to $\frac{5}{9}\gamma H^2 a/g$, acting horizontally at a distance $(4/3\pi)H$ up from the base.

F_I = force of ice on lake surface against dam $\cong 5000\, h_I$, where h_I is the depth of freezing.

R = resultant of foundation shear and bearing pressures: horizontal component, $R_H = F_H + F_S + F_{Q_W} + F_{Q_D} + F_I$ acting along the base; vertical component, $R_V = W + F_v - F_u - F_{Q_D}$ acting at a distance $\bar{x}$ from the toe that can be determined by the requirement for rotational equilibrium of the dam, by equating to zero the sum of the moments of all the foregoing forces about the toe of the dam.

Other forces that may possibly have to be considered include those due to waves, wind, spillway vacuum, pressure of earth fill, tailwater, temperature stresses, superimposed loads on crest highway, etc.

6–8. Stability Analysis of Gravity Dam. A gravity dam may fail by shear along a horizontal plane, or by crushing of the material in the dam or its

Fig. 6–1. Grand Coulee Dam, Washington: a massive straight-gravity dam. Note over-flow spillway and air-entrained flow on spillway. (U.S. Bureau of Reclamation.)

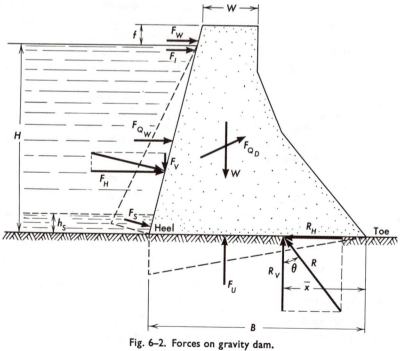

Fig. 6–2. Forces on gravity dam.

foundation on account of combined normal and bending stresses, or by rotation about the toe. The latter possibility is eliminated if the dam is so proportioned that no tension is produced in the section at the heel (or at the toe when the reservoir is empty), a provision which is desirable anyway, since masonry is very weak in tension.

The stability of a dam is ordinarily expressed in terms of its factors of safety against sliding and overturning, respectively. In addition, the maximum foundation pressures are compared with the bearing strength of the foundation. In effect, this means computation of the reactions of the foundation to the applied loads on the dam and a check of the ability of the foundation to supply these reactions.

The factor of safety against sliding is simply the ratio of the total frictional force which the foundation can develop to the force tending to cause sliding. It can be expressed as follows:

$$(FS)_s = \frac{R_V C_f + A_s S_s}{R_H} \tag{6-1}$$

where C_f is the coefficient of friction between the dam and its foundation (usually between $\frac{1}{2}$ and $\frac{3}{4}$), A_s is the shear area and S_s is the average shear strength of any shear areas provided by the keying of the dam to the foundation.

This factor of safety must obviously be greater than 1 for stability; however, it is often quite small. If too small, it may be increased by increasing the base width, stepping the foundation, adding key walls, or other such devices.

The factor of safety against overturning about the toe is the ratio of the resisting moments to the overturning moments:

$$(FS)_o = \frac{Wa + F_v b}{(Wa + F_v b) - R_V \bar{x}} \tag{6-2}$$

where a and b are the distances from the toe to the lines of action of W and F_v respectively. The factor of safety is unity if $\bar{x} = 0$, that is if the resultant foundation reaction passes through the toe. As long as this resultant passes through the base of the dam, it is safe against overturning.

The bearing pressures on the foundation are ordinarily computed from the resultant vertical load on the foundation, R_V, on the assumption that these pressures are distributed linearly across the base of the dam and that the masonry will not resist tension. Two possibilities exist: compression over entire base of dam and compression over a part of the base.

Assume compression exists over the entire base, with foundation pressures p_h and p_t at heel and toe respectively, each in lb/sq ft. Assume linear variation in pressure, as shown in Fig. 6-3.

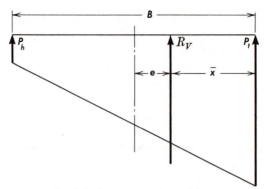

Fig. 6–3. Pressures on base of dam.

Since R_V is the resultant of the distributed foundation pressures, the equations denoting its magnitude and line of action are as follows:

$$R_V = \frac{p_h + p_t}{2} B \tag{6–3}$$

and

$$R_V\,(e) = \frac{p_t - p_h}{2} B \frac{B}{6} \tag{6–4}$$

Solving the two equations for p_h and p_t yields

$$p_t = \left(\frac{R_V}{B}\right)\left(1 + \frac{6e}{B}\right) \tag{6–5}$$

and

$$p_h = \left(\frac{R_V}{B}\right)\left(1 - \frac{6e}{B}\right) \tag{6–6}$$

The eccentricity e of the resultant from the center line is of course computed from the previously known distance $\bar{x}$.

The pressure at the toe, p_t, is of course the critical value, and should be less than the bearing strength of the foundation material, usually with a factor of safety of at least 2. The pressure at the heel is not important, except that it is desired usually to be a positive value to prevent tension cracks from developing and to assure that the entire base width is being utilized to transfer the loads to the foundation.

It will be noted from Eq. (6–6) that, if the resultant passes outside the middle third of the base (i.e., if $e > B/6$), then the pressure at the heel is negative, indicating tension.

It must be assumed that the foundation bond would not be able to resist such tension and cracks would form. Thus the portion of the resultant that theoretically would be carried near the heel in tension will have to be added to the portion that is theoretically being carried in compressive flexure near the toe. This results in a redistribution of foundation pressures as shown in Fig. 6–4. In this case the altitude of the pressure triangle is obviously $3\bar{x}$, and the toe pressure p_t is therefore

$$p_t = \frac{2R_V}{3\bar{x}} = \frac{2R_V}{3(B/2 - e)}$$

$$= \left(\frac{R_V}{B}\right)\left[\frac{1}{(3/2)(1/2 - e/B)}\right] \tag{6-7}$$

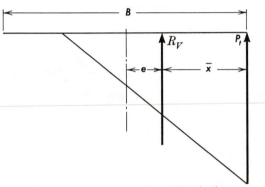

Fig. 6–4. Resultant outside middle third.

Example 6–I

A preliminary design of a masonry gravity dam is based on an assumed triangular section, H feet high and B feet wide at the base, with a vertical upstream face. The masonry has a specific gravity S and a coefficient of friction with the foundation of f. The foundation is fully permeable. Neglect earthquake and other minor forces. Derive formulas for the following, in terms of only H, B, S, f and γ:

(a) Factor of safety against sliding.
(b) Maximum foundation bearing pressure, assuming no tension permitted in masonry or foundation.

Solution.

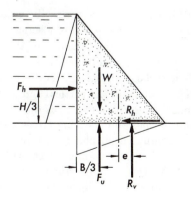

Forces on dam are:

(1) $F_H = \frac{1}{2}\gamma H^2$ (4) $R_H = F_H = \frac{1}{2}\gamma H^2$

(2) $W = \frac{1}{2}\gamma SBH$ (5) $R_V = W - F_u = \frac{1}{2}\gamma HB(S - 1)$

(3) $F_u = \frac{1}{2}\gamma HB$

ΣM about point on base through which pass lines of action of W and F_u:

$$F_H\left(\frac{H}{3}\right) = R_V\left(e + \frac{B}{6}\right)$$

$$\frac{1}{6}\gamma H^3 = \frac{1}{2}\gamma HB(S - 1)\left(e + \frac{B}{6}\right)$$

from which the eccentricity, $e = \dfrac{H^2}{3B(S - 1)} - \dfrac{B}{6}$. However, for no tension at heel (triangular stress distribution throughout base) $e = \dfrac{B}{6}$.

$\therefore B = \dfrac{H}{\sqrt{S - 1}}$, as required minimum base width.

(a) Factor of safety against sliding

$$= \frac{R_V(f)}{R_h} = \frac{\frac{1}{2}\gamma HB(S - 1)(f)}{\frac{1}{2}\gamma H^2} = \frac{B}{H} f(S - 1) \qquad Ans.$$

(b) Maximum bearing pressure, at toe, for triangular stress distribution on base,

$$= \frac{2R_V}{B} = \frac{2}{B}(\tfrac{1}{2}\gamma HB)(S - 1) = \gamma H(S - 1) \qquad Ans.$$

or

$$= \gamma H \frac{H^2}{B^2} = \frac{\gamma H^3}{B^2}$$

6–9. Stress Analysis for Gravity Dam.

The stresses in the interior of a massive body of masonry are difficult or impossible to compute with certainty; however several methods have been used to approximate the correct solution.

The simplest method of analysis is known as the *gravity method*. In this method, the dam is assumed to be composed of a series of vertical cantilever beams, each 1 ft wide and acting independently of the others. This is the same basic assumption as in the stability analysis of the preceding article.

On any horizontal plane through the cantilever it is assumed that the total vertical component of the resultant of the vertical and bending stresses above that plane (computed neglecting uplift forces) is distributed linearly across the plane. The maximum normal and shearing stresses resulting from these vertical stresses are then computed on planes of principal stress as follows. Consider a small slice at the downstream face, as shown in Fig. 6–5, the face making an angle β with the vertical. The vertical stress p_v and the horizontal stress τ act on the horizontal area A. Since there can be no stress on the

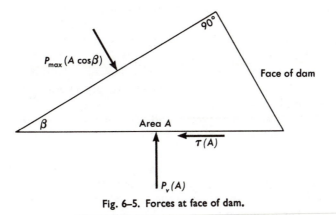

Fig. 6–5. Forces at face of dam.

face of the dam, that face and the plane normal to it represent planes of principal stress, the maximum compressive stress in the masonry at that point therefore acting on the plane normal to the face of the dam. The shearing stress on that plane is zero.

Summing forces in the vertical and horizontal directions yields

$$p_v A = p_{max}(A \cos \beta) \cos \beta$$

and

$$\tau A = p_{max}(A \cos \beta) \sin \beta = \frac{p_v A}{\cos \beta} \sin \beta$$

from which the principal stress at the downstream face of the dam, which will ordinarily be the maximum compressive stress in the masonry at that level, is

$$p_{max} = \frac{p_v}{\cos^2 \beta} = p_v(1 + \tan^2 \beta) \tag{6-8}$$

and the horizontal shear stress resulting from the vertical stress at that point is

$$\tau = p_v \tan \beta \tag{6-9}$$

Similar computations should be made at the upstream face for an empty reservoir. For more accurate studies, as necessary for larger dams, principal and maximum shearing stresses should be computed at other selected points through the body of the dam.

A number of other methods have been developed for determining stresses in large dams more accurately than can be done by the gravity method. One of these is the *trial load-twist method*. In this type of analysis, the structure is conceived as comprising two basic components: (1) a vertical cantilevered structure, with adjacent cantilevers keyed at the construction joints; (2) a twisted structure composed of horizontal elements twisted by virtue of the differential deflections of the cantilevered elements.

The analysis consists essentially of a trial-and-error determination of the division of load necessary to balance homologous deflections as determined by (1) cantilever deflections and (2) deflections of horizontal elements as caused by differential deflection of vertical elements and by shear detrusion. Once these deflections are balanced, moments and shears can be computed from them, and then stresses. The method obviously is very lengthy and laborious and therefore has been used only for large dams.

Another method, still more precise and more lengthy, is the *slab analogy method*. It has the advantage of making allowance for the effect of foundation yielding, but is used only for high and important dams. The dam cross-section is assumed analogous to a slab of the same shape, including part of the foundation. The slab is loaded with edge forces and couples which will cause edge rotations corresponding to the known surface stresses.

The slab is then divided into a grid of horizontal and vertical beams, and the edge forces are distributed arbitrarily between the two systems. Slopes and deflections are computed for each system. The load division must be adjusted by repeated trial until the two systems are brought into slope and deflection agreement. Then, by the analogy, stresses in the vertical and horizontal planes are computed from the moments in the vertical and horizontal beams.

A somewhat similar method is the *lattice analogy*, by which the dam cross-section is replaced by an analogous network of square figures with diagonals. This lattice is assumed to have frictionless joints and to resist the external loads by truss action. The relaxation method is then employed to remove the joint restraints, until terminal joint displacements are obtained. The relative displacements between adjacent joints correspond to strains in the plane section. Elasticity computations then yield the masonry stresses from the strains so determined.

Each of these methods is very tedious and time-consuming unless programed on a high-speed computer. The advantages and need for a detailed stress analysis of course increase with the size of the dam.

Another method that has come into wide use in the last few years, which would have been impossible to use before the advent of the digital computer, is the *finite element* method. By this method, the dam and its foundation are divided into a large number of contiguous triangles, assumed to be interconnected at a finite number of "nodal" points. The elastic properties of each triangle are analyzed and linear equations are written relating the forces and displacements at each nodal point. Prescribed displacements are inserted, as required by foundation and other boundary conditions, as well as the assumed internal displacement pattern, and equilibrium equations written at all the nodes. This process may of course result in several hundred linear simultaneous equations, but these can be programmed for rapid solution on the computer. Once the correct displacements are known at all points, the corresponding stresses can be calculated by elastic theory.

The finite element method is rapidly becoming the preferred analytic method of design for large gravity and arch dams. It may be extended to three-dimensional analysis, using various complex forms of three-dimensional finite elements with many nodes.

Another type of method is the *experimental method*, using structural scale models. It is possible to measure strains at any desired point or points on such a model, loaded proportionally to the loading on the prototype dam, in accordance with principles of dynamic similitude. Of course, masonry cannot be used for the model structure, since the strains would be too small to measure. Some material must be used which possesses a much smaller stiffness but the same value of Poisson's ratio, and which also deforms elastically within the experimental range. Such materials as rubber, celluloid, and plaster of paris have been used, with loads applied by mercury, water, or a system of levers and weights.

Photoelastic models have also been used, not for the entire structure, but for analysis of localized areas, especially regions of stress concentration. The model must be elastic, transparent, isotropic, and free from initial or residual stress. A beam of polarized light is directed through the stressed model, the refraction patterns corresponding to stress conditions. The refracted light is projected on a screen or photographic plate, and then converted to stresses by the techniques of photoelastic stress analysis.

6–10. Earth Dams.   The earth dam is the oldest type of dam, as well as the most common. Dykes and levees also are constructed of earth and are essentially earth dams. It is nearly always the most economical type to construct on soil foundations, particularly if the material for the embankment is conveniently accessible. Often an earth dam is economically competitive even on rock foundations and in canyon sections. A high earth dam is shown in Fig. 6–6.

The design of an earth dam involves both a hydraulic analysis and a structural analysis. In general, several trial designs are necessary before the optimum design is obtained.

The hydraulic analysis primarily involves a determination of the seepage patterns and magnitudes, as well as the internal hydrostatic forces resulting from seepage, for both the foundation and the dam itself. The effect of various adjuncts to the main embankment such as drainage blankets, central zones of impervious materials, cutoff walls, rock toes, aprons, etc., may be determined. Particularly important to investigate is the possibility of removal of the fine particles near the toe by the emerging seepage water, with resultant undermining of the dam, a phenomenon known as *piping*.

The structural analysis involves a study of the stability of the embankment under the given conditions of seepage and other forces. Settlement and stability studies on the foundation are very important also. These are essentially problems in soil mechanics, and are adequately treated in textbooks on this subject under the topic of embankment stability. Consequently, the structural analysis of an earth dam will not be further treated herein.

6–11. Seepage Analysis for Earth Dam.

All earth dams, as well as their foundations, are somewhat permeable and will permit a certain amount of seepage. However, the flow will be slow, and can be assumed laminar, in accordance with the Darcy equation:

$$Q = KiA \tag{6–10}$$

where Q is the seepage, in cfs through a cross-section of A square feet under a hydraulic gradient of i feet per foot. The permeability coefficient K is then in feet per second and depends on the particular material and its condition.

It is usually sufficiently accurate to assume that the seepage flow is two-dimensional. The cross-section varies, however, and therefore also the velocity. The *potential head* at any point is the sum of the elevation and pressure heads at the point, and the hydraulic gradient is the linear rate of potential drop. Flow must everywhere be in the direction normal to lines of equal potential (a special case is the flow of water on the ground surface along lines of steepest slope, perpendicular to contour lines).

Fig. 6–6. Anderson Ranch Dam, Idaho: a zoned earth-fill dam, 456 ft high. Note chute spillway and stilling basin. (U.S. Bureau of Reclamation.)

The seepage velocities in the horizontal and vertical directions can be written as:

$$V_x = K_x \frac{\partial H}{\partial x} \quad \text{and} \quad V_y = K_y \frac{\partial H}{\partial y} \tag{6-11}$$

in which K_x and K_y are the permeabilities in the horizontal and vertical directions, respectively, and $\frac{\partial H}{\partial x}$ and $\frac{\partial H}{\partial y}$ are the corresponding hydraulic gradients. Considering seepage entering and leaving a small element of the dam, (dx long, dy high, one foot thick), the equation of continuity yields

$$dy\left(V_x + \frac{\partial V_x}{\partial x}\, dx\right) + dx\left(V_y + \frac{\partial V_y}{\partial y}\, dy\right) = V_x\, dy + V_y\, dx$$

which reduces to:

$$\frac{\partial V_x}{\partial x} + \frac{\partial V_y}{\partial y} = 0 \tag{6-12}$$

and, combining this equation with Eq. (6-11):

$$K_x \frac{\partial^2 H}{\partial x^2} + K_y \frac{\partial^2 H}{\partial y^2} = 0 \tag{6-13a}$$

Equation (6-13a) is a form of the familiar *LaPlace equation* of mathematical physics.

When the embankment is homogeneous and isotropic, then K_x and K_y are equal and constant, so that:

$$\frac{\partial^2 H}{\partial x^2} + \frac{\partial^2 H}{\partial y^2} = 0 \tag{6-13b}$$

The applicability of Eq. (6-13b) to the flow indicates that the theory of potential flow can in principle be used to determine equations for the family of streamlines and equipotential lines comprising the flow field. These two families form an orthogonal network with the flow lines everywhere perpendicular to the equipotential lines, and vice versa.

Analytical solutions describing this phenomenon can in fact be obtained from potential theory for certain special boundary conditions. In general, however, a graphical solution must be used, involving construction of a *flow net*. A flow net is an orthogonal network of curves consisting of flow lines and equipotential lines, each group perpendicular to the other group (see Fig. 6-7). For convenience, the network is made to consist of figures which are approximately "square"; that is, for each figure the distances between adjacent flow lines and adjacent equipotential lines are approximately the same, with intersections at right angles. The flow lines may be regarded as boundaries of stream tubes, so that the flow per foot of dam width between any two adjacent flow lines, by the Darcy equation, is

$$\Delta Q = Kib = K \frac{\Delta H}{L} b = K \, \Delta H$$

By continuity, ΔQ remains constant regardless of whether the flow lines converge or diverge. Therefore, the potential drop ΔH must also be the same between any two adjacent equipotential lines, since b/L is unity for any one figure. The total seepage is given by

$$Q = \sum \Delta Q = n_f K \, \Delta H = n_f K \frac{H}{n_d} \tag{6–14}$$

where Q is the total seepage per foot of width caused by a total head H, and n_f/n_d is the ratio of flow channels to steps of equal potential drop in the flow net.

A flow net is constructed graphically by trial-and-error sketching of flow lines and equipotential lines to conform to the given boundary conditions and to form as closely as possible a network of true "square" figures. The more closely the lines are spaced, the greater accuracy is attained, particularly in regions of sharp curvature. The boundary conditions normally are two boundary flow lines and two boundary equipotential lines.

In the case of seepage through an earth dam, the boundary flow lines are the contact line with the impervious rock stratum at the base or below the base and the topmost seepage line, or *phreatic line*, corresponding to the water table in groundwater flow. One boundary equipotential line is the portion of the upstream face and reservoir bed through which seepage water can percolate; the other is a horizontal surface below the earth fill where the flow lines all come out at atmospheric pressure (or at constant head in the event of a tailwater pool). These geometrical characteristics are all illustrated in Fig. 6–8.

The ratio n_f/n_d for the flow net of Fig. 6–8 is seen to be $\frac{3}{10}$, so the seepage is $\frac{3}{10} KHL$, where L is the length of dam.

The phreatic line is located by the condition that its intersection with

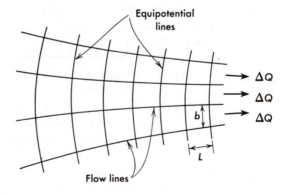

Fig. 6–7. Flow net.

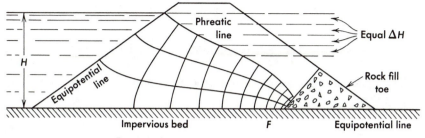

Fig. 6–8. Seepage net through earth dam.

the successive equipotential lines must be at equal increments of elevation (since these intersections are all at the same pressure, atmospheric). This is somewhat difficult to do, usually requiring a number of successive re-adjustments of the flow net to satisfy the requirement.

By inspection, it is evident that the flow net in its lower reaches can be approximated by a set of confocal orthogonal parabolas, with focus at F, the intersection of the bottom flow line and downstream equipotential line. This fact can be used to expedite the location of the phreatic line, otherwise quite a tedious task. The procedure is best illustrated on a dam with impervious foundation and a horizontal drainage blanket near the toe, into which the flow lines enter vertically, as in Fig. 6–9.

It has been found that the phreatic line will closely approximate a parabola whose focus is at F, whose directrix passes through G, and which intersects the water surface at a point A such that $\overline{AB} = 0.3\overline{CB}$. The phreatic line must diverge from the parabola near the upstream face, however, in order to come out perpendicularly at B.

To locate the break out point E of the parabola, the parabolic definition is used, namely, that every point on the parabola is equidistant from the focus and directrix: that is, $\overline{FE} = \overline{EG}$ and $\overline{AF} = \overline{JG}$. From these relations, it follows that

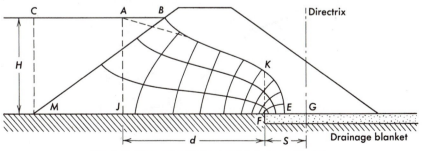

Fig. 6–9. Flow net with horizontal drainage blanket.

$$\overline{FE} = \frac{1}{2} S = \frac{1}{2} [\overline{JG} - \overline{JF}] = \frac{1}{2} [\overline{AF} - \overline{JF}] = \frac{1}{2} [\sqrt{d^2 + H^2} - d] \quad (6\text{--}15)$$

A third point on the parabola (in addition to A and E) is easily located at point K, since $\overline{KF} = \overline{FG} = S$.

The equation of the parabola, in terms of a coordinate system with origin at E is then

$$x = \frac{1}{2S} y^2 \quad (6\text{--}16)$$

as may be determined by insertion of the known coordinates at points A and K.

The slope of the phreatic line at point K is

$$\frac{dy}{dx}\bigg|_{y=S} = \frac{2S}{2y} = 1$$

This is therefore the slope of the potential gradient at a point where the depth of flow channel is S. Consequently, the seepage through this section (the same as at any other section, of course) may be computed, without reference to the flow net, by the Darcy equation:

$$Q = KiA = K(1)(S)(L) = K(S)L \quad (6\text{--}17)$$

In the event the top flow line must emerge from the dam in other than a vertical direction, it will diverge from the computed parabola near the point of emergence. This is illustrated in Fig. 6–10, for a dam with no drainage blanket or toe. In this case, the theoretical parabola is computed exactly as before, with the focus at the emergence point of the bottom flow line. However, the actual phreatic line will diverge from the parabola at some point in order to emerge tangent to the face of the dam at a distance Δa below the theoretical intersection of the parabola and the dam face. The distance Δa can be determined from the approximate empirical relation,

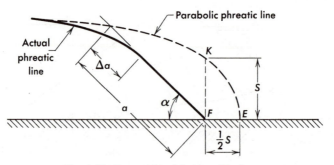

Fig. 6–10. Dam without drainage blanket.

$$\Delta a = a\left(0.5 - \frac{\alpha}{360°}\right) \tag{6-18}$$

where α is the slope in degrees, of the discharge face or of the inner face of a rock toe. A horizontal drainage blanket would correspond to $\alpha = 180°$. Note that the distance a, by the definition of the parabola, is equal to $a \cos \alpha + S$, or

$$a = \frac{S}{1 - \cos \alpha} \tag{6-19}$$

It is impossible to maintain square figures in the flow net at the downstream face, since the flow lines all merge as they come out and flow down the face. The vertical cross-section of seepage at the breakout point of the top flow line is $(a - \Delta a) \sin \alpha$ in height. The slope of the phreatic line as it emerges is probably slightly flatter than the slope of the face, and may be considered approximately equal to $\sin \alpha$, for values of α less than about 60°. By the Darcy equation, the seepage is then given approximately by

$$\frac{Q}{L} = K(\sin^2 \alpha)(a - \Delta a) = K(\sin^2 \alpha)\left(0.5 + \frac{\alpha}{360°}\right)a \tag{6-20}$$

Since $(a \sin^2 \alpha)$, from Eq. (6-19), is equal to $S(1 + \cos \alpha)$, this can also be written:

$$\frac{Q}{L} = KS(1 + \cos \alpha)\left(0.5 + \frac{\alpha}{360°}\right) \tag{6-21}$$

For $\alpha = 60°$, equation (6-21) becomes identical to (6-17). For smaller values of α, the unit seepage is slightly greater than KS. For a dam with a rockfill toe, $90° \leq \alpha \leq$, $180°$ Eq. (6-17) may be used with sufficient accuracy.

An alternate approximate formula for the location of the breakout point of the phreatic line, for a dam without a drainage toe or blanket, is

$$a - \Delta a = \sqrt{d^2 + H^2} - \sqrt{d^2 - (H \operatorname{ctn} \alpha)^2} \tag{6-22}$$

valid for $15° \lesssim \alpha \leq 90°$.

6-12. Special Considerations in Seepage Analysis. It has been assumed in the foregoing that the embankment is homogeneous and isotropic and that, therefore, its permeability is the same in every direction. As a matter of fact, the construction procedures are such that this is seldom, if ever, true. Usually the permeability in the horizontal direction, along the planes of stratification, is materially greater than in the vertical direction.

To allow for this, a *transformed section* is used, whose horizontal dimensions are reduced to correspond to the greater ease of percolation in that direction. Also, for seepage lines with components in both horizontal and vertical directions, an *equivalent permeability* is used, equal to the geometric mean of the vertical and horizontal permeabilities. Thus, Eq. (6–13a) may be rewritten as follows:

$$\frac{\partial^2 H}{\partial \left[\sqrt{\dfrac{K_y}{K_x}}\, x \right]^2} + \frac{\partial^2 H}{\partial y^2} = 0$$

Now, let $x' = \sqrt{\dfrac{K_y}{K_x}}\,(x)$, and transform all horizontal dimensions on the dam in this proportion, leaving the vertical dimensions unchanged. The equation then is of exactly the same form as Eq. (6–13b), and so a true flow net can be drawn on the transformed section. Then, by Eq. (6–10), the flow through one flow channel of the net on the transformed section, at a point where the flow lines have become horizontal, as in Fig. 6–11, is

$$q = K' \frac{\Delta H}{L\sqrt{\dfrac{K_y}{K_x}}}\, b$$

This must be the same as the seepage in the same flow channel on the true section. Thus:

$$K' \frac{\Delta H}{L\sqrt{\dfrac{K_y}{K_x}}}\, b = K_x \frac{H}{L}\, b$$

from which the equivalent permeability on the transformed section K', is equal to $\sqrt{K_x K_y}$. Thus it is legitimate to compute the seepage through such a horizontal square on the basis of the transformed section and the equivalent permeability. Since the seepage quantity through a given flow channel is independent of the direction and spacing of the flow lines, the transformed section may therefore also be used for non-horizontal portions of the flow net.

The procedure is first to draw the flow net and compute the seepage quantities on the basis of the transformed section. Then, if desired, especially for study of seepage pressures in the dam, the flow net may be transferred to the actual section, by proportional placement of net intersection points.

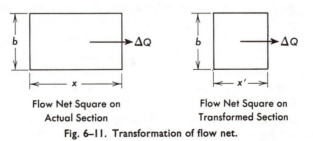

Flow Net Square on   Flow Net Square on
Actual Section   Transformed Section

Fig. 6–11. Transformation of flow net.

Example 6–2

An earth dam has a crown width of 25 feet and a height of 85 feet, with a 10-ft freeboard. The side slopes are at an angle of 20° with the horizontal and the dam has no drainage blanket. The embankment material has a permeability of 2×10^{-5} ft/sec in the horizontal direction and 1×10^{-5} ft/sec in the vertical direction. Determine the quantity of seepage, in cubic feet per day per foot of width of dam, by the equivalent parabola method.

Use transformed section:

$$x' = x \sqrt{\frac{K_y}{K_x}} = x \sqrt{\tfrac{1}{2}} = 0.707x$$

$$\alpha' = \tan^{-1}(\sqrt{2} \tan 20°) = \tan^{-1} 0.515 = 27.25°$$

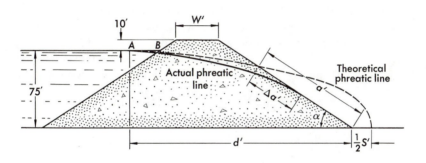

Solution.

$$\overline{AB} = 0.3\,\overline{BC} = 0.3(75)\,\frac{1}{0.515} = 43.7 \text{ ft}$$

$$W' = 0.707(25) = 17.7 \text{ ft}$$

$$d' = W' + \frac{85}{\tan \alpha'} + \overline{AB} + \frac{10}{\tan \alpha'} = 61.4 + \frac{95}{0.515} = 245.7 \text{ ft}$$

$$S' = \sqrt{d'^2 + H^2} - d' = \sqrt{(245.7)^2 + (75)^2} - 245.7 = 10.7 \text{ ft}$$

$$a' = \frac{S'}{1 - \cos \alpha'} = \frac{10.7}{1 - \cos (27.25°)} = 96.4 \text{ ft}$$

$$\Delta a' = a'\left(0.5 - \frac{\alpha'}{360°}\right) = 96.4\left(0.5 - \frac{27.25}{360}\right) = 40.9 \text{ ft}$$

$$K' = \sqrt{K_x K_y} = \sqrt{2}(10^{-5}) \text{ ft/sec}$$

$$Q = K'(\sin^2 \alpha')(a' - \Delta a') = \sqrt{2}(10^{-5})(\sin^2 27.25°)(96.4 - 40.9)(3600)(24)$$

$$= 14.2 \text{ cubic feet per day per foot of dam.} \qquad Ans.$$

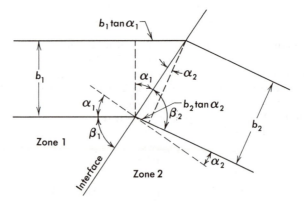

Fig. 6–12. Flow net crossing zone interface.

Another effect of changing permeability is encountered in zoned embankments, when flow lines pass from a zone of one permeability K_1 into a zone of permeability K_2. This effect is illustrated in Fig. 6–12. The heavy dashed lines are equipotential lines representing a certain potential drop ΔH. The seepage ΔQ may be computed on either side of the interface. Thus,

$$\Delta Q = K_1 \frac{\Delta H}{b_1 \tan \alpha_1} b_1 = K_2 \frac{\Delta H}{b_2 \tan \alpha_2} b_2$$

Therefore,

$$\frac{\tan \alpha_1}{\tan \alpha_2} = \frac{K_1}{K_2} = \frac{\tan \beta_2}{\tan \beta_1} \qquad (6\text{-}23)$$

The angles α_1 and α_2 are angles made by the flow lines with the normal to the interface, both of them measured in the same angular direction from that normal.

Once the flow net is carried across the interface, if the same flow lines are retained, the spacing of the equipotential lines must change, so that the flow net becomes a net of rectangles instead of squares, that is,

$$K_1 \frac{(\Delta H)_1}{L_1} b_1 = K_2 \frac{(\Delta H)_2}{L_2} b_2$$

or

$$\left(\frac{b}{L}\right)_2 = \frac{K_1}{K_2}\left(\frac{b}{L}\right)_1 = \frac{K_1}{K_2} \qquad (6\text{-}24)$$

which gives the ratio of the rectangle sides in the flow net in zone 2. Or, if it is desired to retain square figures in the net, the magnitude of the drop in potential between equipotential lines can be changed. Then,

$$(\Delta H)_2 = (\Delta H)_1 \frac{K_1}{K_2} \qquad (6\text{-}25)$$

The flow net may also be used to determine the hydrostatic pressure (also known as "neutral pressure") at any point in the embankment. The total potential is given by the initial potential, H, minus the total potential drop to that point. The static head is the total potential minus the elevation head. Thus,

$$p = \gamma[H - n\Delta H - z] \qquad (6\text{-}26)$$

where p is the hydrostatic pressure at a point in the embankment z feet above the datum, where n potential drops each of magnitude ΔH have occurred, as indicated by the flow net.

If the hydrostatic pressure at a point exceeds the pressure due to the superposed mass of soil and water, then a boiling, or "quick" condition will develop in the mass. For a free water table, as in the embankment, such a condition cannot develop.

However, when seepage water emerges from beneath a dam at the toe or other point of egress, it may well be under a static pressure greater than the exit head, and this will cause the "quick" condition. Furthermore, if these pressures are sufficiently large, they may cause movement of the finer material in the soil, washing it out in the "boil." In effect, this moves the point of egress of the seepage water farther back under the dam where the pressure is still higher. More "fines" will then be eroded and a soil "pipe" may thus quickly be formed. This phenomenon of "piping" may thus undermine the entire dam, and should be prevented by all means. A drainage blanket or toe may be provided, designed as a filter, through which the fines cannot escape but through which the seepage water may flow out easily.

To investigate whether piping may tend to occur, the rate of change of static head (i.e., the hydraulic gradient) at egress, as determined from the flow net, is compared with the hydraulic gradient that would be required to cause the quick condition to occur, and thus to permit movement of the fines in the mass.

The seepage force per unit volume on the sand, in the direction of flow, (upward) is $\gamma(\Delta H)A/AL = \gamma i$, where i is the hydraulic gradient near the point of egress. The force per unit volume resisting this force is simply the *submerged unit weight* of the sand, which is the dry unit weight minus the unit weight of water. This is calculated as follows:

$$\text{Submerged unit weight} = \frac{G\gamma V_s + \gamma V_v}{V_v + V_s} - \gamma = \gamma\left(\frac{G + V_v/V_s}{V_v/V_s + 1} - 1\right)$$

$$= \gamma\left(\frac{G + e}{e + 1} - 1\right) = \gamma\frac{G - 1}{1 + e} \qquad (6\text{-}27)$$

where V_v and V_s are the volumes of voids and solids, respectively, in the mass, G is the specific gravity of the solid material, and e is the "voids ratio" V_v/V_s.

The *critical gradient* is the value of hydraulic gradient at which the seepage force per unit volume becomes equal to the submerged unit weight:

$$\gamma i_c = \gamma \frac{G-1}{1+e}$$

or

$$i_c = \frac{G-1}{1+e} \tag{6-28}$$

The magnitude of i_c is obviously near unity. If it is exceeded, quicksand will form and piping is possible. Either the gradient must be reduced or a filter provided.

The foregoing discussion has assumed fairly simple conditions. For more complex embankment and foundation conditions, the method of finite elements can be used to determine seepage and pressure quantities throughout non-homogeneous regions.

6-13. Arch Dams. The arch dam is subject to essentially the same forces as a gravity dam, but it resists those forces primarily by horizontal arch action. This requires large horizontal reactions to be provided by the abutments. The arch dam, therefore, is restricted to relatively narrow canyon sections with strong rock abutments. Where they can be used, however, arch dams are very effective. The efficiency of the arch as a structural shape is almost intuitively obvious and has been appreciated from antiquity, but this very effectiveness makes arch action extremely intricate and difficult to analyze in detail. An arch dam is especially difficult to analyze, since gravity action is combined with the arch action to a greater or lesser extent.

A very remarkable fact, probably unique in the history of structures, is that, so far as known, there has never been a failure of an arch dam as such. This reflects the inherent ability of the arch stresses to adjust themselves most efficiently to a wide variety of support and loading conditions. The total and essentially uniform crushing of the constituent materials seems to be the only sufficient cause for failure, provided only that the abutments can withstand the total horizontal forces applied.

The stability analysis of an arch dam consists essentially of merely determining the total horizontal loads and comparing them with the abutment strength. There is presumably no danger of sliding or overturning in an arch dam, nor are the foundation pressures critical. The stress analysis, however, is much more difficult. The most desirable method of determining the internal stresses is even yet a matter of considerable difference of opinion, especially as between American and European designers.

For the purpose of preliminary economic studies and for the actual design of many low dams, a thin-cylinder analysis may be sufficient. In this method, all gravity or cantilever action is neglected. The dam is considered as a series of horizontal arches, each 1 ft thick, carrying the horizontal water load to the abutments by arch action. Such a slice is shown in Fig. 6–13.

The rib shown is assumed to be h feet below the reservoir water level, so that the hydrostatic pressure acting radially against the arch is γh. Summing forces in direction parallel to the stream axis yields

$$2R \sin \frac{\theta}{2} = 2\gamma h(r)\left(\sin \frac{\theta}{2}\right)$$

where $2(r)(\sin (\theta/2))$ is the projection of the exposed arch area on a transverse plane. Therefore

$$R = \gamma h(r) \tag{6–29}$$

where R is a single abutment reaction and r is the extrados radius.

If the arch is thin enough to warrant an assumption of uniform stress distribution, then the compressive stress in the masonry, in pounds per square foot, is

$$f = \frac{R}{t} = \gamma h\left(\frac{r}{t}\right) \tag{6–30}$$

The stress permitted has often been taken as one-eighth of the 28-day strength of the concrete. There has been a strong tendency in recent years to increase the allowed stress, 1000 psi being a fairly common figure at present. Some European dams have been designed with stresses allowed up to 1500 psi and higher. For a given stress, the required thickness is

$$t = \frac{\gamma h r}{f} \tag{6–31}$$

indicating that the thickness should increase with depth and with the radius of curvature. It may be noted, too, that the hydrostatic pressure γh may be augmented by earthquake, ice, and other pressures where applicable. A

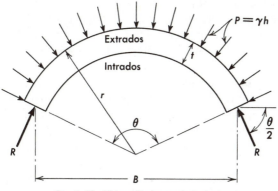

Fig. 6–13. Thin cylinder arch analysis.

general empirical rule is that the thickness should also be at least $(\frac{1}{75})L$ at the crest and at least $(\frac{1}{25})L$ at midheight, where L is the length of the arch ring.

The determination of the radius of curvature (or central angle θ) to use is mainly a matter of economy. The volume of concrete in the rib is

$$
V = \left(r - \frac{t}{2}\right)\theta t = \left[r - \frac{\gamma h r}{2(f)}\right]\theta\left(\frac{\gamma h r}{f}\right)
$$

$$
= r^2\theta\left[1 + \frac{h\gamma}{2(f)}\right]\left(\frac{\gamma h}{f}\right)
$$

$$
= \left[\frac{B/2 + (t/2)\sin(\theta/2)}{\sin(\theta/2)}\right]^2\theta\left[1 + \frac{\gamma h}{2(f)}\right]\left(\frac{\gamma h}{f}\right) \tag{6-32}
$$

This is a rather complicated expression, but essentially it gives the volume of concrete as a function of the central angle θ, for given values of channel width B, depth h, and permitted stress f. It can be simplified materially, without loss of comparative accuracy, by taking the intrados and extrados radii as essentially equal. Then,

$$
V = \frac{\gamma h}{f}\left[\frac{B}{2\sin(\theta/2)}\right]^2\theta \tag{6-33}
$$

The value of θ for which V is minimum is found by setting $dV/d\theta = 0$. Solving the trigonometric equation which results yields $\theta = 133°34'$ for a rib of minimum volume. Other factors, especially topographical conditions, often prevent using this exact value, but the values used are commonly in the range from $100°$ to $140°$.

The optimum value of θ corresponds to a center-line radius of $0.54B$. If the canyon is U-shaped (that is, B does not change significantly with height), then a *constant-radius* dam may be suitable. If the dam is in a V-shaped canyon, then in order to retain the optimum angle (*constant-angle dam*), the radius must decrease from top to bottom.

There are other more precise methods of stress analysis, of course. Arch stresses assuming a uniform arch section can be computed by a method developed by Cain, or for a section with thickened arch-ends by the method of Perkins.

Most large arch dams in this country have been designed by the trial load method developed by the Bureau of Reclamation. By this method the dam is conceived as composed of vertical cantilever elements, with radial sides and horizontal arch elements. The total load is distributed between these two sets of elements by trial, balancing deflections. The cantilever deflections are computed as those due to moment plus shear plus twist, as in the trial load-twist method of gravity dam analysis. Arch deflections also are computed, including effects of temperature. Rotations and deflections at abutments and foundations are assumed also and included in the calculations. All deflections must be balanced radially, circumferentially, and angularly. Once the deflections have all been made to balance, by repeated adjustments of the load distribution, it is relatively straightforward to compute moments, shears, thrusts, and stresses. Both extrados and intrados stresses must be computed, especially at the crown and abutment sections.

Because of the extreme complexity of this type of analysis, designers have been turning more and more to the use of structural models. Models have been used for some time, especially as a check on the stress analysis. Considerable evidence has now accumulated, however, that slight errors in the assumptions as to loading, foundation or abutment conditions, etc., may result in complete redistribution of the computed stresses. Stress measurements on actual structures have often indicated that model measurements are at least as reliable as even the most intricate stress analysis. Accordingly the present trend, especially in Europe, is to base a preliminary design on one of the simpler methods of analysis, and then to make adjustments to this design primarily on the basis of model testing.

Another important recent trend is the use of the finite-element method of analysis in conjunction with the high-speed computer. This technique, together with the development of double-curvature, shell-type, arch structures, may well revolutionize arch-dam design in the future.

6-14. Other Types of Dams.
Quantitatively, the dams already discussed are most important. Rock-fill dams are tending to increase in popularity, however, as are the buttress and multiple-arch types. Rock-fill dams often have an earth core, relatively impervious, and are economical if there exists a convenient supply of good rock and earth for the fill. Usually a rock foundation is desired, because of both seepage and load-carrying characteristics. The dam is analyzed for stability much the same as a gravity dam, the loads being transmitted through the fill by rock-to-rock contacts. Seepage is an important item, in both the dam and the foundation, and is analyzed in the same way as an earth dam. The basic triangular rock section (resting at its natural angle of repose) is provided with a relatively impervious facing of earth or concrete on the upstream slope and by a suitable filter between it and the rock section, for protection from wave action and from drawdown sloughing.

Buttress dams are of several types, as listed in Art. 6-5. Each provides some sort of thin masonry surface transmitting the water loads to spaced supports or buttresses of some type. The structural design of each type is unique to itself but is not usually overly complex. The structural characteristics are indicated by the respective names of the different types. The most popular types are the slab-and-buttress (Ambursen) type and the multiple-arch type. The upstream faces of buttress dams slope rather strongly, in order to utilize the vertical components of hydrostatic pressure. Uplift forces are easily relieved by the gaps between the buttresses. Otherwise the force system is essentially the same as for gravity and arch dams.

The slab-and-buttress type is usually articulated; that is, the slab is not rigidly attached to the buttress. Thus the slab can be designed essentially as a series of horizontally loaded, simply supported beams. Buttresses are massive structures and are therefore difficult to analyze rigorously. They are often approximated as a series of contiguous curved columns. Their design must also include footings to transmit the buttress loads through bearing to the foundation.

The multiple-arch dam is designed by methods of arch analysis, usually neglecting the cantilever action. Larger central angles are used than for the straight arch dam, often 180°, in order to minimize horizontal thrusts between adjacent arches.

6–15. Spillways. Usually the most important of the various appurtenant facilities to a dam is its spillway. The cost of a spillway is one of the major items of expense in a dam construction project and may well determine the type of dam selected.

The function of the spillway is to provide an efficient and safe means of conveying flood discharges, exceeding the capacity of the reservoir to retain them, past the structure to the downstream channel. Ordinary releases are handled through the regular outlet conduits, so that the spillway may actually be used very rarely. However, it is very important that the rare occurrence of an excessive discharge be not permitted to destroy the entire structure, and so a spillway of large capacity must be available.

The capacity is determined by hydrological studies over the drainage area in question; in general, it is desired to provide adequate capacity to transmit the largest flood that may possibly occur at the site. This will involve extrapolation of past flood frequency records and probably studies of a hydrometeorological nature which will indicate the maximum rainfall that can be produced on the area.

The complete spillway will involve several structural elements, each of which requires a particular hydraulic analysis. The *entrance channel* collects and conveys the flood water to a *control structure*, which functions as a weir or other control to measure or determine the quantity of flow passing through the spillway in relation to the reservoir water surface elevation. The *discharge carrier* is a steep channel carrying the flow below the control structure downstream to an *energy dissipator*, which causes a hydraulic jump to form, slowing down the supercritical flow in the discharge carrier to a subcritical velocity. The flow is then carried on out at the lowered velocity through an *outlet channel* into the stream channel below the dam. *Transitions* must also be designed for smooth conveyance of the flow from each spillway element into the next succeeding element. One or more of the above elements may be omitted from certain types of spillways.

Although spillways are normally thought of in relation to dams, similar structures are used also for other purposes. In general, a spillway-type structure is required whenever it is necessary to convey a flow rapidly, safely, and efficiently from a high to a low elevation. Such structures are frequently used in connection with soil conservation, highway drainage, and other hydraulic engineering projects.

6–16. Overfall Spillways. The most common spillway, and usually the most economical for large discharges, is the overfall, or overflow, spillway (see Fig. 6–1). This structure is essentially a large rectangular weir, with crest curved to conform to the natural shape of an overflowing nappe of water. This feature is for the dual purpose of providing the maximum weir discharge coefficient and of preventing the development of a vacuum in the separation zone between the crest and the underside of the nappe. The latter phenomenon would not only cause an additional force on the dam but, more seriously, might lead to damage by cavitation.

Consider the trajectory of a particle of water in the nappe flowing over a sharp-crested rectangular weir, as shown in Fig. 6–14. At time t after passing the crest, the particle's trajectory has coordinates (x, y) such that

$$x = V_0 t \cos \theta \tag{6-34}$$

$$y = y_0 + V_0 t \sin \theta - \tfrac{1}{2} g t^2 \tag{6-35}$$

where V_0, θ, and y_0 are the values of the velocity, direction, and elevation of the particle at the instant it passes the crest. Eliminating t from the two parametric equations yields the following equation for the trajectory of the particle:

$$y = y_0 + x \tan \theta - \frac{g x^2}{2(V_0 \cos \theta)^2}$$

In order to express the equation in dimensionless form, divide through by the total spillway head H, composed of the static head h and the approach velocity head $h_v = V_a^2/2g$:

$$\frac{y}{H} = \frac{y_0}{H} + \frac{x}{H} \tan \theta - \frac{S}{2}\left(\frac{x}{H}\right)^2 \tag{6-36}$$

where, for simplicity, S replaces $gH/(V_0 \cos \theta)^2$. The slope of the trajectory is

$$d\left(\frac{y}{H}\right) \Big/ d\left(\frac{x}{H}\right) = \tan \theta - S\left(\frac{x}{H}\right) \tag{6-37}$$

and the rate of change of slope is

$$d^2\left(\frac{y}{H}\right) \Big/ d\left(\frac{x}{H}\right)^2 = -S \tag{6-38}$$

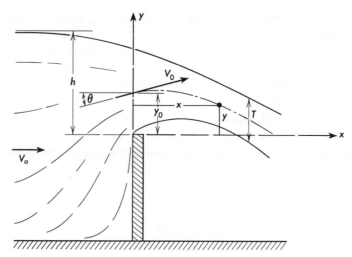

Fig. 6–14. Trajectory of nappe.

The equations thus far are perfectly general equations for any trajectory (neglecting air resistance). For specific application to the overflow nappe, however, the initial conditions must be known (i.e., y_0, V_0, θ). These terms are related in a complex manner to the whole field of approaching flow. The velocity-pressure-direction relationships in such a converging field with a free surface and arbitrary lower boundary have not yet been satisfactorily analyzed mathematically, so that empirical methods must be used.

Model studies conducted by Blaisdell for the most common case, that of a rectangular approach channel of the same width as the spillway and with a vertical upstream spillway face, have yielded the following empirical equations:

$$\frac{y_b}{H} = \left[0.150 - 0.45\left(\frac{h_v}{H}\right)\right]$$
$$+ \left[0.411 - 1.603\left(\frac{h_v}{H}\right) - \sqrt{1.568\left(\frac{h_v}{H}\right)^2 - 0.892\left(\frac{h_v}{H}\right) + 0.127}\right]\left(\frac{x_b}{H}\right)$$
$$- \left[0.425 - 0.25\left(\frac{h_v}{H}\right)\right]\left(\frac{x_b}{H}\right)^2 \quad (6\text{--}39)$$

and

$$\frac{T}{H} = 0.57 - 2\left[\frac{h_v}{H} - 0.208\right]^2 e^{10(h_v/H - 0.208)} \quad (6\text{--}40)$$

where (x_B, y_B) are the coordinates of the bottom side of the nappe, and T is the vertical thickness of the nappe, so that $y_T = y_B + T$.

For the common case of a negligible approach velocity, the equations simplify to

$$\frac{y_B}{H} = 0.150 + 0.055\left(\frac{x_B}{H}\right) - 0.425\left(\frac{x_B}{H}\right)^2 \qquad (6\text{-}41)$$

and

$$\frac{T}{H} = 0.56 \qquad (6\text{-}42)$$

All these equations are actually valid only for $x_B/H \geq 0.5$. They are valid for approach channels of all depths and for all subcritical approach velocities.

The equations are directly applicable to the determination of the point of impact of a free-falling nappe on an apron. In addition the equation of the lower side of the nappe may be used to determine the overfall spillway crest shape, at least for the portion of the spillway downstream from the peak and upstream from the reverse curve leading to the apron.

A simple crest profile which has been found by the U.S. Waterways Experiment Station to give good agreement with actual nappe measurements on prototype weirs is sketched in Fig. 6–15.

The exponential crest curve is continued until it becomes tangent to the slope of the downstream face, the latter having been determined by structural requirements. If the crest requirement is such that a thicker section is required than needed for stability, a reverse curve (of radius at least equal to one-fourth the spillway height) will be needed to bring the crest back to tangency with the downstream face. As an alternative to this, the crest curve can be transposed upstream by the distance necessary to bring it to tangency with the downstream slope, leaving a projecting "corbel" at the lip. This expedient may save a substantial amount of concrete.

The profile shown in Fig. 6–15 is for a vertical upstream face only. For sloping upstream faces, the crest shape may be flattened. The profiles

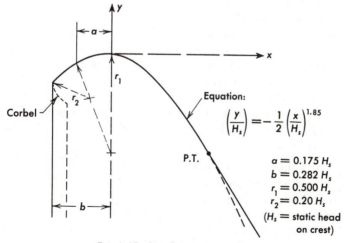

Fig. 6–15. Overflow crest profile.

suggested by the Waterways Experiment Station, based mainly on the Reclamation Bureau data, are indicated by the accompanying table, which refers to Fig. 6–15:

Upstream Slope	$\dfrac{a}{H_s}$	$\dfrac{b}{H_s}$	$\dfrac{r_1}{H_s}$	$\dfrac{r_2}{H_s}$	Crest Equation
$0H:3V$	0.175	0.282	0.50	0.20	$\dfrac{y}{H_s} = -0.500 \left(\dfrac{x}{H_s}\right)^{1.85}$
$1H:3V$	0.139	0.237	0.68	0.21	$\dfrac{y}{H_s} = -0.516 \left(\dfrac{x}{H_s}\right)^{1.836}$
$2H:3V$	0.115	0.214	0.48	0.22	$\dfrac{y}{H_s} = -0.515 \left(\dfrac{x}{H_s}\right)^{1.810}$
$3H:3V$	0	0.199	0.45	∞	$\dfrac{y}{H_s} = -0.534 \left(\dfrac{x}{H_s}\right)^{1.776}$

The equation for discharge over the overflow spillway is essentially the same as for a rectangular weir:

$$Q = CLH^{3/2} \tag{6–43}$$

where Q = design discharge in cfs

L = crest length in feet

H = total head on crest, $H_s + h_v$

C = crest coefficient, ranging from about 3.1 to 4.1, depending on h_v/H and shape of approach

The crest coefficient for a particular spillway is often determined by model tests. If the approach velocity is negligible (usually the case if the static head H_s is less than 75 per cent of the height of the spillway), then the coefficient may be taken as 4.03.

The nappe will conform to the crest as designed for only the design head. As the head decreases, the discharge coefficient will decrease, approaching about 3.1 as the static head approaches zero. The nappe will adhere to the crest and pressure on the crest will be above atmospheric. If the head exceeds the design head, the coefficient approaches a limiting value of about 4.14 when the design head is exceeded by about 40 per cent. Any greater excess than this should not be permitted, as crest pressures become negative, resulting in possible vibration and even cavitation.

6–17. Side Channel Spillways. The side channel spillway combines an overfall section with a channel parallel to it, which carries the spillway discharge away to a chute or tunnel. An outstanding example is the Arizona Spillway at Hoover Dam, shown in Fig. 6–16. The side channel must be sloped sufficiently to accelerate the accumulating flow in the channel. The minimum slope and depth at each point will usually be desired in order to

Fig. 6–16. Side-channel spillway, Hoover Dam. (U.S. Bureau of Reclamation.)

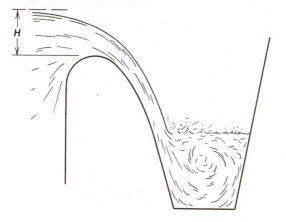

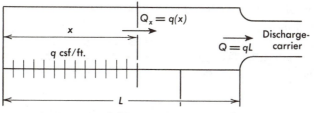

Overfall Section

Fig. 6–17. Side-channel spillway.

259

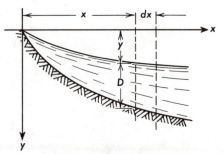

Fig. 6–18. Profile in side channel.

minimize excavation and other construction costs. A schematic plan and transverse section of such a spillway are shown in Fig. 6–17.

The discharge per foot of width over the overfall section is

$$q = CH^{3/2} \tag{6-44}$$

where H is the total reservoir head over the spillway and C is the crest discharge coefficient. At the distance x from the beginning of the side channel, the discharge in the channel is thus

$$Q = qx = CH^{3/2}x \tag{6-45}$$

When x becomes greater than L, the discharge of course then remains constant at qL, usually entering the discharge carrier through some form of control section.

The flow in the channel cannot be analyzed by the energy equation, because of the excess turbulence and energy dissipation in the channel; in fact, one of the main purposes of the control section at the outlet is to maintain sufficient depths in the channel to provide a "water cushion" for the overfall flow, so that the energy of the latter can be substantially reduced before entering the discharge chute or tunnel. However, the following method of analysis,[1] based on the momentum principle, has been adequately validated by both model and prototype measurements.

Consider a longitudinal section through the channel and water flow profile, as shown in Fig. 6–18. At section x, the discharge is $Q = qx$. At section $x + dx$, the discharge is $Q + dQ$ or $Q + q\,dx$.

The body of water between these two sections is now analyzed by the momentum principle, which requires that the resultant force on the body be equal to the time-rate of change of its momentum. The force system is sketched in Fig. 6–19. The forces acting are the hydrostatic forces and the fluid weight. The friction force in the short distance dx can be neglected.

---

[1] This method of analysis can be applied to any open channel in which the discharge increases linearly along the channel, as in gutters, wash water troughs, effluent channels around sewage treatment tanks, etc. See also the discussion of *spatially varied flow* in Article 5–13.

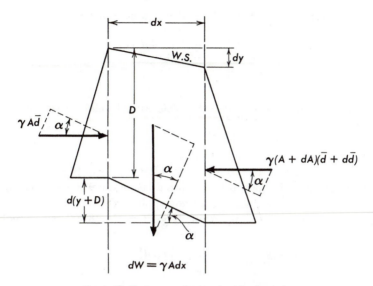

Fig. 6–19. Forces on element in side channel.

It is desired to obtain the equation of the flow profile in the side channel, as measured in terms of the x and y coordinates shown in Fig. 6–18. The bed slope angle is α and $\bar{d}$ is the distance from the water surface to the centroid of the area A. $Q\,dt$ is the volume moving past the section in time dt. The velocity of flow at the section is V.

By the momentum equation:

$$\sum F = \Delta(\rho Q V)$$

Therefore:

$$\gamma A(dx)\sin\alpha + \gamma A\,\bar{d}\cos\alpha - \gamma(A + dA)(\bar{d} + d\bar{d})\cos\alpha$$
$$= [\rho(Q + q\,dx)(V + dV) - \rho Q V]\cos\alpha$$

Simplifying, and eliminating second-order differentials:

$$A\left[(dx)\tan\alpha + \bar{d} - (\bar{d} + dd)\left(1 + \frac{dA}{A}\right)\right] = \frac{1}{g}(Q\,dV + qV\,dx)$$

Now, since $\tan\alpha = \dfrac{d(y + D)}{dx}$ and $q = \dfrac{Q}{x}$, this becomes:

$$A\,dx\left[\frac{d(y + D)}{dx} - \frac{A\,(d\bar{d}) + \bar{d}(dA)}{A\,dx}\right] = \frac{Q}{g}\left(dV + \frac{V\,dx}{x}\right)$$

Replacing $(A\,dx)$ by $(Q\,dt)$:

$$Q\,dt\left[\frac{d(y + D)}{dx} - \frac{d(A\bar{d})}{A\,dx}\right] = \frac{Q\,dt}{g}\left[\frac{dx}{dt}\left(\frac{dV}{dx}\right) + \frac{V}{x}\left(\frac{dx}{dt}\right)\right] \qquad (6\text{-}46)$$

The second term in the left-hand member can be written:

$$\frac{d(A\dot{d})}{A\ dx} = \frac{\dfrac{d}{dx}\left[\displaystyle\int_0^D d(dA)\right]}{\left[\displaystyle\int_0^D dA\right]} \tag{6-47}$$

For any cross-section whose width at any elevation is related to its depth by an equation of the form

$$b = c + K(D - d)^n \tag{6-48}$$

(and this includes trapezoids, rectangles, triangles, parabolas, etc.), this function can be shown[2] also to reduce to dD/dx. In Eq. (6–48) c, K, and n are any constants, b is the width at depth d below the water surface, and D is the total depth. Although it does not apply strictly to all sections, any differences are relatively unimportant. Therefore Eq. (6–46) can be reduced to

$$\frac{d(y + D)}{dx} - \frac{dD}{dx} = \frac{1}{g}\left(V\frac{dV}{dx} + \frac{V^2}{x}\right).$$

Finally, the differential equation of the flow profile becomes:

$$\frac{dy}{dx} = \frac{1}{g}\left(V\frac{dV}{dx} + \frac{V^2}{x}\right) \tag{6-49}$$

The velocity can be assumed to vary with x in any arbitrary manner, provided the channel is then designed to produce such a relation. Assume, therefore, that the velocity is an exponential function of x:

$$V = ax^n \tag{6-50}$$

from which

$$\frac{dV}{dx} = nax^{n-1} = \frac{nV}{x} \tag{6-51}$$

Equation (6–49) can then be written

$$y = \frac{1}{g}\int_0^x (na^2x^{2n-1} + a^2x^{2n-1})\ dx = \frac{a^2(n + 1)}{g}\int_0^x x^{2n-1}\ dx$$

[2] Reduction as follows:

$$\frac{d}{dx}\int_0^D d\ dA \bigg/ \int_0^D dA = \frac{d}{dx}\int_0^D d[c + K(D - d)^n]\ dd \bigg/ \int_0^D [c + K(D - d)^n]\ dd$$

$$= \frac{d}{dx}\left[\frac{cD^2}{2} + K\left(\frac{D}{n + 1}D^{n+1} - \frac{D^{n+2}}{n + 2}\right)\right] \bigg/ \left(cD + K\frac{D^{n+1}}{n + 1}\right)$$

$$= \frac{d}{dx}\left[\frac{cD^2}{2} + \frac{KD^{n+2}}{(n + 1)(n + 2)}\right] \bigg/ \left(cD + \frac{KD^{n+1}}{n + 1}\right)$$

$$= \frac{dD}{dx}\left(cD + \frac{K}{n + 1}D^{n+1}\right) \bigg/ \left(cD + \frac{K}{n + 1}D^{n+1}\right) = \frac{dD}{dx}$$

$$= \frac{a^2(n+1)}{g(2n)} x^{2n} \tag{6-52}$$

This may also be written

$$y = \left(\frac{n+1}{n}\right)\left(\frac{V^2}{2g}\right) \tag{6-53}$$

To satisfy these conditions, the channel cross-section and bottom profile must be designed in such a way as to make the water surface drop proportionately to its velocity head. The channel cross-section is usually trapezoidal, with narrow bottom width and steep side slopes. The constants a and n are arbitrary, and are selected to produce a profile which will most economically conform to the existing topography.

It frequently occurs that the profile elevation at the channel outlet will be fixed by the control section leading to the discharge carrier. This, in effect, specifies the constant a if the exponent n is assumed. In addition, the elevation is determined at the upstream end (at the origin of coordinates) by the overfall crest elevation. Since the overfall flow is essentially at critical depth, which for a rectangular section is at two-thirds the specific energy, the spillway may be as much as two-thirds "submerged" at its upper end without any important effect on its capacity.

The shape of the flow profile between these two limiting points depends on the choice of n. From Eq. (6-52) it will be noted that the profile is rectilinear if n is $\frac{1}{2}$, concave downward if n exceeds $\frac{1}{2}$ and concave upward for n less than $\frac{1}{2}$. The best value is determined by making several trials to find which profile best fits the topography at the site and is most economical to construct.

The equation of the channel bottom is determined by adding the required depth of flow to the coordinate of the water profile. The depth of flow is computed from the required velocity and the given cross-sectional shape. For a rectangular channel, for example, the bottom equation would be

$$y_b = y + D = \frac{a^2(n+1)}{g(2n)} x^{2n} + \frac{qx}{Bax^n} \tag{6-54}$$

The flow in the side-channel is very complex, with transverse vortices, non-uniform velocity distributions, high energy losses and other factors. Attempts to incorporate these into the momentum analysis do not seem at present to warrant the additional complexities thus introduced into the analysis. Experience in both model tests and prototype studies appears to give adequate support to the method of analysis outlined above.

Example 6-3

The overflow spillway leading to a side channel has a discharge coefficient of 3.9, under a design head of 16 feet and discharge of 78,000 cfs. The side channel has a rectangular cross-section, 156 feet wide, with a control section at the outlet. Assume the channel velocity will be made to vary linearly as the distance from the beginning of the channel, at which point the water profile is at the same elevation as the overflow crest. Determine the equation of the channel bed profile.

Solution. Total discharge into the side-channel is:

$$Q = CLH^{3/2} = 3.9L(16)^{3/2} = 78,000 \text{ cfs}$$

from which, required spillway crest length (assumed same as length of side-channel) = 312.5 ft.
Depth of flow at control section at outlet

$$= D_c = \sqrt[3]{\frac{Q^2}{gB^2}} = \sqrt[3]{\frac{(500)^2}{32.2}} = 19.8 \text{ ft}$$

$$\text{Velocity at outlet} = V_c = \frac{Q}{BD_c} = \frac{500}{19.8} = 25.25 \text{ ft/sec}$$

$$\text{Assuming } V = ax \text{ (i.e., } n = 1), \text{ then } a = \frac{V_c}{L} = \frac{25.25}{312.5} = \frac{8}{99}$$

$$\text{Equation of flow profile is: } y = \frac{a^2(n+1)}{2ng} x = \frac{a^2}{g} x$$

At any section, x, depth is

$$D_x = \frac{Q}{BV_x} = \frac{3.9(64)(x)}{156ax}$$

$$= \frac{3.9(64)(99)}{156(8)} = 19.8 \text{ ft}$$

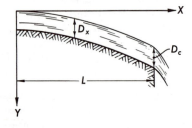

Thus, for the assumed conditions ($n = 1$), depth of flow in the side-channel is constant.
The equation of the bed profile is:

$$y_{bed} = y_{W.S.} + D = \left(\frac{8}{99}\right)^2 \frac{x^2}{g} + 19.8$$

$$= 0.000202x^2 + 19.8 \qquad\qquad Ans.$$

6–18. Siphon Spillways.   The discharge over an overfall spillway is a function of the total reservoir head measured above the spillway crest. This effective head can be materially increased by enclosing the crest and permitting the resulting conduit to flow full. The head on the spillway is then the difference in elevation between the reservoir surface and the spillway outlet.

However, the flow near the crest of the spillway must then be under a negative pressure; in other words the conduit becomes a *siphon*, and all necessary precautions must be taken to ensure that the vacuum is maintained and also that it does not become so strong as to cause trouble. A siphon spillway is sketched in Fig. 6–20.

The maximum vacuum at the spillway crest is theoretically 34 ft of water, at sea level. Actually, it is considered unwise to permit a vacuum of more than about 24 ft of water. Flow will begin when the reservoir level rises above the spillway crest, but siphon action will not start until the level is above the conduit crown at the crest; also air must be prevented from entering at the outlet end. This is sometimes accomplished by placing a constriction or an upward deflecting lip at the outlet.

Once siphoning begins, the conduit will flow full until the vacuum is broken by the dropping of the reservoir level to the elevation of the inlet crown (or of an air vent placed at any arbitrary desired elevation). Since the discharge occurs under a large head, a large quantity of water can be passed in a small conduit and in a short time. Thus, the siphon gives an effective means of automatic control of the reservoir level within the desired range. Its main use is for installations requiring such control or where it is desired to carry the design discharge with a minimum reservoir surface rise.

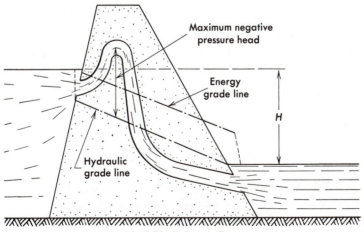

Fig. 6–20. Siphon spillway.

The critical vacuum will form when the reservoir level is just above the inlet crown. The energy equation from the reservoir level to the crest for this case is

$$\frac{p_{atm}}{\gamma} + \frac{(v_{res})^2}{2g} = \frac{p_{crest}}{\gamma} + \frac{v^2}{2g} + \text{(losses and rise from inlet to crest)} \qquad (6\text{--}55)$$

Each term in the first member of Eq. (6–55) is zero. If the inlet is properly rounded, the losses to the crest will be small and their effect can be included in the limit assigned to the negative potential head, which we shall take to be −24 ft water.

Therefore, the velocity át the crest section 'is obtained by setting the velocity head at the crest equal to +24 ft water. Thus,

$$v = \sqrt{2g(24)} = 39.3 \text{ ft/sec}$$

However, the velocity will not be uniform across the section. Since the flow is curvilinear at this section, and since each streamline may be assumed to possess the same total energy (because each started with the same total energy in the reservoir, and subsequent differences in energy losses are small), therefore the motion around the crest may be assumed to be essentially that of a *free vortex*, as described in Eq. (2–19). The horizontal velocity distribution at the crest is therefore as shown in Fig. 6–21. By the vortex equation

$$v_0 r_0 = v_i r_i = vr = \text{constant} \qquad (6\text{--}56)$$

Since $v_i = 39.3$ fps,.the equation is

$$v = 39.3 \frac{r_i}{r} \qquad (6\text{--}57)$$

The maximum discharge which the siphon can pass is limited by these conditions at the crest, and can be computed as follows:

$$Q_{max} = \int_{r_i}^{r_o} v \, dA = \int_{r_i}^{r_o} 39.3 \frac{r_i}{r} \, dA$$

The exact value of course depends on the cross-sectional shape of the siphon. In the simplest and most common case of a rectangular section b feet wide, it becomes

$$Q_{max} = 39.3 r_i b \int_{r_i}^{r_o} \frac{dr}{r} = 39.3 r_i b \log_e \frac{r_o}{r_i} = 90.5 r_i b \log_{10} \frac{r_o}{r_i} \qquad (6\text{--}58)$$

The average velocity at the crest section is then

$$V = \frac{Q}{A} = \frac{90.5 r_i}{r_o - r_i} \log_{10} \frac{r_o}{r_i} \qquad (6\text{--}59)$$

and this will be the average velocity at all sections along the siphon barrel, except where it is necessary to expand or constrict the section.

Since the velocity depends upon the total head H, it may be necessary to limit this head in some manner, so as to prevent V from exceeding the value

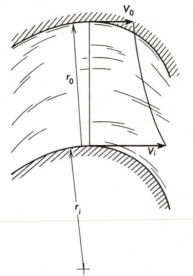

Fig. 6–21. Flow at crest of siphon.

specified in Eq. (6–59). This can be done either by increasing the outlet loss by constricting the outlet section or by decreasing the total head through raising the elevation at the outlet.

If V is the velocity along the barrel and V_o is the outlet velocity, then the energy equation from the reservoir to the outlet is

$$H = \frac{V_o{}^2}{2g} + \frac{V^2}{2g}(k_e + k_f + k_b + \cdots)$$ (6–60)

where k_e, k_f, k_b, etc., represent loss coefficients in the barrel due to entrance, friction, bend, etc.

Since V is determined by the expression of Eq. (6–59) and since the various loss coefficients are determined by the conduit geometry and Reynolds number, either H or V_o must be determined by the value assigned to the other.

If V_o is equal to V (i.e., if the siphon has neither constriction nor expansion at the outlet), then the permitted H is

$$H = \frac{V^2}{2g}(1 + k_e + k_f + k_b + \cdots)$$ (6–61)

If the actual head is greater than this, then the conduit may be aligned so as to emerge above the tailwater, at an elevation H feet below the reservoir level. If the actual head is less than H, then the velocity can be computed from Eq. (6–61); in this case it may be possible to increase the discharge by flaring the outlet, making the outlet loss coefficient less than unity, and by attempting to minimize other barrel losses.

The other alternative, for limiting the velocity at the crest, is to constrict the outlet. The outlet velocity is given by Eq. (6–60) for the given head H. The required outlet area is then

$$A_o = \frac{Q}{V_o} = \frac{AV}{V_o} = \frac{A}{\sqrt{\dfrac{2gH}{V^2} - (k_e + k_f + k_b + \cdots)}} \qquad (6\text{–}62)$$

where A is the normal cross-sectional area of the siphon barrel.

Example 6–4

A reservoir is designed to impound water to a maximum of 60 feet above the tailwater level, for a design discharge of 1600 cfs.

    (a) If an overflow spillway is used, with a crest elevation 5 feet below the maximum headwater elevation and with a well-designed crest shape, what must be the length of the crest?

    (b) If the crest is enclosed to form a siphon spillway, what width of siphon should be used? Assume a constant cross-section 4 feet in depth. The crest can be approximated by a circular arc with a radius of 4 feet, with the crown thus having an 8-ft radius. The total head loss (not including outlet velocity head) is estimated to be 20 feet. How far above or below the tailwater level should the siphon discharge?

Solution.

    (a) For overflow spillway, required crest length,

$$L = \frac{Q}{C(H)^{3/2}} = \frac{1600}{4(5)^{3/2}} = 35.7 \text{ ft} \qquad Ans.$$

    (b) Average velocity allowed through siphon cross-section

$$= V = 90.5 \left(\frac{r_i}{r_0 - r_i} \right) \log \frac{r_0}{r_i} = 90.5 \left(\frac{4}{8 - 4} \right) \log \frac{8}{4} = 27.4 \text{ ft/sec}$$

$$\text{Required crest width} = B = \frac{Q}{VD} = \frac{1600}{27.4(4)} = 14.6 \text{ ft} \qquad Ans.$$

By the energy equation, from the reservoir surface to the siphon outlet, head permitted without exceeding permissible velocity

$$= H = \frac{V_0^2}{2g} + H_L = \frac{(27.4)^2}{64.4} + 20 = 31.6 \text{ ft}$$

    Distance above tailwater at which to place siphon outlet = $60.0 - 31.6 = 28.4$ feet   *Ans.*

6–19. Shaft Spillways. The shaft spillway is simply a closed conduit in which the flood flow is conveyed rapidly from a high to a low elevation. It is similar to the siphon spillway except for the absence of siphon action. Spillways of this general type are used not only for dams (where they are commonly known as morning-glory spillways or glory holes) but also for erosion control structures and as highway culverts (then commonly known as drop-inlet spillways or drop-inlet culverts). For these structures, a rectangular dead-end channel sometimes is used to lead to the shaft inlet, the spillway then being known as a box-inlet spillway. In any case, the essential hydraulic geometry consists of an inlet of some type, a vertical shaft, a bend, and a horizontal section with possibly a flared or bucket-type outlet section. These features are illustrated in Fig. 6–22 (note that the shaft, both vertical and horizontal, may also have a component normal to the stream axis).

As long as the inlet is unsubmerged, it functions as a weir, with the discharge given by

$$Q = cph^{3/2} \tag{6-63}$$

where p is the inlet perimeter and c is the weir coefficient. The amount of flare on the "morning glory" is determined by the required crest length p. A spillway of this type, in operation, is pictured in Fig. 6–23.

However, at discharges large enough to give complete submergence, the inlet ceases to function as a weir and becomes simply a pipe inlet. The head on the entire pipe, H, now controls the discharge:

$$Q = A\sqrt{\frac{2gH}{k_e + k_f + k_b + \cdots + k_o}} \tag{6-64}$$

The various loss coefficients, $k_e, k_f, k_b, \ldots, k_o$ are, however, very difficult to predict, because of the complex nature of the turbulence in the flow. Entrance vortices, entrained air slugs, non-symmetrical approach flows, spiral motion in the vertical shaft, complex eddying at the main bend, and other factors serve to render an accurate hydraulic analysis practically impossible. Model studies are often used for design data on shaft spillways,

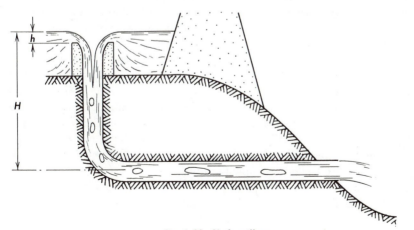

Fig. 6–22. Shaft spillway.

Fig. 6–23. Morning-glory spillway, Hungry Horse Dam, Montana. (U.S. Bureau of Reclamation.)

but it is still uncertain how accurately these conform to prototype behavior (mainly because the effect of entrained air cannot be reduced to model scale).

6–20. Chute Spillways. A chute spillway, also known as a trough spillway, is simply a steep open channel. It is often used as the main spillway structure, as in Fig. 6–6, but is frequently used too as an outlet channel for an overfall, side-channel, or siphon spillway.

For a straight, uniform section, a regular open channel formula such as the Manning formula may be used for the portion of the chute in which the flow is uniform. However, uniform flow will not be attained unless the channel is fairly long. Therefore non-uniform flow profile computations may be necessary, starting with critical depth at the control section at the inlet and proceeding incrementally downstream along the dropdown curve. The capacity of the channel is normally governed by the capacity of its inlet, in accordance with the critical flow formula.

The depth of flow from point to point along the steep channel is determined by the dropdown profile computations. However, the effects of non-hydrostatic pressure distribution and of bulking due to air entrainment may need to be considered.

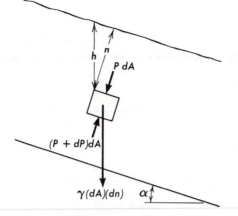

Fig. 6–24. Pressures on steep slope.

The pressure head for flow on a steep slope is not p/γ, as for a static fluid, but $(p/\gamma) \cos^2 \alpha$, as seen from Fig. 6–24. A small fluid element, of volume $dn\, dA$ is in equilibrium under forces due to its own weight and to fluid pressure. Summing forces in the n-direction, normal to the bed, gives

$$dp\, dA = \gamma\, dn\, dA \cos \alpha$$

or

$$dp = \gamma(dn)(\cos \alpha)$$

The pressure at a vertical depth h below the free surface is therefore

$$p = \gamma n \cos \alpha = \gamma h \cos^2 \alpha \qquad (6\text{--}65)$$

Air entrainment is a phenomenon that has received much research attention during the past decade but which is still incompletely understood. Air insufflation apparently begins at that point in the channel at which the turbulent boundary layer thickness becomes equal to flow depth. The flow turbulence entraps air bubbles and mixes them with the flow; at the same time, water droplets separate from the flow and create a region of spray above the main flow. The over-all result is an ill-defined surface of "white water" (see Fig. 6–1).

High-speed photography and electronic measurements have shown that the air concentration increases with distance from the bed, apparently governed by the turbulent mixing process, and that the flow surface is difficult to define, with a region of air filled with water spray above and water filled with air bubbles below.

A considerable "bulking" effect is therefore present, together with significant modification of the turbulence and friction characteristics of the flow. This must be provided for in flow profile calculations, either by an "equivalent Manning roughness coefficient," higher than normally used for the given channel type, or by an air-concentration factor which provides directly for an increased effective depth of flow. An increased flow cross-section of 25 per cent or more due to this factor is not at all uncommon.

The U.S. Corps of Engineers has recommended the following empirical formula for use in design, pending development of a more rational approach:

$$c = 0.70 \log_{10} \frac{\sin \alpha}{q^{1/5}} + 0.97 \tag{6-66}$$

where c is the air-concentration ratio, that is, the ratio of the air volume in a given channel length to the volume of mixed air and water; q is the discharge per unit width; and α is the angle of bed slope. For a rectangular cross-section, the depth of air-entrained flow is then given by

$$D_\alpha = \frac{D}{1-c} \tag{6-67}$$

When the chute involves transitions, bends, or channel obstructions, complex wave patterns may develop in the supercritical flow. These may seriously modify the assumed flow conditions. Analysis of these problems involves wave mechanisms, as discussed in Arts. 5–6 through 5–11. Because of the complexity of supercritical flow phenomena, model studies are often used for their solution.

6–21. Spillway Crest Gates.

Various types of gates have been used to provide control over the discharge on the spillway. The gates may themselves constitute the control structure, or they may be used in connection with an overfall or other spillway structure. On the other hand, many spillways are designed for operation without gates. The most common gate types employed on storage dam projects are vertical-lift gates, tainter gates, rolling gates, and drum gates.

The *vertical-lift gate* or *sliding gate*, is a plane structure made of wood, cast iron, or structural steel, sliding in vertical grooves on piers. Sliding friction is reduced by means of rollers, attached either to the piers (the *Stoney gate*) or to the gate (*fixed-wheel* gate). A sliding gate is sketched schematically in Fig. 6–25. For small gates, the rollers are sometimes omitted. The structural and mechanical design of a sliding gate involves consideration of the hydrostatic force on the gate, the hoisting force, the weight of the gate, and the roller friction. The forces, though few, are very large and require careful attention to stress distributions and concentrations.

The *tainter gate*, called also the *radial gate*, is sketched in Fig. 6–26. It is a steel framework with a circular segment for its face, rotating about its center of curvature. Since all hydrostatic pressures are radial, passing through the trunnion bearing, the thrust on the latter is substituted for the roller friction of the vertical lift gate. Otherwise the forces are the same. The pin friction is usually much less than the roller friction, so that the tainter gate is comparatively light and easy to operate.

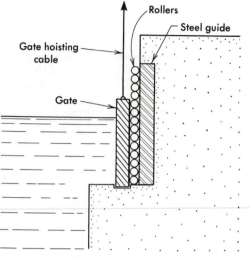

Fig. 6–25. Lift gate.

The *rolling gate* is a steel cylinder spanning between spillway crest piers. It is opened by rolling up an inclined toothed rack on the piers, as sketched in Fig. 6–27. A cylindrical segment is commonly attached to the lower limb of the roller, to give greater height to the gate and to provide a watertight bearing on the spillway crest. Numerous variations of the basic idea have been used.

The *drum gate*, as shown in Fig. 6–28, is of such shape as to conform to the spillway crest itself when open, thus permitting a maximum discharge coefficient to be attained. It is closed by hydrostatic pressure, by admitting water to the recess chamber in the crest, and lowered by draining the chamber. The gate may be hinged at either the upstream or the downstream end, usually the former. The drum gate design is of more recent development than

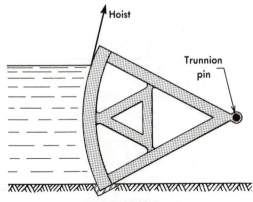

Fig. 6–26. Radial gate.

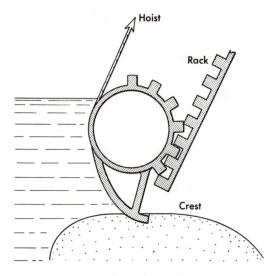

Fig. 6–27. Rolling gate.

the others, but has recently come into rather wide use.

On any of the gates, the hydrostatic forces are easy to determine when the gate is closed. When the gate is open, however, the curvilinear flow past it results in a non-hydrostatic distribution of pressure on the gate. The resulting hydrodynamic forces are determined either by model studies or by flow net analysis.

The weight of the gate is of course mutually interdependent with the design

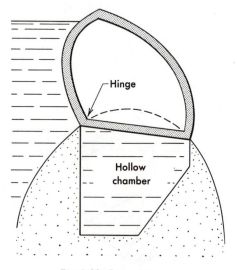

Fig. 6–28. Drum gate.

dimensions. For preliminary analysis, the weight can be computed from the following relation:

$$W = kL^m H^n \tag{6-68}$$

where L is the gate opening and H is the depth of water against the gate, both in feet. k, m, and n are constants for each particular type of gate. The following table gives approximate values for these constants, based on actual designs.

Type of Gate	m	n	Range of k	Mean k
Vertical-lift	1.5	1.75	0.8–2.00	1.20
Tainter	1.9	1.35	0.85–1.45	1.16
Rolling	1.5	1.67	2.40–3.40	2.85
Drum	1.33	1.33	26.0–35.0	31.0

Many other types of gates have been used for various applications. *Flash boards* constitute the simplest type, consisting of horizontal members designed to fail when the static head against it reaches a certain magnitude. *Stop logs* (horizontal) and *needles* (vertical) are members which are removable and replaceable, usually by hand, and are used for small installations.

The *hinged-leaf gate*, known in Europe as the *shutter weir*, is essentially a large flat plate which, when closed, slopes downstream at about 45°. It is prevented from opening by a counterweight system. However, when the stage reaches a determined amount, the hydrostatic force becomes sufficient to open the gate partially and permit water to flow over it; it closes automatically when the head drops sufficiently.

The *bear-trap gate*, or *movable dam*, is formed of two leaves, one hinged at the upstream end, the other at the downstream end. They are joined at the center by a sliding hinge. When the gate is open, the upstream leaf lies partially on top of the other, both lying flat along the spillway crest or stream bed. To close the gate, water is admitted to a chamber beneath it, causing the leaves to rise at the sliding center hinge.

6-22. Stilling Basins. A stilling basin is a channel structure of mild slope, placed at the outlet of a spillway, chute, or other high-velocity flow channel, whose purpose is to dissipate some of the high kinetic energy of the flow in a hydraulic jump. The jump also serves to transform much of the kinetic energy into pressure, i.e., increased flow depth.

Stilling basins or other energy-dissipating devices are almost always necessary in such circumstances, in order to prevent bed scour and undermining of the structure when the high-velocity stream is discharged into the downstream channel. Velocities as low as even 2 or 3 ft/sec are quite capable of causing dangerous erosion. Also, there may be danger of cavitation, particularly at any abrupt changes of channel or bed geometry.

Flow entering the stilling basin is usually at supercritical velocities; the outlet channel, however, is usually on a mild slope and will only support subcritical flow. As shown in Art. 4–6, the transition from supercritical to subcritical flow normally takes place in the form of a hydraulic jump. The stilling basin is designed to ensure that the jump occurs always at such locations that the flow velocities entering the erodible downstream channel are incapable of causing harmful scour.

The design of a particular stilling basin will depend on the magnitude and other characteristics of the flow to be handled, and particularly the Froude number of the approaching flow. Although the jump equation has been verified experimentally for a wide range of Froude numbers, the internal characteristics of the jump vary as the Froude number changes. The U.S. Bureau of Reclamation, on the basis of very extensive and systematic tests, distinguishes five basic types of jump, as follows:

1. *Undular jump* ($1 < N_F < 1.7$), a very weak jump, characterized only by small surface undulations near the jump.
2. *Weak jump* ($1.7 < N_F < 2.5$), a small jump, characterized by a series of small rollers on the otherwise smooth surface, with fairly uniform velocity distribution through the cross-section.
3. *Pulsating jump* ($2.5 < N_F < 4.5$), an unstable jump, characterized by irregular oscillation of the entering jet from bed to surface and back, resulting in generation of strong surface waves of irregular period.
4. *Steady jump* ($4.5 < N_F < 9.0$), a stable jump, with the jet leaving the floor in essentially the same vertical plane as the downstream extremity of the surface roller.
5. *Strong jump* ($9.0 < N_F$), a jump characterized by a very strong jet entering the jump and leaving the floor upstream from the point of termination of the jump. Intermittent mixing of the jet with slugs of water rolling down the face of the jump may create surface waves.

It is convenient to express all relationships involving the hydraulic jump in terms of the approach-flow Froude number. We shall consider first the simplest and most common stilling basin, a horizontal rectangular channel, such as is shown at the base of the spillway in Fig. 6–29.

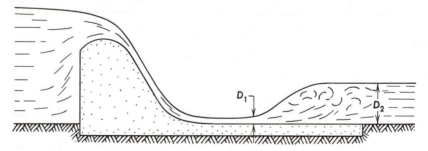

Fig. 6–29. Jump in simple stilling basin.

The hydraulic jump equation for a horizontal rectangular channel has already been derived, Eq. (4–36), on the basis of the pressure-momentum equation for flow through the jump. In dimensionless form, it may be written in terms of the depth ratio D_2/D_1 and the Froude number, as follows:

$$\frac{D_2}{D_1} = \frac{1}{2}(\sqrt{8N_F^2 + 1} - 1) \tag{6–69}$$

where N_F is the Froude number for the upstream rapid flow:

$$N_F = \frac{V_1}{\sqrt{gD_1}} = \frac{q}{\sqrt{gD_1^3}} = \left(\frac{D_c}{D_1}\right)^{3/2} \tag{6–70}$$

Without much loss of accuracy, since this Froude number is always greater than 1, Eq. (6–69) can be simplified to

$$\frac{D_2}{D_1} \cong \sqrt{2}N_F - \frac{1}{2} \tag{6–71}$$

A rather large depth ratio is desirable if a high degree of energy dissipation is to be obtained. This is evident from the jump head loss formula, Eq. (4–38). This equation can be written also in dimensionless form, as follows:

$$\frac{E_j}{D_1} = \frac{(D_2/D_1 - 1)^3}{4D_2/D_1} \tag{6–72}$$

In terms of the Froude number, Eq. (6–72) becomes

$$\frac{E_j}{D_1} = \frac{[(1/2)(\sqrt{8N_F^2 + 1} - 1) - 1]^3}{2(\sqrt{8N_F^2 + 1} - 1)} = \frac{(\sqrt{8N_F^2 + 1} - 3)^3}{16(\sqrt{8N_F^2 + 1} - 1)} \tag{6–73}$$

The efficiency of the jump as an energy dissipator can be expressed as the ratio of the energy dissipated to the energy initially present. The latter is the specific energy, which is

$$E_1 = D_1 + \frac{V_1^2}{2g} = D_1 + \frac{q^2}{2gD_1^2}$$

In dimensionless form, this becomes

$$\frac{E_1}{D_1} = 1 + \frac{q^2}{2gD_1^3} = 1 + \frac{1}{2}N_F^2 \tag{6–74}$$

The efficiency can then be written

$$\frac{E_j}{E_1} = \frac{E_j/D_1}{E_1/D_1} = \frac{(\sqrt{8N_F^2 + 1} - 3)^3}{8(\sqrt{8N_F^2 + 1} - 1)(2 + N_F^2)} \tag{6–75}$$

The efficiency is zero when $N_F = 1$ and becomes about 74 per cent for $N_F = 10$, approaching 100 per cent for very large values of N_F.

The jump height, $D_2 - D_1$, is given in dimensionless form by

$$\frac{D_2 - D_1}{D_1} = \frac{1}{2}(\sqrt{8N_F^2 + 1} - 1) - 1 = \frac{1}{2}\sqrt{8N_F^2 - 1} - \frac{3}{2} \quad (6\text{--}76)$$

The length of the jump has not as yet been adequately calculated from purely theoretical considerations, depending as it does on the internal turbulent mixing and dissipation process. It is a very important factor in stilling basin design, however, since measurements usually indicate it to be six or more times the height of the jump.

The tests conducted by the U.S. Bureau of Reclamation on jump length have been systematized into the form of a design curve, reproduced in Fig. 6–30.

It will be noted that the length of jump has been plotted in terms of the sequent depth D_2, instead of D_1. The reason for this is that the resultant curve is flatter when plotted in this manner.

For a given initial depth D_1 and Froude number N_F, therefore, use of Eqs. (6–69), (6–73), (6–75), and (6–76), along with Fig. 6–30, will quickly yield the sequent depth, head loss, efficiency, height of jump, and length of jump.

The longitudinal position of the jump on the apron must be such that the upstream and downstream depths satisfy the jump equation. It will be recalled that D_1 and D_2 are *conjugate* or *sequent depths*, as defined on the M-diagram (Fig. 4–8). The variation of depth D_1 is determined by energy and friction considerations along the upstream channel and D_2 by conditions in the downstream channel. These flow profiles can be computed for the given discharge. The jump occurs at that point in the channel where the upstream and downstream profiles have respective values of depth which are

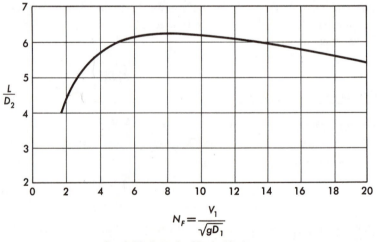

Fig. 6–30. Length of hydraulic jump.

conjugate to each other, as defined by the M-diagram or the jump equation. Generally, the length of jump is neglected in such calculations, but this may lead to substantial errors if the profiles are sharply sloping and the jump is long.

For more precise jump location, a trial-and-error procedure is necessary. The theoretical point location is first determined as outlined above, and this suffices for a tentative determination of the length of jump, from Fig. 6–30.

This jump length is inserted along the two profiles, one supercritical and one subcritical, until the respective depths at the two ends of the jump are found to satisfy the jump equation. This may modify the jump length sufficiently to warrant a recalculation. When finally located correctly, the depths before and after the full jump length should satisfy the jump equation, and the jump length should correspond to the upstream Froude number and the downstream depth as in Fig. 6–30.

The discussion thus far has applied specifically to plain horizontal rectangular basins with no appurtenances. In important structures, especially if subjected to variable discharges, the jump characteristics may make this type of basin excessively long. Flow profile calculations are always somewhat uncertain, owing to imperfect knowledge of bed friction. Furthermore the jump location and length changes as the discharge changes.

It may therefore become advisable to modify the bed or channel geometry in some way so as to shift or shorten the jump, or to stabilize its location. These factors have led to the design of a great many different types of stilling basins, most of them designed primarily by model studies.

The more common expedients which have been used to modify the jump characteristics in stilling basins are the following, used either singly or in combination:

1. Sloping apron, used to decrease the effective tailwater depth on the lower part of the apron to the conjugate value for the rapid flow on the upper part, in order to cause an effective jump; otherwise the tailwater might tend to "submerge" the jump, allowing the high-velocity jet to continue along the bottom a considerable distance before dissipation.

2. Sill, or small dam, at end of apron, used to increase the tailwater depth sufficiently to cause a jump of the desired characteristics to form upstream therefrom.

3. Hydraulic drop, a drop in apron elevation which serves to increase the tailwater depth and thereby decrease the velocity sufficiently to cause a jump to form.

4. Baffles, various kinds and patterns of blocks placed in the basin to increase the bed resistance, causing the flow depth to increase sufficiently to form a jump, and also serving to help dissipate the energy in and following the jump; however, damage to the blocks from ice, debris, and possibly cavitation must be considered carefully in any design of this type.

5. Bucket dissipator, an upturned lip at the bottom of the spillway, serving to

eliminate most of the apron by causing an immediate roller-type jump to form at the spillway toe; a reverse roller is generated below the bucket roller, but this tends to push bed material back toward the bucket and so does not pose a threat of undermining the structure.

6. Stilling pool: a relatively deep basin below the spillway into which the discharge is allowed to fall freely, much of its energy being dissipated by impact with this water cushion; the pool may also be designed to dissipate much of the remaining energy by a jump.

All the above stilling basins can be partially analyzed by the presssure-momentum theory, as in the case of the plain horizontal apron. However, to do so, the hydrostatic and hydrodynamic forces on the baffle blocks, sloping apron, sills, pool, etc., must be included in the analysis, and these are usually unknown. They also cause substantial non-uniformity of velocity distribution, rendering the calculated momentum quantities very uncertain.

Accordingly, most such basins are designed on the basis of model studies, using the Froude number as the correlating parameter. In recent years, a number of systematic testing programs have resulted in standardized designs of stilling basins for specific types of applications. For example the SAF Stilling Basin (so called because of its development at the St. Anthony Falls Hydraulics Laboratory of the University of Minnesota by Blaisdell and others for the U.S. Soil Conservation Service), has come into quite general use in connection with small spillway and drainage structures.

The U.S. Bureau of Reclamation (U.S.B.R.) has developed generalized designs for ten basic types of stilling basins and energy dissipators, very briefly characterized as follows:

1. U.S.B.R. Basin I—Straight, horizontal, plain rectangular basin, as discussed already.

2. U.S.B.R. Basin II—used on high spillways and large canal structures, for $N_F > 4.5$. Contains row of chute blocks at basin inlet and a dentated end sill.

3. U.S.B.R. Basin III—used on small spillways, outlet works, etc., for $N_F > 4.5$. Contains row of inlet chute blocks, row of baffle piers, and solid end sill.

4. U.S.B.R. Basin IV—used for structures with $2.5 < N_F < 4.5$. Uses chute blocks and solid end sill, and may also be followed by rafts or other wave suppressors.

5. U.S.B.R. Basin V—consists mainly of sloping apron, used where economy requires sloping apron, usually on high dam spillways.

6. U.S.B.R. Basin VI—used for pipe or open channel outlets, uses a small box and vertical "hanging baffle wall," with flow energy reduced by impact on wall and flow emerging beneath the baffle.

7. U.S.B.R. Basin VII—used for spillways and other structures with unsubmerged crests. Consists of slotted end bucket-type dissipator, with upsloping bed at outlet.

8. U.S.B.R. Basin VIII—The hollow-jet valve stilling basin, a short basin

used with an outlet works control structure discharging an annular jet of water. Contains a sharply inclined floor followed by a horizontal apron and end sill, with converging side walls and, when two valves are used, a center dividing wall.

9. U.S.B.R. Basin IX—Baffled apron, with large chute blocks or baffle piers staggered in rows down the chute or spillway drop.

10. U.S.B.R. Basin X—Flip bucket spillway, used at outlets of spillway tunnels or large conduits, used to throw water into the air and downstream to concentrate riverbed damage away from the structures. Similar to the "ski-jump" spillway, sometimes used at bottom of overflow reach.

For details of the analysis and design of these and other standardized stilling basins and energy dissipators, the references listed in the bibliography at the end of this chapter should be consulted.

Example 6–5

What minimum length of horizontal apron is needed to contain a hydraulic jump for a discharge of 125 cfs/ft if the initial depth is 3.0 feet? How much energy is dissipated in the jump?

Solution

(a)
$$N_{F_1} = \frac{q}{\sqrt{gD_1{}^3}} = \frac{125}{\sqrt{27g}} = 4.25$$

$$D_2 = \frac{D_1}{2}(\sqrt{8N_{F_1}{}^2 + 1} - 1) = 1.5(\sqrt{8(4.25)^2 + 1} - 1) = 16.6 \text{ ft}$$

From Fig. 6–30, $\dfrac{L}{D_2} = 5.8$, and $L = 5.8(16.6) = 96.5 \text{ ft}$ *Ans.*

(b)
$$E_j = \frac{(D_2 - D_1)^3}{4D_1D_2} = \frac{(16.6 - 4.25)^3}{4(16.6)(4.25)} = 6.7 \text{ ft} \qquad Ans.$$

6–23. Energy Dissipation by Induced Tumbling Flow.

Special stilling basins of the sort described in the preceding article are needed for large hydraulic structures but are relatively expensive to design and build. They are therefore often impracticable economically for smaller structures such as highway drainage chutes and culverts. Nevertheless such smaller structures are of much more common occurrence, and it is very important that they also be protected against destruction by outlet erosion.

An economical alternative in such cases may often be to use simple roughness elements on the bed of the steep channel itself, with size and spacing such as to induce the tumbling-flow regime in the channel. Tumbling flow consists of a cascade of hydraulic jumps, in which the flow oscillates from subcritical to supercritical and then, through the jump, back to subcritical

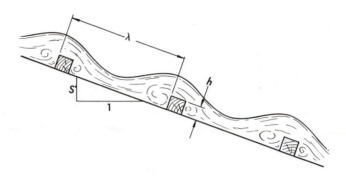

Fig. 6–31. Tumbling flow in open channel.

as it passes each roughness element. This phenomenon is sketched in Fig. 6–31.

Tumbling flow is thus a quasi-critical flow regime, in which the flow is transmitted down the channel under essentially critical flow conditions, which means that the specific energy is maintained at its minimum level for the given discharge. The velocity at the outlet of the channel is then equal to, or near, the critical velocity, $(gq)^{1/3}$, a velocity which can ordinarily be handled safely in the expanding section downstream from the outlet.

Energy dissipation in the tumbling flow is accomplished by the succession of hydraulic jumps. For optimum performance, the proper spacing of elements is essential. Otherwise, good hydraulic jumps will not be maintained and the energy dissipation will be reduced as the flow degenerates to ordinary rapid flow over a rough channel.

Extensive studies on this phenomenon at Virginia Polytechnic Institute have indicated that the optimum spacing of elements is related to the critical spacing for maximum friction factor for flow in closed conduits or tranquil flow in open channels—that is, the spacing marking the boundary between wake-interference flow and isolated-roughness flow. The elements thus should be separated just sufficiently so that each jump is fully developed without interfering with the adjacent jumps.

For practical design purposes, it was found that best results are obtained when the spacing-height ratio, λ/h, is between 7.5 and 12.0, with a normally recommended value of 10.0. For this spacing the empirical design equations for the roughness elements have been recommended as follows:

1. For roughness elements of square cross-section:

$$h = \frac{1}{(3 - 3.7S)^{2/3}} D_c \qquad (6\text{–}77)$$

2. For cubical roughness elements, with cubes in adjacent rows staggered and with a maximum transverse element spacing of $1.5h$:

$$h = 0.7D_c \qquad (6\text{--}78)$$

In the above equations, S is the bed slope and D_c is the critical depth for the given discharge. The value of h thus computed is the minimum element size to assure tumbling flow. The equations apply specifically to rectangular channels but may also be used for trapezoidal channels with reasonable accuracy. In the latter case, the roughness elements should be beveled on their ends in order to fit snugly against the sides of the channel, and the unit discharge q should be calculated on the basis of the width at the bottom of the trapezoidal cross section.

The elements should normally be used throughout the entire length of steep channel. However, if economy dictates, an alternate method is to use only five rows of elements at the downstream end of the channel, preceded by a single large leading element. The large leading element establishes one large hydraulic jump and the five following normal elements are then adequate to establish uniform tumbling flow for the given discharge from the overfall from the leading element. In this case, the leading element size can be calculated from the following semi-empirical equation:

$$h_1 = \left[\sqrt{ \frac{2}{(1-S)^2} \left(\frac{D_c}{D_0} \right) } - (1-S)^2 \right] D_c \qquad (6\text{--}79)$$

In this equation, D_0 is the supercritical depth just upstream from the leading element, as determined by flow profile calculations from the channel inlet.

6–24. Culverts. The term *culvert* is here intended to apply not only to a drainage opening beneath an embankment, as used in highway and railway work, but to any short closed conduit. In this sense, culverts are used as filling and emptying conduits for lock chambers, outlet conduits for dams and tanks, and turnout structures for irrigation canals, and in numerous other situations.

A great variety of flow phenomena can occur in a culvert. It may flow full with either a submerged or unsubmerged outlet, on either mild or steep slopes; it may flow part full, with either tranquil or rapid flow conditions, or a combination of both, with one or more hydraulic jumps in the barrel. Under some conditions, the flow may oscillate from full to part full and back. It may flow full for part of its length, part full for the rest.

The variables that determine the particular flow phenomena are numerous, including culvert size, shape, slope, and roughness, inlet and outlet geometry, capacity of upstream and downstream channels, and culvert discharge. Obviously, the hydraulic design of a culvert cannot correctly be accomplished by any such simple expedient as one of the traditional "waterway-opening" empirical hydrological formulas. A careful hydraulic analysis of the complete flow situation is essential. A great deal of research on culvert hydraulics has been carried out in recent years, but much remains to be done.

The analysis can most conveniently be made in terms of either pipe hydraulics or open-channel hydraulics, depending on whether the culvert flows full or part full. In either case, it is convenient to think in terms of the factors that exert control over the head that will be necessary to force a given discharge through a given conduit opening, as measured by the height of headwater above the inlet center line.

6-25. Culverts Flowing Full. If the culvert flows full, then the "control" will be the barrel friction, entrance and outlet losses, and the tailwater elevation. The last factor depends in turn on the capacity of the downstream channel. This condition is sketched in Fig. 6-32.

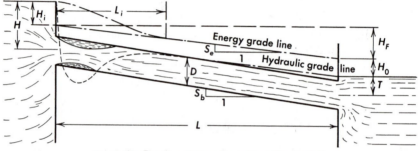

Fig. 6-32. Energy relations for culvert flowing full.

The energy equation for the culvert shown is as follows:

$$H + \frac{V_a^2}{2g} + S_b L - T = \frac{V_e^2}{2g} + H_i + H_f + H_o \qquad (6\text{-}80)$$

in which H and T are the static center-line heads on the culvert inlet and outlet, respectively, V_a and V_e are the velocities in the approach and exit channels, respectively, S_b is the culvert barrel slope, L is the horizontal length of barrel, and H_i, H_f, and H_o are the head losses due to inlet, barrel friction, and outlet respectively.

In terms of the total inlet head, the equation is

$$\left(H + \frac{V_a^2}{2g}\right) = \left(T + \frac{V_e^2}{2g}\right) + \frac{V^2}{2g}(K_i + K_f + K_o) - S_b L \qquad (6\text{-}81)$$

where V is the average velocity in the barrel, and K_i, K_f, and K_o are the loss coefficients. The inlet head is thus seen to depend on the exit channel head, the culvert friction, and the culvert drop.

At the entrance to a conduit, the flow contracts to a certain minimum section, then gradually re-expands to fill the section. At the section of

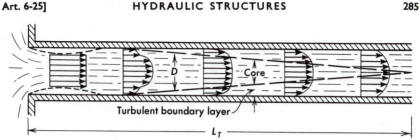

Fig. 6–33. Effect of sharp inlet on flow in culvert.

maximum contraction, the streamlines are all essentially parallel and have the same energy; therefore the velocity distribution is essentially uniform. The velocity distribution for fully developed flow, however, is parabolic for laminar flow and logarithmic for turbulent flow. There must therefore be a transition length, L_T, over which the velocity distribution gradually changes from uniform at the vena contracta to the normal logarithmic distribution (for turbulent flow) at some point far enough downstream.

This transition takes place from the wall outwards, leaving a narrowing core of uniform velocity distribution in the center. The transition process is sketched in Fig. 6–33, for a sharp, flush inlet.

Figure 6–34 shows the same process for a well-rounded inlet, with no re-expansion of the flow after contraction. The transition length, L_T, may be quite long. However, most of the friction loss along the barrel is attributable to the phenomena near the wall rather than in the central core. If the barrel friction loss, H_f, is measured by the slope of the straight-line portion of the energy grade line, S_e, then the length over which the additional energy loss due to the inlet, H_i, is developed, L_i, is substantially less than the total transition length L_T. Approximate values of L_i and L_T, for different entrance and wall conditions,[3] are given below:

Inlet Type	Wall Type	Length-Diameter Ratios	
		Entrance, L_i/D	Transition, L_T/D
Square flush	Smooth concrete	30	60
	Corrugated metal	10	30
Rounded	Smooth concrete	30	100.
	Corrugated metal	10	40
Projecting	Smooth concrete	40	70
	Corrugated metal	15	40

The entrance loss will be fully developed only if the culvert length is

[3] Adapted from measurements made by Morris, at the St. Anthony Falls Hydraulics Laboratory, on smooth pipes and on concrete and corrugated metal pipes.

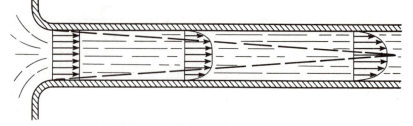

Fig. 6–34. Culvert flow with rounded inlet.

equal to or greater than the inlet length L_i. For a culvert of sufficient length, the inlet loss coefficient, K_i, is the same as for any other pipe inlet of the same geometry. For a shorter culvert, the actual inlet loss coefficient, K'_i, may be approximated by a straight-line relationship:

$$K'_i = K_i \frac{L}{L_i} \quad \text{for} \quad \frac{L}{L_i} < 1 \tag{6–82}$$

The friction coefficient, K_f, is obtained from the Darcy equation, just as for a pipe:

$$K_f = f \frac{L}{D} \tag{6–83}$$

The outlet loss coefficient, K_o, is the same as for any pipe outlet, as long as the outlet is submerged.

If the culvert has a free outlet, the static head on the outlet, $T + V_e^2/2g$, reduces to merely the height of the hydraulic grade line above the outlet center line. Because of the non-hydrostatic pressure distribution in the jet as it emerges, the hydraulic grade line actually is somewhat below the outlet crown, as shown in Fig. 6–35. The outlet head is thus mD, where m is a coefficient that decreases as the Froude number, $N_F = V/\sqrt{gD}$, increases. Experimental results on models yield the following approximate empirical values for the coefficient m:

N_F	0	1	2	>3
m	$\frac{1}{2}$	$\frac{1}{4}$	$\frac{1}{10}$	0

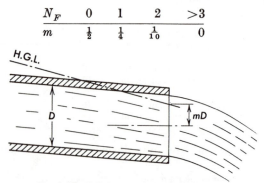

Fig. 6–35. Full flow with free outlet.

The above values of m are approximate only, since their magnitudes depend also to some extent on slope, roughness, entrained air, and other variables.

In general, the design of a culvert will involve the selection of a barrel cross-section to convey a given discharge for a permitted head H over the inlet center line. For a submerged outlet, Eq. (6–81) can be used. For the given conditions, V_e, V_a, and T can be determined. The culvert slope and length will be known at least within definite limits, as controlled by the channel and embankment geometry. Equation (6–81) becomes

$$A = Q\sqrt{\frac{K_i + K_f + K_o}{2g[(H + V_a^2/2g) + S_bL - (T + V_e^2/2g)]}} \qquad (6\text{--}84)$$

which gives the required cross-sectional area. This will have to be solved by trial, since K_f (and, to a lesser extent K_i and K_o) depend on the Reynolds number and therefore on the cross-sectional area.

If the outlet is free, with full flow prevailing in the barrel, then Eq. (6–84) may be modified to replace the term $T + V_e^2/2g$ with the term mD. The factor m obviously is also dependent upon the cross-sectional area, since it depends on N_F. It may be noted that, for either submerged or free outlet conditions, the term in brackets in the denominator of Eq. (6–84) is merely the difference in elevation between energy grade line at entrance and energy grade line at exit, the latter being measured sufficiently away from the outlet for the outlet loss to be complete.

6–26. Culverts Flowing Part Full. If conditions are such that the culvert flows only partly full, then it must be analyzed in terms of the hydraulics of open channels. It becomes important then to determine whether the flow in the culvert is rapid, or tranquil, or a combination.

The flow regime in the culvert depends upon the inlet shape, the barrel slope, and the tailwater level. For supercritical flow to be produced and maintained in the culvert, the inlet must cause a contraction to a smaller area than that of the full cross-section, the slope must be hydraulically steep, and the tailwater must be at a lower level than that corresponding to critical depth at the outlet.

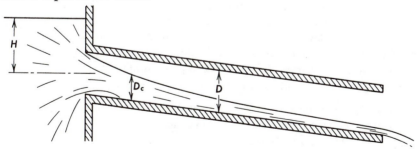

Fig. 6–36. Culvert flow controlled by inlet.

For these conditions, the flow passes through critical depth somewhere near the inlet, so that the inlet functions as a "control section," as shown in Fig. 6–36. The critical depth, however, would be affected by the non-hydrostatic pressure distribution and curvilinear velocity distribution at the control section. Consequently the usual critical depth formula cannot be used to compute the flow. However, the culvert inlet functions in this case essentially as an orifice (or as a contracted weir, if the headwater level is below the culvert crown), and the orifice (or weir) formula çan be used. Thus, Eq. (6–81) simplifies to

$$H + \frac{V_a^2}{2g} = \frac{V_c^2}{2g} = \frac{Q^2}{2g(C_c A)^2} \qquad (6\text{–}85)$$

where C_c is the orifice contraction coefficient and A is the orifice area. If the inlet is a sharp-edged, flush inlet, then C_c can for all practical purposes be taken as 0.62. Therefore the discharge for this case, with the inlet submerged, is given by

$$Q = 0.62 A \sqrt{2g \left(H + \frac{V_a^2}{2g} \right)} \qquad (6\text{–}86)$$

With the headwater below the inlet crown, the weir principle can be applied, but this case is unlikely to be of design significance.

It is possible for a hydraulic jump to form in the barrel if the tailwater level rises above the critical depth level at the outlet. However, this will of course not affect the flow at the inlet unless the jump moves up all the way to the inlet.

If the culvert is on a mild slope, with a free outlet, it is possible that subcritical flow may be established over part or all of the barrel. An accurate analysis would involve computation of a complete flow profile through the barrel, proceeding downstream from the contracted section and proceeding upstream from the outlet flow elevation. It is possible that these two profiles might meet in a hydraulic jump or that supercritical flow would persist all the way to the outlet, in a short culvert, or that subcritical flow would submerge the contraction at the inlet. Only a careful computation of the flow profiles could give a definite answer.

In any case, it is very likely that as long as the culvert flows part full, unless it is very long, the orifice or weir equation at the inlet will serve satisfactorily for design purposes. If conditions develop such that the inlet contraction is submerged by the downstream profile, however, then for practical purposes the culvert can be analyzed as flowing full.

6–27. Criteria for Full and Part-full Flow. In general, it is desirable to design the culvert to flow full at design discharges if possible, since usually a larger discharge can be handled in a given section and with a given inlet head. The effective head producing flow is increased by the amount of the vertical drop in the culvert barrel, though this is partially reduced by the barrel energy losses. Comparison of Eqs. (6–81) and (6–85), both for free outlet, shows the following discharge relationships:

$$\text{Full flow:} \qquad Q = A\sqrt{\frac{2g(H + V_a^2/2g + S_bL)}{K_i + f(L/D) + K_o}} \tag{6–87}$$

$$\text{Part-full flow:} \quad Q = A\sqrt{\frac{2g(H + V_a^2/2g)}{1/C_c^2}} \tag{6–88}$$

Assuming a standard outlet ($K_0 = 1$) these formulas become, for a sharp-edged inlet ($C_c = 0.62$ and $K_i = 0.4$) the following:

$$\text{Full Flow:} \qquad Q = A\sqrt{\frac{2g\left(H + \dfrac{V_a^2}{2g} + S_bL\right)}{1.4 + f\dfrac{L}{D}}} \tag{6–89}$$

$$\text{Part-Full flow:} \quad Q = A\sqrt{\frac{2g\left(H + \dfrac{V_a^2}{2g}\right)}{2.6}} \tag{6–90}$$

For a well-rounded inlet ($C_c = 1$; $K_i = 0$), the formulas become:

$$\text{Full Flow:} \qquad Q = A\sqrt{\frac{2g\left(H + \dfrac{V_a^2}{2g} + S_bL\right)}{1 + f\dfrac{L}{D}}} \tag{6–91}$$

$$\text{Part-Full flow:} \quad Q = A\sqrt{2g\left(H + \frac{V_a^2}{2g}\right)} \tag{6–92}$$

A comparison of culvert capacities can be obtained for the two conditions by setting the corresponding formulas equal to each other. Thus, equating formulas (6–87) and (6–88), it is seen that
Q for full flow $\lessgtr Q$ for part-full flow when

$$S_b \lessgtr \frac{\left(H + \dfrac{V_a^2}{2g}\right)}{L}\left[C_c^2\left(1 + K_i + f\frac{L}{D}\right) - 1\right] \tag{6–93}$$

Whenever the culvert bed slope exceeds the value specified in Eq. (6–93) then the culvert will carry a larger discharge for given head H when it flows full than when it flows part-full. For the sharp-edged flush inlet and well-rounded inlet, respectively, this relation becomes (neglecting the approach velocity head)

$$\text{For } C_c = \tfrac{1}{2}: \quad S_b \lessgtr \frac{H}{D}\left(\frac{f}{4} - \frac{D}{2L}\right) \tag{6-94}$$

$$\text{For } C_c = 1: \quad S_b \lessgtr \frac{H}{D}\,(f) \tag{6-95}$$

It is obvious, from examination of these equations, that the full-flow condition will in most cases yield a larger discharge than the part-full condition.

Part-full flow can best be inhibited by preventing the inlet contraction. Since the contraction creates supercritical velocities, the only way the culvert can then be made to flow full is to submerge the vena contracta by backwater, produced either by a high tailwater elevation or a long mild-sloping barrel.

However, the contraction may be substantially eliminated by rounding the inlet or providing a gradual transition of some other form from the approach channel. A radius of rounding of $\frac{1}{7}$ (pipe diameter) or greater will theoretically eliminate all contraction. Model studies have shown that even on steep slopes with free outlets, the culvert with a rounded inlet will always flow full if H/D exceeds unity. For H/D between 0.75 and 1.00, the culvert flows alternately full and part full, in so-called *slug flow*, with the head H oscillating up and down corresponding to the changing control. If H/D is less than 0.75, the culvert always flows part full.

On the other hand, model tests for the same conditions with a sharp-edged flush inlet showed that, at least for H/D values up to 7.0, and probably indefinitely higher, the culvert would flow only part full. This indicates the very striking advantage to be gained by such a simple expedient as inlet rounding.

Subsequent prototype field tests, however, have not always confirmed the model results on the rounded inlet. The maintenance of full flow requires maintenance of siphon action near the inlet, and this is easier to secure in the laboratory than in the field. Disturbances and asymmetry in approaching flows tend to permit vortex generation at the inlet; the vortex sucks air into the culvert, breaks the siphon, and full-flow conditions deteriorate to part-full flow.

This leads to an establishment of an undesirable unsteady "slug flow" phenomenon, such as described above. Present research is being directed toward design of an inexpensive but efficient "transition piece" which will inhibit the formation of inlet vortices and will establish stable full-flow conditions.

Various forms of "tapered inlets" have been extensively tested by French at the National Bureau of Standards, but the results have indicated such inlets to be unreliable in assuring full flow. Consequently, many engineers prefer to design culverts on the assumption that they will flow part-full unless high tailwater levels can be counted on to maintain full flow.

On the other hand, Blaisdell and others have noted considerable success with the "hooded inlet," which is merely a beveled extension of the crown of the culvert inlet into the headwater pool. Both model and prototpye tests have confirmed that culverts with these inlets will flow full at relatively low heads.

However, to be effective, an "anti-vortex device" of some kind, such as a horizontal circular plate projecting into the pool from the crown of the hood, must be attached to the hooded inlet. This will inhibit the formation of vortices and establishment of intermittent slug flow in the pipe. A disadvantage of this is that the combination of the anti-vortex plate and re-entrant inlet increases the inlet loss coefficient, K_i, almost to 1.0, thus reducing the gain in flow energy due to the siphon action. Perhaps more important is the difficulty of maintaining such a vortex plate in position under field conditions.

Thus, although it is desirable to design for full flow if possible, the problem of a completely reliable rounded or tapered inlet has not yet been satisfactorily resolved. It seems probable that a simple bell-mouth inlet is the most nearly certain device for assuring full flow at reasonably low inlet heads.

Some additional economy in design may be attained by flood-routing procedures, taking advantage of the increased headwater storage due to ponding as the flood crest passes through the culvert. The technique of the "equivalent uniform flood," described in Art. 11–11, has been found useful in this connection.

Example 6–6

A corrugated pipe culvert is laid through a highway embankment, with the pipe inlet flush with the headwall and pipe outlet projecting into the tailwater pool. The culvert is 30 inches in diameter, 100 feet long, and is on a hydraulically steep slope. The Manning coefficient is 0.025 and the design discharge 40 cfs. Compute the head on the culvert, measured above the inlet center-line, when:

(a) the outlet is submerged, with the tailwater elevation at the same level as the inlet center-line.

(b) the outlet is free and the culvert flowing part-full.

Solution

(a) $V = \dfrac{Q}{A} = \dfrac{40(4)}{\pi(2.5)^2} = 8.14$ fps

Inlet loss, $H_i = \dfrac{1}{2}\dfrac{V^2}{2g} = 0.51$ ft

(Note that $L_i = 10(D) = 25$ ft; since $L_i < L$, full H_i applies)

Outlet loss, $H_o = \dfrac{V^2}{2g} = 1.03$ ft

Barrel friction loss, $H_f = \left(\dfrac{nV}{1.49R_h^{2/3}}\right)^2 (L)$

$$= \left(\frac{0.025(8.14)}{1.49(0.625)^{2/3}}\right)^2 (100) = 3.48 \text{ ft}$$

By energy equation, neglecting V_a and V_e.

$$H + S_b L = T + H_i + H_f + H_o$$

Since, in this problem, $S_b L = T$

$$H = H_i + H_f + H_o = 0.51 + 3.48 + 1.03 = 5.02 \text{ ft} \qquad Ans.$$

(b) $H = \dfrac{1}{2g} \left(\dfrac{V}{C_c}\right)^2 = \dfrac{1}{64.4} \left(\dfrac{8.14}{0.62}\right)^2 = 2.66 \text{ ft} \qquad Ans.$

PROBLEMS

6–1. A masonry gravity dam has a 112-ft maximum height with crest width of 30 ft. Its upstream face has a slope of 4 vertical to 1 horizontal and its downstream face a slope of $1\frac{1}{2}$ vertical to 1 horizontal. The reservoir is designed for a maximum depth of 100 ft of water. The dam rests directly on the foundation, without stepping or key walls, the friction coefficient being 0.5, and the foundation being quite permeable. Tailwater depth is negligible. The masonry can be assumed to have a specific gravity of 2.5. Earthquakes are assumed capable of producing accelerations of $\frac{1}{10}g$ in the foundation in both vertical and horizontal directions. Ice on the surface is capable of developing a force of 10,000 lb/ft and sediment at the bottom may accumulate to a depth of 10 ft. Neglect forces due to wind, temperature, spillway or crest loadings, etc. Compute all the forces acting on the dam and the vertical and horizontal components of the resultant foundation reaction, with their lines of action.

6–2. For the dam of Problem 6–1, compute the factors of safety against sliding and against overturning. *Ans.:* 0.93, 1.75.

6–3. Compute the foundation bearing pressures in pounds per square foot at the heel and toe respectively. *Ans.:* 2,700 psf, 11,600 psf.

6–4. Compute the heel and toe pressures with the reservoir empty (i.e., due to weight of dam only).

6–5. Compute the heel and toe pressures on the assumption that the foundation will be made impervious but that the other forces continue to act.

6–6. Compute the heel and toe pressures, assuming uplift is present, but neglecting earthquake, wave, ice, and sediment forces.

6–7. Compute the vertical stress, maximum normal stress, and horizontal shearing stress in the masonry at the toe of the dam, for all forces acting. *Ans.:* 11,500 psf, 16,700 psf, 7,700 psf.

6–8. Compute the same stresses at the downstream face at an elevation of 50 ft above the base. Neglect uplift and quake forces.

6–9. Compute the stresses at the heel, with reservoir empty, due to weight of dam only.

6–10. Compute the stresses at the upstream face at 50 ft elevation, due to weight of dam only.

6-11. A masonry gravity dam (specific gravity $= 2.5$) has a 20-ft crest width, a 75-ft base width, and a vertical upstream face. The dam is 100 ft high, has a full reservoir, and an ice thrust at the surface of 12,480 lb/lin ft. The foundation is permeable, so that uplift is assumed to vary from full hydrostatic head at the heel to zero at the toe. Earthquake and other minor forces are neglected. Determine the foundation bearing pressures at heel and toe.

6-12. A masonry gravity dam is to be designed for a preliminary study with a triangular cross-section, H ft high and B ft in base width. The upstream face is vertical and the masonry has a specific gravity S and coefficient of friction f with the foundation. Assume full reservoir and uplift, but neglect earthquake and other forces. Derive formulas for the following:

  a. Factor of safety against sliding.
  b. Maximum normal stress in masonry at toe.
  c. Horizontal shearing stress in masonry at toe.

6-13. A masonry gravity dam has a vertical upstream face and a 20-ft crest width. The dam is 60 ft high and has a coefficient of 0.5 with its pervious sandstone foundation. Neglecting earthquake and other minor forces, determine the minimum base width:

  a. To prevent sliding. *Ans.:* 46.7 ft.
  b. To prevent development of tension cracks. *Ans.:* 47.4 ft.

6-14. The gravity dam cross section approximates design conditions at Elephant Butte Dam in New Mexico. The masonry weighs 135 lb/cu. ft and the foundation is fully pervious, with a coefficient of friction of 0.75. Neglect ice, earthquake, wind, and other secondary forces. Compute

  a. Factor of safety against sliding
  b. Bearing pressures at heel and toe.

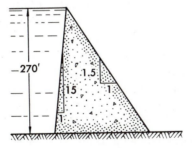

Prob. 6-14.

6-15. A gravity dam is constructed of masonry weighing 150 lb/cu. ft and is 300 ft high. Neglect the crest width, assuming the cross-section to be triangular with an upstream slope of 10 vertical to 1 horizontal and a downstream slope of 1.5 vertical to 1 horizontal. The foundation is fully pervious, with a friction coefficient of 0.75. Neglect ice, earthquake, wind, and other secondary forces. Assume reservoir is full to crest elevation. Compute:

  a. Factor of safety against sliding.
  b. Maximum normal stress in masonry along base of dam, in psi.

6–16. For the concrete gravity dam shown, compute the foundation bearing pressures at the heel and toe. Assume the reservoir full, no freeboard, and masonry of specific gravity 2.5. Neglect uplift and other forces, considering only the hydrostatic forces and force of gravity.

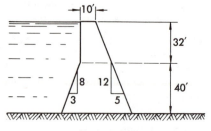

Prob. 6–16.

6–17. An earth dam is 90 ft high with a 10-ft freeboard and 20-ft crest width. The side slopes are 1 vertical to 3 horizontal, and it rests on an impervious foundation. Assume the embankment is isotropic with a permeability of 0.0001 cm/sec. It is provided with a horizontal drainage blanket extending back 200 ft from the toe. Compute the quantity of seepage, in cfm per ft of dam width, by the basic parabola method.

6–18. Sketch the flow net for the dam of Problem 6–17, and compute the seepage, in cfm, from the flow net. Use a scale of 1 in. = 100 ft.

6–19. Repeat Problems 6–17 and 6–18, assuming no underfilter.

6–20. Assume the embankment of Problem 6–17 to be stratified, with a permeability of 0.0003 cm/sec in the horizontal direction and 0.0001 cm/sec in the vertical direction. Sketch the flow net on both the actual and transformed sections. Use a scale of 1 in. = 50 ft on the transformed section. Compute the seepage from the flow net, in cfm per ft.   *Ans.:* 0.0091 cfm per ft.

6–21. Compute the seepage from the dam of Problem 6–20 from the formula based on the equivalent parabola.   *Ans.:* 0.0081 cfm per ft.

6–22. Plot diagrams showing the distribution of hydrostatic pressure along the base of the dam and along a surface through the midheight of the dam.

6–23. Repeat Problems 6–20 and 6–21 for a dam with no underfilter. *Ans.:* 0.00582 cfm per ft.

6–24. Repeat Problem 6–22 for a dam with no underfilter.

6–25. In the dam of Problem 6–17, if a central core of less permeable material were inserted, having a permeability one-fourth that of the main body of the embankment, compute the deflection angle of the flow lines as they enter and leave the core. Sketch a typical flow net sequence along one channel for the flow entering, traversing, and leaving the core. Assume the central core is symmetrical about the center of the dam, and that the two interfaces are at an angle of 60° with the horizontal base. Assume that the flow lines approaching the upstream interface are horizontal and that, after being deflected at the interface, they continue in straight lines until they reach the downstream interface.

6-26. If the dam of Problem 6–17 is on an isotropic pervious foundation, with no underfilter, draw the flow net for underseepage and compute the factor of safety against piping at the toe. Assume the foundation and embankment flow nets are independent of each other, so that the base of the dam is a flow line for each. The foundation material has a specific gravity of 2.60 and a voids ratio of 0.60.   *Ans.:* 6.

6-27. An earth dam has a crown width of 30 ft and a height of 60 ft, with upstream and downstream slopes of 4 horizontal to 1 vertical. Water in the reservoir is 50 ft deep. The material in the dam has a permeability of 0.0004 cm/sec horizontally and 0.0001 cm/sec vertically. The dam has an underfilter extending back 50 ft from the toe. Determine the quantity of seepage in gallons/day per ft.   *Ans.:* 32.6 gal/day per ft.

6-28. Underseepage occurs beneath a masonry dam of 60-ft height and 46.7-ft base width. The permeable foundation material is 40 ft thick (above an impermeable stratum) and has a permeability of 4 ft/day horizontally and 1 ft/day vertically. Construct a flow net for underseepage, assuming full reservoir, and determine:

a. Quantity of seepage per foot. *Ans.:* 669 gal/day per ft.

b. Factor of safety against piping. *Ans.:* 0.7.

6-29. An earth dam on an impervious foundation has a crest width of 20 ft upstream and downstream slopes of 45°, a total height of 85 ft and a freeboard of 10 ft. It has a uniform permeability coefficient of 0.2×10^{-4} cm/sec. Compute the seepage in cfm per foot of dam, for:

a. Drainage blanket, extending back 40 ft from toe. *Ans.:* 0.000985 cfm per ft.

b. No drainage blanket. *Ans.:* 0.00075 cfm per ft.

6-30. An earth dam is 75 feet high, with a 25-ft crown width. The upstream and downstream slopes are 2 horizontal to 1 vertical. The dam rests on an impervious foundation and has a 75-ft horizontal drainage blanket at the downstream end. The compacted embankment is isotropic, with a permeability of 0.0002 cm/sec in all directions. The water depth in the reservoir is 60 feet. Determine the seepage through the dam in cubic feet per day per foot of width.

6-31. An earth dam is 75 feet high with a 25-ft crown width. The upstream and downstream slopes are at an angle of 30° with the horizontal and the freeboard is 15 feet. The dam rests on an impervious foundation and has a permeability of 0.00394 ft/min in all directions. There is no drainage blanket. Determine the seepage through the dam, by the equivalent parabola method, in cu ft/day/ft of width.

6-32. An earth dam has a crown width of 25 ft and a height of 85 ft, with a 10-ft freeboard. The side slopes are at an angle of 20° with the horizontal and the dam has no drainage blanket. The embankment material has a permeability of 2×10^{-5} ft/sec in the horizontal direction and 1×10^{-5} ft/sec in the vertical direction. Determine the quantity of seepage, in cu ft per day per ft of width of dam.

6-33. A constant-center arch dam is to be designed to span a canyon 400 ft wide. The height of dam is to be 270 ft, including a 10-ft freeboard. Assume a central angle for the arch of 120° and a permitted masonry stress of 800 psi. Using the thin-cylinder method for an approximate analysis, determine:

    a. Thickness of dam at crest.

    b. Thickness of dam at midheight.

    c. Thickness of dam at base.

    d. Total reaction at one abutment.

6–34. Design an approximate section for a gravity dam for the conditions of Problem 6–33, assuming a coefficient of friction with the base of 0.5. Permitted bearing pressure on the foundation is 15 tons/ft^2. Neglect uplift and all other forces except the hydrostatic forces and weight of dam. Assume a vertical upstream face and a triangular cross-section. Compare volumes of masonry required for the dam with that for the arch dam of Problem 6–33.

6–35. A gravity dam is constructed with an overflow spillway to handle a maximum discharge of 5000 cfs. The stream bed elevation is 830 and the spillway crest elevation is to be 850, the maximum allowable water surface elevation being 855. Assume the approach channel to be of rectangular cross-section with a breadth of 125 ft. Assuming the upstream face to be vertical, and the discharge coefficient to be 3.9, determine the required crest length. Design the crest shape to conform to the hydraulic overfall requirements at the design head, and plot to scale.

6–36. A reservoir has a normal pool elevation of 146, a maximum allowable elevation of 150 and a design flood of 5000 cfs. A side channel spillway 160 ft long is built to handle a maximum flow of 5000 cfs. The side channel is to have a base width of 20 ft and side slopes of 2 vertical to 1 horizontal. Conditions downstream fix the water surface elevation at the channel outlet at 143.50. At the upstream end of the channel, assume that the spillway may be two-thirds submerged without appreciably affecting its discharge. Derive equations for the water-surface and channel-bottom profiles, assuming channel velocity varies as the $\frac{1}{3}$ power of the distance from its beginning point. Plot the computed profiles.

6–37. The overflow spillway leading to a side channel has a discharge coefficient of 3.5, with a design head of 9 feet and discharge of 35,000 cfs. The side channel is of rectangular cross-section, with a width of 100 feet. Design a channel bed profile such that the velocity in the channel will vary linearly as the distance from the beginning of the channel, at which point the water profile in the channel is at the same elevation as the overflow crest.   *Ans.:* $y = 0.000117x^2 + 15.6$.

6–38. The overflow spillway leading to a side channel has a discharge coefficient of 4.0, under a design head of 16 ft and discharge of 64,000 cfs. The side channel is rectangular in form, 128 ft wide, with a control section at its outlet. Assume the channel velocity will vary as the $\frac{3}{4}$ power of the distance from the beginning of the channel, at which point the water profile is at the same elevation as the overflow crest. How far vertically below the spillway crest is the control section bed, and what is the equation of the channel bed?

6–39. A side-channel spillway has a crest coefficient of 3.6, with a design head of 9.0 ft and a design discharge of 6400 cfs. The side channel is rectangular in form, 32 ft wide, with a control section at the end of the spillway. The water profile is to be a straight line, beginning at the same elevation as that of the spillway crest.

   a. Calculate the required spillway length.   *Ans.:* 66 ft.

   b. How far below the spillway crest is the control section bed level?
Ans.: 26.9 ft.

6–40. A side-channel spillway is rectangular in cross-section, of horizontal bed slope, and has a free overfall at its outlet. Water enters it from an overflow structure, at the rate of q cfs/ft of length of L ft. Neglecting friction, and assuming the overfall occurs at the end of the length L, derive a differential equation for the flow profile in the channel, giving flow depth as a function of only the horizontal distance from the channel inlet, the unit discharge q, the flow depth at the overfall, and the channel length L.

6–41. A side-channel spillway is designed of such form that the velocity of flow at any point equals the natural logarithm of the distance from the beginning of the channel to that point. The channel is rectangular, 20 ft wide. The overfall spillway leading to the channel is 4 ft wide, with a crest coefficient of 3.6, and is under a head of 4.0 ft. What should be the equation of the channel bed (in x and y only), with the origin taken at the bed at its upstream end?

6–42. A siphon is to discharge 180 cfs with a crest and reservoir elevation of 60 and an outlet at elevation 0. The radii of the tube at the crest are 3 ft and 5 ft, and the minimum permissible gage pressure at this section is minus 24 ft of water. The throat section is followed, in order, by a vertical section, a 90° bend with a center-line radius of curvature of 8 ft, a horizontal section at elevation 0, and a reducing section, if necessary. The horizontal distance from the face of the dam to the siphon crest is 8 ft., and from the vertical section to the outlet 40 ft. The inlet should have an area approximately twice that of the throat and its top should be submerged several feet. Provide a uniform taper from the inlet to the throat and make the conduit rectangular in cross-section throughout, with an area equal to that at the throat except at the beginning and end. Assume an n value of 0.012 in the Manning formula for computing the conduit loss, and assume that other losses may be expressed in the form kh_v where k has the following values:

$$\text{Entrance:} \quad k = 0.25$$
$$\text{Contraction:} \quad k = 0.05$$
$$90° \text{ bend:} \quad k = 0.25$$

Determine the proper dimensions of the various sections to satisfy the given requirements.

6–43. A reservoir impounds water to a maximum of 50 ft above the tailwater level. The spillway must handle a peak flow of 1000 cfs under these conditions, with crest elevation placed 4 ft below the peak water level.

   a. If an overflow spillway is used, with crest coefficient of 4.0, what spillway length is required?   *Ans.:* 31.2 ft.

   b. If a siphon is used, with crest radii of 2.0 ft and 5.0 ft, what width is required?   *Ans.:* 13.9 ft.

c. For this constant cross-section, with a total head loss (not including outlet velocity head) of 10 ft, how far above the tailwater should the siphon discharge?   *Ans.:* 31.1 ft.

6–44. A rectangular siphon spillway through a dam is designed for the following conditions:

Inlet ₵ elevation	424.0 ft	Crest ₵ radius	6 ft
Outlet ₵ elevation	364.0 ft	Friction factor	0.025
Siphon depth	6 ft	Inlet loss coefficient	0.10
Siphon length	160 ft	Outlet loss coefficient	1.0
Crest ₵ elevation	436 ft	Σ(Bend loss coefficients)	(0.80)

a. What maximum discharge can theoretically be passed through the siphon, assuming headwater does not rise above level of inlet crown?   *Ans.:* 233 cfs/ft.

b. What maximum discharge can actually be passed, without cavitation danger?   *Ans.:* 129 cfs/ft.

6–45. A siphon spillway has a throat section of inner radius 1 ft and outer radius 3 ft, with a uniform width of 4 ft, discharging at elevation 100. The siphon is designed to prime when the headwater reaches elevation 140. The total energy loss in the conduit, exclusive of outlet velocity head, is 10 ft. What is the discharge, and what must be the area of outlet section, for:

a. Tailwater elevation 110 ft?

b. Tailwater elevation 90 ft?

6–46. A siphon spillway has a throat section of inner radius 2 ft and outer radius 4 ft. The conduit has a uniform cross-section 2 ft × 4 ft throughout its length. The total head loss in the system under design conditions is 5 ft, when the reservoir surface level is 50 ft above the conduit outlet. In order to protect against possible undesirable conditions at the throat, what should be the dimensions of the conduit at the outlet, assuming the width is kept at twice the depth?

6–47. A chute spillway is on a slope of 30° with a Manning coefficient of 0.0149 and is long enough for uniform flow to be attained. The discharge is 10,000 cfs and the channel is rectangular, of 20-ft width. Determine the probable air-entrainment concentration ratio and the depth of air-entrained flow. Also compute the pressure on the bottom of the chute.

6–48. Spillway crest gates for a dam are the vertical-lift type, of average size and design. Each gate spans an 8-ft clear opening between piers and may carry a 9-ft head of water between spillway and dam crest elevations. If the coefficient of friction between the gate and seats is 0.3, what lifting force is required to open the gate?   *Ans.:* 7355 lb.

6–49. A horizontal rectangular stilling basin, 40 ft wide, is provided at the outlet of a steep channel for energy dissipation. The discharge is 200 cfs and, at the point where the hydraulic jump begins, the velocity is 20 fps. Compute:

a. Depth at end of jump.   *Ans.:* 2.37 ft.

b. Head loss in jump.   *Ans.:* 4.01 ft.

c. Efficiency of jump.   *Ans.:* 62%.

d. Length of jump.   *Ans.:* 14.6 ft.

6–50. A simple stilling basin at the base of an overflow spillway causes a hydraulic jump to form, with the flow jumping from a depth of 1 ft to a depth of 4 ft. What is the horsepower dissipated in the jump?   *Ans.:* 3.44 hp/ft.

6–51. Determine the location and dimensions of the hydraulic jump of Problem 4–55, allowing for the effect of length of jump.

6–52. A hydraulic jump forms on a simple horizontal rectangular stilling basin. When the discharge is 40 cfs/ft and the depth just upstream from the jump is 2 ft, calculate:
  a. Downstream depth, ft.
  b. Head loss, ft.
  c. Efficiency of jump, per cent.
  d. Length of jump, ft.

6–53. What should be the dimensions of large roughness elements on the bed of a steep channel for most effective energy dissipation? The discharge is 60 cfs and the channel is 10 ft in width, with a slope of 30°. Calculate the following:
  a. Height and longitudinal spacing of square roughness bars if placed throughout length of channel.
  b. Height and transverse spacing of cubical elements used throughout length of channel.
  c. Height of leading element for only five rows of elements at downstream end. Assume the supercritical depth just before the first hydraulic jump is 3 inches.

6–54. A "smooth" concrete culvert 3 ft in diameter and 300 ft long is to discharge 140 cfs at a temperature of 73°F. Assume that the pipe is straight and uniform throughout its length and that there is no approach transition. The elevation of the culvert invert at the entrance is 100 ft. Determine the headwater elevation for the following conditions:
  a. Culvert on horizontal slope, sharp flush inlet, tailwater elevation 104.0 ft.
  b. Culvert on slope of 0.0500, bell-mouthed inlet, tailwater elevation 104.0 ft.
  c. Culvert on slope of 0.0500, sharp flush inlet, tailwater elevation 80.0 ft.
  d. Culvert on slope of 0.0500, bell-mouthed inlet, tailwater elevation 80.0 ft.

Repeat the above computations for a culvert only 30 ft long. Assume the following loss coefficients:

$$\text{Sharp flush entrance} = 0.50$$
$$\text{Bell-mouthed entrance} = 0.05$$
$$\text{Outlet} = 1.00$$

6–55. An outlet conduit through a dam has a diameter of 6 ft and a length of 50 ft and discharges under a head of 100 ft measured above the center line. The pipe is horizontal and discharges freely with a flared outlet (loss coefficient 0.60). The inlet is rounded (loss coefficient 0.05), and the material has an equivalent sand roughness of 0.01 ft. The water temperature is 60°F. The reservoir surface area is 100,000 ft².
  a. What is the discharge?   *Ans.:* 2480 cfs.
  b. What is the pressure just below the inlet?   *Ans.:* −10.8 psi.
  c. Assuming no further inflow to the reservoir, how long would it take to drop the reservoir level 19 ft through two outlet conduits of this size?   *Ans.:* 6.8 min.

6–56. A circular corrugated metal highway culvert is 200 ft long and is laid on a grade of 10 per cent, with a free outlet. Assume the Manning coefficient for corrugated metal to be 0.025 and the inlet contraction coefficient to be 0.75. The approach channel velocity head can be neglected. The permissible head on the culvert above its inlet center line is 4.0 ft and the design discharge is 40 cfs.

    a. What commercial-size pipe should be used?   *Ans.:* 30 in.

    b. If a special inlet piece is used which will cause the culvert to flow full and will eliminate entrance loss, what size pipe may be used?   *Ans.:* 24 in.

6–57. A square concrete box culvert is 100 ft long and is on a 5 per cent slope, with a free outlet. The Manning coefficient is 0.013, and the design discharge 60 cfs. The permitted head on the culvert is 4.0 ft above the inlet center line. Approach and exit channel velocity heads are negligible. Determine the minimum required culvert size:

    a. For sharp entrance, contraction coefficient $\frac{2}{3}$.   *Ans.:* 2.37 ft square.

    b. For inlet transition piece eliminating all contraction and loss at entrance. *Ans.:* 1.94 ft square.

6–58. A concrete pipe culvert is 6 ft in diameter, 400 ft long and on a 5 per cent slope, with a straight, free outlet. The inlet is set flush with the headwall, and the contraction coefficient resulting from the tongue-and-groove lip on the pipe is $\frac{3}{4}$. The pipe friction factor can be assumed to be 0.015. Compute the head over the culvert inlet center-line for a design discharge of 300 cfs (neglecting approach velocity head):

    a. Assuming outlet control.

    b. Assuming inlet control.

Which condition is more likely to prevail? Why?

6–59. A box culvert is 6 ft × 6 ft in cross section, 100 ft long and on a 2 per cent slope. It has a sharp, flush inlet and outlet. The outlet is submerged at the design discharge of 1000 cfs, with the tailwater level 6 ft above the outlet center-line. The concrete surface can be considered to have an equivalent sand roughness of 0.001 ft and the water to have a kinematic viscosity of 10^{-5} ft²/sec. How high is the headwater above the inlet center-line? Neglect velocity heads in approach and exit channels.

6–60. A horizontal concrete pipe culvert 100 ft long is laid through a highway embankment, with the pipe inlet flush with the headwall and the pipe outlet projecting into the tailwater pool. The concrete surface has an equivalent sand roughness of 0.01 ft and the pipe diameter is 4.0 ft. If the permitted head on the culvert is 16 ft (difference between headwater and tailwater pool elevations), what is the culvert discharge? The kinematic viscosity of the water is 10^{-5} ft²/sec. Neglect approach and exit velocity heads.   *Ans.:* 275 cfs.

6–61. A corrugated metal culvert is 6 ft in diameter, 400 ft long, and on a 5 per cent slope, with a straight, free outlet. The inlet is set flush with the head-wall, and the contraction coefficient resulting from the initial corrugation is $\frac{3}{4}$. The pipe friction factor can be assumed at 0.060.. The design discharge is 600 cfs. Neglect approach velocity head. Compute the head over the culvert inlet center-line:

    a. Assuming outlet control and full flow.

b. Assuming inlet control and part-full flow.
Which condition will prevail? Why?

6-62. An outlet conduit through a dam is of concrete, 8 ft in diameter, 200 ft long, and on a horizontal slope. It has a sharp, flush inlet and outlet, and the friction factor can be assumed to be 0.020. The outlet is free and the head over the inlet center-line is 64 feet. Assume the conduit flows full throughout its length.

  a. What is the discharge in cfs?
  b. What is the pressure, in psi, at the vena contracta near the inlet? What inference can be drawn from the magnitude of this pressure?

6-63. A culvert consists of twin barrels, each 6 ft × 6 ft in cross-section, 500 ft long and on a 2 per cent slope. The culvert is concrete, with a Manning coefficient of 0.012. The inlet is well rounded and the outlet discharges onto a paved apron which is at the same elevation as the culvert floor. At a certain discharge, the depth of flow on the apron is 8 ft, and the velocity is 6 ft per second. The downstream channel is of rectangular shape, 30 ft wide. Neglecting the approach velocity head, how high is the headwater above the center-line of the culvert inlet?

7

HYDRAULIC POWER CONVERSION

7–1. Introduction. The conversion of hydraulic energy to other forms of energy, and vice versa, on an efficient and economical basis, requires the coordination of the efforts of many types of specialists. Hydropower developments are of great scope and involve the design and operation of many and diverse structures, machines, processes, and systems.

This chapter is limited to the hydraulics of the actual conversion process. Many topics already discussed (dams, spillways, open channels, pipes, etc.) deal with elements associated with hydropower plants, but these elements are also associated with many other types of hydraulic engineering projects. This chapter will be concerned primarily with draft tubes, hydraulic machinery, and other hydraulic topics identified especially with hydropower developments. The two major categories of hydropower developments in view are hydroelectric plants, for the conversion of hydraulic energy into electrical energy through turbines, and pumping plants, for elevating water or increasing its hydraulic head by pumps.

The basic hydropower equation is

$$P = Q\gamma H \qquad (7\text{--}1)$$

in which P is the power, in foot pounds per second, being transmitted by water flowing at the rate of Q cubic feet per second under a total head of H feet, γ being the specific weight in pounds per cubic foot. In horsepower units, Eq. (7–1) is

$$\text{HP} = \frac{Q\gamma H}{550} = \frac{QH}{8.8} \qquad (7\text{--}2)$$

and in kilowatts

$$\text{KW} = \frac{Q\gamma H}{737} = \frac{QH}{11.8} \qquad (7\text{--}3)$$

The general power conversion equation for steady flow is

$$Q\gamma\left(\frac{p_1}{\gamma} + Z_1 + \frac{V_1^2}{2g} + H_P\right) = Q\gamma\left(\frac{p_2}{\gamma} + Z_2 + \frac{V_2^2}{2g} + H_T + H_L\right) \quad (7\text{--}4)$$

where H_P is the head added to the flow by pumps between points 1 and 2, H_T is the head extracted from the flow by turbines in the same reach, and H_L is the head lost in friction in the reach.

The term $p/\gamma + Z + V^2/2g$ represents the total head of the flow. The change in total head, $\Delta(p/\gamma + Z + V^2/2g)$, represents the energy available to, or required from, the power conversion system taken as a whole. Thus, from Eq. (7-4),

$$H_T + H_L - H_P = \left(\frac{p_1}{\gamma} + Z_1 + \frac{V_1^2}{2g}\right) - \left(\frac{p_2}{\gamma} + Z_2 + \frac{V_2^2}{2g}\right) = \pm H$$

$$(7\text{-}5)$$

Normally, either H_T or H_P would be dropped from the equation. The overall hydraulic efficiency of the system is then either

$$e_h = \frac{H_T}{H} = \frac{H - H_L}{H} \qquad (7\text{-}6)$$

or

$$e_h = \frac{H}{H_P} = \frac{H_P - H_L}{H_P} \qquad (7\text{-}7)$$

If the mechanical efficiency of the equipment is e_m and the total efficiency is $e = e_h e_m$, then the horsepower conversion for the entire system is

$$\text{HP} = \frac{Q\gamma H_T}{550} e_m = \frac{Q\gamma H}{550} e = \frac{QHe}{8.8} \qquad (7\text{-}8)$$

or

$$\text{HP} = \frac{Q\gamma H_P}{550 e_m} = \frac{Q\gamma H}{550 e} = \frac{QH}{8.8e} \qquad (7\text{-}9)$$

In Eq. (7-8), the horsepower is that which can actually be placed on the transmission lines; in Eq. (7-9), it is that which must be supplied to the pumping plant.

The design of a hydropower installation largely devolves about planning equipment which will accomplish the indicated power transformations with as high efficiencies as practicable. This requires not only careful selection of the proper pumping or turbine units, but also optimum design of the entire intake and outlet systems, for both steady and unsteady flow conditions.

7-2. Power Available and Required. The planning of a hydroelectric plant requires first a determination of the power available in the stream. This is essentially a product of the available discharge and head. If the plant efficiency is about 88 per cent, Eq. (7-8) yields a quick determination of the possible power available, as about $\frac{1}{10}QH$, where Q is the average usable discharge and H is the average head to be provided, this of course depending largely on the reservoir design and operation.

More accurate estimates of power available may be obtained by detailed flow routing through the proposed reservoir, as discussed in Chapter 11. Thus, a month-by-month tabulation of QH quantities can be synthesized. *Firm power* is that amount of power which can be supplied virtually at all times.

Surplus power is all available power in excess of the firm power. The relative amounts of firm and surplus power are often affected by the other water uses in a multiple-purpose reservoir.

A similar analysis may be made of the anticipated demands on a pumping installation. These should be more nearly constant, although there will usually be diurnal, weekly, and seasonal variations to take into account.

7–3. Types of Turbines and Pumps. Following is an outline list of the major categories of hydraulic turbines and pumps:

Turbines

1. Pressureless (impulse)—buckets on periphery of wheel, moved by jet; all available head converted into kinetic energy, some of which is used by changed velocity causing wheel torque. Wheel and buckets enclosed in case, but under essentially atmospheric pressure. Used for high heads, above 800 ft. Pelton type used exclusively.
2. Pressure (reaction)—flow directed through stationary guide vanes onto runner, entering with both radial and tangential components. Energy imparted to runner by reduction of both pressure and velocity. Casing is filled and all buckets in operation simultaneously, more power developed for given diameter than in Pelton wheel. Shaft usually vertical, and water discharges downward at center into expanding draft tube. Francis type is used, mixed flow, radial and axial. Heads from 15 ft to 1100 ft.
3. Propeller (also a form of pressure turbine)—flow is axial, uses principle of lift on airfoil to impart energy to propeller. Heads less than 60 ft.

Pumps

1. Centrifugal—uses principle of forced vortex (increased pressure from centrifugal force) combined with radial flow to give spiral flow from impeller through vanes to casing and discharge pipe. Maximum head per stage about 100 ft (possibly up to 500 ft) but can have any number of stages. Pump size usually in terms of discharge pipe diameter (velocity about 10 ft/sec, economical for long pipe lines).
   a. Volute pump—spiral casing, with area changing to correspond to increasing Q and to convert velocity into pressure energy.
   b. Turbine pump—has guide vanes to take water from impeller and convert velocity into pressure energy. Economical only for very large installations.
2. Propeller—reverse of propeller turbine, useful mainly for large capacities at low heads, up to 40 ft.
3. Displacement.
   a. Reciprocal (piston).
   b. Rotary.
   c. Screw.
4. Ejector (jet pump).
5. Air lift pump.
6. Hydraulic ram.
7. Diaphragm pump.
8. Steam vacuum pump.

The displacement pump, as well as the other special types following it on the above list, are of interest for special applications only and will not be discussed here. The centrifugal volute pump and the Francis turbine are by far the most commonly encountered types, and propeller pumps and turbines also find frequent use. The impulse turbine, or water wheel, is now used very infrequently except for high head installations.

7–4. The Centrifugal Pump. The most frequently used hydraulic machine is the centrifugal pump, which will therefore be first considered. Furthermore, much of the basic theory of centrifugal pump operation is essentially identical with that of the Francis turbine, which operates basically as a centrifugal pump in reverse.

A cross-section through a typical centrifugal pump of the volute type is shown in Fig. 7–1.

The volute casing is proportioned to keep velocities approximately equal around the circumference, as well as to provide a gradual conversion of kinetic head to pressure head in the discharge pipe.

Centrifugal pumps may be further identified as

1. Single-suction or double-suction.
2. Single-stage or multistage.
3. Horizontal-mounting or vertical-mounting.
4. Driven by steam, motor, gas, gasoline, Diesel, wind, etc.
5. Used for low lift, high lift, fire, well, sump, sludge, etc.

From Eq. (7–5), the net head developed by a pump (after losses) may be expressed as

$$H = \left(\frac{p_d}{\gamma} - \frac{p_s}{\gamma}\right) + (Z_d - Z_s) + \left(\frac{V_d^2}{2g} - \frac{V_s^2}{2g}\right) \qquad (7\text{–}10)$$

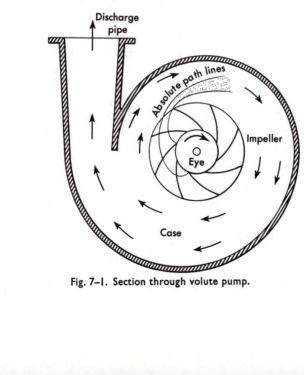

Fig. 7–1. Section through volute pump.

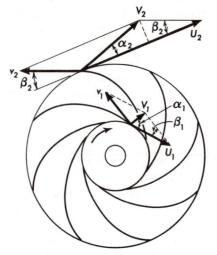

Fig. 7–2. Centrifugal pump impeller.

where the subscripts d and s stand for discharge pipe and suction pipe, respectively.

The suction pressure head, p_s/γ, may be either positive or negative. In order to avoid possible cavitation and other problems, a negative pressure head of more than 20 to 24 ft should be avoided. The equation is simplified, of course, if the suction and discharge pipes are at the same elevation or if they have the same diameter.

A pump impeller is sketched in Fig. 7–2, showing velocity vector relations at inlet and outlet of the impeller. The symbol U represents impeller tangential velocity, v represents the relative velocity of water to the blade, and V, the vector sum of U and v, is the absolute velocity of the water. The subscript 1 represents impeller inlet and 2 impeller outlet. The angle α is the angle between U and V, and the angle β is the blade angle, or the angle between v and $-U$.

From the equation of continuity, it is obvious that

$$2\pi r_1 b_1 (V_1)_r = 2\pi r_2 b_2 (V_2)_r = Q \tag{7–11}$$

and therefore that

$$(V_2)_r = (V_1)_r \frac{r_1 b_1}{r_2 b_2} \tag{7–12}$$

where b is the width of impeller passageway, r the radius, and the subscript r denotes radial component. From the vector diagram, if the subscript t denotes tangential components,

$$(V_1)_t = U_1 - (V_1)_r \operatorname{ctn} \beta_1 \tag{7–13}$$

It is desirable that the water enter the impeller radially, at least at the design discharge, in order to minimize shock at entrance. Thus α_1 can be assumed to be $90°$, and $(V_1)_t = 0$. Then,

$$(V_1)_r = U_1 \tan \beta_1 = \omega r_1 \tan \beta_1 \qquad (7\text{--}14)$$

ω being the angular speed, in radians per second.

At the impeller outlet, the vector diagram similarly indicates that

$$(V_2)_t = U_2 - (V_2)_r \operatorname{ctn} \beta_2 = \omega r_2 - \frac{(V_2)_r}{\tan \beta_2} \qquad (7\text{--}15)$$

Thus, if the pump impeller dimensions and speed are known, then the radial components of V can be determined from Eqs. (7–12) and (7–14). The tangential component of V_2 is obtained from Eq. (7–15).

The momentum flux entering the impeller is $\rho Q V_1$, with $\rho Q V_2$ at the exit. By the impulse-momentum principle, the net force on the impeller is

$$\vec{F} = \rho Q(\vec{V}_2 - \vec{V}_1) \qquad (7\text{--}16)$$

The torque exerted on the impeller is

$$T = (F_t)_2 r_2 - (F_t)_1 r_1 = \rho Q[V_2(\cos \alpha_2)r_2 - V_1(\cos \alpha_1)r_1] \qquad (7\text{--}17)$$

The head imparted to the flow by the impeller is

$$H = \frac{P}{Q\gamma} = \frac{T\omega}{Q\gamma} = \frac{V_2(\cos \alpha_2)U_2}{g} - \frac{V_1(\cos \alpha_1)U_1}{g} \qquad (7\text{--}18)$$

Since in each vector diagram, by the law of cosines, $2UV \cos \alpha$ is equal to $U^2 + V^2 - v^2$, Eq. (7–18), known as the Euler equation, can also be written

$$H = \frac{U_2^2 - U_1^2}{2g} + \frac{V_2^2 - V_1^2}{2g} - \frac{v_2^2 - v_1^2}{2g} \qquad (7\text{--}19)$$

The pump characteristics and efficiency largely depend upon the particular design of blades, casing, etc. The energy relationships for flow through the impeller are shown in Fig. 7–3. The shutoff head is approximately $U_2^2/2g$, resulting from the parabolic rise in pressure distribution obtained in a fluid in a rotating cylinder. For this condition, no flow occurs and the suction head, H_s, is entirely pressure head.

When flow begins, the suction head is in two parts, p_s/γ and $V_s^2/2g$. Similarly the pressure head at point 2, the outside edge of the impeller blades, drops by an amount $v_2^2/2g + k(v_2^2/2g)$, where k is the loss coefficient for the reach from point s to point 2, and v_2 is the relative velocity of flow to the blades.

The pressure continues to rise in the casing, reaching a final value p_d at the discharge pipe, where the velocity also remains constant at V_d. If V_2 is the absolute velocity at point 2 the losses between points 2 and d will be a function of $V_2^2/2g$, equal, say, to $(1 - m)V_2^2/2g$.

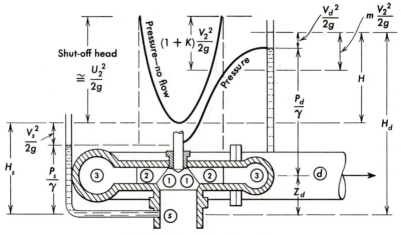

Fig. 7–3. Flow through impeller.

Then, as seen from Fig. 7–3, the net head added to the flow by the pump, measured by the difference in heads between suction and discharge points, is

$$H = H_d - H_s = \frac{U_2^2}{2g} - (1 + k)\frac{v_2^2}{2g} + m\frac{V_2^2}{2g} \qquad (7\text{-}20)$$

This equation gives the head H in terms of the three components, U_2, v_2, and V_2, of the velocity vector diagram at the impeller exit.

Obviously, this head is increased when the pump speed (therefore U_2) is increased, and when V_2 is increased (therefore when the blade angle β_2 is high). It is decreased when v_2 is high (therefore when the discharge is high).

The loss coefficients, k and $1 - m$, depend mainly on the blade design and the discharge. In particular, they are minimum values when the design discharge occurs, for which the impeller and casing proportions have been correlated with the velocity vector relations. At other conditions, separation occurs and the efficiency drops.

For best efficiency, V_2 should be small, cutting down losses in the casing and facilitating conversion into pressure energy. This normally means that the blade should curve back and the exit blade angle β_2 should be small. On the other hand, special designs for other requirements are sometimes needed, with radial or even forward-curving vanes.

Typical pump characteristic curves are shown in Fig. 7–4. It is obvious from Eq. (7–20) that a rising characteristic corresponds to a blade design such that $m\,V_2^2/2g > (1 + k)\,v_2^2/2g$. This implies forward curving vanes. A steep characteristic requires that $(1 + k)\,v_2^2/2g > m\,V_2^2/2g$, which implies back-curving vanes, as in the usual centrifugal pump. A flat characteristic is obtained by radial vanes. In any case, losses increase materially when the discharge exceeds the design discharge, so that the head developed also drops off rapidly.

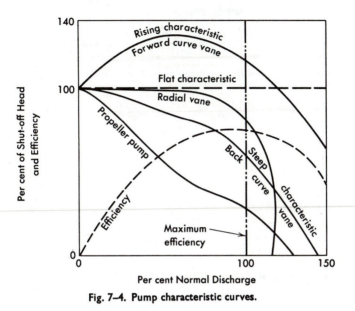

Fig. 7–4. Pump characteristic curves.

Example 7–1

A centrifugal water pump has the following characteristics:

$$r_1 = 4 \text{ inches} \qquad Q = 10 \text{ cfs} \qquad N = 1800 \text{ rpm}$$
$$r_2 = 12 \text{ inches} \qquad b_1 = b_2 = 2 \text{ inches} \qquad e_h = 90\%$$
$$\alpha_1 = 90° \qquad \text{Net power added to flow} = 300 \text{ horsepower.}$$

What are the inlet and outlet blade angles β_1 and β_2?

Solution. Peripheral velocities:

$$U_1 = \frac{\pi N}{30} r_1 = \frac{\pi(1800)}{30} \frac{4}{12} = 20 \text{ fps}$$

$$U_2 = \frac{r_2}{r_1} U_1 = 3U_1 = 60 \text{ fps}$$

Head on impeller, $H = \dfrac{P}{Q\gamma e} = \dfrac{300(550)}{10(62.4)(0.9)} = 294 \text{ ft}$

Tangential component of absolute velocity of water at exit of impeller =

$$(V_2)_t = \frac{gH}{U_2} = \frac{32.2(294)}{60\pi} = 50.1 \text{ fps}$$

Absolute velocity at entrance to impeller $=$

$$(V_1)_r = \frac{Q}{2\pi r_1 b} = \frac{10}{2\pi(\frac{1}{3})(\frac{1}{6})} = \frac{90}{\pi} = 28.6 \text{ fps}$$

Radial component of absolute velocity at exit of impeller $=$

$$(V_2)_r = (V_1)_r \frac{r_1 b_1}{r_2 b_2} = (V_1)_r \frac{r_1}{r_2} = 28.6(\tfrac{4}{12}) = 9.53 \text{ fps}$$

(a) Blade angle at inlet, $\beta_1 = \tan^{-1} \frac{(V_1)_r}{U_1} = \tan^{-1} \frac{28.6}{20\pi}$

$$= 24.5° \text{ (from back tangent)}. \quad Ans.$$

(b) Blade angle at exit, $\beta_2 = \tan^{-1} \frac{(V_2)_r}{U_2 - (V_2)_t} = \tan^{-1} \frac{9.53}{60\pi - 50.1}$

$$= 3.94° \text{ (from back tangent)}. \quad Ans.$$

7-5. Similarity Relationships. Equation (7–20) may be written in dimensionless form as follows:

$$m \frac{V_2^2/2g}{H} - (1 + k)\frac{v_2^2/2g}{H} + \frac{U_2^2/2g}{H} = 1 \qquad (7\text{–}21)$$

The dimensionless ratios $(V_2^2/2g)/H$ and $(U_2^2/2g)/H$ are constants if dynamic similarity can be maintained. However, this can be accomplished only by homologous impellers operating under homologous conditions.

If true similarity is attained, then the ratios are constants. That is,

$$U_2 = \phi\sqrt{2gH} = \omega r_2 = \frac{\pi DN}{720} \qquad (7\text{–}22)$$

and

$$V_2 = c\sqrt{2gH} = \frac{Q}{kD^2} \qquad (7\text{–}23)$$

where D is the impeller diameter in inches. Then,

$$N = 1840\phi \frac{\sqrt{H}}{D} \qquad (7\text{–}24)$$

Substituting D from Eq. (7–23) in Eq. (7–24) yields

$$N = 1840\phi \frac{\sqrt{H}(ck\sqrt{2gH})^{1/2}}{\sqrt{Q}} = N_s \frac{H^{3/4}}{Q^{1/2}} \qquad (7\text{–}25)$$

where the constant N_s is the specific speed for the particular pump and operating conditions. The constant ϕ is called the *speed factor*.

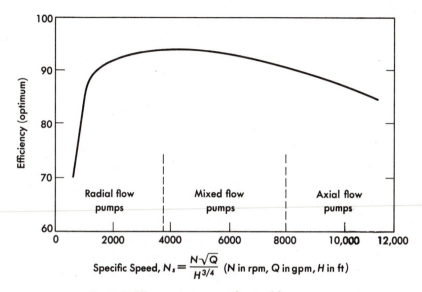

Fig. 7–5. Efficiency versus specific speed for pumps.

As shown in Art. 8–13, the specific speed is a similitude parameter related to the Euler number. It is usually defined at the condition of greatest efficiency for the given design. Designs which yield high discharges for given heads give high specific speeds, and vice versa. Figure 7–5 shows the approximate range of specific speeds corresponding to various types of pumps. Since the power developed is $Q\gamma H$, the specific speed may be written alternatively as

$$N_s = \frac{NQ^{1/2}}{(P/Q)^{3/4}} = \frac{NQ^{5/4}}{P^{3/4}} \tag{7–26}$$

or as

$$N_s = \frac{N(P/H)^{1/2}}{H^{3/4}} = \frac{NP^{1/2}}{H^{5/4}} \tag{7–27}$$

The last form is commonly used in connection with turbines. In any case, it is not dimensionally a "speed." Therefore its specific numerical value depends on the units assigned to H, P, N, and Q. Usually, P is in horsepower, H in feet, N in revolutions per minute, and Q in gallons per minute.

Kinematically similar conditions may be obtained for different discharges and heads on a given pump, provided the speed is adjusted correspondingly. Similarly, homologous impellers will give similar conditions if they all run at the same speed.

Assuming that full similarity is maintained, then all homologous velocity vector diagrams will be geometrically similar. Thus the speed factor ϕ and the coefficient c are constants. Thus:

$$V_2 = \frac{c}{\phi} U_2 \propto U_2 = \frac{2\pi N}{60} \frac{D}{2} \propto ND \qquad (7\text{–}28)$$

Also, since any area is proportional to the impeller cross-sectional area:

$$Q \propto V_2 D_2^2 \qquad (7\text{–}29)$$

Then, combining Eqs. (7–28) and (7–29):

$$Q \propto ND^3 \qquad (7\text{–}30)$$

Since $V_2 = c\sqrt{2g}\, H^{1/2}$, it also follows that

$$H \propto N^2 D^2 \qquad (7\text{–}31)$$

Now, since $P = Q\gamma H$, then

$$P \propto N^3 D^5 \qquad (7\text{–}32)$$

Equations (7–30), (7–31), and (7–32) specify for similar conditions the variation of Q, H, and P with impeller speed and diameter. Various other relations are implicit in these. For example, combining Eqs. (7–30) and (7–31):

$$Q \propto H^{1/2} D^2 \qquad (7\text{–}33)$$

Combining (7–31) and (7–32):

$$P \propto H^{3/2} D^2 \qquad (7\text{–}34)$$

Or, again combining (7–30) and (7–31):

$$Q \propto \frac{H^{3/2}}{N^2} \qquad (7\text{–}35)$$

Also,

$$P \propto \frac{H^{5/2}}{N^2} \qquad (7\text{–}36)$$

It is obvious from Eq. (7–35) that the pump specific speed is merely the square root of the similitude constant in the equation. Thus

$$Q = (N_s)^2 \frac{H^{3/2}}{N^2} \qquad (7\text{–}37)$$

which is the same as Eq. (7–25).

Similarly, the turbine specific speed is merely the square root of the similitude constant in Eq. (7–36):

$$P = (N_s)^2 \frac{H^{5/2}}{N^2} \qquad (7\text{–}38)$$

which is the same as Eq. (7–27).

7–6. The Radial-Flow Turbine. The counterpart of the centrifugal (radial-flow) pump is the radial-flow turbine, commonly known as the Francis turbine. A schematic cross-section through the runner of the Francis turbine is shown in Fig. 7–6. The water enters the runner at point 1, with an absolute

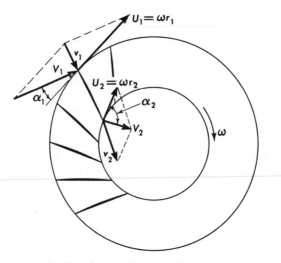

Fig. 7–6. Section through turbine runner.

velocity V_1, its direction controlled by the setting of the fixed blades on the *guide ring* outside the runner (not shown). The relative velocity v_1, at design conditions, has a direction corresponding to the blade angle on the runner, which has a peripheral velocity U_1. Similar relations apply at the exit point 2, where the water leaves the runner to enter the draft tube.

Application of the angular momentum principle yields exactly the same equation as obtained for the pump, except for interchange of subscripts. Thus,

$$T = \rho Q(r_1 V_1 \cos \alpha_1 - r_2 V_2 \cos \alpha_2) \tag{7–39}$$

and

$$H = \frac{U_1 V_1 \cos \alpha_1 - U_2 V_2 \cos \alpha_2}{g} \tag{7–40}$$

where T and H are the torque and head, respectively, imparted to the turbine runner by the flow. If the angular momentum is zero at exit ($\alpha_2 = 90°$) as usually designed, then the second term in each of the above equations vanishes.

In the region outside the runner, as the flow nears the runner, $T = 0$ and the free vortex equation applies:

$$rV \cos \alpha = \text{constant} \tag{7–41}$$

Just as Eq. (7–18), for the centrifugal pump, is equivalent to Eq. (7–40) for the Francis turbine, with only the subscripts interchanged, so likewise

Eq. (7–19) can be applied to the turbine merely by interchanging subscripts. It may also be written as:

$$H = \frac{U_1^2 + V_1^2 - v_1^2}{2g} \cdot - \frac{U_2^2 + V_2^2 - v_2^2}{2g} \qquad (7\text{–}42)$$

The same equation can also be derived from the energy relationships between relative flows at points 1 and 2, neglecting energy losses in the runner.

Equations (7–28) through (7–38), giving relations between discharge, head, power, speed, and diameter for homologous impellers for centrifugal pumps, can also be applied to turbine runners.

The speed factor, as defined in Eq. (7–22), commonly has a value from about 0.55 to 0.90 in turbines operating at good efficiencies. The specific speed, as defined for turbines in Eq. (7–27), ranges from about 10 to about 100 for Francis turbines. It is also defined as the speed necessary to operate a turbine to develop 1 hp under a head of 1 ft. Equation (7–27) can be conveniently utilized to select a suitable turbine for given requirements for power, head, and speed.

7–7. Axial-Flow Pumps and Turbines. In axial-flow (propeller) pumps and turbines, the flow has no radial component, but is over the blades in a direction parallel to the axis of rotation of the runner or impeller. The principle of lift on an airfoil is employed for the transmission of force and torque. A schematic diagram of a propeller turbine setting is shown in Fig. 7–7.

Although the flow has no radial component as it passes through the runner, it may have a tangential component. The latter is determined by the setting

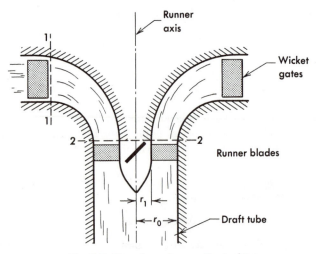

Fig. 7–7. Flow through propeller turbine.

of the wicket gates in the guide ring and by the free vortex relation, Eq. (7-41), in the region between sections 1—1 and 2—2.

In this same region, the initial radial flow components are converted to axial flow by the boundary formed by the casing. The axial flow velocity distribution can ordinarily be assumed uniform at section 2—2 because of the contracting approach:

$$V_a = \frac{Q}{\pi(r_o{}^2 - r_i{}^2)} \tag{7-43}$$

where V_a is the axial component of the absolute velocity and Q is the total discharge. This combines with the tangential component for the absolute water velocity. The peripheral velocity of the blade, U, of course depends on the radial distance of the particular element. The tangential component of absolute velocity, as determined from the free vortex relation at the leading edge of the blade, is reduced by the blade curvature, so that it should be small or zero at the trailing edge. These relationships are illustrated in Fig. 7-8, which shows a cross-section through the blade element at some arbitrary radius r.

The blade angles β_1 and β_2 are designed to correspond with the required directions of the relative velocities v_1 and v_2 for the design conditions. Since the direction of v depends on the magnitude of U, which in turn depends on the radius, it is obvious that each element of the blade must have a different shape corresponding to the local requirements for β. Since V_a and U are constant from the leading to trailing edge for each element, the vector diagrams of Fig. 7-8 may be superposed, as in Fig. 7-9. As the radius decreases, it is clear that U must decrease but the tangential component of V_1 must increase. Therefore the blade angle β_1 for the leading edge must

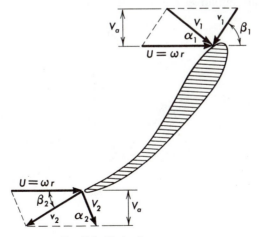

Fig. 7-8. Blade element.

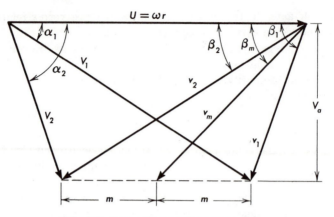

Fig. 7–9. Vector diagrams for runner blade.

decrease as the radius of the element increases. If it is required that V_2 be fully axial, then,

$$\tan \beta_2 = \frac{V_a}{U} = \frac{Q}{\omega \pi r (r_o^2 - r_i^2)} \tag{7-44}$$

Hydraulic force and torque are imparted to the runner through the mechanism of lift and drag forces on the propeller blade. Each element of the blade may be considered as an airfoil section, upon which flow impinges at some angle of attack θ, as sketched in Fig. 7–10. The resultant force on the airfoil can be considered in terms of its lift and drag components (normal and parallel to the direction of the relative flow velocity v_0). The angle of attack is measured to the *chord*, a straight line usually drawn connecting the leading and trailing edges of the airfoil.

The lift and drag forces are computed by the relations,

$$F_L = C_L A \frac{\rho v_o^2}{2} \tag{7-45}$$

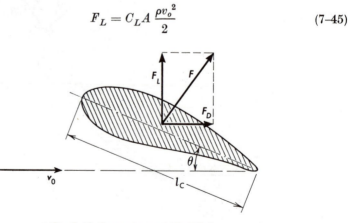

Fig. 7–10. Forces on propeller blade.

and

$$F_D = C_D A \frac{\rho v_o{}^2}{2} \qquad (7\text{--}46)$$

in which A is the effective area, usually defined as (length of chord) × (span of airfoil), and C_L and C_D are coefficients of lift and drag respectively. These coefficients are determined empirically and are dependent primarily on angle of attack and airfoil shape.

In the case of the turbine runner, the effective relative flow velocity is defined as v_m, at an angle β_m such that, as shown on Fig. 7–9,

$$\text{ctn}\,\beta_m = \frac{1}{2}\,(\text{ctn}\,\beta_1 + \text{ctn}\,\beta_2) \qquad (7\text{--}47)$$

Similarly, it is obvious from Fig. 7–9 that:

$$V_a = v_1 \sin \beta_1 = v_2 \sin \beta_2 = v_m \sin \beta_m \qquad (7\text{--}48)$$

The lift and drag forces can then be evaluated for a span of blade dr, from Eqs. (7–45) and (7–46), using v_m as the effective velocity v_0. The total

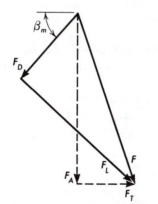

Fig. 7–11. Force components on blade element.

force F on the span dr can then be resolved into axial and tangential forces, as in Fig. 7–11. The axial component causes a thrust on the bearing, and the tangential component is effective in producing torque on the shaft. By combining Eqs. (7–45) and (7–46) with the vector relationships of Fig. 7–11, expressions for thrust and torque are obtained, all referring only to a single blade and to a single element on the blade, of thickness dr and radial distance r:

$$F_A = \frac{\rho v_m{}^2}{2}\, l_c\, dr(C_L \cos \beta_m + C_D \sin \beta_m) \qquad (7\text{--}49)$$

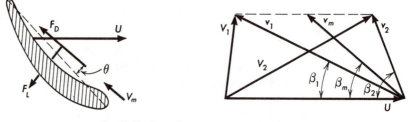

Fig. 7–12. Propeller pump vector relations.

$$T = F_T r = \frac{\rho v_m^2}{2} l_c r \, dr (C_L \sin \beta_m - C_D \cos \beta_m) \qquad (7\text{--}50)$$

where l_c is the chord length, and v_m, C_L, C_D, β_m and l_c are all functions of r. Total thrust and torque could in principle be obtained by integrating with respect to r and multiplying by the number of blades.

The relationships described above all apply to design conditions of operation. If conditions change, serious energy losses due to shock and separation may occur. This may be avoided to considerable extent by changing the blade angles correspondingly by changing the blade setting. A turbine with adjustable blades, usually governor-operated, is known as a Kaplan turbine.

The axial-flow pump is essentially an axial-flow turbine in reverse. The primary difference is that the peripheral velocity of the pump impeller is directed opposite to the tangential component of the lift force, as shown in Fig. 7–12.

The similarity relationships discussed above in connection with radial-flow pumps and turbines are also valid for axial-flow turbomachinery. The *specific speed* is defined in exactly the same way, although numerical magni-

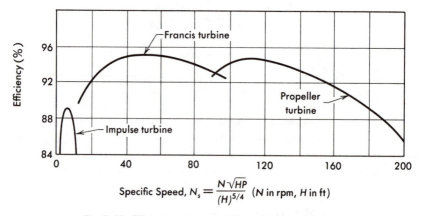

Fig. 7–13. Efficiency versus specific speed for turbines.

tudes of specific speed are quite different. Figure 7–13 indicates the range of specific speeds for Francis and propeller turbines operating at conditions of optimum efficiency, for fairly large turbine units.

It is seen that Francis turbines are most effective at specific speeds less than about 100, propeller turbines at specific speeds greater than about 100. For specific speeds less than about 10 (corresponding to high heads and small power potential), the impulse turbine (Pelton water wheel) is most suitable.

Approximate ranges of specific speed for pumps have already been shown, in Fig. 7–5.

7–8. Draft Tubes. The draft tube is ordinarily considered as an integral part of a pressure turbine, either axial-flow or radial-flow. It consists of a conduit of gradually expanding cross-section, conveying the water discharged from the turbine runner to the tailrace below the power plant. Its function in the complete hydropower operation is best understood in terms of a schematic representation of the plant as shown in Fig. 7–14.

The effective head on the turbine is usually considered in hydropower plant practice to be the difference in elevations of the energy grade line at points 1 and 4, which are at the entrance to the turbine casing and in the tailrace channel below the draft tube outlet, respectively. This is the head

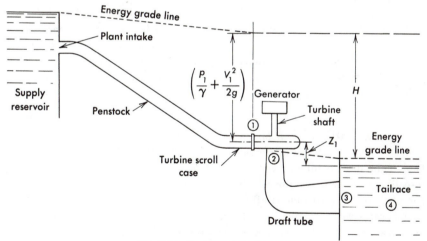

Fig. 7–14. Turbine plant elements.

H shown in Fig. 7–14. The head actually transmitted to the shaft is reduced by losses in the turbine runner and casing, in the draft tube, and at the draft tube outlet, all of which are minimized by careful design of the turbine and draft tube.

A draft tube performs two primary functions: (1) to permit utilizing at least a part of the difference in turbine and tailrace elevation as effective head on the turbine; and (2) to convert the high kinetic energy at the runner outlet into pressure energy with a minimum of losses. These functions require the maintenance of a suction head on the runner outlet.

A draft tube may be either a vertical tube or an elbow tube. The former is more efficient but ordinarily requires more vertical space than is feasible, so that the latter is more common. The design of the tube is usually supplied by the turbine manufacturer as an integral part of the turbine design.

If the tailwater level is taken as datum, the energy equation can be written as follows:

$$Z_1 + \frac{p_1}{\gamma} + \frac{V_1^2}{2g} = H + \frac{V_4^2}{2g} \qquad (7\text{--}51)$$

The energy equation can also be written from point 2, at the runner outlet, to point 4, this equation applying then to the draft tube itself:

$$Z_2 + \frac{p_2}{\gamma} + \frac{V_2^2}{2g} = \frac{V_4^2}{2g} + H_f + H_o \qquad (7\text{--}52)$$

where H_f is the friction loss in the draft tube and H_o is the outlet loss approximately $(V_3 - V_4)^2/2g$.

The suction head, p_2/γ, may be computed from Eq. (7–52) if the other terms are known. Its magnitude will of course be of prime concern with respect to the possibility of cavitation on the runner blades. The terms on the right-hand side of the equation are usually small, but V_2 will be high, probably between 20 and 40 fps. The static head Z_2 is limited to a maximum of 15 ft, and usually is less than 12 ft.

7–9. Cavitation.
The phenomenon of cavitation becomes a potential danger whenever the fluid pressure drops substantially below atmospheric. One of the most common occurrences of cavitation is on the blades of pump impellers and turbine runners. Not only is the pressure in the pump suction chamber and the turbine draft tube usually subatmospheric, but the high peripheral speeds of the blades may set up regions of highly excessive velocities, and therefore very low pressures, in certain segments of the blade passageways.

Cavities of vapor tend to form in the flow wherever the absolute pressure drops to the fluid vapor pressure. Collapse of the cavities when the pressure rises in the downstream direction causes serious problems of noise, vibration, and pitting of surfaces. Turbine and pump blades, especially on high specific speed units, have in the past often been seriously damaged and even destroyed by this phenomenon.

The *cavitation index*, representing the cavitation-producing characteristics of the flow, is defined as

$$\sigma = \frac{p_0 - p_v}{(\tfrac{1}{2})v_0^2 \rho} = \frac{p_0/\gamma - p_v/\gamma}{v_0^2/2g} \tag{7-53}$$

where p_0 and v_0 are the absolute fluid pressure and the flow velocity, usually taken at some representative reference point in the flow field, and p_v is the vapor pressure.

The parameter σ is a similarity parameter and, if made equal in model and prototype, it can be used to estimate cavitation probabilities in a given hydropower plant. That is, σ is determined experimentally on a model pump or turbine of a given specific speed at the point where cavitation actually begins. Assuming the prototype will be operated under homologous conditions, then cavitation in the prototype should begin at approximately the same value of σ. The permissible value of p_0 can then be determined from Eq. (7-53) and this in turn utilized in the general energy equation for the draft tube to select the permissible turbine setting. A similar analysis can be made to determine a permissible pump setting. This subject is discussed further in Art. 8-10. It should be noted, however, that many factors can affect the cavitation phenomenon and considerable uncertainty still exists regarding the best procedures for predicting prototype cavitation from model testing.

7-10. Impulse Turbines. The impulse turbine is sometimes used in high-head installations and differs from pressure turbines primarily in being open to atmospheric pressure throughout the wheel and casing. In its usual form, it consists of a series of "buckets" attached to the periphery of a wheel (usually rotating on a horizontal shaft) which may be up to 15 ft or more in diameter. The energy available in the water is converted into kinetic energy exclusively, by means of a nozzle at the end of the penstock, creating a high-speed jet to be directed on the buckets. The jet is split into two parts by a "splitter" ridge on each bucket, thus eliminating the bearing thrust that would otherwise be caused by diversion of the jet. A wheel containing such split buckets is known as a Pelton water wheel and is the only type of impulse turbine in commercial use.

The operation of an impulse turbine can be explained in terms of the hydrodynamic force produced by a jet of water on a moving blade, as shown in Fig. 7-15. The blade is one of a series, so that the entire jet is used continuously to drive the wheel.

If the blade is moving to the right with an absolute velocity U and the jet has an absolute velocity V_1, then the relative velocity of the jet to the blade is $V_1 - U = v$. As the jet is deflected by the blade, through a total deflection angle θ, it retains a relative velocity of the same magnitude v (neglecting retardation by friction), although its direction is gradually changed.

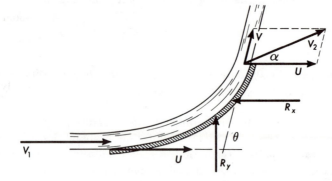

Fig. 7–15. Action of jet on moving blade.

The jet has an absolute velocity V_2 at its exit from the blade, obtained by vector summation of velocities U and v, the latter at angle θ with U. The angle between V_2 and U is denoted by α.

In order to divert the jet, the blade must exert a resultant force on it, of components R_x and R_y. By the momentum principle,

$$R_x = \rho Q(V_1 - V_2 \cos \alpha) \qquad (7\text{–}54)$$

and

$$R_y = \rho Q(V_2 \sin \alpha) \qquad (7\text{–}55)$$

These forces are equal and opposite to the force components F_x and F_y exerted by the water on the blade. If the blade is attached to the periphery of a rotating wheel, F_y is an axial thrust and can be eliminated by the expedient of a "splitter." See Fig. 7–16. The force F_x, however, is a tangential force and thus produces a torque which serves to rotate the wheel.

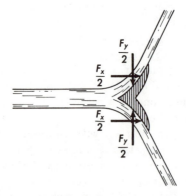

Fig. 7–16. Splitter on blade.

The vector relationships yield

$$V_2 \cos \alpha = U + v \cos \theta = U + (V_1 - U) \cos \theta$$
$$= U(1 - \cos \theta) + V_1 \cos \theta$$

Therefore, substituting in Eq. (7–54) and replacing R_x by its numerical equivalent F_x gives

$$F_x = \rho Q(V_1 - U)(1 - \cos \theta) \qquad (7\text{–}56)$$

The torque developed on the blade is seen to be

$$T = F_x r \qquad (7\text{–}57)$$

Therefore,

$$T = \rho Q(V_1 - U)r(1 - \cos \theta) \qquad (7\text{–}58)$$

In these equations, r is the wheel radius, measured to the jet center line. The power developed is

$$P = T\omega = T\,\frac{U}{r} = F_x U = \rho Q U(V_1 - U)(1 - \cos \theta) \qquad (7\text{–}59)$$

On the assumption that the jet area and velocity are held constant, the power that can be transmitted varies with U and θ, which can be adjusted by wheel design. For the best values, make $\partial P/\partial U$ and $\partial P/\partial \theta$ equal to zero.

$$\frac{\partial P}{\partial U} = \rho Q(1 - \cos \theta)(V_1 - 2U) = 0$$

$$\frac{\partial P}{\partial \theta} = \rho Q U(V_1 - U) \sin \theta = 0$$

From which, the turbine should be designed so that

$$U = \tfrac{1}{2}V_1 \qquad (7\text{–}60)$$

and

$$\theta = 180° \qquad (7\text{–}61)$$

If the speed N (in rpm) of the wheel is fixed by power plant design, then the water wheel diameter should be

$$d = \frac{30}{\pi}\,\frac{V_1}{N} \qquad (7\text{–}62)$$

The above relations are for ideal conditions. Substitution in Eq. (7–59) indicates the power transmission to be

$$P = \rho Q\,\frac{V_1}{2}\left(\frac{V_1}{2}\right)[1 - (-1)] = \tfrac{1}{2}\rho Q V_1{}^2 \qquad (7\text{–}63)$$

Similarly the kinetic energy flux in the impinging jet is

$$\frac{E}{T} = \frac{(1/2)M V_1{}^2}{T} = \tfrac{1}{2}\rho Q V_1{}^2$$

Thus there is a theoretically 100 per cent efficient transmission of power by a wheel of this design. If only one blade were involved, the effective Q is $A(V_1 - U)$ instead of AV_1. The efficiency is then $100(1 - U/V_1)$.

Actually there is substantial energy dissipation by friction with the blades and by interference of the diverted jet with the approaching flow. More precise theoretical analyses allowing for friction, supported by empirical tests, indicate the best design to be such that θ is about 165° to 175° and U about $0.43V_1$ to $0.48V_1$. Efficiencies at design discharge are usually about 80 per cent, and may be as high as 90 per cent, but drop substantially for other discharges. However, if the load on the power plant changes, the speed N of the shaft must be kept constant by changing the discharge. This is usually accomplished by a governor-operated needle valve in the nozzle leading to the buckets.

It may be noted that the torque equation for an impulse wheel, Eq. (7-58), can be written

$$T = \rho Qr[V_1 - (V_1 \cos \theta + U - U \cos \theta)]$$
$$= \rho Qr(V_1 - V_2 \cos \alpha) \tag{7-64}$$

This is simply the forced vortex equation for the special case of *tangential flow*, with radius constant at r and $\alpha_1 = 0°$.

The "speed factor" as previously defined for pressure turbines may also be used for impulse turbines, for which it is essentially the ratio of U to V_1. Thus,

$$\phi = \frac{U}{\sqrt{2gH}} = \frac{U}{V_1/C_v} \cong 0.98 \frac{U}{V_1} \tag{7-65}$$

Here H is the total head at the base of the nozzle, consisting of the penstock velocity head plus terminal pressure head. The velocity coefficient C_v is a measure of the small energy loss in the nozzle. It is noted that the head H does not include the position head of the turbine setting above the tailrace. The latter is not utilized in energy conversion, but since impulse turbines are limited to high-head installations, this is not of great importance.

The similarity parameters and relationships developed previously for other turbomachines may also be applied to the impulse turbine. The specific speed for the latter is quite low, ranging from about 3 to 10, for wheels with one jet, up to about 14, for two jets on a single runner.

The jet dimensions and velocity obviously exert a controlling influence on the wheel design. In general, the jet velocity depends on the available head at the nozzle, which in turn depends on the gross head in the supply reservoir and on losses in the penstock. Thus,

$$V_1 = C_v\sqrt{2g(H_t - H_f)} \tag{7-66}$$

Neglecting minor losses, the friction loss H_f can be computed from the Darcy equation. H_t represents the difference in elevation between the reservoir surface and nozzle center line. Since H_t may be very large for an impulse turbine installation, the penstock may be quite long and the friction loss therefore quite large unless a large diameter pipe is used. In any case, the power in the jet is given by

$$P_1 = Q\gamma(H_t - H_f) = \gamma A_1 V_1 \left(H_t - f\frac{L}{D}\frac{V^2}{2g}\right)$$

where V is the penstock velocity and L/D its length-diameter ratio, with f the Darcy friction factor. By continuity, $A_1 V_1 = A\,V$. Therefore,

$$P_1 = \gamma A_1 V_1 \left(H_t - f\frac{L}{D}\frac{V_1^2}{2g}\frac{A_1^2}{A^2}\right) \qquad (7\text{-}67)$$

For maximum power in the jet,

$$\frac{dP_1}{dA_1} = \gamma V_1 \left(H_t - f\frac{L}{D}\frac{V_1^2}{2gA^2}3A_1^2\right) = 0$$

from which,

$$H_t = f\frac{L}{D}\frac{V^2}{2gA^2}\frac{A^2}{A_1^2}3A_1^2 = 3H_f \qquad (7\text{-}68)$$

Since the nozzle head, $V_1^2/2g$, is equal to $(H_t - H_f)$, neglecting nozzle losses, it follows from continuity that:

$$\frac{V_1^2}{2g} = 2H_f = \frac{2fL}{D}\frac{V_1^2}{2g}\left(\frac{D_1}{D}\right)^4$$

from which, simplifying, the following is obtained:

$$D_1 = \left(\frac{D^5}{2fL}\right)^{1/4} = D\left(\frac{D}{2fL}\right)^{1/4} \qquad (7\text{-}69)$$

as the best relation between jet diameter and penstock characteristics.

Example 7–2

An impulse turbine of 4 ft diameter is driven by a water jet 2 inches in diameter, with a discharge of 4 cfs. The blade angle is 150° and the wheel speed is 250 rpm.

Calculate (a) tangential force on the blades

(b) horsepower developed on the turbine.

Solution.

Blade velocity, $U = \dfrac{2\pi N r}{60} = \dfrac{2\pi(250)(2)}{60} = 52.4$ fps

Jet velocity, $V_1 = \dfrac{Q}{A_1} = \dfrac{4(12)^2}{\pi} = 183.2$ fps

(a) Tangential force on blades $= F_x$

$= \rho Q (V_1 - U)(1 - \cos \theta) = 1.94(4)(130.8)(1 + \cos 30°) = 1890 \text{ lb.}$

Ans.

(b) Power developed $= \text{HP} = \dfrac{1}{550} \rho Q U (V_1 - U)(1 - \cos \theta)$

$$= \frac{F_x U}{550} = \frac{1890(52.4)}{550} = 180 \text{ hp} \qquad \textit{Ans.}$$

7-11. Water Hammer. A problem of considerable importance in turbine penstocks, as well as in other pipe lines subject to unsteady flow conditions, is the phenomenon of water hammer. This is the graphic though somewhat misleading name given to the transient motion of a pressure wave through a pipe line occasioned by a change in valve opening. The effects associated with this phenomenon depend on the celerity of the pressure wave in relation to the rapidity of the rate of valve closure or opening, and may under some conditions produce severe stresses in the pipe walls.

In an extensive field of fluid, the transmission of a pressure increment wave of magnitude dp through a given area dA can be described by the energy and continuity equations. Thus, by the energy equation,

$$d\left(\frac{c^2}{2g}\right) + d\left(\frac{p}{\gamma}\right) = 0 \tag{7-70}$$

and by the continuity equations,

$$d(\rho c)\, dA = 0 \tag{7-71}$$

By solving for dc from each equation, and setting the resulting expressions equal to each other, the following is obtained:

$$dc = -\frac{dp}{\rho c} = -\frac{c\, d\rho}{\rho}$$

Then, since the modulus of elasticity E of a fluid is defined as the ratio of its unit stress (i.e., pressure change) to unit strain (i.e., change in density per unit of density), this can be solved for the celerity c as follows:

$$c = \sqrt{\frac{dp}{d\rho}} = \sqrt{\frac{E}{\rho}} \tag{7-72}$$

This is the familiar expression for "sonic velocity," or the celerity of a pressure wave in a fluid medium of unlimited extent and density ρ. For water at ordinary temperatures, c is about 4720 ft/sec.

However, this must be modified if the transmission of the pressure wave is inhibited, rather than free in all directions. Thus, if a pressure wave is transmitted in a pipe line, some of the energy is utilized in stretching the pipe walls.

Consider first the case of an instantaneous valve closure in a pipe in which the flow velocity under steady conditions had been V_0. As the flowing mass of fluid is suddenly brought to rest, a sharp rise in pressure at the valve occurs, and is transmitted upstream at celerity c. By the momentum principle,

$$F = \Delta p\, A = \rho Q\, \Delta V \tag{7-73}$$

where Δp is the pressure rise, ΔV is the velocity change. ρQ is the mass per unit time affected, and A is the cross-sectional area. Therefore, since $Q = A(c)$,

$$\Delta p = \rho V_0 c \cong V_0 \sqrt{\rho E} \tag{7-74}$$

The rise in pressure head is

$$\Delta H = \frac{\Delta p}{\gamma} = \frac{V_0 c}{g} \tag{7-75}$$

This pressure rise is transmitted upstream at celerity c. If the total length of pipe is L, the pressure wave reaches the supply reservoir in a time L/c.

In order to determine the value of the celerity under these conditions, it may be assumed that the kinetic energy of the flowing column of water is converted into pressure energy by compression of the column when it is stopped, and into elastic strain energy by stretching of the pipe walls. If dL is the length affected in time dt, the total kinetic energy of flow is $\gamma A\, dL\, (V^2/2g)$. The work done in compressing the fluid is equal to the reduction in volume multiplied by the average pressure increase, or

$$A\, dL \left(\frac{\Delta p}{E}\right) \left(\frac{\Delta p}{2}\right)$$

since $A\, dL$ is the initial volume and since $E = \Delta p\ (\text{volume})/\Delta\ (\text{volume})$. The work done in stretching the walls is equal to the average force in the wall multiplied by the elongation of the circumference. The latter in turn is the initial circumference times the unit strain. Since the tensile force per unit length in the walls is $\Delta p(D/2)$, the work done is

$$\frac{1}{2}\left(\frac{\Delta p\ D}{2}\, dL\right) \left(\frac{\pi D}{E_p}\right) \left(\frac{\Delta p\ D}{2t}\right)$$

where t is the wall thickness and E_p is the modulus of elasticity for the pipe material. By energy conservation,

$$\gamma A\, dL\, \frac{V_0^2}{2g} = A\, dL \left(\frac{\Delta p}{E}\right)\left(\frac{\Delta p}{2}\right) + \frac{1}{2}\left(\frac{\Delta p\ D}{2}\, dL\right)\left(\frac{\pi D}{E_p}\right)\left(\frac{\Delta p\ D}{2t}\right)$$

Simplification gives

$$\gamma \frac{V_0^2}{2g} = \frac{(\Delta p)^2}{2E} + \frac{(\Delta p)^2}{2E_p}\left(\frac{D}{t}\right)$$

Solving for Δp yields

$$\Delta p = V_0 \sqrt{\frac{\rho E}{1 + (E/E_p)(D/t)}} \tag{7-76}$$

By substituting in Eq. (7–74) the celerity is seen to be

$$c = \frac{\sqrt{E/\rho}}{\sqrt{1 + (E/E_p)(D/t)}} \tag{7-77}$$

This is less than the previous expression $\sqrt{E/\rho}$, but reduces to this for an infinitely rigid pipe.

The rise in pressure head occasioned by the valve closure is

$$\Delta H = \frac{\Delta p}{\gamma} = \frac{V_0}{g} \sqrt{\frac{E/\rho}{1 + (E/E_p)(D/t)}} \tag{7-78}$$

and this rise is transmitted upstream at the rate indicated by Eq. (7–77). As the wave moves upstream, the water velocity is reduced to zero. When it reaches the reservoir, the entire column of water in the pipe is at rest.

However, upon reaching the reservoir the wave is reflected as a negative wave, or wave of rarefaction. Pressure waves are true waves, conforming to the principles of wave mechanics, and may thus be reflected in either of two ways: (a) from a solid boundary, at which the velocity becomes zero but at which the pressure may vary; (b) from a fluid boundary, at which the pressure must be constant but the velocity may change. In like manner, when two waves meet, their velocities and pressures may be superposed. When equal and opposite compression waves meet, the result is analogous to that occurring when a single wave meets a solid boundary: the velocity becomes zero and the pressure doubles. When opposite compression and expansion waves meet, the pressure reduces to zero and the *relative* velocity doubles; this is analogous to reflection from a boundary of constant pressure.

Thus, when the pressure wave reaches the reservoir, it is reflected as an expansion wave of zero pressure and equal, but opposite, velocity. Looking at the phenomenon in a different way, there is instantaneously a sharp differential of pressure from the normal (zero, relatively speaking) pressure in the reservoir and the high pressure of the elastically compressed water in the pipe. This condition cannot remain stable, so water starts to flow out of the pipe at its previous forward velocity V. This relieves the stressed condition and the expansion wave reaches the valve, at which time the entire water column has assumed a zero pressure and a negative velocity V.

The expansion wave is now reflected from the solid boundary, the pressure dropping to $-\Delta p$ and the velocity returning to zero. This wave moves upstream to the reservoir, where its negative pressure is obliterated by the reflection of an equal and opposite positive wave, of forward velocity V.

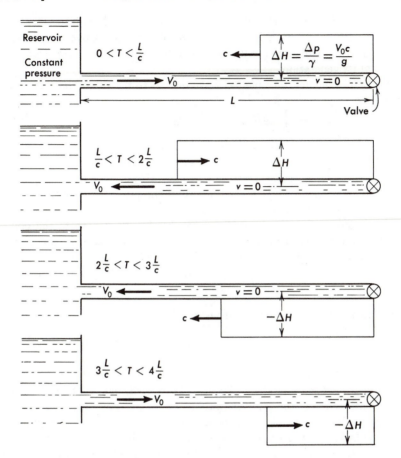

$$0 < T < \frac{L}{c}$$

$$\Delta H = \frac{\Delta p}{\gamma} = \frac{V_0 c}{g}$$

$$\frac{L}{c} < T < 2\frac{L}{c}$$

$$\Delta H$$

$$2\frac{L}{c} < T < 3\frac{L}{c}$$

$$-\Delta H$$

$$3\frac{L}{c} < T < 4\frac{L}{c}$$

$$-\Delta H$$

Fig. 7–17. Water hammer cycle.

This latter wave reaches the valve after a time $4L/c$ from the instant of valve closure, where L is the total length of pipe. At this time the pressure is again zero and the forward velocity is V, which is exactly the situation when the valve was first closed. Consequently, the cycle will now repeat itself. This cycle is sketched in Fig. 7–17.

The same cycle continues until it is eventually damped out by friction. After a sufficient number of cycles, the oscillation stops and the entire system is brought permanently to rest.

Actually, it is not possible for the valve to be closed instantaneously. Rather, the closure must occupy some finite time period T_c, which may be of any arbitrary magnitude.

If the valve closure is visualized as occurring in a series of small increments, each producing a velocity decrease dV, the corresponding head rise is

$dH = -(c/g)\,dV$, from Eq. (7–75). The negative sign indicates that H increases with decreases in V. Each incremental rise will move up the pipe, returning to the valve via a reflected expansion wave in time $2L/c$ from the instant of its generation.

Just before the first incremental wave returns to the valve, the total head rise will be

$$H' = \int_0^{H'} dH = -\frac{c}{g}\int_{V_0}^{V'} dV = \frac{c(V_0 - V')}{g} \tag{7–79}$$

where V' is the pipe velocity at $T = 2L/c$. Thus, if T_c is less than this time, V' will be zero, and there will be a pressure rise in the pipe near the valve, just as if the valve had closed instantaneously. However, this maximum rise will occur over only a part rather than all of the pipe, and over only a part rather than all of the time period $2L/c$.

The part of the pipe subjected to peak head is the length to the point where the wave of full head rise meets the returning initial expansion wave. Thus, equating travel times gives

$$T_c + \frac{x}{c} = \frac{2L - x}{c}$$

from which the length of pipe x (measured from the valve) over which peak head will occur is

$$x = L - \tfrac{1}{2}cT_c \tag{7–80}$$

Obviously, $x = L$ when T_c is zero.

If it is desired to know the variation of head with time, at the valve or at other locations along the pipe, the problem is one of unsteady flow and an exact analysis is impossible. The variation of valve opening as a function of time may be known or assumed, and also the valve discharge coefficient must be known, either as a variable function of the valve position or as an assumed constant quantity.

By continuity across the valve, it can be assumed that

$$AV = C_d A_v \sqrt{2gH} \tag{7–81}$$

where V is the pipe velocity, H the head across the valve, and A_v the valve opening area at time T. After a further time increment dt,

$$A(V - dV) = C_d(A_v - dA_v)\sqrt{2g(H + dH)} \tag{7–82}$$

By subtracting Eq. (7–81) from Eq. (7–82), and combining with Eq. (7–75) in differential form, the following differential equation results:

$$dH = \frac{cC_d}{Ag}[A_v(\sqrt{2g(H + dH)} - \sqrt{2gH}) - dA_v\sqrt{2g(H + dH)}] \tag{7–83}$$

In principle, this could be solved to give H as a function of time for any given functional relation of A_v to time. Actually, it would normally have to be solved by arithmetic integration, that is, by determining successive incremental values of ΔH corresponding to assumed values of ΔA_v, through trial-and-error solution of the equation.

Another factor which has been ignored in the discussion thus far is the boundary friction loss in the pipe. That is, when the valve is closed, not only is there a pressure rise due to water hammer but also one due to elimination of the friction loss when the velocity is reduced to zero. This is usually small compared to the water-hammer pressure and might be ignored. If the friction head loss, when the normal velocity V_0 prevails, is

$$H_f = f\left(\frac{L}{D}\right)\left(\frac{V_0^2}{2g}\right)$$

it may be shown from energy considerations, that its gradual elimination by the moving pressure wave will finally cause an additional head rise at the valve equal to $H_f/\sqrt{2}$. Thus the total rise in head at the valve for any time of closure less than $2L/c$ is

$$(\Delta H)_t = \frac{V_0 c}{g} + \frac{f}{\sqrt{2}}\frac{L}{D}\frac{V_0^2}{2g}$$

$$= \frac{V_0^2}{2g}\left[2\frac{c}{V_0} + \frac{fL}{\sqrt{2}D}\right] \tag{7–84}$$

In order to keep water-hammer pressures within reasonable limits, it is common practice to design gates and valves for closure times considerably greater than $2L/c$, the exact time depending on the permissible pipe wall stress. If $T_c > 2L/c$, then reflected expansion waves will mitigate the pressure waves generated by valve closing.

A comprehensive analysis of this phenomenon, yielding water-hammer pressure at any selected point in the pipe as a function of time, is tedious and will not be given here. In general, the principles are identical with those outlined above. That is, at any instant the flow in the pipe is equated with the flow out of the valve, by using the orifice equation for the latter. This equation is solved simultaneously with the equation relating pressure rise to decrease in velocity. Incremental calculations can be made, taking due account of both positive and negative wave fronts at each instant, which will yield paired values of velocity and pressure rise for each increment.

An alternative procedure is to set up the basic partial differential equations describing the transient phenomena, and then to adapt the equations for digital computer programming. The basic equations, those of energy and continuity, are derived below, for a segment dx in a straight horizontal pipe experiencing a pressure surge. Both V and H are functions of both x and T.

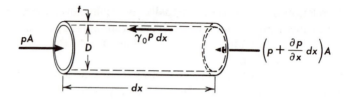

Fig. 7–18 Force system for water-hammer flow.

The energy equation may be derived from the Newtonian equation $F = Ma$. Referring to Fig. 7–18,

$$p(A) - \left(p + \frac{\partial p}{\partial x}\, dx\right)A - \tau_0 P\, dx = \rho A\, dx\, \frac{dV}{dT} \tag{7–85}$$

from which

$$\frac{\partial p}{\partial x} + \frac{\tau_0}{R_h} = -\rho\, \frac{dV}{dT}$$

But since

$$\frac{dV}{dT} = \frac{\partial V}{\partial T} + \frac{\partial V}{\partial x}\left(\frac{dx}{dT}\right) = \frac{\partial V}{\partial T} + V\, \frac{\partial V}{\partial x} \tag{7–86}$$

and since $p = \gamma H$ and $\tau_0 = \rho V^2 = \rho f V^2/8$, Eq. (7–85) becomes

$$\frac{\partial H}{\partial x} + \frac{f V^2}{D(2g)} = -\frac{1}{g}\left(\frac{\partial V}{\partial T} + V\, \frac{\partial V}{\partial x}\right) \tag{7–87}$$

The equation of continuity follows from the requirement that the excess mass entering the segment dx in a given time increment must equal the increment added to the total mass already within the segment. The latter increment is made possible by: (1) stretching of the pipe walls, and (2) compression of the fluid. Thus,

$$-\rho A\, \partial V\, \partial T = \rho\, dx\, P\, \frac{\partial D}{2} + \rho\, dx\, A\, \frac{\partial p}{E} \tag{7–88}$$

From the formula for stress in a thin ring, $s = pD/2t$, the increase in diameter due to the pressure increment ∂p is

$$\partial D = \frac{D^2}{2t E_p}\, \partial p \tag{7–89}$$

Substituting in Eq. (7–88) and simplifying gives

$$-\partial V\, \partial T = \frac{dx\, D\, \partial p}{t E_p} + \frac{dx\, \partial p}{E}$$

$$= \frac{dx\, \partial p}{E}\left[\left(\frac{D}{t}\right)\left(\frac{E}{E_p}\right) + 1\right] \tag{7–90}$$

Upon replacing ∂p by $\gamma \, \partial H$ and $(D/t)(E/E_p) + 1$ by $E/\rho c^2$, Eq. (7–90) becomes

$$\frac{\partial H}{\partial T} = -\left(\frac{c^2}{g}\right)\frac{\partial V}{\partial x} \qquad (7\text{–}91)$$

This is the usual form of the continuity equation for water-hammer flow. Equation (7–87) is the energy equation. In many cases, the friction term is small enough to neglect, as compared with the head caused by water hammer. Also $V(\partial V/\partial x)$ is usually quite small compared with $\partial V/\partial T$. If these two terms are neglected, Eq. (7–87) becomes

$$\frac{\partial H}{\partial x} = -\left(\frac{1}{g}\right)\frac{\partial V}{\partial T} \qquad (7\text{–}92)$$

These two partial differential equations in effect give the dependent variables H and V as functions of the independent variables x and T. They are the fundamental equations for surging flows in pipe lines.

Numerous methods have been proposed for their solution, all of which are approximations to some degree. In general, numerical methods are employed, with simultaneous solutions of the equations for ΔH and ΔV for assumed values of Δx and ΔT. The boundary values must be known as a starting point in the computation. For example, the velocity and head at the valve can be related by the orifice equation, as discussed previously. Reference may be made to the work of Parmakian, Rich, Streeter, Wood, and others as listed in the references at the end of the chapter, for the detailed techniques employed.

A widely used equation for approximate determination of the maximum water-hammer head that will be developed at the valve is based on the assumption that the product of the time of closure and the maximum pressure developed is a constant. This relation was suggested by Joukowsky, who was apparently the first to develop an essentially correct theory of water hammer. This assumption yields the following relation, for closure times in excess of $2L/c$:

$$p \cong p_{\max}\frac{2L/c}{T_c} \qquad (7\text{–}93)$$

or in terms of head rise,

$$H \cong \left(\frac{V_0 c}{g}\right)\left(\frac{2L/c}{T_c}\right) = \frac{2V_0 L}{gT_c} \qquad (7\text{–}94)$$

However this approximation should only be used with caution, and as a first estimate, as it may sometimes give seriously incorrect results, depending upon the manner of valve closure.

A sufficiently accurate, yet mathematically simple, technique for computing the head rise with time as the valve slowly closes in any known fashion has been developed by Wood and is known as the method of "wave plan analysis." In this method, it is assumed that the valve closes in a series of small increments, each during a time ΔT. Each incremental closure generates a pressure wave, which travels up the pipe and then is reflected negatively at the reservoir. The effect of the returning wave on the pressure at the valve and the flow through the valve is duly noted in the next step calculation.

The time increment ΔT is arbitrarily chosen such that the total time for wave travel and return, $2L/c$, is an even multiple of ΔT. Thus each wave returns to the valve just as one of the incremental changes in valve coefficient occurs.

The flow out of the valve is defined in terms of the usual orifice equation:

$$Q = AV = C_D A_V \sqrt{2g(H - H_{\text{ext}})} \tag{7-95}$$

The terms are as previously defined for Eq. (7–81) except that H represents the pressure head in the line just upstream from the valve and H_{ext} the pressure head in the external medium just downstream from the valve.

Now let V_1 and V_2 be the line velocities at the valve just before and just after an incremental valve closure, with H_1 and H_2 the corresponding line heads. The water hammer wave generated is ΔH_2; at the same instant a returning wave of magnitude ΔH_1 reaches the valve. The undisturbed line velocity is V_0. From Eq. (7–75)

$$\Delta H_1 = \frac{c}{g}(V_0 - V_1)$$

and

$$\Delta H_2 = \frac{c}{g}(V_0 - V_2)$$

When these equations are combined to eliminate V_0,

$$\Delta H_2 = -\Delta H_1 + \frac{c}{g}(V_1 - V_2) \tag{7-96}$$

By superposition of the pressure waves

$$H_2 = H_1 - \Delta H_1 + \Delta H_2 \tag{7-97}$$

Equations (7–96) and (7–97) are simultaneous equations in V_2 and ΔH_2, for known values of V_1 H_1 and ΔH_1. H_2 is replaced by its equivalent in terms of V_2, from Eq. (7–95). Their combination yields a quadratic equation in V_2, as follows:

$$V_2 = cK_2\left[-1 + \sqrt{1 + \frac{2g}{K_2 c^2}\left(H_1 - 2\Delta H_1 + \frac{V_1 c}{g} - H_{\text{ext}}\right)}\right] \tag{7-98}$$

In the above equation, K_2 is $\left(C_D \dfrac{A_v}{A}\right)^2$ computed just after the incremental valve closure.

In the calculations, $2L/c$, the wave travel time to and from the reservoir, is an even multiple of ΔT. For each increment ΔT, the corresponding valve coefficient and opening, C_D and A_v, are specified, this being a function of the type of valve and method of closure.

At the end of the first increment, the velocity at the valve is determined from Eq. (7–98). The magnitude of the outgoing pressure head wave is computed from Eq. (7–96) and the new pressure head from Eq. (7–97). The velocity V_2 and head H_2 become V_1 and H_1 for the next step calculation.

The term ΔH_1 in the analysis is the head increment of the returning pressure wave, which was therefore the increment generated at a time $2L/c$ earlier. For the first several calculations, ΔH_1 is zero, until the first pressure wave has time to return as the impinging wave.

Example 7–3

A valve at the end of a 2000 ft pipe closes in five seconds. The initial line velocity at the valve is 10 ft/sec and the initial head at the valve is 25 feet. Assume the celerity of a pressure wave in the pipe is 4000 ft/sec. The valve closes such that the term $(K)^{1/2}$—that is, $\left(C_D \dfrac{A_v}{A}\right)$—decreases linearly in five seconds from $(0.0623)^{1/2}$ to zero. Use time increments of $\dfrac{1}{2}\left(2\,\dfrac{L}{c}\right)$ in the calculations. Compute the head-time variation.

Solution.

$$\Delta T = \frac{1}{2}\left(2\,\frac{2000}{4000}\right) = 0.5 \text{ seconds. } H_{\text{ext}} = 0.$$

The calculations are summarized in the table below.

Time (secs)	K_2	V_1 ft/sec	H_1 ft	ΔH_1 ft	V_2 (Eq. 7–98)	ΔH_2 (Eq. 7–96)	H_2 (Eq. 7–97)
0.5	0.0504	10	25	0	9.96	5.59	30.59
1.0	0.0398	9.96	30.59	0	9.89	7.65	38.23
1.5	0.0305	9.89	38.23	5.59	9.72	15.60	48.25
2.0	0.0224	9.72	48.25	7.65	9.48	21.29	62.52
2.5	0.0156	9.48	62.52	15.60	9.08	35.51	82.43
3.0	0.0100	9.08	82.43	21.29	8.48	52.05	112.55
3.5	0.0056	8.48	112.55	35.51	7.55	81.38	158.43
4.0	0.0025	7.55	158.43	52.05	6.12	127.43	233.81
4.5	0.0006	6.12	233.81	81.38	3.80	209.32	360.7
5.0	0.0000	3.80	360.7	127.43	0	347.41	580.7

Thus, the maximum head produced by the valve closure is 580.7 ft. In comparison, the approximate Joukowsky relation would give only:

$$H = \frac{2V_0 L}{gT_c} = \frac{2(10)(2000)}{32.2(5)} = 250 \text{ ft}$$

7-12. Surge Tanks. Water-hammer pressures may be either resisted by pipe walls designed strong enough to allow them or relieved by surge tanks designed to accommodate the surge of water resulting from gate closure. Surge tanks are commonly employed on penstocks close to the turbine gates and in pumping plants on the discharge pipe lines. Consider a simple surge tank near a turbine gate, as sketched in Fig. 7–19.

Initially water is flowing in the penstock at velocity V_0, experiencing a friction loss (H_{f_0}). When the gate is suddenly closed, water will rise in the surge tank to a level $S + H_f$ ft above its initial running level. Elastic effects in the pipe and water may be neglected.

The basic unsteady-flow energy equation, from the equation $F = Ma$, is

$$-\gamma(S + H_{f_0})A = \rho A L \frac{dV}{dT}$$

or

$$\frac{dV}{dT} = -\frac{g}{L}\left[S + f\left(\frac{L}{D}\right)\left(\frac{V_0{}^2}{2g}\right)\right] = -\left(\frac{Sg}{L} + \frac{fV_0{}^2}{2D}\right) \qquad (7\text{--}99)$$

and by continuity requirements,

$$\frac{dS}{dT} = \frac{AV}{A_s} \qquad (7\text{--}100)$$

where A and A_s are the areas of the penstock and the surge tank, respectively.

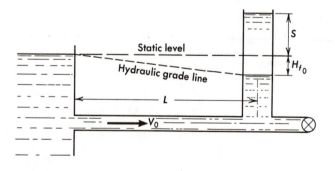

Fig. 7–19. Surge tank operation.

If friction losses are neglected, Eq. (7–99) simplifies to

$$\left(\frac{dV}{dS}\right)\left(\frac{dS}{dT}\right) = -\frac{Sg}{L} \tag{7–101}$$

and this may be combined with Eq. (7–100) to yield

$$\frac{dV}{dS} = -\left(\frac{A_s}{AV}\right)\left(\frac{g}{L}\right)S \tag{7–102}$$

from which, by integration,

$$V^2 = V_0{}^2 - \left(\frac{A_s}{A}\right)\left(\frac{g}{L}\right)S^2 \tag{7–103}$$

which gives the relation between pipe velocity and surge at any instant. The latter is maximum when V becomes zero. Then,

$$S_{\max} = V_0\sqrt{\left(\frac{A}{A_s}\right)\left(\frac{L}{g}\right)} \tag{7–104}$$

The relation between surge and time is obtained by combining Eq. (7–100) and (7–103):

$$T = \frac{A_s}{A}\int\frac{dS}{V} = \left(\frac{A_s}{A}\right)\frac{dS}{\sqrt{V_0{}^2 - (A_s/A)(g/L)S^2}}$$

$$= \frac{A_s}{A}\left[\sin^{-1}\frac{S}{V_0\sqrt{(A/A_s)(L/g)}}\right]\sqrt{\left(\frac{A}{A_s}\right)\left(\frac{L}{g}\right)} = \sqrt{\left(\frac{A_s}{A}\right)\left(\frac{L}{g}\right)}\sin^{-1}\left(\frac{S}{S_{\max}}\right) \tag{7–105}$$

The time required to reach the maximum surge is then obviously

$$T_{\max} = \frac{\pi}{2}\sqrt{\left(\frac{A_s}{A}\right)\left(\frac{L}{g}\right)} \tag{7–106}$$

The above equations are fairly accurate and give a conservative value for maximum surge. In long pipe lines, however, the friction loss should be included in the analysis. Equation (7–100) may be differentiated and combined with Eq. (7–99):

$$\frac{d^2S}{dT^2} = \left(\frac{A}{A_s}\right)\frac{dV}{dT} = -\frac{A}{A_s}\left(\frac{Sg}{L} + \frac{fV^2}{2D}\right)$$

$$= -\frac{A}{A_s}\left[\frac{g}{L}S + \frac{f}{2D}\left(\frac{A_s}{A}\right)^2\left(\frac{dS}{dT}\right)^2\right]$$

This gives the following differential equation relating surge and time:

$$\frac{d^2S}{dT^2} + \left(\frac{f}{2D}\right)\left(\frac{A_s}{A}\right)\left(\frac{dS}{dT}\right)^2 + \left(\frac{g}{L}\right)\left(\frac{A}{A_s}\right)S = 0 \tag{7–107}$$

This equation can be solved only by numerical and graphical methods, of which a number have been developed. Some of these are discussed in the references listed at the end of the chapter.

For high heads, the simple surge tank as discussed so far may be impracticable, and modifications may be desirable. Damping of surge action may be aided by a restricted entry to the surge chamber, for example. The tank may be enclosed at the top, providing an air cushion to absorb a portion of the water-hammer pressure. A *differential surge tank* has a small-diameter vertical riser in the middle of the large tank. Openings in the riser admit water to the main tank; however, oscillations in the riser have a greater amplitude than those in the main tank. Because the two are out of phase, surge damping is accomplished more rapidly.

PROBLEMS

7-1. A flow of 2 cfs is conveyed by a centrifugal pump of the following impeller dimensions:

$$r_1 = 2 \text{ in.} \qquad \alpha_1 = 90°$$
$$r_2 = 6 \text{ in.} \qquad \beta_1 = 75°$$
$$b_2 = \tfrac{1}{2} \text{ in.} \qquad \beta_2 = 60°$$
$$\omega = 100 \text{ rad/sec}$$

Determine the torque and horsepower added to the flow by the impeller. *Ans.:* 80 ft-lb, 14.5 hp.

7-2. The impeller of the pump of Problem 7-1 is modified to make $\beta_1 = 60°$ and $\beta_2 = 30°$, with $b_1 = \tfrac{3}{4}$ in. The pump is then used to draw water from a reservoir into a discharge pipe 6.0 in. in diameter. At a point 6.0 ft above the reservoir water surface, the discharge pipe pressure is measured to be 45 psi. Neglecting suction pipe friction losses, and assuming shockless entrance to the impeller, for a pump speed of 1800 rpm, determine the following:

   a. Entrance and exit velocity vector diagrams.
   b. Discharge. *Ans.:* 3.56 cfs.
   c. Shaft horsepower. *Ans.:* 55.7 hp.
   d. Pump hydraulic efficiency. *Ans.:* 83.3%.

7-3. A centrifugal pump operates at 150 rad/sec and requires 294 hp. The exit dimensions are $r_2 = 8$ in., $b_2 = 1$ in., $\beta_2 = 45°$. Neglecting any tangential component of entrance velocity, compute the discharge.

7-4. A pump tested at 2000 rpm discharges 6.0 cfs against a head of 340 ft. For similar hydraulic conditions, yielding the same efficiency, with a similar pump twice the size of the first, what discharge, head, and horsepower would be obtained at a speed of 1500 rpm.

7-5. The following test data apply to a centrifugal water pump having impeller diameter 0.875 ft and running at 24.2 rps:

Q (cfs)	H (ft)	Efficiency (%)
2.23	70	72
2.67	67	77
3.12	64	84
3.57	60	88
4.02	55	86
4.46	44	78

Geometrically similar pumps of diameters 0.750 ft and 0.667 ft are available. Each of the three can run at 19.2, 24.2, or 29.2 rps. Select from these the most suitable pump and speed to use for an application requiring 1.90 cfs to be pumped against a head of 51 ft of water. Compute the input horsepower required for the pump and speed selected.

7–6. A pump is to be selected which will pump 10 cfs against a head of 150 ft, with a shaft speed of 2400 rpm.
 a. What should be its specific speed? *Ans.:* 3750.
 b. What type of pump should be used? *Ans.:* Radial or mixed-flow pump.
 c. For a geometrically similar pump, operating at 3600 rpm, with a diameter one-half that of the first pump, what would be the rated head and discharge? *Ans.:* 85 ft, 1.88 cfs.

7–7. A centrifugal pump impeller has the following dimensions:

$b_1 = 1.5$ in. $r_1 = 3$ in. $\beta_1 = 50°$ $Q = 3$ cfs
$b_2 = 0.75$ in. $r_2 = 8$ in. $\beta_2 = 40°$ $\alpha_1 = 90°$

Neglecting losses, compute the following:
 a. Pump speed. *Ans.:* 490 rpm.
 b. Head added by impeller. *Ans.:* 21.8 ft.
 c. Torque added by impeller. *Ans.:* 79.6 ft-lb.

7–8. A centrifugal water pump has the following characteristics:

$r_1 = 4$ in. $\beta_1 = 20°$
$r_2 = 12$ in. $\beta_2 = 10°$
$b_1 = 2$ in. $N = 1800$ rpm
$b_2 = \frac{3}{4}$ in. $e_h = 88\%$

Assuming shockless entrance, compute:
 a. Pump discharge, cfs.   b. Net head added by pump, ft.

7–9. A centrifugal water pump has the following characteristics

$r_1 = 3$ in. $b_1 = 2$ in. $\beta_1 = 45°$ $N = 1500$ rpm.
$r_2 = 9$ in. $b_2 = 1$ in. $\beta_2 = 30°$ $\alpha_1 = 90°$

Compute the design head and discharge, in ft and cfs, respectively.

7–10. A pump must be selected which will pump 12 cfs against a head of 256 ft, with a shaft speed of 2400 rpm.

a. What is the specific speed and what type of pump should be selected?

b. If conditions change such that it becomes necessary to pump 24 cfs against a head of 128 ft, with the same pump, what shaft speed should be used?

c. If a $\frac{1}{10}$ scale model of the pump is tested in the laboratory, using a speed of 2400 rpm, what discharge ratio (model to prototype) should be used?

7–11. A pumping system is shown in the accompanying figure, Part A. Its design flow is to be 360,000 gal/day, with the pump operating 6 hours each day and 300 days per year. It is desired to select a pump size and pipe size combination for optimum economy.

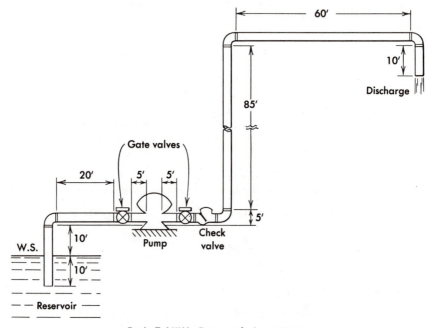

Prob. 7-11(A). Pump and pipe system.

Five pumps (A, B, C, D, E) are available with operating characteristics as shown in Part B. Pipes with diameters 6 in., 8 in., and 10 in. are available, of material such that the Hazen-Williams coefficient can be assumed to be 100.

First costs are as follows:

Pump	Cost	Motor	Cost
A	$710	10 hp	$ 80
B	650	20	100
C	600	30	120
D	550	40	140
E	510	50	190

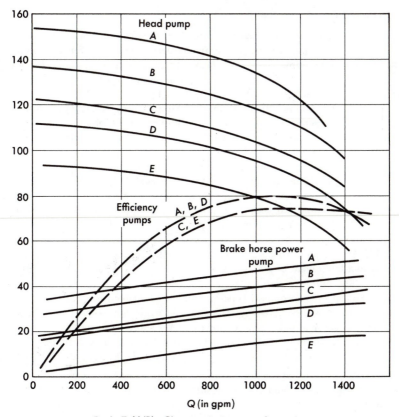

Prob. 7-11(B). Characteristic curves for system.

	6″ diameter	8″ diameter	10″ diameter
Pipe (100′)	$120	$150	$192
Elbow	5	8	16
Gate value	20	32	54
Check value	24	42	70

Cost of power is $0.02/kw-hr, and the motor efficiency is 90 per cent in each case.

Minor losses may be calculated in terms of equivalent lengths of straight pipe, as follows:

	6″ diameter	8″ diameter	10″ diameter
Elbow	18 ft	24 ft	32 ft
Gate value	7	10	13
Check value	52	75	100

Required:

     a. Curves of total head on pump for each of the various pipe diameters, superimposed on the pump characteristic curves, for flows from 0 to 1400 gpm.

     b. Selection of pump to satisfy the discharge and head requirements for each pipe size.

     c. For the rated condition (1000 gpm), pump efficiency, motor brake horsepower, and annual pumping cost for each case.

     d. Selection of best pump and pipe system on the basis of (1) first cost; (2) annual cost of pumping; (3) combination.

     e. Costs for a combination of the 8 in. pipe and the pump selected for the 10-in pipe (note effect of increased daily pumping time).

     f. Determination of pumps that should be selected on the basis of the assumption that each pipe has its effective diameter reduced by $\frac{1}{2}$ in. by encrustations.

7–12. A turbine is to be selected which will operate with a flow of 200 cfs with an available head of 600 ft. Assume an efficiency of 90 per cent and shaft speed of 1000 rpm.

     a. What specific speed and type of runner should be used?

     b. Assuming a model test on the selected type of turbine has indicated a speed factor of 0.6, what runner diameter should be used?

     c. If the discharge through the turbine as selected increases to 400 cfs, with no increase in head, what should be the new shaft speed and horsepower?

7–13. At a hydropower installation, the difference in elevation between reservoir water surface and tailwater surface is 1010 ft, and the turbine setting is 10 ft above the tailwater. At the design discharge of 100 cfs through each of the three turbines in the power plant, with a generator shaft speed of 400 rpm, a penstock head loss of 100 ft, and a hydraulic efficiency of 90 per cent:

     a. What specific speed and type of turbine should be used?

     b. What runner diameter should be used, assuming a speed factor of 0.5?

     c. What torque is developed on the turbine shaft?

7–14. At a hydro-electric plant, there is an available head of 300 ft and an average discharge of 400 cfs. The generator speed is 200 rpm and the turbine efficiency is 92 per cent. At the entrance to the draft tube the velocity is 30 ft/sec for vapor pressure of 1 ft of water and barometric pressure of 32 ft of water. The draft tube expands from 6 to 18 ft diameter, with the expansion head loss assumed equal to $0.5 \dfrac{(V_1 - V_2)^2}{2g}$

     a. What specific-speed turbine is needed for this installation?

     b. How high above tailwater level may the turbine be set without cavitation occurring?

7–15. A propeller turbine is driven by a flow of 1000 cfs at a speed of 240 rpm. The flow enters through wicket gates set at an angle of 19.4° with the tangential direction. The diameter of opening just inside the wicket gates is 12 ft and the height 4 ft. Assuming constant axial velocity just above the turbine runner, which has a hub radius of 1 ft and outer radius of 3 ft, determine the required blade angles at the outer radius, at entrance and exit, for efficient operation and removal of all torque from the flow. *Ans.:* 46.6°, 27.8°.

7–16. An impulse turbine of 6-ft diameter is driven by a water jet 2 in. in diameter moving at 200 ft/sec. Calculate the tangential force on the blades and the horsepower developed at 250 rpm. The blade angles are 150°.

7–17. The buckets on a Pelton turbine have a blade angle of 120°, and are driven by a jet of 2-in. diameter and 48-ft/sec velocity. For theoretical maximum power output, what is the resultant force on the blades? What wheel diameter should be specified for a shaft speed of 180 rpm?

7–18. A single blade is moved by a jet of water 3 in. in diameter with a velocity of 110 ft/sec. The blade is moving in the direction of the jet with a velocity of 70 ft/sec. The blade has one cusp only, with a deflection angle of 150°. Neglecting friction, calculate the x and y components of the force exerted by the water on the blade.

7–19. A 3-ft-diameter penstock delivers water to an impulse turbine through a nozzle of velocity coefficient 0.98. The penstock has a friction factor of 0.030 and is 6000 ft long. The total head, from nozzle level to reservoir surface, is 600 ft. The turbine has a speed factor of 0.45 and blade angles of 170°, and runs at a speed of 600 rpm. What jet diameter and what wheel diameter should be specified?

7–20. A 48-in. steel pipe, $\frac{3}{8}$ in. thick, carries water at a velocity of 6 fps. If the pipe line is 10,000 ft long, and if a valve at the discharge end is shut in 2.50 sec, what water-hammer force would be exerted on the valve just after closure? Assume the moduli of elasticity for water and steel to be respectively 313,000 and 30,000,000 psi. *Ans.:* 461,000 lb.

7–21. A 36-in. diameter penstock 8000 ft long carries water at a discharge of 70 cfs. A valve at the pipe terminus is closed in a time of 4 seconds. The pipe is $\frac{1}{2}$ in. thick and has a modulus of elasticity of 30,000,000 psi. The water modulus is 300,000 psi. What is the maximum rise in pressure head due to water hammer and where does it occur?

7–22. A 6-ft diameter penstock is made of steel plate $\frac{1}{2}$ in. thick. The velocity of flow is 8 ft/sec and the pipe is 15,200 ft long. The modulus of elasticity of the water is $\frac{1}{100}$ that of the steel. Using the Joukowsky approximation, how much time should be taken to close a valve near the end of the pipe if the excess pressure is not to exceed 100 psi?

7–23. A 24-in. steel pipe is $\frac{1}{4}$ in. thick and 6000 ft long. The modulus of elasticity for steel is 30,000,000 psi, and for water 300,000 psi. The pipe is carrying a flow of 20 cfs. Assume the Joukowsky approximation to be adequate.

  a. What is the wave celerity? *Ans.:* 3880 fps.
  b. What rise in water pressure is occasioned by instantaneous valve closure? *Ans.:* 332 psi.
  c. About how long should the closure time be to reduce this pressure by a factor of 4?

7-24. If a simple surge tank of 4-ft diameter is placed near the terminus of the pipe of Problem 7-23, what maximum surge will be experienced? How soon after the assumed instantaneous valve closure will this maximum surge be attained? Neglect the effects of pipe friction. *Ans.:* 43.5 ft; 42.9 sec.

7-25. Solve the problem of Example 3-3 assuming that all conditions are the same except that the valve (a spring-loaded check valve) closes in such a way that the rate of reduction of the open area increases linearly with time. *Ans.:* 912 ft.

8

HYDRAULIC MODEL STUDIES

8–1. Use of Hydraulic Models. Because of the large number of variables in many hydraulic problems, together with the infinite variety of boundary configurations in hydraulic structures and conduits, it is often impossible to develop comprehensive rational relationships to use as the basis of design. Consequently many hydraulic phenomena are studied by means of models, using the basic principles of similitude to correlate model and prototype behavior. The design of hydraulic structures and machinery today is commonly either accomplished or checked by model measurements.

Consequently, it is very important that the hydraulic engineer acquire a good understanding of both the advantages and the limitations of model analysis. The principles of similarity on which model studies are based are, of course, valid for other phenomena as well as hydraulic phenomena. Structural model analysis has become an important tool of the structural designer, and models are increasingly used in other fields as well. However, it is probably in hydraulic studies that they have the greatest utility and economic value.

8–2. Principles of Similitude. In a hydraulic model study, it is desired that the physical behavior of the model simulate in a known manner the physical behavior of the prototype, so that the latter can be predicted from the former. Several kinds of similarity are defined, as follows:

1. Geometric similarity exists when the ratios of all homologous dimensions on the model and prototype are equal.
2. Kinematic similarity exists when the ratios of all homologous velocities and accelerations are equal in the model and prototype.
3. Dynamic similarity requires that the ratios of all homologous forces be the same in the model and prototype.

Thus, geometric similarity is similarity of form, kinematic similarity is similarity of motion, and dynamic similarity is similarity of force system. In general, the attainment of dynamic similarity also requires and implies kinematic similarity (since motion patterns are established by the force system), and kinematic similarity likewise implies geometric similarity.

If the model and prototype are geometrically similar, then the *scale ratio* (or length ratio) is denoted by $L_R = L_P/L_M$, where the subscripts P and M

refer to prototype and model, respectively. Thus,

$$L_R = \frac{x_P}{x_M} = \frac{y_P}{y_M} = \frac{z_P}{z_M} = \cdots \tag{8–1}$$

where $x, y, z, \ldots$ refer to particular dimensions. Geometric similarity also implies similarity of areas and volumes:

$$A_R = \frac{A_P}{A_M} = L_R^2 \tag{8–2}$$

$$(\text{vol})_R = \frac{(\text{vol})_P}{(\text{vol})_M} = L_R^3 \tag{8–3}$$

Kinematic similarity is expressed in terms of the velocity ratio, V_R, or the time ratio, T_R. Thus,

$$V_R = \frac{V_P}{V_M} = \frac{L_R}{T_R} = \frac{L_P T_M}{L_M T_P} \tag{8–4}$$

The acceleration ratio, a_R, is defined by

$$a_R = \frac{a_P}{a_M} = \frac{V_R}{T_R} = \frac{L_R}{T_R^2} \tag{8–5}$$

Another important kinematic term is the discharge. For kinematic similarity, the discharge ratio is given by

$$Q_R = \frac{Q_P}{Q_M} = A_R V_R = \frac{L_R^3}{T_R} \tag{8–6}$$

Dynamic similarity implies that the ratio of the inertial forces $(F_i)_P/(F_i)_M$, is the same as the ratio of each pair of homologous forces in the system:

$$F_R = \frac{(F_i)_P}{(F_i)_M} = \frac{(F_p)_P}{(F_p)_M} = \frac{(F_g)_P}{(F_g)_M} = \frac{(F_v)_P}{(F_v)_M} = \frac{(F_t)_P}{(F_t)_M} = \frac{(F_e)_P}{(F_e)_M} \tag{8–7}$$

where F_p, F_g, F_v, F_t, F_e are, respectively, the forces due to pressure, gravity, viscosity, surface tension, and elasticity.

Each force is related to the geometry and motion of the flow, and to some fluid property or other factor. The inertial force can be considered as the vector sum of all the others:

$$F_i = F_p \leftrightarrow F_g \leftrightarrow F_v \leftrightarrow F_t \leftrightarrow F_e \tag{8–8}$$

If other types of force are present, they should be included in the equation, but usually these suffice to describe hydraulic phenomena. Equation (8–7) can actually be considered as the statement of five independent conditions, or equations, that must be satisfied by the model fluid and behavior if true dynamic similitude is attained. However, this number can be reduced to four by means of Eq. (8–8), which permits one of the forces to be regarded as a function of the others. Usually the pressure force, F_p, is taken as the dependent variable;

$$\frac{(F_p)_P}{(F_p)_M} = \frac{(F_i)_P \to [(F_g)_P \leftrightarrow (F_v)_P \leftrightarrow (F_t)_P \leftrightarrow (F_e)_P]}{(F_i)_M \to [(F_g)_M \leftrightarrow (F_v)_M \leftrightarrow (F_t)_M \leftrightarrow (F_e)_M]} \qquad (8\text{–}9)$$

8-3. Equations for Dynamic Similarity. The nature of the conditions required for dynamic similitude can be better appreciated if the respective force quantities are written in terms of basic units of length and time, and of the pertinent fluid properties. The inertial force is given by Newton's second law of motion:

$$F_i = Ma = \rho(\text{vol})a \qquad (8\text{–}10)$$

The inertial force ratio is therefore

$$(F_i)_R = M_R a_R = \rho_R L_R{}^3 \frac{L_R}{T_R{}^2} = \frac{\rho_R L_R{}^4}{T_R{}^2} \qquad (8\text{–}11)$$

This equation is sometimes known as the Bertrand equation.

In similar fashion, the other force ratios can be expressed as follows:

$$(F_p)_R = (pA)_R = p_R L_R{}^2 \qquad (8\text{–}12)$$

$$(F_g)_R = (\gamma \cdot \text{vol})_R = \gamma_R L_R{}^3 \qquad (8\text{–}13)$$

$$(F_v)_R = \left(\mu \frac{V}{L} \cdot A\right)_R = \mu_R \frac{V_R}{L_R} L_R{}^2 = \frac{\mu_R L_R{}^2}{T_R} \qquad (8\text{–}14)$$

$$(F_t)_R = (\sigma L)_R = \sigma_R L_R \qquad (8\text{–}15)$$

$$(F_e)_R = (EA)_R = E_R L_R{}^2 \qquad (8\text{–}16)$$

By equating each of these expressions in turn to the inertial force ratio, the following equations result:

$$\frac{\rho_R L_R{}^4}{T_R{}^2} = p_R L_R{}^2; \qquad \frac{\rho_R V_R{}^2}{p_R} = 1 \qquad (8\text{–}17)$$

$$\frac{\rho_R L_R{}^4}{T_R{}^2} = \gamma_R L_R{}^3.; \qquad \frac{V_R{}^2}{g_R L_R} = 1 \qquad (8\text{–}18)$$

$$\frac{\rho_R L_R{}^4}{T_R{}^2} = \frac{\mu_R L_R{}^2}{T_R}; \qquad \frac{\rho_R V_R L_R}{\mu_R} = 1 \qquad (8\text{–}19)$$

$$\frac{\rho_R L_R{}^4}{T_R{}^2} = \sigma_R L_R; \qquad \frac{\rho_R V_R{}^2 L_R}{\sigma_R} = 1 \qquad (8\text{–}20)$$

$$\frac{\rho_R L_R{}^4}{T_R{}^2} = E_R L_R{}^2; \qquad \frac{\rho_R V_R{}^2}{E_R} = 1 \qquad (8\text{–}21)$$

These expressions will be recognized as the flow parameters derived by dimensional analysis as Eqs. (2–43) through (2–47). Thus, Eqs. (8–17) through (8–21) can be stated as requiring that, for dynamic similarity between a model and its prototype, the Euler number, Froude number, Reynolds number, Weber number, and Mach number must respectively be equal in the model and prototype.

As discussed above, only four of these are independent. That is, the Euler numbers will automatically be equal in the model and prototype if the others are. But this still leaves four independent equations:

$$\frac{\rho_R V_R{}^2}{\gamma_R L_R} = \frac{\rho_R V_R L_R}{\mu_R} = \frac{\rho_R V_R{}^2 L_R}{\sigma_R} = \frac{\rho_R V_R{}^2}{E_R} = 1 \qquad (8\text{–}22)$$

It is immediately apparent that no single model fluid of the requisite properties (γ_M, μ_M, σ_M, E_M) could be found which would permit all four of these equations to be satisfied at once. Therefore, it seems impossible that true dynamic and kinematic similarity could ever be obtained between a model and a prototype.

However, it is often the case that one or more of the specific types of forces will be either absent or negligible. If this is so, the number of equations to be satisfied is reduced correspondingly. In fact, in most cases, it will be found that the phenomena in a particular instance result mainly from the effect of only *one* of the forces, and that others are negligible. The flow in the great majority of hydraulic structures, for example, is determined primarily by the effect of gravity. In others, viscosity, surface tension, or elasticity is paramount, but seldom is more than one determinative. Consequently, the use of hydraulic models becomes entirely practicable and, in fact, a very powerful tool of analysis.

8–4. Hydraulic Phenomena Controlled by Gravity. If the flow phenomena are determined primarily by gravitational forces, so that the others (except pressure and inertia) can be neglected, then Eq. (8–18) is the only one that must be satisfied for dynamic similarity. The parameter V^2/gL, or more commonly $V/\sqrt{gL}$, is known as the *Froude number* (after William Froude, an English engineer who first utilized a parameter of this form in ship model studies), and Eq. (8–18) as the *Froude model law*. This law is applicable to model studies of most types of hydraulic structures, including spillways, orifices, chutes, gates, weirs, stilling basins, (see Fig. 8–1), locks, intakes, wasteways, transitions, and many others. The model must be planned and operated in such a way that the Froude number in the model is equal to the Froude number in the prototype.

There are various physical interpretations that may be given the Froude number. For example, it is most fundamentally the ratio of inertial to gravitational forces. This is best illustrated by flow out of an orifice, as in Fig. 8–2. The ratio of inertial and gravitational forces acting on the jet is

$$\frac{F_i}{F_g} = \frac{Ma}{Mg} = \frac{V/T}{g} = \frac{V}{gL/V} = \frac{V^2}{gL} \qquad (8\text{–}23)$$

Fig. 8–1. Model study of stilling basin for John Redmond Dam. (U.S. Army Engineer District, Tulsa.)

In more general form, it is evident that the actual gravitational effect is measured by the difference in the weight of the emerging fluid and that of the surrounding fluid, the latter manifesting itself as a buoyant force. In this case, the Froude number would be written

$$N_F = \frac{F_i}{F_g} = \frac{Ma}{W} = \frac{\rho(\text{vol})V/T}{(\gamma_j - \gamma_0)(\text{vol})} = \frac{V^2}{(\Delta\gamma/\rho)L} \qquad (8\text{–}24)$$

Equation (8–24) shows that, if the jet emerges into a reservoir of the same fluid as the jet, there is no net effect of gravity on the motion—the jet simply dissipates in the reservoir. The phenomenon then becomes one in which presumably viscosity and the Reynolds number would control.

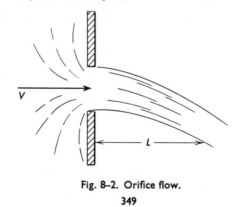

Fig. 8–2. Orifice flow.

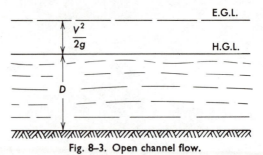

Fig. 8-3. Open channel flow.

Another physical interpretation of the Froude number is in terms of relative kinetic and potential heads in an open channel, as in Fig. 8-3. Obviously the Froude number as defined above is twice this ratio, but the factor of two has no bearing on its basic physical significance. Thus,

$$N_F = \frac{V^2}{gD} = 2\frac{V^2/2g}{D} \qquad (8\text{--}25)$$

The Froude number can thus be interpreted either as the ratio of two dynamic terms or as the ratio of two geometric terms (the latter being also the ratio of two energy terms). Finally, it can be visualized as the ratio of two kinematic terms. Since the celerity of a gravity wave, c, is $\sqrt{gD}$, Eq. (4-29), the Froude number, when written in its alternate form, $V/\sqrt{gD}$, becomes

$$\frac{V}{\sqrt{gD}} = \frac{V}{c} \qquad (8\text{--}26)$$

In this form it is the ratio of the flow velocity to the velocity of a small gravity wave on the free surface. (See Fig. 8-4).

It should be realized also that any representative velocity and any representative length can be used for the terms in the Froude number, so long as dynamic similarity is maintained, and homologous terms are used in the model and prototype. Say, for example, it is desired to use a certain dimension L' and velocity V' instead of L and V as before. Dynamic similarity implies that

$$L' = K_1 L \quad \text{and} \quad V' = K_2 V \qquad (8\text{--}27)$$

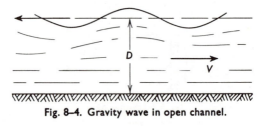

Fig. 8-4. Gravity wave in open channel.

The ratio of Froude numbers in model and prototype is

$$(N_F)_R = \frac{V_R^2}{g_R L_R} = \frac{V'^2_R (K_1)_R}{(K_2)_R^2 g_R L'_R} = \frac{V'^2_R}{g_R L'_R} \qquad (8\text{--}28)$$

since $(K_1)_R$ and $(K_2)_R$ must each be unity.

A model operated by the Froude law will usually have a scale ratio, L_R, of about 5 to 25 for intricate structures such as gates, about 30 to 100 for spillways and about 100 to 1000 for river models. Conversions from model measurements to prototype quantities are made by means of Eq. (8–24):

$$(N_F)_R = \frac{V_R^2}{(\Delta\gamma/\rho)_R L_R} = \frac{V_R^2}{g_R(1 - \rho_0/\rho_j)_R L_R} = 1 \qquad (8\text{--}29)$$

Usually, $g_R = 1$ and $\rho_0 \cong 0$, so that

$$V_R = \sqrt{L_R} \qquad (8\text{--}30)$$

Some common conversion factors are as follows:

Discharge: $\quad Q_R = A_R V_R = (L_R)^2 \sqrt{L_R} = L_R^{5/2}$ $\qquad (8\text{--}31)$

Pressure: $\quad p_R = \gamma_R L_R = L_R$ (if same fluid in model $\qquad (8\text{--}32)$
and prototype)

Force: $\quad F_R = \gamma_R(\text{vol})_R = \gamma_R L_R^3 = L_R^3$ $\qquad (8\text{--} `3)$

Energy: $\quad E_R = F_R L_R = \gamma_R L_R^4 = L_R^4$ $\qquad (8\text{--}34)$

Power: $\quad P_R = E_R/T_R = F_R V_R = \gamma_R L_R^3 \sqrt{L_R} = \gamma_R L_R^{7/2} = L_R^{7/2}$
$\qquad (8\text{--}35)$

Momentum: $\quad M_R = m_R V_R = \rho_R(L_R)^3 \sqrt{L_R} = \rho_R L_R^{7/2} = L_R^{7/2}$ $\qquad (8\text{--}36)$

It is pertinent to note that the Euler number, in the case of a model operated in accordance with the Froude law, can be written as follows (since pressure in a gravity-controlled phenomenon is computed in terms of the hydrostatic equation):

$$N_E = \frac{\rho V^2}{p} = \frac{\rho V^2}{\gamma H} = k \frac{V^2}{gL} = k N_F \qquad (8\text{--}37)$$

where k is a constant, the ratio of the two dimensions L and H. Thus, if the Froude number is the same in the model and prototype, the Euler number ratio will also be unity. In fact, if the length parameter in the Froude number is chosen as the pressure head, the Euler number and the Froude number are identical.

Example 8-1

A horizontal rectangular stilling basin has a design discharge of 30 cfs per foot of width and is to be studied by a 1:36 scale model.

(a) What model discharge per unit width should be used?

(b) The model hydraulic jump occurs when the upstream depth becomes 0.05 ft. What will be the probable jump *length* in the prototype?

(c) The head dissipated in the model jump is measured as 0.1 ft. If the prototype channel is 20 feet wide, what horsepower is dissipated in the prototype jump?

Solution.

(a) Unit discharge ratio, $q_R = \dfrac{Q_R}{L_R} = L_R^{3/2} = (36)^{3/2}$

$$q_M = \frac{1}{(36)^{3/2}} q_P = \frac{30}{216} = 0.139 \text{ cfs/ft.} \qquad Ans.$$

(b) For prototype basin, $D_1 = 36(0.05) = 1.80$ ft.

$$V_1 = \frac{q}{D_1} = \frac{30}{1.8} = 16.67 \text{ ft/sec}$$

$$N_{F_1} = \frac{V_1}{\sqrt{gD_1}} = \frac{16.67}{\sqrt{1.8g}} = 2.19$$

$$D_2 = \tfrac{1}{2}D_1\left[\sqrt{8N_{F_1}^2 + 1} - 1\right] = 0.9(5.27) = 4.75 \text{ ft.}$$

From Fig. 6–30, $\dfrac{L}{D_2} = 4.6$

$\therefore$ Length of jump, $L_j = 4.6(4.75) = 21.8$ ft.   *Ans.*

(c) For prototype, head loss, $H_L = 36(0.1) = 3.6$ ft.

$$\text{Power loss} = \frac{Q\gamma H_L}{550} = \frac{30(20)(62.4)(3.6)}{550} = 245 \text{ horsepower} \qquad Ans.$$

8–5. Phenomena Controlled by Viscosity.

When the flow is fully enclosed, gravity and surface tension will have no net effect on the flow, so that either viscosity or elasticity will control the velocity and pressure distributions. Assuming inelastic flow, viscosity alone, as manifested in internal and boundary friction, will determine the flow phenomena. For this condition, the Reynolds number, N_R, is the characteristic flow parameter. For similitude, the model and prototype Reynolds numbers must be equal.

The Reynolds number, in like manner to the Froude number, may be conceived in terms of several kinds of physical ratios. Fundamentally it is the ratio of inertial and frictional forces in the flow. Dimensionally,

$$N_R = \frac{F_i}{F_v} = \frac{Ma}{\tau A_s} = \frac{\rho(\text{vol})(V/T)}{\mu(V/y)A_s} = \frac{\rho L^2}{\mu T} = \frac{\rho LV}{\mu} \qquad (8\text{–}38)$$

It may be transformed as the ratio of two velocities:

$$N_R = \frac{\rho L^3 V^2/L}{\tau L^2} = \frac{V^2}{\tau/\rho} = \left(\frac{V}{v^*}\right)^2 \tag{8-39}$$

where v^* is the *friction velocity* or *shear velocity*, defined as

$$\sqrt{(\text{boundary shear stress})/(\text{fluid density})},$$

a parameter very important in studies of boundary resistance to flow.

For many commercial pipe surfaces, it is known that the friction factor almost becomes a constant and is independent of the Reynolds number when the latter becomes very large, say 10^6 or more. In this case, the friction head, as calculated from the Darcy equation, is directly proportional to the velocity head. Since the prototype-model head ratio is merely the length ratio, L_R, this means that L_R becomes proportional to $V_R{}^2$, and this is essentially the same as in the Froude law.

Occasionally, therefore, it may be possible when designing a particular closed conduit system (such as a short reach of pipe, a valve, a fitting, etc.) to use the V^2 law as the basis for converting from model results to prototype. However, the model Reynolds number must be kept high, over 10^6, and this means a fairly large model must be used with L_R between about 5 and 30 usually. Also the particular pipe and its geometry and boundary surface must be such as to give reasonable assurance that the boundary friction phenomena will be independent of Reynolds number for the design range.

Another important application of the Reynolds law is in the study of drag forces on immersed objects. The Reynolds number is then usually written as $V_0 H/\nu$, where V_0 is the velocity of the undisturbed approach flow and H is the height of the immersed object. In the case of a streamlined body, for which friction drag is more important than form drag $V_0 L/\nu$ is used, L being the length of the object.

The Euler number ratio, in the case of flow controlled by viscosity, becomes

$$(N_E)_R = \left(\frac{\rho V^2}{\Delta P}\right)_R = \frac{\rho_R V_R{}^2}{\left(\dfrac{\mu_R V_R}{L_R}\right)} = \frac{V_R L_R}{\nu_R} = (N_R)_R$$

Thus the Euler number is the same in model and prototype if the Reynolds number is the same in model and prototype.

In energy terms, the Reynolds number may be further transformed, as follows:

$$N_R = \frac{V^2/2g}{(\tau/\rho)(1/2g)} = \frac{V^2/2g}{(1/2)(\tau/\gamma)} = \frac{(V^2/2g)}{(1/2)H_f(A/A_s)} = 2\frac{(V^2/2g)/A}{H_f/A_s} \tag{8-40}$$

since, for equilibrium of forces on a flowing body of fluid $\Delta p\, A = \tau A_s$, and $H_f = \Delta p/\gamma$. In this form, the Reynolds number is seen to be proportional to the ratio of (velocity head per unit cross-sectional area) to (friction head per unit shear area). If the Darcy relationship for friction head is inserted, the expression becomes

$$N_R = 2\,\frac{A_s}{AfL/4R_h} = 2\,\frac{p(4R_h)}{fA} = \frac{8}{f} \propto \frac{1}{f} \tag{8-41}$$

where f is the Darcy friction factor. Equation (8–41) of course applies directly to viscous flow only, since the Newtonian equation for viscous force was used in its derivation. In any case the Reynolds number is thus seen to be intimately related to the friction factor.

Common conversion factors used in connection with the Reynolds model law are as follows:

Velocity ratio: $\quad V_R = \dfrac{\nu_R}{L_R} = \dfrac{1}{L_R}$ (if same fluid is used in model) $\quad$ (8–42)

Discharge ratio: $\quad Q_R = A_R V_R = L_R{}^2\dfrac{\nu_R}{L_R} = \nu_R L_R = L_R \tag{8-43}$

$$\text{(if same fluid is used in model)}$$

Pressure ratio: $\quad p_R = \tau_R = \mu_R\dfrac{V_R}{L_R} = \dfrac{\mu_R \nu_R}{L_R{}^2} = \dfrac{1}{L_R{}^2} \tag{8-44}$

$$\text{(if same fluid is used in model)}$$

Force ratio: $\quad F_R = \rho_R L_R{}^2 V_R{}^2 = \rho_R L_R{}^2\,\dfrac{\mu_R{}^2}{\rho_R{}^2 L_R{}^2} = \mu_R \nu_R \tag{8-45}$

Energy ratio: $\quad E_R = F_R L_R = \mu_R \nu_R L_R \tag{8-46}$

Power ratio: $\quad P_R = F_R V_R = \mu_R \nu_R\,\dfrac{\nu_R}{L_R} = \dfrac{\mu_R{}^3}{\rho_R{}^2 L_R} \tag{8-47}$

Momentum ratio: $\quad M_R = m_R V_R = \rho_R L_R{}^3\,\dfrac{\nu_R}{L_R} = \mu_R L_R{}^2 \tag{8-48}$

A number of difficulties are encountered in Reynolds law model studies if the Reynolds number is high. Since the velocity ratio is to be inversely proportional to the length ratio, model velocities have to be high. This often requires use of a wind tunnel, with air as the model fluid.

Also, true geometric similarity requires similarity of the boundary roughness elements, since the viscous and turbulence phenomena associated with the boundary layer and boundary separation are directly dependent on these. At low Reynolds numbers and on smooth surfaces, this is not serious, and the Reynolds criterion is quite adequate as a similitude requirement.

For true dynamic similarity, of course, the flow structure should be similar in model and prototype. The question of whether the flow is laminar or turbulent was shown by Reynolds to be determined by a specific numerical value of a *bulk Reynolds number*; for example, in the case of a circular pipe, flow is always laminar if the parameter DV/ν is less than 2000.

For conduit inlets, or other sections where the flow patterns are changing in the direction of flow, it is necessary for this *zone of flow establishment* to be simulated in the model. Studies of this sort are frequently made in terms of an *entrance Reynolds number*, VX/ν, where X is the longitudinal distance from the initial point of the flow establishment zone. Since the boundary layer thickness δ is a function of X, this Reynolds number may alternatively be written as $V\delta/\nu$.

The important question of boundary roughness is the most difficult of the Reynolds-law problems to treat by means of model studies. Complete geometric similarity of roughness elements and separation points needs to be obtained, as well as full kinematic similarity of velocity distributions, and turbulence patterns. Nikuradse and von Karman used the similarity parameter v^*K/ν, where K was the roughness element height. For smooth walls, the parameter $v^*\delta/\nu$ may be used. Probably a more generally significant *wall Reynolds number* is the parameter $v^*\lambda/\nu$, where λ is the spacing of roughness elements.

Example 8–2

A 12-in. pipe line which carries oil (sp. gr. = 0.85, viscosity = 10^{-4} lb-sec/sq ft) is to be modeled using a 2-in. pipe carrying water, with kinematic viscosity of 10^{-5} sq ft/sec. The prototype design discharge is 1000 gpm. Compute:

(a) Model velocity, in ft/sec.

(b) Model discharge, gpm.

(c) The friction factor for the prototype, for a measured friction factor of 0.025 in the model.

Solution. The bulk Reynolds number for the design flow is

$$N_R = \frac{DV\rho}{\mu} = \frac{4Q\rho}{\pi\gamma D} = \frac{4\left(\dfrac{1000}{449.6}\right)(0.85)(1.94)}{\pi(10^{-4})(1)} = 46{,}700$$

The flow therefore is turbulent, since $N_R > 2000$, but is definitely in the transition range for which the friction factor varies with both pipe roughness and Reynolds number. Consequently it is necessary to test the model at the same Reynolds number and also to be sure that the boundary roughness is geometrically similar in model and prototype and, further, that the turbulent flow regimes are the same. Assuming that the latter requirements are satisfied, the problem is solved as follows:

(a) $\quad V_M = \dfrac{N_R}{D_M}\nu_M = \dfrac{46{,}700}{\frac{1}{6}(10^5)} = 2.8$ ft/sec.   *Ans.*

(b) $Q_M = A_M V_M = \pi(\tfrac{1}{12})^2 (2.8)(449.6) = 27.5$ gpm   *Ans.*

(c) Friction factor is dimensionless and thus is independent of L_R.

$\therefore f_P = f_M = 0.025$ *Ans.*

8–6. Model Studies Involving Both Gravity and Viscous Forces.

If both viscous and gravity forces are important, as in the case of open channel flow on mild slopes and the motion of surface vessels through water, then theoretically the Froude and Reynolds laws would both have to be satisfied simultaneously. That is,

or

$$\frac{L_R V_R}{\nu_R} = \frac{V_R}{\sqrt{L_R}}$$

$$\nu_R = L_R^{3/2} \tag{8–49}$$

This requirement can be met only by choosing a special model fluid such that the kinematic-viscosity ratio is equal to the three-halves power of the scale ratio. This is obviously all but impossible.

Two other expedients may be available, however. An approximate similarity requirement may be used, based on empirical relationships which include the major effects of frictional forces (such as the Manning formula). Another approach may be to attempt to correct both model and prototype measurements for the forces due to friction, while operating the model in accordance with the Froude law. The first approach is commonly used in studying "fixed-bed" open channel problems, the second in studying ship resistances.

8–7. Fixed-Bed Open Channel Studies.

If a relatively long reach of river channel is to be modeled (see Fig. 8–5), particularly if effects due to changes in bed configuration are only of secondary concern, a *fixed-bed* study may be made (as distinguished from a *movable-bed* study, in which sedimentation effects are important). Problems of this type are concerned mainly with velocity and slope patterns and therefore the effect of bed roughness is very important.

Since fairly high Reynolds numbers are usually associated with river flows, the shear stresses are primarily determined by form drag phenomena rather than friction drag at a boundary layer. Consequently, an empirical equation of the Manning type may usually be used. Assuming similarity gives

$$V_R = \frac{R_R^{2/3} S_R^{1/2}}{n_R} = \frac{R_R^{2/3}(y_R/L_R)^{1/2}}{n_R} \tag{8–50}$$

where y is the drop in elevation in length L. If an undistorted model is used, $S_R = 1$ and $R_R = L_R$, so that

$$V_R = \frac{L_R^{2/3}}{n_R} \tag{8–51}$$

Fig. 8–5. Model of Mississippi River at Old River diversion. (U.S. Army Engineers Waterways Experiment Station.)

and

$$n_R = L_R^{1/6} \quad (\text{since } V_R = \sqrt{L_R}) \tag{8–52}$$

Frequently, however, the model velocity will be so small (or the model roughness relatively so large) as to make accurate measurements difficult or even to cause laminar flow in the model. For such conditions a distorted model may be used, with the vertical scale ratio y_R smaller than the horizontal scale ratio L_R. In effect, this means utilizing a slope ratio, $S_R = y_R/L_R$, determined from Eq. (8–50) as follows:

$$S_R = \frac{n_R^2 V_R^2}{R_R^{4/3}} \tag{8–53}$$

The velocity ratio must vary with the square root of the depth ratio, according to the Froude law. Thus,

$$S_R = \frac{y_R}{L_R} = \frac{n_R^2 y_R}{R_R^{4/3}} \tag{8–54}$$

If the roughness coefficient is known for both model and prototype, Eq. (8–54) may be used to determine the distortion required. If, on the other hand, the distortion ratio is fixed by space considerations, the model roughness required is given by

$$n_R = \frac{S_R^{1/2} R_R^{2/3}}{y_R^{1/2}} = \frac{R_R^{2/3}}{L_R^{1/2}} \tag{8–55}$$

In general this means that the model roughness must be adjusted by trial and error until the required model flow is obtained. The discharge ratio is obtained as

$$Q_R = A_R V_R = L_R y_R V_R = L_R \sqrt{V_R}\, V_R = L_R V_R^{3/2} \qquad (8\text{–}56)$$

The use of the Manning formula as a similarity criterion requires that the flow be fully turbulent in both model and prototype. This is one of the reasons for vertical distortion. Certain studies have indicated that the following criterion assures fully turbulent flow in the model:

$$\frac{V^*\epsilon}{\nu} \left(= \frac{\sqrt{gRS_\epsilon}}{\nu} \right) \geq 100 \qquad (8\text{–}57)$$

where ϵ is the equivalent sand roughness. A bulk Reynolds number, defined as RV/ν, should be greater than 1800 for assured turbulence.

Vertical distortion must not be made so large, however, as to cause strongly curvilinear flow in the model, since this will seriously alter the velocity and pressure distributions.

8–8. Ship Model Studies. The resistance to motion of a vessel moving along a water surface is caused by three factors: (1) friction drag along the boundary; (2) form drag due to separation zones behind the vessel; and (3) force expended in generation of gravity waves. The first two are viscous phenomena and therefore are functions of the Reynolds number; the last, usually the largest, is a gravity phenomenon and therefore related to the Froude number.

In ship model studies, the Froude law is used as the basis of operation. However the drag force as measured on the model is first corrected for the computed model viscous and form drag, using known formulas and drag coefficients. The remaining force is the wave resistance and is scaled up to prototype values by the Froude law. The total prototype drag is then obtained by adding computed values of friction and form drag for the prototype vessel.

8–9. Movable-Bed Studies. Open channel studies involving problems of sediment erosion, transportation, or deposition require a *movable bed*, that is, a bed composed of sand or other material which can be moved in response to tractive stress at the bed. It is essentially impossible to attain quantitative similarity in such a study, but qualitative similarity can often be produced and this may be all that is necessary. That is, it may be possible to determine locations of scouring or other sedimentary action, and to design methods of modifying it, without actually having to determine specific quantities of sediment moved.

It is usually impossible to scale the bed material down to model size. Consequently, a vertical scale distortion is usually employed on movable-bed models, in order to provide sufficient tractive force to cause bed movement.

Because of the absence of quantitative similarity, a verification study of a movable-bed model is always very important. That is, the model and its operation must be adjusted until it accurately reproduces bed configuration changes that are known to have occurred in the prototype. When this is accomplished, there is then some ground for confidence that the model can predict further prototype behavior that may occur in the future.

A somewhat different approach to river model studies, not used much as yet in this country, is in terms of the so-called *regime theory*. The model river is regarded simply as a miniature river, subject to the same regime principles as its prototype. It is constructed to computed regime dimensions (width, slope, depth, and meander size) as determined from the regime formulas for the assumed dominant discharge scale ratio. Any changes then introduced in the model presumably would be at least qualitatively reproduced in the prototype. The regime equations are discussed in Chapter 13.

8-10. Cavitation Studies. The phenomenon of cavitation (formation and collapse of small vapor cavities in a fluid) may occur whenever local velocities become very high. Since energy conservation requires that increased kinetic energies be accompanied by reduced potential energy, low pressures tend to accompany high velocities, and if these approach the absolute vapor pressure for the fluid, cavitation may occur. Since collapse of the vapor cavities adjacent to the flow boundaries causes very high point stress intensities on those boundaries and possible local structural failure, cavitation is generally a highly undesirable phenomenon.

Since calculation of a complete velocity field is quite difficult and often impossible, especially when the boundaries are irregular, model studies are often made, on both hydraulic structures and hydraulic machines, in order to determine their cavitation potentialities. For such studies, it is desirable to establish a similarity criterion which will correlate the cavitation-producing characteristics of the model and prototype.

For this purpose a *cavitation index*, σ, is defined such that

$$\sigma = \frac{\text{Forces tending to prevent cavitation}}{\text{Forces tending to cause cavitation}} = \frac{\text{Pressure forces}}{\text{Inertial forces}} = \frac{1}{N_E} \quad (8\text{-}58)$$

The net pressure inhibiting cavitation is $p_0 - p_v$, where p_v is the vapor pressure of the fluid and p_0 is the fluid pressure at some reference point. The inertial force per unit area is proportional to $\frac{1}{2}\rho V_0^2$. Thus,

$$\sigma = \frac{p_0 - p_v}{(1/2)\rho V_0^2} = \frac{p_0/\gamma - p_v/\gamma}{V_0^2/2g} \quad (8\text{-}59)$$

For cavitation similitude,

$$\sigma_M = \sigma_P$$

This criterion may also be written as follows, since both pressure head and velocity head are linear terms:

$$L_R = \frac{(p_0/\gamma - p_v/\gamma)_P}{(p_0/\gamma - p_v/\gamma)_M} = \frac{(V_0^2/2g)_P}{(V_0^2/2g)_M} = V_R^2 \tag{8-60}$$

This is the same as the Froude law, which must also be satisfied if the problem is one involving free-surface flow. In this case, the model velocity is determined by the Froude law, and the model pressure, $(p_0)_M$, is maintained at the value computed from

$$(p_0)_M = \frac{(p_0 - p_v)_P}{L_R} + (p_v)_M = \frac{(p_0)_P}{L_R} + p_v\left(\frac{L_R - 1}{L_R}\right) \tag{8-61}$$

Under these conditions, the model should simulate any behavior that might cause prototype cavitation. The minimum pressure measured on the model can be converted to the corresponding prototype pressure; if this is less than one-half atmosphere, it is usually considered that cavitation might develop if small changes in boundary geometry or roughness occur. Generally the reference pressure, p_0, will be atmospheric pressure, p_A, and this may have to be reduced for model testing by a vacuum-tank enclosure.

8–11. The Weber Model Law.   If the force of primary importance in the study is that of surface tension, then the Weber model law is employed, as defined from Eq. (8–20):

$$\frac{\rho_M V_M^2 L_M}{\sigma_M} = \frac{\rho_P V_P^2 L_P}{\sigma_P} \tag{8-62}$$

where the parameter $\rho V^2 L/\sigma$ is the Weber number, representing the ratio of inertial to surface tension forces. The unit surface tension, σ, in pounds per foot, unfortunately has the same symbol as the cavitation index, although the two quantities are unrelated.

The Weber number can also be written

$$\frac{V}{\sqrt{\sigma/L\rho}} \cong \frac{V}{\sqrt{2\pi\sigma/\lambda\rho}} \tag{8-63}$$

where $\sqrt{2\pi\sigma/\lambda\rho}$ can be shown to be the celerity of a capillary wave, of wave length λ. Capillary waves are surface ripples, whose form is controlled by capillary forces, and may be of importance in some model studies.

In general, surface tension effects are of importance only when sharply curvilinear free surfaces occur, as for flow under low heads over weirs, capillary movements of soil water, and roll waves in steep channels. If needed, conversion factors from model to prototype can be worked out in a manner similar to those for viscous and gravity forces.

8-12. The Cauchy Model Law. When fluid elasticity is the predominant physical force, then the ratio of inertial to elastic forces yields the parameter known as the Cauchy number:

$$N_C = \frac{\rho L^2 V^2}{EL^2} = \frac{\rho V^2}{E} \qquad (8-64)$$

E being the bulk modulus of elasticity of the fluid, in pounds per square foot. When this parameter is expressed as the ratio of two velocities,

$$N_M = \frac{V}{\sqrt{E/\rho}} = \frac{V}{c} \qquad (8-65)$$

it is known as the Mach number, where $c = \sqrt{E/\rho}$ is the *celerity*, or the velocity of propagation of a pressure wave in the fluid.

Similarity under conditions where elasticity predominates in the phenomenon requires equality of Mach numbers in the model and prototype. This type of study is of particular importance in aerodynamics, particularly when the Mach number approaches or becomes greater than unity. At low values of N_M, elasticity is usually of lesser importance than other forces, and this is the case in most hydraulic engineering problems.

Problems involving water hammer, underwater and water-entry ballistics, and some other problems in unsteady flow constitute the range of situations in hydraulic engineering where elasticity may be a controlling factor.

8-13. Model Studies of Hydraulic Machinery. Flow in turbine runners and pump impellers is enclosed flow, so that pressure, elasticity, viscosity, and inertia are the pertinent forces. Cavitation may be important, and special tests of the sort already mentioned may be necessary if the possible effects of compressibility or cavitation are to be studied. Gravity is important in the case of impulse turbines.

Ordinary performance tests, however, involve only the efficient transfer of energy between the fluid and the rotating blades. Effects of viscosity and surface roughness, though definitely present and important, do not usually exercise much influence on the velocity patterns. Kinematic similarity for both moving machine parts and homologous points in the fluid, is very important, however, and this is primarily conditioned by geometric similarity of model and prototype and by the mechanism of conversion of pressure energy into kinetic energy and vice versa.

Essentially, this requires equality of the Euler numbers in model and prototype, since the problem reduces to one in which pressure and inertia are the dominant forces. The Euler number can be modified as follows:

$$N_E = \frac{\rho V^2}{\Delta p} \propto \frac{\rho V^2}{\gamma (N^2 D^2/g)} = \frac{V^2}{N^2 D^2} \propto \frac{(Q/D^2)^2}{N^2 D^2} = \frac{Q^2}{N^2 D^6} \qquad (8-66)$$

where D is the diameter, Q the discharge, and N the shaft speed. The change in pressure resulting from forced rotation is assumed proportional to $\gamma(\omega^2 r^2/2g)$, from the principle of the forced vortex.

Similarity of model and prototype therefore specifies that

$$\frac{Q_R}{N_R D_R{}^3} = 1 \tag{8-67}$$

If it is desired to obtain a similarity parameter involving the head on the pump, or turbine, instead of the discharge, the energy equation for the system specifies that

$$H = \Delta\left(\frac{V^2}{2g} + \frac{p}{\gamma} + Z\right) = k_1 \frac{V^2}{2g} = k_2 \frac{Q^2}{D^4} \tag{8-68}$$

Equation (8–68) is justified by the fact that the phenomenon is one of energy conversion; therefore, it is reasonable to regard the net added (or extracted) flow energy, H, as proportional to the kinetic energy evaluated at any convenient location in the system. The coefficients k_1 and k_2 are not truly constants, since they depend on friction loss, but they will not vary appreciably over the ordinary range of values. Thus the similarity requirement can also be stated as

$$\frac{H_R^{1/2} D_R{}^2}{N_R D_R{}^3} = \frac{H_R}{N_R{}^2 D_R{}^2} = 1 \tag{8-69}$$

Or, if it is desired to obtain a similarity parameter involving only the head, speed and discharge, then the diameter can be eliminated from Eqs. (8–67) and (8–69) as follows:

$$\frac{Q_R}{N_R D_R{}^3} = \frac{Q_R}{N_R(H_R^{3/2}/N_R{}^3)} = \frac{N_R{}^2 Q_R}{H_R^{3/2}} = 1 \tag{8-70}$$

The square root of this parameter is commonly called the specific speed, N_s. Thus,

$$N_s = \frac{N Q^{1/2}}{H^{3/4}} \tag{8-71}$$

where N is the pump speed in rpm, Q the pump discharge in cfs, and H the net head on the pump, in feet. Although not dimensionally a "speed" it can be thought of numerically as the speed necessary on a given pump to convey one cubic foot per second against a head of one foot. For two geometrically similar pump impellers, the specific speed must be the same for kinematic similarity.

The specific speed may be modified to a form involving horsepower instead of discharge as follows, since $P = Q\gamma H/550$:

$$N_s = \left(\frac{N}{H^{3/4}}\right)\left(\frac{550P}{\gamma H}\right)^{1/2}$$

By dropping the constant terms, the specific speed becomes

$$N_s = \frac{NP^{1/2}}{H^{5/4}} \qquad (8\text{--}72)$$

where P is the horsepower. This is the form of the parameter as commonly used in turbine studies. The above relations are further discussed in Art. 7–5.

PROBLEMS

8-1. A stilling basin at the outlet of a steep chute consists of a horizontal apron with baffles placed to stabilize the location of the hydraulic jump, and is designed by a model study, on a 25:1 scale. The design discharge on the apron is 20 cfs per foot of width, and the velocity just before the jump is 25 ft/sec. Compute:

    a. Depth upstream and downstream from the jump, neglecting force on the baffle blocks. *Ans.:* 0.8 ft, 5.19 ft.

    b. Froude number upstream and downstream from the jump. *Ans.:* 4.91, 0.296.

    c. Horsepower dissipated in the jump per foot of width. *Ans.:* 11.52 hp/ft.

    d. Discharge to use in the model study per foot of width. *Ans.:* 0.16 cfs/ft.

    e. Dynamic force on a prototype baffle block, if the measured force on the model block is 2 lb. *Ans.:* 31,250 lb.

8-2. A model of a reservoir is drained in 4 min. when its gates are opened. If the model scale is 1:225, how long should it take to drain the actual reservoir? *Ans.:* 60 min.

8-3. What force in pounds per foot would be exerted against a sea wall if a 1:36 model, which was 3 ft long, experienced a total wave force of 27.0 lb?

8-4. An overflow spillway is to be 200 ft high, with a 400-ft crest length, carrying a design discharge of 40,000 cfs, under a permitted maximum head of 9.0 ft. The spillway operation is to be studied by means of a 1:49 scale model.

    a. What should be the model discharge? *Ans.:* 2.38 cfs.

    b. If the model discharge coefficient is found to be 3.85, what is the corresponding prototype discharge coefficient?

    c. If the velocity at the toe of the model spillway is 15 ft/sec, what is the homologous prototype velocity?

    d. What are the model and prototype Froude numbers, as measured at the spillway toe?

    e. If the force on a bucket dissipator at the toe is measured to be 40 lb in the model, what is the corresponding prototype force? *Ans.:* 4.71×10^6 lb.

8-5. A horizontal rectangular stilling basin, 20 ft wide, has a design discharge of 400 cfs. It is to be studied by means of a 1:25 scale model.

    a. What model discharge should be used? *Ans.:* 0.128 cfs.

    b. The model hydraulic jump is found to occur when the flow depth becomes 0.04 ft. What will be the probable length of the jump in the prototype? *Ans.:* 25.0 ft.

c. If baffle blocks are then installed on the model to stabilize the jump and a dynamometer assembly measures a drag force of 0.2 lb per unit width of channel, what is the *total* drag force on the prototype baffle blocks? *Ans.:* 2500 lb.

d. The head dissipated in the model at the jump is measured as 0.1 ft. What horsepower will be dissipated in the prototype jump? *Ans.:* 114 hp.

8–6. A hydraulic model study is made of an overflow spillway and the appurtenant stilling basin. The prototype discharge is 100 cfs per ft of width and the spillway crest is 600 ft long. The model spillway is 10 ft long.

a. What discharge per ft of width should be used in the model?

b. The model crest discharge coefficient is found to be 3.90. What will be the design head on the prototype spillway?

c. In the stilling basin, the depth just before the jump in the model is measured to be $\frac{1}{2}$ in. For no baffle blocks in the stilling basin, what is the length of jump and head loss in the prototype?

d. If baffle blocks are used and the total drag force on all blocks in the model is found to be 4 pounds, what will be the average drag force, per ft of width, in the prototype basin?

8–7. An overflow spillway is to be studied by means of a $1:10$ scale model. The design discharge is 50 cfs per ft of crest length. The prototype spillway is 40 ft high, has a crest length of 60 ft, and operates under a design head of 5 ft. Determine:

a. Dimensions of the model.

b. Required pump capacity to operate the model, in gpm.

c. Length of time the model must be operated to check the equivalent of 36 hours of prototype operation.

8–8. Water at 60°F flows in a 2-in. smooth pipe at 3 ft/sec. What velocity must glycerine at 80°F have in a 6-in. smooth pipe for the two flows to be dynamically similar?

8–9. A submerged submarine moves at 10 mph. At what theoretical speed must a $1:20$ model be towed for dynamic similarity between model and prototype, assuming that sea water and towing-tank water are the same, and that separation zones behind the model and prototype are made geometrically similar? *Ans.:* 293 fps.

8–10. The moment exerted on a submarine by its rudder is to be studied with a $1:100$ scale model in a water tunnel. If the torque measured on the model is 3.50 ft-lb for a tunnel velocity of 50 ft/sec, what is the corresponding torque and speed for the prototype?

8–11. A fully submerged body is to move horizontally through oil ($\gamma = 52$ lb/cu ft, $\mu = 0.0006$ lb-sec/ft^2) at a velocity of 30 ft/sec. The characteristics of this phenomenon are to be studied by an enlarged model of the body, eight times larger than the actual body, tested in water, of kinematic viscosity 0.00001 ft^2/sec.

a. At what velocity must the model be towed for dynamic similarity?

b. If the drag force on the model is measured to be 2.0 lb, what would be the drag force on the prototype?

8–12. The wave resistance of a boat is to be tested by a 1:49 scale model. The skin drag and form drag forces are assumed to be negligible and the resistance to motion due entirely to generation of gravity waves. The wave resistance on the model is measured to be 0.10 lb when operating at 3 ft/sec.

a. Calculate the wave resistance for the actual boat.

b. Calculate the horsepower requirements for the actual boat.

8–13. Assuming a model study is to be made on a phenomenon in which surface tension is the controlling force, determine conversion ratios from prototype to model for the following quantities:

a. Discharge. b. Pressure.

c. Force. d. Energy.

e. Power. f. Momentum.

Express all ratios in terms of the model scale only, assuming same fluid is used in model and prototype.

8–14. Repeat Problem 8–13 for a phenomenon in which elasticity predominates, expressing the conversion ratios in terms of the model scale, ratio of moduli of elasticity, and density ratio.

8–15. A pump with a 48-in. impeller is to discharge 200 cfs at 200 rpm. What discharge would a 1:4 scale model pump discharge at 1800 rpm, assuming dynamic similarity?

9

PRINCIPLES OF HYDROLOGY

9–1. Introduction. Central to the application of hydraulic engineering is the science of hydrology, which deals with the occurrence and distribution of the waters of the earth. It is important for the engineer to understand the principles that constitute the science of hydrology, and for him to be able to use the methods of analysis that serve to quantify hydrologic principles. In this chapter we shall deal primarily with the occurrence and distribution of water on the land—that which we will call the land phase of the hydrologic cycle—and with the precipitation processes that initiate the land phase.

The hydrologic cycle is the name given to the combination of transport of water from one location to another and its residence and actions in the various locations. The hydrologic cycle includes the complex interconnections of transport, locations, and actions. Water in the atmosphere undergoes condensation, evaporation, nucleation and precipitation. When water precipitates onto a land surface it may run off in a stream, over the land, or infiltrate into the ground to reappear later in a stream or in a ground-water aquifer. Some of the water that reaches a land surface evaporates or goes into plant structure to then transpire to the atmosphere. Water that reaches the ocean, either in a stream or through ground-water flow, evaporates to the atmosphere, in effect completing the cycle. Finally, some of the precipitated water remains in place for untold centuries in the ocean deeps, glaciers and snow packs. The alternative actions that result in the many diverse paths and the diverse paths themselves make up the hydrologic cycle.

9–2. Precipitation. Precipitation is not in the land phase of the hydrologic cycle, but nevertheless the engineering hydrologist should know something of precipitation processes and distribution, both temporal and spatial, in order to properly assess their actions and effects on the hydrologic processes in the land phase.

All precipitation results from condensed water in the atmosphere. Water in the gaseous state is always present in the air as a result of transpiration from plants and evaporation from the oceans and land surfaces. A parcel of air can hold a certain amount of water vapor in its composition, depending upon temperature. That amount can be expressed as the partial pressure of the water vapor, and is known as the *saturation vapor pressure* when expressed for pure water vapor over a plane surface of pure water at the same temperature as the air. More useful for this discussion is the *absolute humidity*, the density of the water vapor. The saturation vapor pressure and the saturation absolute humidity are related by the equation of state written for water vapor.

The total amount of water that exists in the atmosphere above a unit area of surface is called the *precipitable water*. The determination of the precipitable water between two levels, z_1 and z_2, is accomplished by the integration of the equation

$$W_p = \int_{z_1}^{z_2} \rho_w \, dz \qquad (9\text{--}1)$$

where W_p is the precipitable water and ρ_w is the absolute humidity. If the value used for ρ_w is the saturation absolute humidity, the value of W_p obtained will be the maximum precipitable water. The value of maximum precipitable water changes little by including the part of the atmosphere above 40,000 feet. If maximum precipitable water is calculated with z_1 equal to sea level and z_2 equal to 40,000 feet for a surface temperature (dewpoint) of 20°C, the result is an equivalent depth of water of about 2.0 inches. Obviously mechanisms which cause the inflow of moist air to a region must be present, since observed rainfalls for even short times often exceed that amount.

Since the saturation absolute humidity is an increasing function of temperature, it follows that water vapor in the air must condense out of the vaporous state when the air is cooled. This condensation process is the first in precipitation. A parcel of air that has a certain absolute humidity, less than saturation, will reach saturation when cooled to a low enough temperature. Since temperature decreases with height in the atmosphere due to turbulent convection processes, cooling of a parcel of air can be effected by lifting it to a higher elevation. Accordingly, the lifting of air must be considered in the precipitation phenomenon.

Lifting of the atmosphere can occur in three ways. The first two methods we will call mechanical, the third thermodynamical. When winds carry moist air (the term given to a parcel of air near saturation) into a region where the surface of the earth is at a higher elevation, such as from the ocean to a mountain range, the air is forced to a greater altitude in order to go over the highlands. The increase in altitude causes an expansion to a lower pressure which, as shown by the equation of state, results in a lower temperature. The lowering of temperature results in an increase in relative humidity (the ratio of absolute humidity to saturation absolute humidity), and when the air reaches the saturation point the water vapor can condense. This lifting process, often called *orographic*, results in extremely high local values of annual precipitation—the Pacific Northwest coast of the United States is a good example.

Another mechanical process causing lifting is the expansion that air undergoes as it moves from an area with low surface friction to one of high surface friction—as from the region over an oceanic surface to the region over a land surface. The increased drag due to the rougher surface causes growth of the boundary layer. The growth of the boundary layer results in part of the air being lifted by displacement to a higher elevation with the consequent cooling and condensation.

Thermodynamical lifting is the name given to the process of lifting of moist air that occurs when the moist air is lighter than its surroundings. When water vapor condenses in the atmosphere, release of the latent heat of condensation causes the parcel of air to be warmed relatively. Because this air is warmer than its surroundings it tends to be buoyed up and continues to rise. Since the rising air is saturated it will continue to have the moisture condense and will realize more heat from the condensation. When a condition such as is described is reached the moist air is said to be unstable. instability of moist air can be ascertained by comparison of lapse rates. (The *lapse rate* is the gradient of temperature with height.) The lapse rate for the adiabatic process—no heat added or taken away, essentially the unsaturated lifting—is $10°C$ km^{-1}; the pseudo-adiabatic lapse rate—the lapse rate occurring at saturation with condensation—is variable, but a representative figure is about $6.8°C\,km^{-1}$. Obviously, if air is started on an ascent and reaches condensation, it will become lighter than the surrounding atmosphere, and it will be lighter the higher it goes.

Thermodynamical lifting, the instability of the moist air, can be initiated in several ways. Even the orographic situation can cause instability, but we consider other causes of instability as significant in regions of the land where orographic precipitation is not dominant. Unequal heating of the surface of the land can result in convective cells wherein moist air rises to heights that permit condensation. Precipitation resulting from such situations can be significant locally in terms of large, flood-producing rains. During certain seasons of the year, primarily summer and early fall, large regions in the central portion of the country may receive a major amount of their rainfall from events resulting from convective processes.

Large-scale storms account for much of the precipitation in the midcontinental and eastern United States. These storms are large air masses associated with low pressure centers and fronts. The name given to the air masses, which rotate in a counter-clockwise direction, is cyclones. In general, the cyclones have cold fronts extending to the southwest from their low center. The general path of cyclones is from the northwestern U.S., across the mid-continent, roughly at latitude 45°N, and then out into the Atlantic generally following the St. Lawrence River valley. The counter-clockwise circulation results in moist air being brought into the region of the low pressure center from the south when the cyclone is north of the Gulf of Mexico. The geographical location of the Gulf of Mexico in relation to the normal cyclonic path is the reason for the relatively moist, or humid climate of the eastern United States. Figure 9–1 is a replica of the surface weather map of

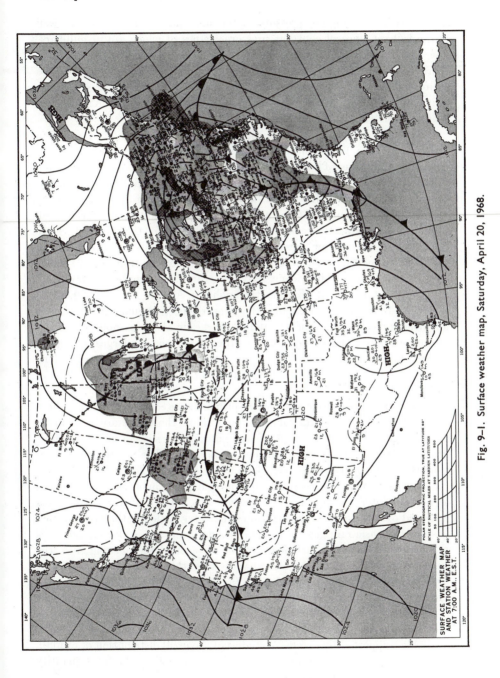

Fig. 9–I. Surface weather map, Saturday, April 20, 1968.

April 2, 1970, showing low pressure centers and the associated fronts as well as the isobars, lines of constant pressure, values of which are in millibars. The cold fronts are distinguished by heavy lines marked with triangles. Warm fronts are similarly marked with semi-circles and occluded fronts have both symbols alternating along the length. The direction of travel of the front is in the direction of the side of the line the symbols are on. A stationary front is depicted with the symbols on both sides of the line.

The engineer often needs to estimate rainfall magnitudes for design purposes. If the values needed are estimates of maximum rainfalls for particular durations of rain, it is apparent that the type of storm studied will be different depending upon the length of rainfall duration that is being considered. As illustration, it is noted that short-duration maxima recorded for the United States are generally the result of convective cell lifting. Medium-duration maximum values recorded for the central U.S. are the result of intense low pressure systems that are cyclonic in nature. In the Atlantic coastal region record storms with durations from 6 to 24 hours are often the result of hurricane activity. Hurricane Camille, which crossed the Piedmont region of Virginia the night of August 19–20, 1969, is an example. In rural Nelson County, that storm deposited quantities that were unofficially recorded at approximately 30 inches in about six hours. Precipitation values for long periods, on the order of months, are largest in regions where orographic influences result in constant lifting of moist air. The previously cited example of that condition is the Pacific northwest coast of the United States.

Precipitation measurements that we primarily are concerned with are those that are made in the traditional collection fashion, those measurements made in rain gages. The basic, common feature of rain gages is the collector, which has a horizontal opening eight inches in diameter. The water is conveyed from the opening to a receptacle where it is measured. The water content is recorded as it enters the receptacle in some rain gages; in others it is measured at periodic intervals. In all cases, the result is a depth of water which was collected over the area of the eight-inch diameter circle. If the precipitation is snow the basic rain gage is still the standard measuring device, but it is modified to reduce the turbulence of the air around the opening since the turbulence would affect the quantity of snow that would enter the gage. Gages are also equipped with devices (heaters) or materials (antifreezes) to melt the snow and prevent clogging of the opening with bulked snow.

In hydrologic applications we are interested in the amount of precipitation that falls on the watershed, or catchment area. Techniques are generally available, supported by extensive studies of the areal distribution of rainfall, that permit reduction of rain gage data to an estimate of the total quantity of water that fell on the watershed.

The basic precipitation data are found in the publication of the Environmental Sciences Services Administration, U.S. Department of Commerce, called *Climatological Data*. The publications are issued monthly for each state, summarizing precipitation data as well as other information. Most precipitation data are given for 24-hour intervals. If more complete data are desired they can often be obtained from the National Weather Records Center, Asheville, North Carolina.

Precipitation is obviously the "input" to the land phase system in the hydrologic cycle. The importance of considering that input in civil engineering activities, such as drainage design, is quite apparent. Another reason for consideration of precipitation is the matter of data, with regard to both length of record and homogeneity. In general, precipitation records have been kept longer than streamflow records. Streamflow data also are more susceptible to environmental changes than are precipitation data (however, there is evidence that annual rainfall values are increased in urban situations over what they would be in rural, or more natural conditions).

Several types of analysis of precipitation data are made for hydrologic purposes. For example, it is often necessary to estimate the homogeneity of data. One method is the double-mass analysis for consistency, in which mass values (accumulated chronologically) of the data in question are plotted against mass values of data that are known to be consistent but subject to the same general environmental factors. Consistent data would plot as a straight line relationship in an arithmetic graph. If at a point in time the questionable data had undergone a change in statistical parameters (particularly the annual average), the straight line relation would exhibit a break in the linearity. Where the values of the questionable data are yearly rainfall totals at a gage, and data resulting from summing the yearly totals of a number of gages in the proximate area are taken as the data of known consistency. (See Example 9–1, with Fig. 9–2). The summing of a number of gages can be assumed to be consistent since the sum will be little influenced by the inconsistency of one of its contributors. The inconsistent data on either side of the change in slope can be reduced to consistency with the data of the other side by multiplying the values themselves by a number which is the ratio of the slopes of the two segments of the line. It must be determined whether the inconsistency is significant.

Example 9–1

Analyze the annual rainfall at Roanoke, Va. by the double-mass technique. See Fig. 9–2 and Table 9–1.

Solution. See Table 9–1 and Fig. 9–2.

In studies of the relation between rainfall and runoff it is necessary to have an estimate of the volume of rainfall on a watershed. Since the

TABLE 9-1

Data for Double–Mass Analysis of Annual Rainfall at Roanoke, Va.

Station	Annual Precipitations—Inches															
	1953	1954	1955	1956	1957	1958	1959	1960	1961	1962	1963	1964	1965	1966	1967	1968
Bedford	34.81	47.31	36.90	38.02	43.88	46.90	39.85	37.06	49.67	45.44	27.17	39.11	33.45	44.24	41.15	34.06
Huddleston	36.19	46.19	37.65	38.43	39.61	40.18	37.27	35.30	44.81	49.77	28.55	48.11	33.73	39.33	33.84	32.50
Rocky Mount	31.73	47.68	34.40	36.29	51.13	45.99	41.14	35.68	50.80	44.63	31.90	49.02	35.74	44.74	48.45	46.00
Buchanan	35.18	49.80	36.97	39.43	42.20	40.79	41.66	35.57	46.14	43.70	27.96	34.41	31.81	40.98	43.82	33.93
Catawaba	31.59	44.49	35.67	39.43	46.94	39.74	43.41	39.99	49.29	48.02	25.32	38.29	33.12	42.51	39.08	37.75
New Castle	32.52	43.05	35.85	32.03	44.75	42.21	38.86	35.36	39.34	40.23	26.78	36.64	36.33	39.32	40.22	34.96
Blacksburg	34.60	36.10	33.37	39.60	49.02	36.95	39.30	33.87	41.55	54.39	26.95	36.62	35.42	37.07	34.48	33.45
Copper Hill	35.38	46.44	42.69	39.75	60.43	42.21	41.93	49.75	50.01	44.94	28.38	45.32	33.82	42.39	35.83	38.24
Lafayette	33.07	39.05	34.15	38.71	41.87	46.45	46.49	33.07	43.03	41.66	23.42	38.73	28.04	40.86	33.23	37.23
Pilot	25.60	37.78	37.57	43.15	46.68	37.06	42.37	37.72	41.05	39.67	24.36	36.64	29.02	41.09	37.83	36.87
Average	33.07	43.79	36.52	38.48	46.65	41.85	41.23	37.34	45.57	45.25	27.08	40.29	33.05	41.25	38.79	36.50
Accumulations	33.07	76.86	113.38	151.86	198.51	240.36	281.59	318.93	364.50	409.75	436.83	477.12	510.17	551.42	590.21	626.71
Roanoke	33.42	44.87	38.11	38.33	51.37	39.89	42.40	38.95	49.46	43.23	25.67	37.87	31.72	40.12	36.65	35.08
Accumulations	33.42	78.29	116.40	154.73	206.10	246.08	288.48	327.43	376.89	420.12	445.79	483.66	515.38	555.50	592.15	627.23

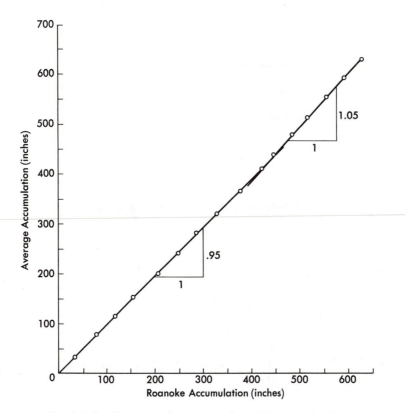

Fig. 9–2. Double-mass analysis of annual rainfall at Roanoke, Va.

measured rainfall is the volume that falls on the gage, there must be a method for reducing the data of the gage to represent areal data. If the gages are evenly distributed on the watershed a strict arithmetic average can suffice for the average depth estimate for the volume of the precipitation on the watershed. If the areal distribution of the gages is uneven, several methods of weighting the precipitation methods are available. The simplest method of weighting is through the Thiessen diagram, in which the depth of precipitation assigned to an element of area is the depth of rainfall which fell on the gage closest to that element of area. This is effected on a map of the watershed by drawing lines connecting rain gages and then constructing the perpendicular bisectors to those lines, the bisectors being the bounds of the area around the particular gage. In complex situations, it can be difficult to draw the Thiessen diagram, but there is only one correct solution and it can be found by care in construction. Figure 9–3 illustrates the construction of the

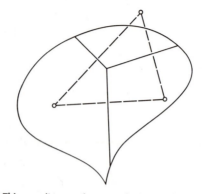

(a) Thiessen diagram for a simple three-station case.

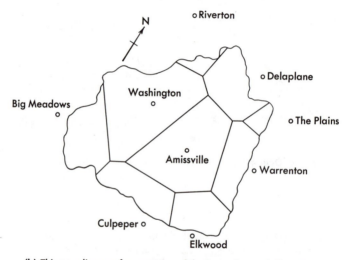

(b) Thiessen diagram for a portion of the Rappahannock River Basin.

Fig. 9–3. Thiessen diagram construction.

Thiessen diagram in a simple case and an example of the resulting diagram in an actual, more complex situation. The precipitation values are weighted by area according to the rule

$$P_t = \frac{\sum\limits_{i=1}^{i=n} P_i A_i}{A_t} \qquad (9\text{–}2)$$

where P_t is the average precipitation on the total area, P_i is the precipitation at the gages $i = 1, 2, \ldots, n$, A_i is the subarea associated with each gage (found by planimetry or some other suitable method) and A_t is the total area of the watershed. The computation for the case illustrated in Figure 9-3 is performed in Table 9-2.

Example 9-2

Weight the observed precipitation for the storm on the Rappahannock River at Remington, Va., by the Thiessen method. (See Figure 9-3b.)

Solution. See Figure 9-3b and Table 9-2.

TABLE 9-2

Computations for Thiessen-Weighting of Precipitation for Rappahannock River at Remington, Va.

Station	Observed Precipitation (inches)	Percent Total Area (by planimeter)	Weighted Precipitation (inches)
Riverton	2.35	2	0.047
Delaplane	2.14	10	0.214
The Plains	1.75	2	0.035
Washington	1.41	26	0.367
Amissville	1.35	26	0.351
Warrenton	1.48	8	0.118
Big Meadows	0.65	8	0.052
Culpeper	0.77	10	0.077
Elkwood	0.84	8	0.067
		Total =	$\overline{1.328}$

In some situations, the experience of the hydrologist and his knowledge of topographical features and hydrologic processes enable a better estimate of the total volume of precipitation by application of the isohyetal method. This method considers the values of precipitation at the gages as "contour points" and constructs, by interpolation between those points and experience gained on the watershed, isohyetal lines—lines of equal rainfall depth. The average depth on the area between two of the isohyetal lines can be taken as their average, and the same sort of weighting performed as described in the case of the Thiessen diagram.

Area weighting methods, particularly the isohyetal method, are used in conjunction with information on the temporal distribution of rainfall, to develop depth-area-duration curves. Depth-area-duration curves give the variation of depth of rainfall with area for a given duration. The curves, an example of which is shown in Figure 9-4, are useful in prediction of rainfall

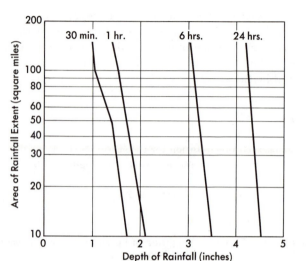

Fig. 9–4. Depth-area-duration curves, derived for the point where Indiana, Kentucky, and Illinois meet for a 10-yr storm. (Data from the *Handbook of Hydrology*, p. 9–51 to 9–57.)

input to watersheds for drainage design studies. The frequency of the precipitation event described by the depth-area-duration curve should be explicitly stated. Methods of determining frequency of hydrologic events will be discussed later in this chapter. Related to depth-area-duration curves are the diagrams developed by the U.S. Weather Bureau which have depth of rainfall lines on a map of the United States for selected values of rainfall duration and frequency. These diagrams, in connection with generalized area-depth ratio curves (Fig. 9–5), can also be used to construct depth-area-duration curves for a selected locale for specified frequencies.

9–3. Runoff. Runoff is the name given to flow on the land's surface. Runoff can be derived from one or more of several sources, but generally from surface accumulations and from sources of water in the ground. Runoff which is from the accumulation of water on surfaces is called *surface runoff*. Sometimes we refer to flow over the surface but not in a defined channel as *overland flow*, or *sheet flow*. Flow in a defined channel is called *streamflow*, and is the flow in rivers, streams, and other drainage channels.

Runoff is most often described analytically as the time distribution of discharge for a point in the flow path. The particular form of description most used is a graphical presentation, called a hydrograph. The instantaneous discharge is plotted on the ordinate and the time scale is on the abscissa. Obviously, such a representation must refer to a particular point in a stream, or on the flow path in the case of overland flow. An example of a hydrograph is shown in Fig. 9–6, plotted from data given in the Water

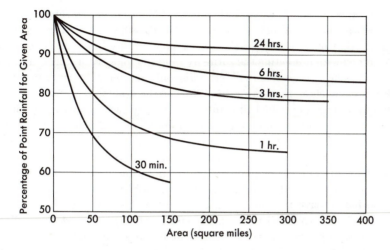

Fig. 9–5. Area-depth curves. (By the U.S. Weather Bureau.)

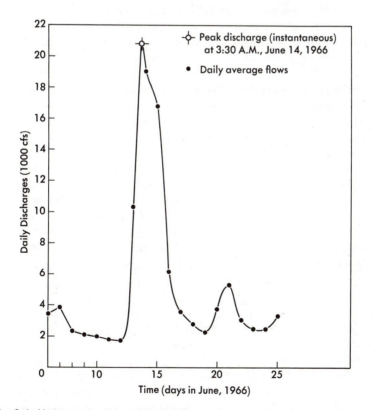

Fig. 9–6. Hydrograph of the Roanoke (Staunton) River at Randolph, Virginia. Area is 3,000 square miles, approximately. (Data from U.S. Geodetic Survey.)

Supply Papers of the U.S. Geodetic Survey. The time period represented on the abscissa is not always so short. Also classified as hydrographs are diagrams that may show a year's flow or more. Many times the hydrologist is concerned with predicting the entire hydrograph for a given storm situation. Other times he is interested in determining only specific values relating to the hydrograph, such as the peak rate of flow or the volume represented by the hydrograph over a specified time interval. In any case, the hydrograph is a central concept to the study, analysis, and prediction of runoff.

When rain falls to the ground some of the water is *intercepted* by plants and their foliage. Thus, a portion of the total rainfall is lost to runoff. In some light rainfalls, the quantity of precipitation lost this way will be significant on a percentage basis. In heavy rainfalls the amount lost can be disregarded relative to other quantities. Some of the water which falls to the ground goes to fulfill the requirement called *depression storage*. Water in this category, such as that trapped in puddles, does not contribute to runoff, but is eventually either evaporated from the storage or infiltrates into the ground.

Some of the moisture which falls on the ground will infiltrate into the ground even though the location has a slope which would permit drainage. The quantity of water which infiltrates into a specified area goes to one of several locations. Some of this water goes to make up the *field moisture deficiency*—field moisture being the water that the soil holds in affinity against the action of gravity. The water in soil moisture is variable, and so the water that is necessary to meet the soil moisture deficiency is variable. Depletion of soil moisture is accomplished through evaporation and transpiration by plants that have their roots in the soil. The water that goes to soil moisture is also lost to runoff.

Water that infiltrates into the ground in excess of the soil moisture deficiency will flow through the ground in response to the forces acting on it. The primary force is gravity, but other forces, such as surface tension, can affect the path that the water takes. Water sometimes percolates through the ground roughly parallel to the surface. Such flow is called *subsurface flow*, or *interflow*. The force is gravity; the reason for the lateral component of velocity can be the presence of a relatively impervious layer beneath the immediate surface or a non-isotropicity of permeability. In the latter case the horizontal permeability could be much higher than the vertical.

Overland flow, or *sheet flow*, is that part of runoff which is diffused during its motion over the surface of the ground before it reaches the stream channel. In many instances, it is of major concern to the engineer. In situations where the surface is impermeable, overland flow is the primary contributor to streamflow. When the rate of rainfall exceeds the capacity of the ground to infiltrate water, or in snow melting situations, the phenomenon of overland flow can also occur. To a considerable extent, the analysis of overland flow is amenable to hydraulic methods, for either the laminar or turbulent regimes.

Finite-difference methods of computation, based upon the method of characteristics, have also been used to develop overland flow hydrographs.

It can be seen that the runoff process is an extremely complex one, combining time-wise averages with space-wise averages. The time and space distribution of rainfall and the resulting contributions to runoff are blended together by the drainage basin to produce the complex description of runoff that the engineer uses, called the hydrograph. In analyzing the response of a drainage basin, it is useful to separate the flow resulting from overland flow from the flow resulting from groundwater contribution. It is generally considered that the flow resulting from precipitation on the stream surface is small enough that it can be included in surface runoff. In a similar fashion, subsurface flow occurs so closely in time to surface runoff that it, too, is included in the quantity of the hydrograph that is called surface runoff. The phenomenon of bank storage can complicate the process of separating surface runoff from groundwater

For the purposes of engineering analysis which lead to reconstruction (synthesis) of the hydrograph for design or prediction purposes, the method of separation of surface runoff from groundwater is considered to be important only in that it be reasonable, and be applied in a consistent manner. A method which can meet these requirements is the construction of a straight line from the point of rise to the point of greatest curvature on the falling side. The point of greatest curvature can be considered to be the point where the overland flow contribution ceases. The area above that line represents the volume of surface runoff (plus subsurface flow and precipitation on the stream). The portion of the discharge below that line represents the contribution to streamflow from groundwater, called *base flow*. Figure 9–7 shows a hydrograph with the separation made by means of a straight line.

Separation of surface runoff from base flow results in a time distribution of surface runoff, the values of surface runoff as functions of time. The volume of surface runoff can be determined from integration of the surface runoff portion of the separated hydrograph. This volume is often expressed as an equivalent depth on the area of the watershead. Such considerations lead to the concept of the *unit hydrograph.*

A unit hydrograph is a hydrograph of surface runoff of volume equivalent to one-inch depth on the watershed resulting from a uniform rate of precipitation excess over a specified duration. Precipitation excess is the portion of precipitation which results in overland flow, and ultimately surface runoff. Since the length of time that precipitation excess persists affects the time that surface runoff continues, when a unit hydrograph is described for a basin, the duration of the precipitation excess which produced the runoff must be specified.

The definition of the unit hydrograph is one which describes its ordinates by a continuous function. Since the methods of determining unit hydrographs for natural basins are primarily graphical and discrete, the result is a

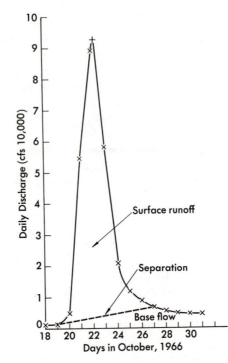

Fig. 9–7. Hydrograph of James River at Cartersville, Va., area 6242 square miles flood of October 22, 1962.

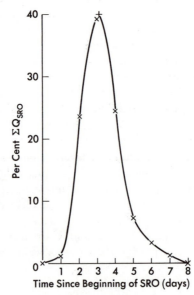

Fig. 9–8. Distribution graph of surface runoff for the James River at Cartersville, Va. flood of October 22, 1962.

description of a unit hydrograph where the values given for the ordinates, the runoff, are really averages over a time interval. Accordingly, in practice what is called the unit hydrograph is really a form of distribution graph. A *distribution graph* has as its ordinate the percentage of the total volume that passed during the time interval. Since the percentage could be expressed as the ratio of the flow passing to the total surface runoff, the ordinates can be related to the depth in inches of runoff for a unit time, or discharge, or any other flow unit. Obviously, as the time interval of the distribution graph is decreased, and with the appropriate transformation of ordinate units, the distribution graph approaches the unit hydrograph.

Figure 9–8 shows a distribution graph developed from the hydrograph of the James River at Cartersville, Va. (Fig. 9–7). Note that the value of the instantaneous peak flow is carried along in the tabular computation in Example 9–3. Its values are italicized to emphasize that they are not included in the summing of the ordinates of the distribution graph since the peak value does not represent an average over a time interval.

Example 9–3

Given the hydrograph of the James River at Cartersville, Va., develop the distribution graph for the flood of October 22, 1962.

Solution.　See tabulation below and Fig. 9–8.

Day	Q_{total}	Groundwater Separation	Q Surface Runoff (cfs-days)	Q Unit Hydrograph (cfs-days)	$\% \Sigma Q$ Surface Runoff
19	1,320	1320	0	0	0
20	4,340	1980	2,360	1,798	1.07
21	54,700	2640	52,060	39,676	23.64
22	89,600	3300	86,300	65,771	39.19
Peak (1330)	*91,200*	*3341*	*87,859*	*66,959*	*39.89*
23	58,100	3960	54,140	41,261	24.58
24	20,300	4620	15,680	11,950	7.12
25	12,200	5280	6,920	5,274	3.14
26	8,710	5940	2,770	1,730	1.26
27	6,600	6600	0	0	0

$$\Sigma Q = 220{,}230 \text{ cfs-days}$$
(Surface Runoff)

$$\text{Depth of storm runoff} = \frac{220{,}230 \,\dfrac{ft^3}{sec} - days \times 86{,}400\,\dfrac{sec}{day} \times 12\,\dfrac{in}{ft}}{6242 \text{ sq. mi.} \times (5280)^2 \text{ ft}^2/\text{mi}^2} = 1.312 \text{ in.}$$

The synthesis of streamflow with a unit hydrograph, or a derivative, is done by assuming that the ordinates of surface runoff are proportional to the volume of runoff—the distribution of volume in time is unchanged. This assumption of superposition, or linearity, is the cause of the main criticism of the unit hydrograph method, since real hydrologic systems do not exhibit linearity. If, however, the unit hydrograph is used to synthesize a flow that is not too different in·magnitude from the one used in derivation of the unit hydrograph, the watershed can be expected to respond in a fashion similar to that described by the synthesis. In practice the use of a unit hydrograph means that the ordinates of the graph are multiplied by the total volume of surface runoff resulting from the precipitation excess. Total flow values are estimated by adding the ordinates computed through the unit hydrograph to assumed base flows, representing the ground water contribution. Thus, one can construct the entire hydrograph.

Several parameters are considered important in hydrographs of watersheds. The duration of rain is one already mentioned. Another is the length of time of surface runoff. A third parameter that is often referred to is lag time, most often described as the time between the center of mass of the rainfall rate-time graph (hyetograph) and the center of mass of the surface runoff hydrograph.

Because of the widespread use of unit hydrograph concepts, and their acceptance in practice, there has been considerable development of synthetic unit hydrographs, the name given to those that are not based on measured and recorded flows at the point on the stream in question. There have been a number of methods developed for estimating the unit hydrograph using such features of the watershed as area, length of stream channel, slope, etc. Nearly all such relationships describing synthetic unit hydrographs are constructed from examining many actual hydrographs and relating their characteristics to quantifiable characteristics of the watershed.

Many references have been made to stream flow measurements. Since that aspect of hydrologic data collection is so important, a description of the method of stream flow measurement is useful. The measuring element that makes the continuous record is, in most natural-stream-gaging situations, a float whose motion is recorded in either a graphical form or in a code (such as on paper tape). Sometimes a pressure transducer is used to sense water depth. The record that is produced is what is called a stage hydrograph, a graph of stage height (the height of the water surface above an arbitrary datum) versus time. In order to produce a hydrograph from the stage record, a rating curve is needed. A rating curve is the relation between gage height, or stage, and discharge. A rating curve is shown in Figure 9–9. Obviously, the rating curve is unique to a particular point in the stream. The rating curve itself must be constructed by stream gaging. Data points on the rating curve are found by calculating discharge from current meter measurements. These are devices which measure velocity at a point, by measuring

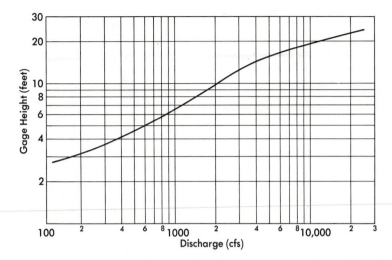

Fig. 9-9. Rating curve for the Nottoway River near Stoney Creek, Va.

the rotational speed of a cup or propeller assembly placed in the stream. Discharge is calculated by integrating the velocity distribution measured by the current meter over the area of the stream.

Stream flow data gathered as described provide the hydrologist with much raw material for analysis. Some of the common engineering analyses, such as the unit hydrograph, have already been mentioned. Other analyses often used are flow-duration curves, essentially cumulative frequency-of-flow curves, where the ordinate scale is that of flow values over the range of experience and the abscissa has units of the percent of flows greater than certain values. A mass curve is constructed of flow data; its use is described in Art. 11-3.

Since stream flow and precipitation data exhibit many stochastic characteristics, it is often useful to examine the data in a statistical reference. In a real sense, the flow duration curve is a distribution function. Standard statistical tests on the mean, variance, sample size, etc. may be made on hydrologic data. It must be remembered that the standard tests assume normality and independence of variables. Hydrologic data do not necessarily exhibit these properties. If we are considering precipitation or streamflow, we find that the data are bounded by zero, destroying an assumption of normality. Persistence is observed in streamflow data, particularly daily flows and, depending on basin size, in other flows too. Such an observation does not permit an assumption of independence. Nevertheless, there are useful statistical analyses that can be made since the results are not always sensitive to the assumptions.

One interesting example of an application of statistics to a hydrologic problem is Gumbel's extreme value analysis. Gumbel considers the distribution of the largest values each from a sub-sample of equal size, where the values making up the sub-samples are exponentially distributed and unbounded. The probability p of an event of magnitude X not being equalled or exceeded (the probability of non-occurrence), as the sub-samples and the total sample both become large, is

$$p = \exp\left\{-\exp\left[-\sigma_n \frac{(X - \bar{X})}{\hat{S}} - \bar{y}_n\right]\right\} \qquad (9\text{--}3)$$

where σ_n and $\bar{y}_n$ are functions of the sample size only (they are given in Table 9–3), $\bar{X}$ is the average of the sample, and $\hat{S}$ is the standard deviation of the sample. The probability of non-occurrence p is not normally used in hydrologic practice. Instead it is related to the return period, or recurrence interval T, defined as the length of time in years, on the average, for an event of a given magnitude to be equalled or exceeded. The probability of occurrence q of an event being equalled or exceeded in any one year is given by the reciprocal of the return period, and thus the probability of non-occurrence p is

$$p = 1 - \frac{1}{T} \qquad (9\text{--}4)$$

The return period of an event of the magnitude of one of the sample data can be estimated by the relation

$$T = \frac{n + 1}{m} \qquad (9\text{--}5)$$

where n is sample size, and

m is the rank of the event; for the largest value, m equals one; and for the smallest value, m equals n.

Thus, a technique is available to compute either the return period for an event of given magnitude or the magnitude of an event of given return period.

The probability of non-occurrence in a period of t years, p_t, is given by

$$p_t = p^t \qquad (9\text{--}6)$$

The probability of occurrence at least once in t years is then given as

$$q_t = 1 - p_t = 1 - p^t \qquad (9\text{--}7)$$

This equation can be used for computation of the probability of observing an event in an interval equal to its own return period as well as the probability of observing the event in other intervals. For example, the probability of observing an event with a return period of 25 years in a 25-year period is

$$q_{25} = 1 - (1 - \tfrac{1}{25})^{25} = 0.640$$

TABLE 9–3

Values of $\bar{y}_n$ and σ_n

n	$\bar{y}_n$	σ_n
20	0.5236	1.0628
21	0.5252	1.0695
22	0.5268	1.0755
23	0.5282	1.0812
24	0.5296	1.0865
25	0.5309	1.0915
26	0.5230	1.0961
27	0.5332	1.1004
28	0.5343	1.1047
29	0.5353	1.1086
30	0.5362	1.1124
31	0.5371	1.1159
32	0.5380	1.1193
33	0.5388	1.1226
34	0.5396	1.1255
35	0.5403	1.1285
36	0.5410	1.1313
37	0.5418	1.1339
38	0.5424	1.1363
39	0.5430	1.1388
40	0.5436	1.1413
41	0.5442	1.1436
42	0.5448	1.1458
43	0.5453	1.1480
44	0.5458	1.1499
45	0.5463	1.1519
46	0.5468	1.1538
47	0.5473	1.1557
48	0.5477	1.1574
49	0.5481	1.1590
50	0.5485	1.1607
55	0.5504	1.1681
60	0.5521	1.1747
70	0.5548	1.1854
100	0.5600	1.2065
∞	0.5772	1.2826

Source: After Gumbel.

We may also use the equation to seek answers to such questions as; What is the probability of observing the design flood of a culvert, say the 50-year flood, during the projected economic life of the culvert. If the economic life of the culvert is estimated at 20 years,

$$q_{20} = 1 - (1 - \tfrac{1}{50})^{20} = 0.332$$

The use of extreme value analysis is not limited to floods. The theory described above can be applied to distributions of other hydrologic phenomena, such as maximum rainfall, wind speeds, etc. There are also theories for minimum values; they are of special interest to hydrologists concerned with droughts.

Example 9–4

Annual maximum instantaneous discharges for the Clinch River at Speer's Ferry, Va., are given in Table 9–4. Also tabulated is the rank of each flow in descending order, its return period, and the "reduced variate" or plotting position. Plot the data and the theoretical relationship on an extreme-value grid.

Solution. The plotting position is found by solving

$$1 - \frac{1}{T} = \exp\left(-\exp\left[-y\right]\right) \tag{9–8}$$

for y, the reduced variate. Since the reduced variate is linear with the variate (X) itself,

$$y = \sigma_n \frac{(X - \bar{X})}{\hat{S}} + \bar{y}_n \tag{9–9}$$

the relationship is most often plotted on a graph with X on the ordinate and y on the abscissa. Figure 9–10 shows the data from Table 9–4 plotted along with the curve of Eq. (9–9). The scales of the abscissa were computed from Eq. (9–8).

9–4. Evaporation and Transpiration.

Evaporation is the name given to the process by which water goes from the liquid state to the vaporous state. The process by which plants give off the water that passes through their structure is called *transpiration*. The latter represents the major water "loss" between precipitation and runoff. The percentage of precipitation that goes to transpiration is difficult to estimate accurately, but it is very high, ranging up to perhaps 80 per cent in humid regions. Evaporation losses are high in arid regions where water is impounded, such as the western United States. There evaporation losses can be of the magnitude of the water use. *Sublimation* is the name given to the change of state of water from the solid state to the vaporous state; it is convenient in hydrologic considerations to combine sublimation and evaporation since they are similar in response to environmental conditions and since sublimation is not a high-rate process and opportunity for large quantities of water to undergo sublimation is not often present. Because the processes of transpiration and evaporation are so closely linked to common environmental factors and because both represent in the same way a loss to streamflow, the two processes are considered combined and the resulting loss is given the name *evapotranspiration*.

TABLE 9–4

Data and Computations for Extreme Value Flood Analysis for the Clinch River Speer's Ferry, Va.

Year	Annual Maximum Discharge	Rank	Return Period	P	y
1921	10,500	45	1.07	0.0654	−1.00
22	19,200	28	1.71	0.4152	0.128
23	37,200	3	16.00	0.9375	2.74
24	20,400	27	1.78	0.4382	0.204
25	25,400	15*	3.20	0.6875	0.984
1926	14,700	35	1.37	0.2701	−0.269
27	35,000	4	12.00	0.9167	2.44
28	9,560	46	1.04	0.0385	−1.18
29	25,400	16	3.00	0.6667	0.904
30	13,600	41	1.17	0.1453	−0.656
1931	11,500	44	1.09	0.0826	−0.911
32	30,500	10	4.80	0.7917	1.45
33	18,000	30	1.60	0.3750	0.020
34	17,000	31	1.55	0.3548	−0.034
35	20,500	26	1.85	0.4595	0.251
1936	20,700	24	2.00	0.5000	0.367
37	22,700	22	2.18	0.5413	0.487
38	13,900	37	1.30	0.2308	−0.382
39	23,900	19	2.53	0.6047	0.689
40	25,100	18	2.67	0.6255	0.757
1941	8,240	47	1.02	0.0196	−1.37
42	18,600	29	1.66	0.3976	0.079
43	20,700	25	1.92	0.4792	0.306
44	30,700	9	5.33	0.8124	1.57
45	16,600	32	1.50	0.3333	−0.095
1946	31,000	8	6.00	0.8333	1.70
47	29,400	12	4.00	0.7500	1.24
48	30,000	11	4.36	0.7706	1.34
49	13,900	38	1.26	0.2064	−0.456
50	31,100	7	6.86	0.8542	1.85
1951	13,900	39	1.23	0.1870	−0.517
52	15,300	34	1.41	0.2908	−0.206
53	23,700	20	2.40	0.5833	0.619
54	11,600	43	1.12	0.1071	−0.805
55	25,600	14	3.43	0.7085	1.07
1956	28,400	13	3.69	0.7290	1.15
57	45,300	2	24.00	0.9583	3.16
58	33,100	6	8.00	0.8750	2.02
59	15,800	33	1.45	0.3103	−0.157
60	14,600	36	1.33	0.2481	−0.329
1961	23,700	21	2.29	0.5633	0.555
62	22,600	23	2.09	0.5215	0.429
63	46,800	1	48.00	0.9792	3.85
64	13,300	42	1.14	0.1228	−0.740
65	25,300	17	2.82	0.6454	0.826
1966	13,700	40	1.20	0.1667	−0.581
67	33,500	5	9.60	0.8958	2.20

$$\hat{S} = 9020 \text{ cfs}, \qquad \overline{X} = 22,300 \text{ cfs}$$

* For uniformity, identical flows are ranked consecutively. Thus the flows of 1925 and 1929, both equal to 25,400 cfs, are ranked 15 and 16 respectively. No significant error is introduced by this assignment.

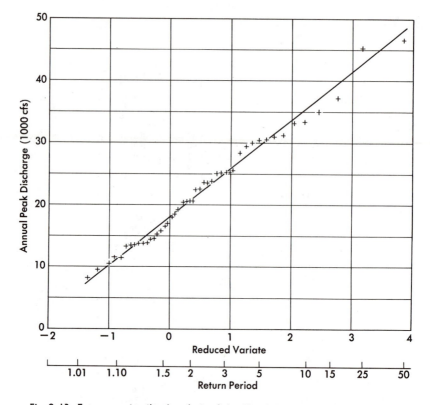

Fig. 9–10. Extreme-value flood analysis of the Clinch River at Speer's Ferry, Va.

The determination of the rates of evapotranspiration, or the total quantity of water going that route, is extremely difficult in field conditions. The primary problem is the inability to measure directly the quantity of evapo-transpiration. Thus, we can only calculate, or estimate, evapotranspiration from measured quantities which are related in some way to the process.

The basic relation which governs the rate of evaporation from a free surface of water into the air is Dalton's law:

$$E = C(e_w - e_a) \qquad (9\text{--}10)$$

where E is the rate of evaporation, usually expressed as a depth per unit time

  e_w is the partial pressure of water vapor as the temperature of the water body

  e_a is the partial pressure of the water vapor in the air directly above the water surface

  C is a constant of proportionality of appropriate units

It must be recognized that Dalton's law is essentially a rate equation, and as such must be integrated to obtain quantities. The practice is to accomplish the integration by taking averages of the vapor pressures over a time interval; a commonly used interval is one month.

No similar relation is available for transpiration. Through careful observation and measurement it has been established that transpiration has a strong dependence on temperature. Since the transpiration process is partly a component of the action of plant growth (transpiration also serves to cool the plant) its rate is dependent upon the presence of light. Obviously, quantities of transpiration are dependent on availability of moisture in the ground.

In many regions of the western United States evapotranspiration losses from the water available for irrigation are very great. One primary source of such loss is found in phreatophytes, particularly the salt cedar, a water-loving plant that grows in the low-lying lands near rivers, using large quantities of water from the water table and having no commercial value. Another large loss is the evaporation from reservoirs. The estimate has been made that the loss of water from man-made reservoirs in the western United States is equal to the consumption of water from them. Water management under these conditions includes then, control of the phreatophytes (with due consideration to ecological balances) and methods of evaporation reduction. Evaporation reduction can be accomplished in several ways; all other things being equal, it is desirable to put the reservoir at a higher elevation thus keeping its contents at a lower temperature. Sometimes it is feasible to reduce the area of a reservoir without significantly reducing its volume. Since the evaporation rate is proportional to area, some saving perhaps can be realized without great costs. In recent years there has been considerable interest in the development of evaporation retardants, particularly monomolecular films on the surface of the water which inhibit the escape of the water molecule with high kinetic energy. Another way of reducing evaporation is to seek out and use suitable underground reservoirs, since the evaporation from a water table is very much less than the evaporation from a water surface.

9–5. Social Factors. As there are social limitations on many engineering activities, there are also restraints on the application of hydrologic knowledge that are not technical. In the general fields of study making up the area of study known as water resources, one of the primary considerations is where the right rests to use the natural waters—the factors of water law. Two basic doctrines of water law exist in the United States—riparian and prior appropriation.

"Riparian" is a word that means "pertaining to the bank of a river or stream," and that meaning extends to the definition of rights under riparian doctrine. The right to use the water is a right of the owners of riparian land.

It is a right that may not be lost by non-use; the right exists whether it is exercised or not. The definition of riparian land limits the meaning to land within the watershed. In some states, land can be severed from riparian status by sale or transfer of title, and land once severed cannot reattain riparian status through purchase by the owner of adjacent, riparian land. In other jurisdictions, all contiguous land under a single owner is considered one tract, and can all have riparian status.

The riparian owner is restricted to making a reasonable use of the water; he may be enjoined from a practice which is considered to be an excessive use. He is also protected from unreasonable attacks on his use. Riparian rights are subject to prescription by nonriparians. A nonriparian who uses water for a period of time defined by the statute of limitations establishes a power to take water which is superior to the right of downstream riparians. In some states the diminution of flow for the stated period is sufficient to establish the prescriptive right. It may be more difficult to establish prescriptive rights where the limitation of reasonable use is applied. The riparian owner must be on guard against prescriptive rights being established above him, but in most situations a riparian owner can complain only when there is clear injury to him if the use above him—either by a riparian or a nonriparian— is reasonable.

A municipality bordering a watercourse is not riparian in a sense that permits it to extract water for its inhabitants that do not own riparian land. The city can obtain water by prescription or by permit from downstream riparians.

The doctrine of prior appropriation gives the right to use water to a beneficial use, not to land ownership, and priority of use is the basis for allocation during times of shortage. The use of the water does not have to be on riparian land; in some places the use is not even in the watershed. Similarly, the transfer can be across political boundaries, but some states are now restricting exportation of water from within their boundaries. The water right may be lost by nonuse; in most states a particular interval of nonuse can be a cause for termination of the right. Under appropriation doctrine the water right may be sold to someone who intends a different use; however, such use may not harm junior appropriators (ones with a more recent right).

Beneficial use is a primary requirement for a right under appropriation doctrine; most states recognize domestic uses as beneficial, including in some places watering stock and irrigating gardens. The application of water to agricultural lands—irrigation—is also considered a beneficial use, and no preference is given to the type of crop. Appropriation doctrine arose in the mining camps of the west, so it is not surprising to find that mining is considered a beneficial use. Industrial, power and municipal uses are all considered beneficial in most jurisdications. Generally, domestic uses are higher uses than others, and in most cases irrigation comes next. In some states preferences are stated, allowing a junior appropriator to establish a right over a senior appropriator. In some places this is a true preference, in others it permits the junior to appropriate with compensation to the senior appropriators that are damaged. Municipalities often have preference.

As mentioned earlier, both doctrines of water law exist in the United States. In general, riparian doctrine modified by reasonable use is found in the states east of the Mississippi. Combinations of riparian and appropriation doctrines are found in Pacific coast states and in the states bordering the Mississippi on the west. The combinations generally recognize early riparian claims, but new claims have essentially the factors of appropriation. The remaining western states (including Alaska but excluding Hawaii) have appropriation doctrine.

No discussion on hydrology can be considered complete today without considering the impact of manmade hydrological changes on the ecological system, including human society. Decisions on water-resource development have tended to be irrevocable, destroying forever some other, perhaps unrecognized, facet of our surroundings. An interesting example is the Aswan High Dam on the Nile. It provides power, flood control, and irrigation on the upper Nile. However, even the ancients realized that the river, in its annual flood, imparted to the flood plain those nutrients that made the Nile delta so productive. Now fertilizer plants must be built and commercial fertilizer (which we are beginning to identify as a pollutant) is being applied to the agricultural lands. The ancients did not realize that the annual flooding kept water-loving snails close to the river bed, snails which are the vector for the organism that causes schistosomiasis. The snails are increasingly found in cultivated fields, which are now irrigated (with attendant brackish return flows) because the flooding no longer provides the necessary moisture; the snails follow the irrigation process, causing an increasing incidence of schistosomiasis in humans.

The concern in the western United States for the increasing salinity of irrigation return flows is another example, as is the increasing depth of the Salton Sea in the Imperial Valley. Perhaps we, as a society, are willing to pay these costs, but they were not apparent to us as costs when the projects were begun. It will be increasingly important for the engineer-hydrologist to better anticipate these effects on the environment, and include them in his considerations when designing and planning.

PROBLEMS

9-1. For a watershed assigned by the instructor, construct the Thiessen diagram and compute the weighted average rainfall.

9-2. Using the results of Example 9–4 on the Clinch River, determine the return periods for flows of 35,000 cfs and 55,000 cfs. Find the discharge of floods which have return periods of 25 yrs and 60 yrs. What is the return period of the average flood?

9-3. Given below are the annual maximum instantaneous flows of the Cowpasture River near Clifton Forge, Va. for the years 1926 through 1965. Compute the mean and standard deviation of the extreme values, and estimate the return period of a flow of 45,000 cfs. Plot the data on a graph of discharge versus reduced variate.

Year	Q cfs × 1000	Year	Q cfs × 1000	Year	Q cfs × 1000
1926	5.97	1940	9.32	1954	12.1
1927	6.69	1941	3.38	1955	11.6
1928	6.63	1942	18.5	1956	4.35
1929	5.86	1943	11.0	1957	12.5
1930	6.85	1944	6.28	1958	5.85
1931	2.66	1945	6.93	1959	8.95
1932	7.12	1946	8.86	1960	11.1
1933	6.28	1947	6.12	1961	6.32
1934	5.80	1948	7.61	1962	13.1
1935	13.7	1949	10.4	1963	11.4
1936	34.2	1950	9.49	1964	8.57
1937	10.1	1951	13.4	1965	8.95
1938	10.1	1952	13.9		
1939	9.32	1953	14.8		

9-4. Tabulated below are the values of annual instantaneous maximum discharge for the Mattaponi River at Beulahville, Va., 619 sq mi in watershed area. Compute the mean and standard deviation of the extreme value distribution, and determine the magnitude of flows with return periods of 10, 15 and 25 years. Plot the data on a graph of discharge versus return period.

Year	Q cfs	Year	Q cfs
1942	2700	1954	1130
1943	7250	1955	7120
1944	1730	1956	2000
1945	8580	1957	2620
1946	2800	1958	4690
1947	1890	1959	2350
1948	4770	1960	4280
1949	7280	1961	5900
1950	4360	1962	5320
1951	1590	1963	4280
1952	5210	1964	2800
1953	5410	1965	1780

9-5. Using the distribution graph derived for the James River at Cartersville, Virginia, in Example 9–4, compute the average daily ordinates and the peak ordinate of surface runoff for a precipitation excess (precipitation less infiltration) of 0.85 inch depth on the basin.

9–6. Given below are 6-hr averages of the flow of the Jackson River at Falling Spring, Virginia for the storm of December, 1967. Compute the distribution graph from the data. The area of the watershed is 409 sq mi. Compute the surface runoff, and put the distribution graph ordinates into discharge value form, as in a unit hydrograph.

Q (6-hr average) cfs		
945	1756	834
959	1588	785
1401	1437	741
2471	1293	702
3316	1181	676
3027	1094	646
2577	1016	619
2224	949	591
1955	888	572

10

GROUNDWATER HYDRAULICS

10–1. General. The extent of groundwater use in the United States varies from region to region. In many places, such as the southwestern United States, the use of groundwater supplies for irrigation is extensive. Many moderately-sized municipalities use wells as the primary source of water. In this chapter we shall look at the engineering aspects of the occurrence and recovery of the water in the ground.

The water that is recoverable by engineering methods is that which can be drained from the void spaces in the soil matrix by the action of gravity. All the void spaces in the matrix can be filled, in which case the soil is said to be saturated. The unsaturated condition, where only part of the void spaces are filled with water, is common in the upper soil zone during and after precipitation. The motion of the water in the unsaturated zone is extremely complex, and although it is important in the hydrologic cycle, it is not of major importance in water supply considerations. In this chapter we shall consider only saturated flows.

10–2. Darcy's Law. The basic flow relation in cases of flow through porous media is Darcy's Law. The formulation is an empirical one, arising from experiment.

$$Q = -KA\frac{\Delta h}{\Delta l} \tag{10–1}$$

where Q = the discharge
A = the cross-sectional area of the flow
$-\Delta h$ = the change in piezometric head over the length of flow
Δl = the length of the flow
K = the coefficient of permeability, or hydraulic conductivity
The dimensions of K are derived to be those of velocity, or length divided by time. When Darcy's law is modified by dividing the discharge by the area, the result is

$$v = -K\frac{\Delta h}{\Delta l} \tag{10–2}$$

The velocity v is only an average velocity over an area. In the medium itself the fluid follows a very tortuous path around the particles. Nevertheless, the apparent or average velocity, v, is a useful concept in the analytic development which follows. The substitution of Q divided by A by the velocity does not create serious problems in mathematical simulation of groundwater flows since the cross-sectional areas of flow are large. The gradient of piezometric head is used rather than the energy gradient because the velocities are so small that the kinetic energy head can be neglected relative to changes in piezometric head.

Further modification of Darcy's law is made by considering the medium as a continuum (really, the assumption is made with regard to the velocity) and taking the length of flow to the limit.

$$v = -K \frac{dh}{dl} \tag{10–3}$$

This is the form of Darcy's Law used in the development which is to follow.

10-3. Differential Equation of Groundwater Flow. Consideration of continuity of fluid in a unit volume of saturated aquifer routinely yields

$$-\left[\frac{\partial(\rho u)}{\partial x} + \frac{\partial(\rho v)}{\partial y} + \frac{\partial(\rho w)}{\partial z} \right] \Delta x \cdot \Delta y \cdot \Delta z = \frac{\partial(\Delta M)}{\partial t} \tag{10–4}$$

where x, y and z are the rectangular coordinates (z vertical)

u, v and w are the velocities in the x, y and z directions respectively

ρ is the density

ΔM is the mass of fluid in the unit volume of aquifer

t is the time variable

Development of the terms in the left side of the equation will be dealt with first. The definition of piezometric head when differentiated gives

$$dp = \rho g \, dh - \rho g \, dz + \frac{p}{\rho} \, d\rho \tag{10–5}$$

The definition of fluid compressibility, β, is

$$\beta = -\frac{1}{\zeta} \frac{d\zeta}{dp} \tag{10–6}$$

where ζ is specific volume. In the case of a confined aquifer,

$$\rho\zeta = \text{constant} \tag{10–7}$$

Thus,

$$\frac{d\zeta}{\zeta} = -\frac{d\rho}{\rho} \tag{10–8}$$

and substituting into Eq. (10–6),

$$dp = \frac{1}{\beta\rho} \, d\rho \tag{10–9}$$

Combining Eqs. (10–5) and (10–9), we get

$$\frac{\partial \rho}{\partial x} = \frac{\rho^2 \beta g}{1 - \beta \rho}\frac{\partial h}{\partial x} \cong \rho^2 \beta g \frac{\partial h}{\partial x} \tag{10–10}$$

The approximately equal sign is permissible, because if the magnitude of p is of order unity, the magnitude of β is generally of the order 10^{-3} or smaller.
Similarly

$$\frac{\partial \rho}{\partial y} \cong \rho^2 \beta g \frac{\partial h}{\partial y} \tag{10–11}$$

Differentiation of the piezometric head with respect to z leads to a result different from differentiation with respect to the other dimensions.

$$\frac{\partial p}{\partial z} = \rho g \frac{\partial h}{\partial z} + \frac{p}{\rho}\frac{\partial \rho}{\partial z} - \rho g \tag{10–12}$$

Combining Eq. (10–12) with an equation of the form of Eq. (10–9), but written with respect to z, results in

$$\frac{\partial \rho}{\partial z} \cong \rho^2 \beta g \left(\frac{\partial h}{\partial z} - 1\right) \tag{10–13}$$

Expansion of the left side of Eq. (10–4) will permit substitution of Eqs. (10–10), (10–11) and (10–13), resulting in

$$-\left[\frac{\partial(\rho u)}{\partial x} + \frac{\partial(\rho v)}{\partial y} + \frac{\partial(\rho w)}{\partial z}\right] = -\left[u\rho^2 \beta g \frac{\partial h}{\partial x} + v\rho^2 \beta g \frac{\partial h}{\partial y} + w\rho^2 \beta g \left(\frac{\partial h}{\partial z} - 1\right)\right]$$
$$- \rho\left[\frac{\partial u}{\partial x} + \frac{\partial v}{\partial y} + \frac{\partial w}{\partial z}\right] \tag{10–14}$$

Writing Darcy's law for each of the three directions, viz.,

$$u = -K_x \frac{\partial h}{\partial x}, \quad v = -K_y \frac{\partial h}{\partial y}, \quad w = -K_z \frac{\partial h}{\partial z} \tag{10–15}$$

and substituting into Eq. (10–14) provides

$$-\left[\frac{\partial}{\partial x}(\rho u) + \frac{\partial}{\partial y}(\rho v) + \frac{\partial}{\partial z}(\rho w)\right]$$
$$= \left[K_x \rho^2 \beta g \left(\frac{\partial h}{\partial x}\right)^2 + K_y \rho^2 \beta g \left(\frac{\partial h}{\partial y}\right)^2 + K_z \rho^2 \beta g \left\{\left(\frac{\partial h}{\partial z}\right)^2 - \frac{\partial h}{\partial z}\right\}\right]$$
$$+ \rho\left[K_x \frac{\partial^2 h}{\partial x^2} + K_y \frac{\partial^2 h}{\partial y^2} + K_z \frac{\partial^2 h}{\partial z^2}\right] \tag{10–16}$$

We can neglect the terms which are products of differentials as of higher order.

The term
$$K_z \rho^2 \beta g \frac{\partial h}{\partial z}$$

we will assume to be small relative to the remaining terms. If

$$K_x = K_y = K_z = K \tag{10-17}$$

that is, if the medium is isotropic with respect to permeability, Eq. (10–4) can be written

$$\rho K \left(\frac{\partial^2 h}{\partial x^2} + \frac{\partial^2 h}{\partial y^2} + \frac{\partial^2 h}{\partial z^2}\right) \Delta x \cdot \Delta y \cdot \Delta z = \rho K \nabla^2 h = \frac{\partial(\Delta M)}{\partial t} \tag{10-18}$$

The right-hand side of Eq. (10–4) or (10–18) expresses the change in the fluid mass in the unit volume of saturated aquifer. It is

$$\frac{\partial(\Delta M)}{\partial t} = \frac{\partial(n\rho \, \Delta x \cdot \Delta y \cdot \Delta z)}{\partial t} \tag{10-19}$$

where n is the porosity, the ratio of the volume of voids to the total unit volume of aquifer.

Assuming that Δx and Δy are restrained from changing size by lateral action in the aquifer

$$\frac{\partial(\Delta M)}{\partial t} = \frac{\partial \rho}{\partial t} \, n \, \Delta x \cdot \Delta y \cdot \Delta z + \frac{\partial(n \, \Delta z)}{\partial t} \, \rho \, \Delta x \cdot \Delta y \tag{10-20}$$

Examining the contributions to the mass change in the right-hand side of Eq. (10–20) term-by-term, we consider the term expressing the change in density of the fluid first. Substituting into that first term the relation of Eq. (10–9), we obtain

$$\frac{\partial \rho}{\partial t} \, n \, \Delta x \cdot \Delta y \cdot \Delta z = n \beta \rho \, \frac{\partial \rho}{\partial t} \, \Delta x \cdot \Delta y \cdot \Delta z \tag{10-21}$$

In examining the second term, of Eq. (10–20), we first consider the volume of the solids in the unit volume of aquifer:

$$V_s = (1 - n) \, \Delta x \cdot \Delta y \cdot \Delta z \tag{10-22}$$

The change in this volume of solids is small compared to the change in the unit volume itself. Therefore we can assume

$$d(V_s) = [d(\Delta z) - d(n \, \Delta z)] \, \Delta x \cdot \Delta y = 0 \tag{10-23}$$

and

$$d(n \, \Delta z) = d(\Delta z) \tag{10-24}$$

Substituting Eq. (10–24) into the second term of Eq. (10–20) results in that term becoming

$$\frac{\partial(n \, \Delta z)}{\partial t} \, \rho \, \Delta x \cdot \Delta y = \frac{\partial(\Delta z)}{\partial t} \, p \, \Delta x \cdot \Delta y \tag{10-25}$$

Introducing the compressibility of the soil matrix as the reciprocal of its elasticity gives

$$\frac{1}{\alpha} = E_s = \frac{d(\sigma_z)}{d(\Delta z)/\Delta z} \qquad (10\text{--}26)$$

where α is the compressibility of the soil matrix

E_s is the elasticity of the soil matrix

σ_z is the stress in the grains in the matrix

Further we recognize that the stress, σ_z, and the pressure of the fluid in the voids, p, must combine to support the total mass above the unit area if arching of the overlying strata is neglected, resulting in

$$d(\sigma_z) = -dp \qquad (10\text{--}27)$$

Combining Eqs. (10–25), (10–26) and (10–27), we see that the second term in Eq. (10–20) becomes

$$\frac{\partial(n\,\Delta z)}{\partial t}\,\rho\,\Delta x \cdot \Delta y = \rho\alpha\frac{\partial p}{\partial t}\,\Delta x \cdot \Delta y \cdot \Delta z \qquad (10\text{--}28)$$

Then, upon substitution of Eqs. (10–21) and (10–28) into Eq. (10–18) we obtain the equation

$$\frac{\partial^2 h}{\partial x^2} + \frac{\partial^2 h}{\partial y^2} + \frac{\partial^2 h}{\partial z^2} = \nabla^2 h = \frac{(\alpha + n\beta)}{K}\frac{\partial p}{\partial t} \qquad (10\text{--}29)$$

Use of Eqs. (10–5) and (10–9) result in

$$\frac{\partial p}{\partial t}\frac{\rho g}{1 - \beta\rho}\frac{\partial t}{\partial h} \cong \rho g\frac{\partial h}{\partial t} \qquad (10\text{--}30)$$

This result leads to Eq. (10–29) being written

$$\frac{\partial^2 h}{\partial x^2} + \frac{\partial^2 h}{\partial y^2} + \frac{\partial^2 h}{\partial z^2} = \nabla^2 h = \frac{(\alpha + n\beta)}{K}\rho g\frac{\partial h}{\partial t} \qquad (10\text{--}31)$$

Then we multiply top and bottom of the right side of Eq. (10–31) by b, the vertical thickness of the confined aquifer, and define

$$S = (\alpha + n\beta)\rho g b \qquad (10\text{--}32)$$

and

$$T = Kb \qquad (10\text{--}33)$$

where S is the coefficient of storage, the water volume obtained from a unit volume of aquifer per unit decline in head (S is dimensionless), and T is the transmissivity of the aquifer.

The final result is then

$$\nabla^2 h = \frac{S}{T}\frac{\partial h}{\partial t} \qquad (10\text{--}34)$$

the equation of piezometric head for unsteady flow in a saturated, confined, isotropic aquifer. For steady flows, the equation becomes the familiar Laplace equation,

$$\nabla^2 h = 0 \tag{10-35}$$

10-4. Solutions of the Steady Flow Differential Equation . Techniques of solution of the Laplace equation are well known, and are to be found many places in potential theory literature. Unfortunately, for the hydraulic engineer, most methods of solution deal with either two-dimensional flows or axi-symmetric conditions. The effect of some geologic formations, such as intrusions into the aquifer by igneous material, on the problem is to create boundary conditions which demand the solution of the Laplace equation in a three-dimensional field. In this section we will consider solutions to problems that describe confined aquifers that are infinite in areal extent.

For steady flows the equation of piezometric head was given as

$$\nabla^2 h = 0 \tag{10-35}$$

If an aquifer of infinite areal extent is assumed, the flow field has radial symmetry about the well. Further, we assume that the well fully penetrates the aquifer, and that all streamlines of the apparent velocity are horizontal. Then the Laplace equation becomes

$$\frac{d^2 h}{dr^2} + \frac{1}{r}\frac{dh}{dr} = \frac{1}{r}\frac{1}{dr}\left(r\frac{dh}{dr}\right) = 0 \tag{10-36}$$

Thus,

$$r\frac{dh}{dr} = \text{constant} \tag{10-37}$$

The constant in Eq. (10-37) can be evaluated by considering Darcy's law for radial flow.

$$Q = K2\pi r b \frac{dh}{dr} \tag{10-38}$$

Then,

$$r\frac{dh}{dr} = \frac{Q}{2\pi K b}$$

and

$$\int_{h_w}^{h} dh = \frac{Q}{2\pi K b} \int_{r_w}^{r} \frac{dr}{r} \tag{10-39}$$

results in

$$h - h_w = \frac{Q}{2\pi K b} \ln\left(\frac{r}{r_w}\right) \tag{10-40}$$

Equation (10-38) was derived from considerations of continuity in the flow field, and itself could be the starting point in the analysis (see Fig. 10-1).

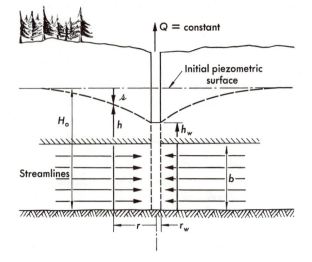

Fig. 10-1. Steady, radial flow to a well in a confined aquifer.

Equation (10-40) has h increasing without limit as r increases. Obviously then, the equation has validity only in the region of the aquifer near the well where steady flow is established. In addition, the limits of integration do not have to be the head at the well and the radius of the well, h_w and r_w respectively, but any pair of values of head and radius. Observation wells can be drilled, the drawdown, $s = H_0 - h$ noted, and Eq. (10-40) can be solved for K or Kb—in effect, the aquifer can be calibrated.

By starting with the continuity principle and considering steady flow, the differential equation for head as a function of radial distance under symmetrical conditions for pumping from an unconfined aquifer can be shown to be

$$Q = K2\pi r h \frac{dh}{dr} \tag{10-41}$$

Integrating as before

$$h^2 - h_w{}^2 = \frac{Q}{\pi K} \ln \frac{r}{r_w} \tag{10-42}$$

The essential difference between confined and unconfined flow is apparent upon examination of Eqs. (10-40) and (10-42). The equation for unconfined flow is nonlinear in h. This nonlinearity makes flow in unconfined aquifers—flows with a free surface—more difficult to analyze.

10-5. Solutions of the Differential Equation, Unsteady, Confined Flows. Radial flow to a well in a confined aquifer permits the transformation of Eq. (10-34) to the polar-coordinate form

$$\frac{\partial^2 h}{\partial r^2} + \frac{1}{r}\frac{\partial h}{\partial r} = \frac{S}{T}\frac{\partial h}{\partial t} \tag{10-43}$$

The solution of Eq. (10–43), when referred to an aquifer of infinite extent, is given by

$$h = H_0 - \frac{Q}{4\pi T}\int_{r^2 S/4Tt}^{\infty}\frac{e^{-u}}{u}\,du \tag{10-44}$$

The variables, r and t, and the aquifer constants, S and T, are in the lower limit of the integral. The dummy variable u is often written

$$u = \frac{r^2 S}{4Tt} \tag{10-45}$$

The evaluation of the integral can be found in mathematical tables under $-Ei(-u)$. It is common in well hydraulics to refer to $-Ei(-u)$ as the "well function," and to write it $W(u)$. A plot of the well function versus its argument is given in Fig. 10-2. Such a diagram is called a type curve.

Solution of a problem describing the unsteady flow in practice is the determination of the aquifer constants from pumping tests. Pumping tests are conducted at a constant pumping rate, observing the drawdown in the piezometric level in an observation well some distance r away from the primary well. The drawdown and the time of observation of the drawdown with the distance from the pumped well are the data for analysis of the aquifer.

For determination of the aquifer constants we have two methods available.

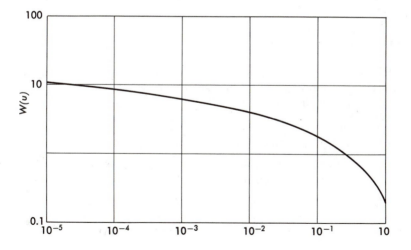

Fig. 10–2. Type curve for the well function.

We first write Eq. (10–44) as

$$s = H_0 - h = \frac{Q}{4\pi T}\, W\!\left(\frac{r^2 S}{4Tt}\right) \tag{10–46}$$

and we see that the drawdown varies with the well function as the argument of the well function varies with the variables. That is,

$$s = \frac{Q}{4\pi T}\, W(u) \tag{10–47}$$

and

$$u = \frac{r^2 S}{4Tt}, \qquad \text{or} \qquad \frac{r^2}{t} = \frac{4T}{S}\, u \tag{10–48}$$

where u is the argument of the well function. The relationships above mean that a plot of drawdown versus the variable form r^2/t should be similar in shape and slope to a plot of $W(u)$ versus u. Since the slope of $W(u)$ versus u is monotonically decreasing with increasing u, and is never constant, there will be only one region of the $W(u)$ versus u curve that is similar in shape to the s versus r^2/t curve. The region of similarity can be found by plotting the data of s versus r^2/t on log-log graph paper of the same type and physical size as the type curve. (The type curve for this purpose can be plotted from the tables of $-Ei(-u)$ previously mentioned.) When the curve representing s versus r^2/t is held over the type curve and made to coincide with a portion of the type curve while keeping the axes of the graphs parallel, the conditions of Eqs. (10–47) and (10–48) are met. Any common point on the two graphs gives corresponding values of s, r^2/t, $W(u)$ and u that can be used in Eqs. (10–47) and (10–48) to compute the values of T (Eq. [10–47]) and S (Eq. [10–48]).

Example 10–1

Values of drawdown and observation time are given below. They are the data from an observation well located 500 ft from the primary well. The primary well is pumped at the rate of 2 cfs. Find the transmissibility T and the storage coefficient S.

Time, hours	Drawdown, feet
0.25	1.09
0.5	2.1
1.0	3.2
2.0	4.5
3.0	5.4
6.0	6.9
9.0	7.7
12.0	8.4
24.0	9.7
48.0	11.1

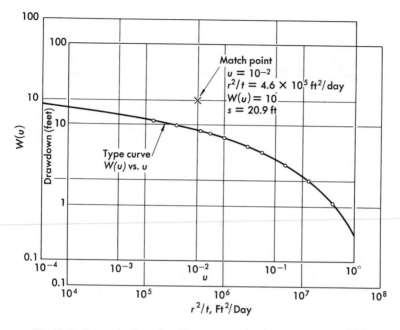

Fig. 10-3. Determination of aquifer constants by the type curve method.

Solution. The data are plotted on log-log paper, and then superimposed on the type curve as shown in Fig. 10–3 and described previously. The match point, common to both graphs, is selected in this case to provide convenient values of $W(u)$ and u.

The match point values are:

$$u = 10^{-2}, \qquad r^2/t = 4.6 \times 10^5 \text{ ft}^2/\text{day}$$
$$W(u) = 10, \qquad s = 20.9 \text{ ft.}$$

Then, solving for T, Eq. (10–47) gives

$$T = \frac{Q}{4\pi s}\, W(u) = \frac{(2.)(10)}{(20.9)(4)\pi} = 0.0762 \text{ ft}^2/\text{sec} = 6580 \text{ ft}^2/\text{day} \qquad Ans.$$

Solving Eq. (10–48) for the storage coefficient, S,

$$S = \frac{4Ttu}{r^2} = \frac{(4)(6579)(10^{-2})}{4.6t \times 10^5} = 0.000572 \qquad Ans.$$

A second method of solution is the so-called approximate method, developed from consideration of the series expansion of the well function.

$$W(u) = -0.5772 - \ln(u) + u - \frac{1}{2.2!}\, u^2 + \frac{1}{3.3!}\, u^3 - \cdots \qquad (10\text{–}49)$$

For small values of u, for example $u = 0.01$, the series can be approximated by the first two terms. Since

$$u = \frac{r^2S}{4Tt} \tag{10-48}$$

all that is necessary for satisfactory approximation of the well function by the first two terms is for the time from beginning of pumping to become large—simply to continue pumping. Then

$$s = \frac{Q}{4\pi T}\left[-0.5772 - \ln\left(\frac{r^2S}{4Tt}\right)\right] - \frac{Q}{4\pi T}\ln\left(\frac{2.25Tt}{r^2S}\right) = \frac{2.30Q}{4\pi T}\log_{10}\left(\frac{2.25Tt}{r^2S}\right)$$

$$\tag{10-50}$$

Reduction of pumping data to determine the aquifer constants can be accomplished by using the form of Eq. (10–50) to advantage. The data are plotted on semi-logarithmic paper: the drawdown on the arithmetic scale, and the variable r^2/t on the logarithmic scale. A straight line tangent to the curve through the data points is drawn, and two values of drawdown are determined for two values of r^2/t differing by a power of 10. If the difference between the two values of drawdown is called Δs, Eq. (10–50) reduces to

$$\Delta s = \frac{2.30Q}{4\pi T} \tag{10-51}$$

and can be solved for T. Any other point on the tangent line can be used to evaluate Eq. (10–50) for S; a convenient point is the extension of the straight line to the point where s equals zero. Then Eq. (10–50) becomes

$$\frac{2.25T}{S}\left(\frac{t}{r^2}\right)_0 = 1 \tag{10-52}$$

and can be solved for S. ($[t/r^2]_0$ is the value of the variable at $s = 0$ ft, a point on the line not represented by data.)

Example 10–2

Using the data of Example 10–1 determine the values of the aquifer constants, S and T, by the approximate method.

Solution. The data are plotted in Fig. 10–4, and the tangent line drawn. Note that since r is constant, the form of the variable plotted on the logarithmic axes of the graph is t. The values from the curve are

$$\Delta s = 4.8 \text{ ft}$$

$$t_0 = 0.225 \text{ hours}$$

Then

$$T = \frac{2.30Q}{4\pi\Delta s} = \frac{(2.30)(2)}{4\pi(4.8)} = 0.0763 \text{ ft}^2/\text{sec}$$

$$= 275 \text{ ft}^2/\text{hr} \qquad\qquad Ans.$$

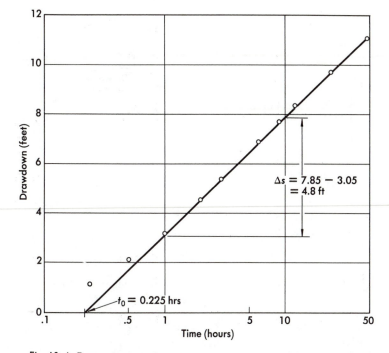

Fig. 10–4. Determination of aquifer constants by the approximate method.

and

$$S = \frac{2.25Tt_0}{r^2} = \frac{(2.25)(275)(0.225)}{(500)(500)} = 0.000556 \qquad Ans.$$

10–6. Boundary Conditions, Imagery. As implied previously, practical well problems involving the determination of the drawdown resulting from a specified pumping schedule and in an aquifer configuration of other than infinite lateral extent requires modification of the principles outlined in the previous article. The practical constraints can be considered as boundary conditions on the problem. Pumping schedules result in constraints on the value of discharge relative to the time variable. Dimensions of the aquifer are constraints on the space variables and their effect on the piezometric head.

The drawdowns resulting from a variable pumping schedule can be determined by superposition, or addition, of the computed drawdown from each steady segment of pumping rate. For example, if a well is pumped for time t_1, at the rate of Q_1 cfs, and then the pumping rate is increased to Q_2 cfs for an additional time t_2, the drawdown at the end of time $t_1 + t_2$ would be (at a distance r from the well)

$$s = \frac{Q_1}{4\pi T}\left[W\left\{\frac{r^2S}{4T(t_1 + t_2)}\right\}\right] + \frac{Q_2 - Q_1}{4\pi T} W\left(\frac{r^2S}{4Tt_2}\right) \qquad (10\text{–}53)$$

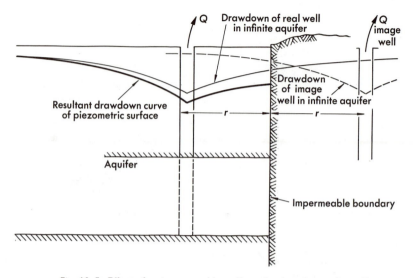

Fig. 10–5. Effect of an impermeable wall on the drawdown of a well.

The recovery of piezometric head in a well can be determined by considering a negative discharge (recharge, in a sense) so that if a well stops pumping after a time t_1 and it is desired to compute the residual drawdown after time $t_1 + t_2$, then

$$s = \frac{Q}{4\pi T} W\left(\frac{r^2 S}{4T[t_1 + t_2]}\right) - \frac{Q}{4\pi T} W\left(\frac{r^2 S}{4T t_2}\right) \qquad (10\text{--}54)$$

The concept of superposition is also applied for the solution of groundwater flow problems with physical boundaries. For example, a vertical, impervious wall restricting the lateral dimension of an aquifer can be simulated mathematically by an imaginary well placed on the opposite side of the boundary from the real well. Figure 10–5 illustrates the drawdown of the piezometric head if the aquifer was infinite, the drawdown of the image well, and the final drawdown, the sum of the two, which satisfies the conditions at the wall. The method can be applied to both steady and unsteady solutions. Other configurations of the aquifer can be obtained by combinations of real wells and image wells. Hydrodynamic literature describes the method of images in detail.

PROBLEMS

10–1. Assuming conditions of steady flow, determine the drawdown at the face of a well of diameter 1 ft in a confined aquifer of thickness $b = 70$ ft and $K = 15$ ft/day. The drawdown should be computed as that from the piezometric level at $r = 1000$ ft. The pumping rate is 200 gpm.

10-2. Find the drawdown in an unconfined aquifer using $K = 15$ ft/day and a pumping rate of 20 gpm. The piezometric level at a radius of 1000 ft is 37.3 ft.

10-3. The piezometric head at an observation well 770 ft from the pumped well in a confined aquifer is 44.3 ft. In another observation well 210 ft away the head is 35.8 ft. If the levels have remained essentially steady for a period of time, determine the coefficient of transmissibility. The pumping rate is 0.5 cfs.

10-4. Pumping test data for a constant rate of 2 cfs are given below. Determine the aquifer constants T, in ft²/day, and S by the type-curve method. The observation well is 300 ft away.

Time, hours	Drawdown, feet
4	0.67
6	0.90
8	1.08
12	1.35
18	1.62
24	1.85
30	2.00
36	2.18
48	2.40
72	2.70
96	2.90

10-5. Use the information of Problem 10-4 to determine T and S by the approximate method.

10-6. A well in a confined aquifer is found to have $T = 2500$ ft²/day and $S = 0.0007$. The well is pumped at the rate of 2 cfs for 3 days, and then pumping stops. What would be the residual drawdown 200 ft from the well 60 hours after pumping stopped?

10-7. The well described in Problem 10-6 is pumped for 5 days at 1.5 cfs, 2 days at 2.0 cfs and then allowed to recover (no pumping) for 7 days. What will be the residual drawdown 200 ft from the well at the end of that time?

10-8. A well in an unconfined aquifer is located 300 ft from an impervious vertical wall and pumped at 0.06 cfs. The permeability of the aquifer is $K = 23$ ft/day. Assuming steady conditions, and further assuming that the piezometric head is 42 ft at a distance of 2000 ft,

    a. What is the drawdown at the well (radius 1 ft)?

    b. At the face of the vertical wall?

10-9. A well in a confined aquifer is pumped at the rate of 0.1 cfs. An essentially straight vertical wall is formed by an intrusion through the aquifer 450 ft from the well. What is the drawdown in piezometric head at the wall after 36 hours of pumping, $T = 1800$ ft²/d, $S = 0.0005$.

10-10. The well in Problem 10-9 is pumped for 10 hours at 0.5 cfs, and then is allowed to recover for 24 hours. What would the residual drawdown be at the well (radius 1 ft) at the end of the period?

11

RIVER ENGINEERING

11-1. Definition and Scope. From the very beginning of human history, rivers have been of high importance in the life and activities of man (note Genesis 2: 10–14). The two most ancient civilizations were built along the valleys of great rivers, Sumeria-Babylonia along the Tigris and Euphrates, and Egypt along the Nile, and depended for their sustenance on the waters flowing in them. The River Jordan is of key significance in the development of both ancient and modern Israel. One also thinks of the Ganges and Indus Rivers in southern Asia, the Yangtze in China, the Rhine and Danube in Europe, and others, as illustrations of the great importance of the role played by rivers in human history.

The total annual discharge in the rivers of the world is about 8,200 cubic miles, representing about one-third of the annual precipitation. The ten largest rivers are probably the following:

Rank	River	Average Discharge (cfs)
1	Amazon (Brazil)	4,000,000 [1]
2	LaPlata-Parana (Argentina)	1,600,000
3	Congo (Africa)	1,400,000
4	Yangtze (China)	1,000,000
5	Ganges-Brahmaputra (India)	800,000
6	Mississippi (United States)	620,000
7	Yenisei (Siberia)	610,000
8	Orinoco (Venezuela)	600,000
9	Mekong (Thailand)	560,000
10	Lena (Siberia)	540,000

The above figures are, in many cases, only rough estimates. Stream gaging even today is systematically carried out on only a relatively small number of the world's rivers. The measurement of river discharges began, for practical purposes, only in 1880, with the efforts of the British naturalist, Dr. H. B. Guppy, to obtain information on the quantity of water flowing in the Yangtze and other large rivers. Soon after this, the U.S. Geological Survey began its stream gaging program, mainly for the benefit of western settlers.

At present, the Survey operates over 6000 gaging stations, on all the

---

[1] Some studies, including a recent U.S.G.S. study, indicate this may be closer to 8,000,000 cfs.

major streams in the United States. The largest of these rivers are indicated in the accompanying table, arranged in order of magnitude of average discharge. Of the 26 large streams listed, only ten are independent, but these ten carry over 75 per cent of the total water (the total being approximately 1,800,000 cfs) drained from the United States in its river system.

Rank	River	Length (miles)	Drainage Area (square miles)	Average Discharge (cfs)
1	Mississippi	3,892	1,243,700	620,000
2	St. Lawrence	2,100	565 000	500,000
3	Ohio (T)	1,306	203,000	255,000
4	Columbia	1,214	258,200	255,000
5	Mississippi (T)	1,170	171,600	91,300
6	Missouri (T)	2,714	529,400	70,100
7	Tennessee (TT)	900	40,600	63,700
8	Mobile	758	42,300	59,000
9	Red (T)	1,300	91,400	57,300
10	Arkansas (T)	1,450	160,500	45,200
11	Snake (T)	1,038	109,000	44,500
12	Susquehanna	444	27,570	38,000
13	Alabama (T)	720	22,600	31,600
14	White (T)	690	28,000	31,000
15	Willamette (T)	270	11,250	30,700
16	Wabash (TT)	475	33,150	30,400
17	Cumberland (TT)	720	18,080	27,800
18	Illinois (T)	420	27,900	27,400
19	Tombigbee (T)	525	19,500	27,000
20	Sacramento	382	27,100	26,000
21	Apalachicola	500	19,500	25,000
22	Pend Oreille (T)	490	25,820	24,000
23	Colorado	1,450	243,900	23,000
24	Hudson	306	13,370	21,500
25	Allegheny (TT)	325	11,700	19,200
26	Delaware	390	12,300	19,000

(T)—first-order tributary.   (TT)—second-order tributary.

Engineering works for the control or use of these rivers have been constructed on all of them in greater or lesser degree. Many smaller rivers, in terms of average flow, have also been of considerable importance, especially in the West and Southwest. The following have large drainage areas but carry less than 10,000 cfs each, on the average:

River	Drainage Area (square miles)
Rio Grande	171,585
Platte	90,000
Kansas	61,300
Gila	58,100
Brazos	44,500
Green	44,400
Colorado (Texas)	41,500
Pecos	38,300
Canadian	29,700

The waters of a river are useful to man in many ways, notably the following:

1. Water supply
   a. Agricultural
   b. Municipal
   c. Industrial
   d. Domestic
2. Power generation
3. Navigation and water transportation
4. Recreation
5. Waste disposal
6. Land drainage
7. Fish and wildlife conservation

On the other hand, rivers are also potentially very harmful in certain respects. These dangers include

1. Floods
2. Erosion, including channel shifting
3. Pollution

The most effective utilization of the river's resources and mitigation of its dangers requires careful engineering planning and, often, many types of structures and controls. With the present rapidly rising populations and increasing industrialization of the world, these demands are growing everywhere.

Furthermore, since uses or controls of river waters at any point in its course may have significant effects at all other points, there has developed a strong present-day trend to coordinate all such works into a basinwide program of development. This of course often requires either governmental supervision or some other form of cooperative action by all parties with an interest in the river resources of the basin. In either case, a high degree of truly professional engineering planning, design, construction, and operation is required for optimum development.

All of this can be properly included within the scope of river engineering. River engineering also can be defined as the planning and implementation of works for the purpose of (1) control of river flow; (2) control of river stage; (3) control of river sediment regimen; (4) control of river quality; or (5) a combination of the foregoing.

In this chapter, we shall be concerned particularly with basic concepts in the planning of works for control of river flows and stages. Sedimentation problems will be discussed in the next chapter. Stream quality studies and controls require specialized biochemical applications and are treated in sanitary engineering publications.

11-2. Stream Regulation by Reservoirs. The key factor in control of the flow or stage of a stream is usually a reservoir or system of reservoirs, which serves to retain and store high flows, later to be released as needed to augment low flows. Lakes along a stream course perform essentially this function as natural reservoirs. An artificial reservoir usually is created by constructing a dam across the stream valley at some feasible dam site.

The design of dams has been discussed in the chapter on hydraulic structures. Here we are concerned with the selection of a reservoir capacity to meet specific stream regulation requirements. This is probably the most important single decision to be made in the development of a river control project, but it must often be based on admittedly inadequate data and criteria.

The first step necessary in arriving at such a decision is to obtain a representative sequence of anticipated actual flows on which to base the study. This normally must be based on records of past flows. However, such records are often very inadequate and must be modified and extended by some means in order to be more realistically representative of future conditions.

It is desirable to have a record extending over at least fifty years' duration if possible. Shorter periods may well miss a long-term cyclic period of abnormally high or low runoff and thus not be truly representative. If only a short-term record is available, an attempt should be made to extend it synthetically, by establishing a correlation with a long-term record of rainfall on the watershed or with a long-term record of runoff on some nearby watershed of similar hydrologic characteristics. If the flow record is for some point on the stream other than the proposed dam site, it may be modified by estimated accretions or depletions of flow in the intervening reach.

The ultimate purpose of this phase of the study is to arrive at a representative sequence of flows as they may be expected to occur during the future life of the reservoir. Therefore the past record must be still further modified to allow for estimated future upstream modifications. Any future upstream regulation works, augmented irrigation withdrawals, etc., will of course have an effect on the flow, and so should be anticipated and applied as corrections to the past flow record, in so far as possible.

The second major step in a river regulation study is to determine the anticipated demands on the reservoir or reservoir system. In other words, a hydrograph of *desired outflows* from the system must be developed. This will depend of course upon the uses to which the water will be put, and will likely vary seasonally. Allowance must again be made for future conditions and needs for water, anticipating future populations, irrigation developments, etc.

Economic studies must also be made of the value of the benefits to be derived by provision of the full demand and of various fractions of the full demand, in order to make comparison with the amortized cost of the reservoir and appurtenant facilities. A major factor in the latter, in addition to the direct relationship with the storage capacity provided, is the location of the dam site. The best location must be chosen, as based primarily on geologic and topographic investigations, in order to minimize the project cost.

Site studies are also necessary to determine the hydrologic factors of evaporation and seepage losses from the reservoir when in operation. Thus, a topographic survey will determine the relation between reservoir surface level, stored volume, and water surface area. A geologic survey can determine the relation between reservoir level and seepage losses, as well as data needed for structural design of the dam.

These preliminary studies provide information as to expected inflows, desired outflows, and probable losses. The problem then is to determine the required reservoir capacity to provide the desired outflows and the necessary losses from the expected inflows. The solution of this problem involves a detailed routing of flows through the projected reservoir or reservoirs, using the basic "hydrologic bookkeeping" equation:

$$\text{Initial storage} + \text{Inflow} - \text{Outflow} - \text{Losses} = \text{Final storage} \quad (11\text{--}1)$$

The equation is applied, month by month, through the entire period of study, thus developing a complete hydrograph of each term in the equation.

Such a study may be made for several reservoir sizes or desired outflows, in order to permit comparative economic studies of costs and benefits.

11–3. River Operation Study. In a preliminary analysis of the effect of a single reservoir on the river flow, a graphical procedure utilizing the mass curve technique is often suitable. The mass curve is a chronological graph of *cumulative* flows, the ordinate corresponding to any given date representing the total accumulated volume of flow up to that time, from the beginning of the period of study. If mass curves of inflow to the reservoir and outflow from the reservoir are superimposed, it is obvious that the vertical intercept between the two at any time must represent the net drawdown from reservoir storage up to that time.

In Fig. 11–1 the operation of a reservoir is shown by such superimposed mass curves of inflow and outflow. In the illustration, the reservoir is assumed full at the beginning of the period of study. The slope of the mass curve over any time increment represents the average flow rate during that time. The normal controlled outflow is indicated by the slope of the outflow mass curve throughout all periods except when the reservoir is either full or empty, deficiencies of inflow being supplied by reservoir drawdown and excesses of inflow being balanced by storage increases. At times when the reservoir is filled, any further excess inflows must be wasted, as there is no longer any storage capacity for their retention. In this case, inflow and outflow are equal, and the mass curves of inflow and outflow are coincident. The vertical distance between the normal outflow line extended and the actual outflow line is the volume of water wasted.

In dry periods, the reservoir drawdown is indicated by the vertical distance between inflow and outflow mass curves. The maximum such drawdown is the reservoir capacity required to maintain a minimum flow equal to the assumed normal outflow. Similarly, for a given reservoir capacity, the

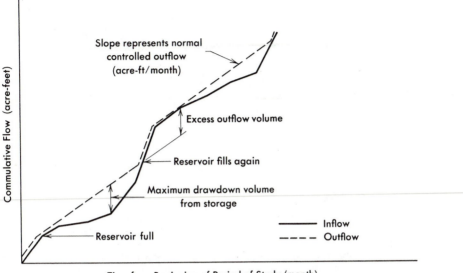

Time from Beginning of Period of Study (month)

Fig. 11-1. Mass diagram of reservoir study.

maximum outflow line slope which will never cause a drawdown greater than the given capacity is the *dependable* flow that can be provided by that capacity. If the normal rated outflow (as based on demand and economic studies) is to exceed this dependable flow, then the reservoir will go dry on occasions, as illustrated in Fig. 11-2. On such occasions, the inflow and outflow are equal, and the inflow and outflow mass curves are therefore parallel. The intercept between normal and actual outflow curves gives the volume of deficiency.

If the reservoir were assumed empty at the beginning of the period of study, instead of full, the outflow mass curve is begun at a point vertically above the origin of the inflow mass curve a distance equal to the reservoir

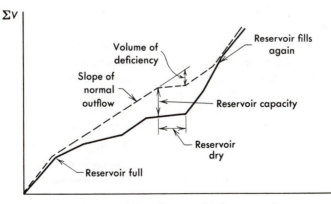

Fig. 11-2. Mass diagram with dry reservoir.

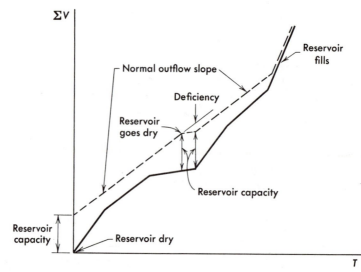

Fig. II–3. Reservoir empty at beginning of period of study.

capacity, thus retaining the condition that the intercept between curves represents the net drawdown from full storage up to that time. (See Fig. 11–3.) The remainder of the curves is constructed exactly as before.

A complete diagram of inflow and outflow mass curves, showing periods and quantities of surplus and deficiency, is called a *regulation diagram*.

The mass curve, by definition, must continually rise, and therefore may require either an inconveniently large paper or an inconveniently small scale. This difficulty may be alleviated by using instead a *residual mass curve*, for which the ordinates are the cumulative flow departures from normal. In effect the mass curve is plotted about a horizontal axis obtained by rotating the average slope on the mass curve to the horizontal. Thus, the ordinates on the residual mass curve may be either positive or negative, with the first and last points on the residual mass curve of inflow falling on the axis. This type of plot also has the advantage of accentuating more clearly the peaks and valleys of the flow record.

II–4. Multiple-Purpose Operation. The graphical analysis as above is convenient only for a simple regulation type of operation. However, most reservoirs today are designed for two or more purposes, and the method of operation must be modified accordingly.

Certain portions of the available storage capacity must be designated for each purpose. For example, a typical division might be as follows, proceeding vertically upward through the reservoir:

1. Sediment storage—volume set aside to provide for gradual deposition of silt on the reservoir bed, regarded as essentially dead storage in the operation analysis.

2. Power storage—volume necessary to maintain minimum head for contracted power generation, also regarded as useless for regulation purposes.

3. Ration storage—volume of storage within which releases are rationed to a specified fraction of the demand, in order to prevent or postpone full depletion of available and usable water.

4. Normal storage—volume within which releases are kept equal to the demands; the latter may be for irrigation, municipal supply, navigation, power, or other purposes.

5. Flood storage—volume of storage normally kept empty in order to be able to catch large flood runoffs when they occur and to control them to some specified flood release; the latter is normally made as large as possible without causing significant downstream flood damage.

11-5. Multiple-Reservoir Systems. A very strong trend in modern planning, necessitated by increasing water demands, requires that each river basin be developed as a unit. That is, each reservoir in the basin must be designed and operated so as to coordinate most effectively with other reservoirs in the basin. Furthermore, the amount of control needed on the river can frequently not be obtained at any single reservoir site; thus, two or more reservoirs are built to provide this control.

A river operation analysis for a multiple-reservoir, multiple-purpose system is necessarily more complex than for a single reservoir. The graphical method is impracticable, and an arithmetical tabulation of all pertinent quantities is required.

The location of the various reservoirs depends primarily on available dam sites, with hydrologic and economic factors also to be considered. In the usual multiple-reservoir development, the best dam sites will probably be found in the upstream reaches of the river, where there are more likely to be gorges and better foundation materials. On the other hand, the demands on the reservoir system are more likely to be concentrated somewhat in the downstream sections, where the flood plains are broader and more fertile, and where the needs for hydroelectric works and municipal water are greater. Similarly, it is often the flood plain areas which are in greatest need of adequate flood protection. A multiple-reservoir development serves in considerable measure to reconcile these conflicting situations.

In such a situation as just outlined, the reservoir farthest upstream could be located at a site permitting economic development of a large storage capacity. It should, of course, be located as near to the downstream areas which it will serve as the existence of good dam sites will permit.

However, there will be still needed a regulatory and storage reservoir closer to the actual areas served. Also, there may be substantial tributary inflows downstream from the large upstream reservoir, which would need to be controlled. Consequently the multiple-reservoir development will also contemplate one or more reservoirs near the areas of greatest demand and below all important tributaries which enter the stream above the areas served by the system.

Further, it may be deemed necessary to build flood control reservoirs on tributary streams on whose watersheds large floods may occasionally originate which contribute considerably to flood stages on the main stream.

11–6. Regulation Storage. Regulation storage is that portion of the total capacity of the reservoir or system of reservoirs from which are supplied those amounts of water necessary to satisfy the downstream water demands, of whatever type they may be. In general, it would be considered as distinct from both sediment storage and flood storage, although some types of operation might conceivably employ either or both of the two latter storage capacities for partial stream flow regulation.

It would seldom be economically feasible to provide sufficient regulation storage capacity to satisfy all demands at all times. There may be certain critical drought periods of such duration and severity that only an extremely large and probably financially impracticable reservoir system could eliminate or even alleviate them. The exact amount of expected future deficiencies in supply remaining under various possible combinations of reservoir capacities is one of the important items which is to be determined by the river operation analysis.

In a particular study, the amount of regulation storage will be selected somewhat arbitrarily, perhaps partially on the basis of a previous mass curve analysis. The total quantity of surpluses and deficiencies for the period studied will then be determined by the river operation study. The same thing will be done for other amounts of regulation storage. For any given quantity of regulation storage, it will probably be desirable also to consider several possible apportionments of this total regulation storage among the several reservoirs of the system. The economic implications for each combination, comparing the cost of the system and its operation with the economic benefits derived, must then be examined in detail in order to arrive finally at the best design.

If the demand is made up largely of irrigation and municipal requirements, further economies can generally be realized by dividing the regulation storage into two parts, which may be called normal storage and ration storage. When water in the reservoir system falls below the volume considered as normal storage, the release can be reduced to some fraction of the demand, thus conserving the available water and stretching out its usefulness over a longer period. The exact fraction employed would depend upon the uses to which the water is put, and the amount of curtailment of the pertinent activities that would be engendered by the fractional supply. In any case, it would almost certainly be desirable to regard a certain amount of the regulation storage as ration storage. This would considerably alleviate the very critical drought periods which cannot be economically eliminated by the reservoir system.

It is obvious that any number of subdivisions of the regulation storage may be possible, and also it is possible that several fractional releases might be studied. Thus, still more river operation studies are indicated before the optimum design is obtained.

It is evident that to arrive at the truly optimum design of a multiple-purpose reservoir system and its operation requires a really tremendous amount of study. Nevertheless, the great importance of such projects undoubtedly warrants this detailed study and planning. Of course, preliminary/graphical studies involving mass hydrograph analysis can probably be used to obtain approximate values for at least the range of storage volumes that should be analyzed in more detail by a computational procedure.

11-7. Method of Operation. The river operation analysis consists essentially of a month-by-month operation of the postulated reservoirs, using the historical record of inflows, as modified by estimated future conditions of upstream regulation and depletion, and impressing upon the system the monthly estimated demands.

Normally, the sediment storage will be reserved for the expected deposition of sediment during the life of the reservoir, and will not be considered available for regulatory purposes. It will then affect the operation only as it affects reservoir stage and consequently evaporation losses. An alternative method would be to allow for an annual reduction in available regulation storage corresponding to the estimated annual accumulation of sediment.

The flood storage in each reservoir is reserved for flood control and is not encroached upon except in times of flood. A maximum release is set. If the inflow exceeds that maximum release and if the reservoir stage is at the upper limit of the regulation storage, then the excess of the inflow over the specified maximum release is detained in the flood storage. As long as water remains in the designated flood storage, the reservoir outflow is held at the maximum release. When the reservoir stage drops to the level of the regulation storage, the release is then reduced to equal the net inflow or the current demand, whichever is greater.

It is understood that this flood storage operation is described only on a monthly basis, which is all that is necessary for the purpose of the general river operation analysis. The routing of the actual design flood must of course be on a more detailed basis, for purposes of design of gates, spillways, outlet conduits, etc.

When the total volume in all reservoirs is within the limits of the total normal regulation storage, releases from the system are equal to the demands. If the storage drops to within the limits of the ration storage, the releases are then reduced to the specified fractional part of the demand. If a prolonged drought completely depletes even the ration storage, then of course the release will be equal to the net inflow until the latter again exceeds the ration release.

In addition to monthly inflows and demands, it is necessary to include in the computations monthly estimates of reservoir losses. This can be done by use of an area-capacity curve for each reservoir, together with estimates of average month-by-month evaporation. The evaporation loss is computed by multiplying the evaporation rate for the particular month by the surface area corresponding to the mean reservoir storage for the month. The area-capacity curve is prepared from the topographic map of the reservoir site. Seepage losses at a particular dam site depend upon the reservoir stage somewhat, and may also be estimated from the mean stage for the month. The net inflow is then the actual inflow reduced by the evaporation and seepage losses for the month. A slightly more accurate computation of evaporation loss, though perhaps less conservative, would be based on the difference between the average evaporation and precipitation rates for the month.

For purposes of the computations, a surplus is defined as the excess of the actual inflow over the demand. A deficiency is the excess of the demand over the inflow. Surpluses must either be stored in the reservoirs, or used to supply reservoir losses, or released as flood releases. Deficiencies are met, as far as possible, by withdrawals from reservoir storage. In periods of deficiency, reservoir losses would also have to be met by withdrawals from reservoir storage.

It will be economical in general, particularly if water conservation is the chief purpose of the reservoirs, to store surpluses in upstream reservoirs first; then, as the regulation storage in the upstream reservoir is filled, further surpluses may be discharged from the upstream reservoir and stored in the next reservoir downstream, and so on.

Similarly, all deficiencies and reservoir losses are borne by the reservoir farthest downstream as far as possible, then the next reservoir upstream, and so on. Thus, as long as any regulation storage remains unused in an upstream reservoir, the downstream reservoir will be kept dry except for interreservoir tributary inflows. This is usually the most efficient method of operation, since it assures that a maximum of the interreservoir inflows can be controlled by one or more reservoirs in the system, and that there will be the maximum likelihood that excess releases necessary at an upstream reservoir can be retained in storage at a downstream reservoir. However, this procedure conceivably would need to be modified if there are large differences in reservoir losses at the different locations.

In carrying through the computations, the basic principle is enunciated in the familiar storage equation, as follows:

$$S_1 + I - O - L = S_2 \qquad (11\text{--}2)$$

where S_1 = stored volume of water at beginning of month
   S_2 = stored volume of water at end of month
   I = total inflow during the month
   O = total outflow during the month
   L = total reservoir losses during the month

The above equation can be applied to the monthly operation of each reservoir individually, or to that of the system as a whole. Let M represent the net modification of the natural river flow for the month, equalling the total change in storage plus reservoir losses. Then,

$$M = I - O = S_2 - S_1 + L \qquad (11\text{--}3)$$

Note that M can be positive or negative, and careful attention must be paid to algebraic sign. Subscripts a, b, c, etc., can be appended to each of the terms in Eqs. (11–2) and (11–3) to denote the proportions of the corresponding quantities without subscripts which are applicable to the several reservoirs of the system, A, B, C, etc.

Let D represent the demand on the reservoir system for the month. For simplicity let it be assumed that the entire demand applies to areas below the lowermost reservoir. If there are important interreservoir demands, the analysis will be considerably complicated, but the general principles and methods will still be applicable. Then, let E represent the excess of the unregulated inflow to the lowermost reservoir over the demand. If E is positive, it is a surplus as previously defined; if negative, it is a deficiency. Then,

$$E = M = M_a + M_b + M_c + \cdots \qquad (11\text{--}4)$$

The breakdown of M among the several reservoirs is made according to the principle already mentioned. Surpluses are applied first to the upstream reservoirs, deficiencies to the downstream reservoirs.

The loss at any reservoir, say reservoir A, is determined from the average storage in that reservoir for the month. It would ordinarily be sufficiently accurate to take the average storage for the month as being equal to the initial storage and monthly modification average. That is,

$$S_a = \frac{1}{2}(S_{1a} + S_{2a}) \cong \frac{1}{2}(2S_{1a} + M_a) \qquad (11\text{--}5)$$

The actual end-of-month storage can then be computed by deducting the losses from the sum of the initial storage and the net modification for the month.

Equation (11–4) is applicable only when the storage volume is within the limits of the normal regulation storage volume. When rationed releases are necessary, then the total M must be made equal to the difference of the unregulated inflow to the lowermost reservoir and the fractional demand specified. When flood releases are necessary, M is made equal to the difference between the unregulated inflow and the actual flood release. In any case, M must always be equal to the unregulated inflow at the lowermost reservoir minus the actual release from that reservoir, and thus represents the over-all effect of the reservoir system.

As a measure of the effectiveness of the entire system, the difference between the actual release from the last reservoir and the actual demand can also be computed for each month. When the monthly release is exactly equal to the demand for that month, then the system is completely effective. When the release exceeds the demand, the difference can be described as a residual surplus. When the demand is greater than the release, the difference is a residual deficiency. The total of all residual surpluses and deficiencies is a measure of the effectiveness of the system of reservoirs.

The regulated inflow to any reservoir, if desired, is equal to the unregulated inflow at that reservoir minus, algebraically, the sum of the net stream modifications at each upstream reservoir. The release from any upstream reservoir may be computed as the inflow to that reservoir minus the modification at that reservoir.

If the storage and inflows are such that the total storage changes from normal to ration storage, or vice versa, during the month, releases from the lowermost reservoir are based on the assumption that the inflow and demand are constant throughout the month. For example, if the initial storage is in the normal limits and the demand sufficiently greater than the supply to pull the storage down into the ration storage during the month, the release would be normal for the portion of the month required to deplete the normal storage and would be rationed for the remainder of the month. A similar procedure is used when flood releases are necessary for a part of the month only.

The computations for the river operation analysis can be most efficiently carried through by means of a systematic tabulation, with the aid of a calculating machine. Graphical methods are not sufficiently accurate and, in fact, are more time-consuming than the actual computation when there is more than one reservoir included in the development.

Example 11-1

Assume three reservoirs, whose operation is primarily for the twofold purpose of water conservation and flood control. The available storage capacities will be assumed to be those given in the tabulation shown. Initial storage quantities as given are taken to be the volumes in the reservoirs at the beginning of the particular period being analyzed.

| Reservoir | Storage Capacity Available | | | | Initial Storage |
	Flood	Regulation	Sediment	Total	
A	800,000	1,000,000	200,000	2,000,000	800,000
B	800,000	500,000	200,000	1,500,000	500,000
C	800,000	500,000	200,000	1,500,000	200,000

Solution. All the quantities in the table are in units of acre-feet. Assume that, of the total regulation storage of 2,000,000 acre-feet, 500,000 acre-feet is ration storage and 1,500,000 acre-feet is normal storage. When the total storage is in the range of the normal storage, releases will be equal to the demands. When the storage is within the ration storage range, releases will be two-thirds of the demand. When the total storage enters the flood storage range, releases will be set at 10,000 cfs, which is equal to 603,000 acre-feet per month. For simplicity, the evaporation loss at each reservoir will be assumed to be constant and equal to 5000 acre-feet per month. Seepage losses will be neglected. A period of two years will be studied.

For this computation, a tabulation of some 25 columns will be needed, with the following suggested arrangement. Note that the data in Columns 1 through 6 will have been prepared before-hand, and these columns may be completely filled in before beginning the actual study.

1. Unregulated inflow at site of Reservoir A.
2. Unregulated inflow at site of Reservoir B.
3. Unregulated inflow at site of Reservoir C.
4. Demand on reservoir system (assumed all required below Reservoir C).
5. Surpluses (E) in unregulated inflow at Reservoir C (Column 3 minus Column 4).
6. Deficiencies $(-E)$ in unregulated flow at Reservoir C (Column 4 minus Column 3).
7. Total modification in unregulated flow at Reservoir C (M) equal to total change in storage plus evaporation at all reservoirs.
8. Modification at Reservoir A.
9. Modification at Reservoir B.
10. Modification at Reservoir C.
11. Storage plus evaporation at Reservoir A.
12. Evaporation at Reservoir A.
13. Storage at Reservoir A.
14. Releases from Reservoir A.
15. Storage plus evaporation at Reservoir B.
16. Evaporation at Reservoir B.
17. Storage at Reservoir B.
18. Releases from Reservoir B.
19. Storage plus evaporation at Reservoir C.
20. Evaporation at Reservoir C.
21. Storage at Reservoir C.
22. Releases from Reservoir C.
23. Residual surpluses in regulated outflow from Reservoir C (Column 22 minus Column 4).
24. Residual deficiencies in regulated outflow from Reservoir C (Column 4 minus Column 22).
25. Total storage in all reservoirs.

Assume for the moment that the total storage at the beginning and at the end of a certain month is within the normal storage. Then the release from the lowermost reservoir (Column 22) would equal the demand (Column 4).

If there were a surplus that month, the quantity of surplus would be either stored or lost through evaporation. Consequently, Column 7 would equal Column 5. A plus sign is arbitrarily assigned to increases in storage. If a deficiency existed during that month, it would have to be met by available stored water. Evaporation losses would also have to be met from storage. Consequently, Column 7 would then equal Column 6. Decreases in storage are denoted by a minus sign.

Columns 8, 9, and 10 indicate the distribution of the quantity in Column 7 among the three reservoirs. In making this distribution, the method of operation described previously must be adhered to. Surpluses are stored in Reservoir A if the capacity is available and if there is sufficient inflow there; if not, Reservoir B is used, then Reservoir C. Deficiencies and evaporation losses in all reservoirs are met from Reservoir C if possible, then from Reservoir B and Reservoir A in order.

The quantity in Column 8 is added to the storage in Reservoir A at the end of the preceding month (Column 13) and the sum entered in Column 11. The estimated evaporation loss for the month in Reservoir A is deducted from Column 11 to get the storage in Reservoir A at the end of the month. The storage contents in Reservoirs B and C are obtained in a similar way.

If the storage is within the limits of the ration storage, releases from Reservoir C (Column 22) are equal to two-thirds of the demand for the month. Consequently, the total modification (Column 7) is the difference between the rationed release and the unregulated inflow at Reservoir C. This quantity is then distributed among the several reservoirs in accordance with the same principles outlined above.

When it is necessary to ration releases, either for a full month or for part of a month (in accordance with the principle of linear proportional distribution described previously), the difference between the demand and the actual release is a residual deficiency and is entered in Column 24.

If the storage in any reservoir reaches flood stage, it may be necessary to maintain the flood capacity by excess releases at that reservoir. Flood storage may only be used when necessary to control outflows to the specified maximum release of 10,000 cfs. If excess releases become necessary at Reservoir C, they cannot be recaptured at a lower reservoir and consequently are wasted. The difference between the actual release at Reservoir C in this case, and the demand, is a residual surplus and is entered in Column 23.

The total storage at the end of the month is obtained by adding the quantities in storage in each of the reservoirs, as given in Columns 13, 17, and 21. This total is entered in Column 25. It is used to indicate the type of operation necessary during the month following. The total storage can never drop below the gate level, which is assumed to be at the top of the sediment storage, which in this example is 600,000 acre-feet. When the storage is less than 1,100,000 acre-feet, releases are rationed. When the storage rises above 2,600,000 acre-feet, the flood release of 10,000 cfs is scheduled until the storage drops below this figure again. Finally, a brief notation is made at the end of the row concerning the type of operation employed during the month.

A sample sheet of computations for two years is given in Table 11–1.

11–8. Degree of Control. The effectiveness of the regulation that the reservoirs provide for the stream depends upon how well the demands on the system are met. By building sufficiently large reservoirs, it would be possible to control the river completely, but such a degree of regulation would usually be economically unjustifiable.

The degree of control afforded by any given system of reservoirs oan be judged both in terms of duration and in terms of quantity. The *degree of duration control* may be defined as the per cent of the total period of study during which releases are at a rate equal to or exceeding the rate of demand. The *degree of discharge control* may be defined as the total *usable* release in per cent of the total demand. These two measures of degree of control would almost always be of the same order of magnitude, though the degree of discharge control would usually be higher than the degree of duration control, especially under the system of rationed releases. It might be desirable to compute degrees of control in terms of ration releases also.

11–9. Other Methods of Operation. The example given above was based on a method of reservoir operation with water conservation and flood control as primary purposes. Silt storage was also contemplated.

However, the general principles and computation procedure may easily be adapted to other methods of operation. If power development is included, then it will be desirable to set a minimum storage and discharge from each reservoir where a hydroelectric installation is planned. The product of the minimum head and discharge will then represent the primary or firm power available. The maintenance of such minimum head and discharge will of course reduce the degree of control afforded as far as water conservation purposes are concerned, and it will be a matter of economic study to balance the cost of the power generated against the cost of the water lost thereby. However, there is no difficulty in making the river operation analysis itself when provisions for power development are included. The tabulations of actual reservoir stage and release may of course be examined to determine the total secondary power that the actual reservoir operation can develop.

Operation of a reservoir for navigation maintenance is somewhat parallel to that for power. A certain minimum release must be set, sufficient to maintain navigable depths of water in the downstream channels. Again, this will tend to decrease the degree of control for water conservation purposes that the system can provide. Provision of recreational facilities would mostly involve maintenance of a minimum lake water level, as does the operation of a reservoir with the purpose of benefiting fish and wildlife. These operations are also somewhat parallel and consistent with operation for purposes of power generation.

It should also be noted that a systems analysis approach involving dynamic programming techniques, may sometimes be employed effectively in the design of a system of multi-purpose reservoirs, provided all the necessary

	1	2	3	4	5	6	7	8	9	10	11	12	13	14
													800	
Jan.	60	125	199	150	49	—	+49	+60	−16	+5	860	5	855	0
Feb.	45	62	150	300	—	150	−150	+45	−200	+5	900	5	895	0
March	32	60	124	350	—	226	−226	−162	−69	+5	733	5	728	194
April	54	83	116	250	—	134	−67	−77	+5	+5	651	5	646	131
May	90	115	145	240	—	95	−15	−25	+5	+5	621	5	616	115
June	85	111	128	180	—	52	+8	−2	+5	+5	614	5	609	87
July	70	81	130	150	—	20	+30	+20	+5	+5	629	5	624	50
Aug.	60	89	188	165	23	—	+78	+60	+13	+5	684	5	679	0
Sept.	100	215	364	300	64	—	+73	+71	−3	+5	750	5	745	29
Oct.	122	223	325	264	61	—	+61	+51	+5	+5	796	5	791	71
Nov.	144	198	302	197	105	—	+105	+95	+5	+5	886	5	881	49
Dec.	130	204	284	243	41	—	+41	+31	+5	+5	912	5	907	99
Total	992	1566	2455	2789	343	677	−13	+167	−240	+60		60		825

	1	2	3	4	5	6	7	8	9	10	11	12	13	14
Jan.	254	424	501	125	376	—	+376	+254	+117	+5	1161	5	1156	0
Feb.	242	430	462	120	342	—	+342	+49	+288	+5	1205	5	1200	193
March	465	595	725	120	605	—	+605	+5	+110	+490	1205	5	1200	460
April	282	315	405	280	125	—	+30	+5	+5	+20	1205	5	1200	277
May	708	925	988	110	878	—	+385	+105	+217	+63	1305	5	1300	603
June	410	480	520	200	320	—	−83	−95	−28	+40	1205	5	1200	505
July	322	415	482	182	300	—	−121	+5	−174	+48	1205	5	1200	317
Aug.	185	224	287	177	110	—	−121	+5	+5	−131	1205	5	1200	180
Sept.	160	218	276	204	72	—	+15	+5	+5	+5	1205	5	1200	155
Oct.	132	153	175	208	—	33	−33	+5	+5	−43	1205	5	1200	127
Nov.	122	137	164	253	—	89	−89	+5	+5	−99	1205	5	1200	117
Dec.	125	142	180	225	—	45	−45	+5	+5	−55	1205	5	1200	120
Total	3407	4458	5165	2204	3128	167	+1261	+353	+560	+348		60		3054

economic data and their functional relationships to the hydrologic and hydraulic parameters are adequately defined. For information on this approach, refer to the bibliography at the end of the chapter.

It should be stressed again that this discussion has been concerned primarily with long-period reservoir operation analyses made in connection with preliminary studies for the purpose of evaluating proposed water control developments of one kind or another. Details of operation involving short-term studies are not primarily in view, although the computation method described can be adapted to these as well.

11–10. Flood-Routing Through Reservoir.

The amount of storage capacity to provide for flood detention should be determined by the process of *routing* the design flood through the reservoir. The design flood recurrence

11-1
tion Study
sands of acre-feet)

15	16	17	18	19	20	21	22	23	24	25	Operation
		500				200				1500	
484	5	479	81	205	5	200	150	0	0	1534	Release = Demand
279	5	274	217	205	5	200	300	0	0	1369	Release = Demand
205	5	200	291	205	5	200	350	0	0	1128	Release = Demand
205	5	200	155	205	5	200	183	0	67	1046	$\frac{1}{6}$Mo.R = D. $\frac{4}{5}$Mo.R = $\frac{2}{3}$D.
205	5	200	135	205	5	200	160	0	80	1016	Release = $\frac{2}{3}$ Demand
205	5	200	108	205	5	200	120	0	60	1009	Release = $\frac{2}{3}$ Demand
205	5	200	56	205	5	200	100	0	50	1024	Release = $\frac{2}{3}$ Demand
213	5	208	16	205	5	200	110	0	55	1087	Release = $\frac{2}{3}$ Demand
205	5	200	147	205	5	200	291	0	9	1145	$1\frac{1}{2}$Mo.R = $\frac{2}{3}$D. $1\frac{1}{2}$Mo.R = D.
205	5	200	167	205	5	200	264	0	0	1191	Release = Demand
205	5	200	98	205	5	200	197	0	0	1281	Release = Demand
205	5	200	168	205	5	200	243	0	0	1307	Release = Demand
	60		1639		60		2468		321		

15	16	17	18	19	20	21	22	23	24		Operation
317	5	312	53	205	5	200	125	0	0	1668	Release = Demand
600	5	595	93	205	5	200	120	0	0	1995	Release = Demand
705	5	700	480	690	5	685	120	0	0	2585	Release = Demand
705	5	700	305	705	5	700	375	95	0	2600	$\frac{1}{8}$Mo.R = D. $\frac{7}{8}$Mo.(I.F. − Evap.)
917	5	912	603	763	5	758	603	493	0	2970	Release = 10,000 cfs.
884	5	879	603	798	5	793	603	403	0	2872	Release = 10,000 cfs.
705	5	700	584	841	5	836	603	421	0	2736	Release = 10,000 cfs.
705	5	700	214	705	5	700	408	231	0	2600	$\frac{2}{5}$Mo.R = 10,000 cfs. $\frac{3}{5}$Mo.R = I.F. − Evap.
705	5	700	208	705	5	700	261	57	0	2600	R = I.F. minus Evaporat.
705	5	700	143	657	5	652	208	0	0	2552	Release = Demand
705	5	700	127	553	5	548	253	0	0	2448	Release = Demand
705	5	700	132	493	5	488	225	0	0	2388	Release = Demand
	60		3545		60		3904	1700	0		

interval is selected on the basis of economic considerations, probably that expected to occur about once or twice during the anticipated economic life of the reservoir. The spillway is designed, of course, to pass much larger floods, approaching in magnitude the maximum *possible* flood at the site, although this again must be based on a careful cost-benefit analysis.

Flood-routing through reservoirs thus serves to aid in the determination of the size of reservoirs, spillways and other outlet works. It also provides information regarding the timing of flood flows—it permits calculation of the time delay of the peak of the flow as it passes through the reservoir. The prediction of the timing of floods thus obtained is part of flood forecasting along with the prediction of peak flow and stage.

In either case, the design flood hydrograph is first determined, by means of the unit hydrograph principle or by some other method, and then routed, incrementally, through the proposed spillway or outlet conduits, obtaining a hydrograph of flood releases and reservoir stages during the passing of the flood. If necessary, the reservoir storage capacity or the outlet works or both are redesigned to yield desired values of these data.

The routing procedure is again based on the hydrologic bookkeeping equation:

$$S_1 + I\,\Delta T - \frac{1}{2}\,(O_1 + O_2)\,\Delta T = S_2 \qquad (11\text{--}6)$$

in which S_1 and S_2, O_1 and O_2, represent the volume in storage and the reservoir release rate, at the beginning and end of the time increment ΔT, respectively. The average flood inflow rate during ΔT is $I = (I_1 + I_2)/2$. The increment ΔT is commonly taken as about one-twentieth of the total base length of the flood hydrograph; this division will normally provide sufficient detail in the outflow hydrograph. The flood is assumed as occurring on top of a reservoir already filled to the top of its regulation storage, if any, so that the designated flood release operation may begin simultaneously with the onset of the flood hydrograph.

The flood release may be either controlled or uncontrolled. If controlled, usually by gate or valve structures, the release is determined by the pre-established operation schedule. The analysis, in this case, is identical with that in the preceding article, except that the time increment used is much shorter.

If the outflow is uncontrolled (except by the hydraulic factors), then O is a function of (1) the stage in the reservoir and (2) the hydraulics of the outlet system. Thus,

$$O = C_1 E^m \qquad (11\text{--}7)$$

in which E is the reservoir surface elevation in feet, measured above a datum plane such that E is also the effective hydraulic head producing flow; C is a coefficient determined from the area, geometrical shape, and friction characteristics of the outlet structure; and m is an exponent similarly determined by the hydraulics of the outlet.

Thus, if the outlet system is a typical overfall spillway, E is measured above the spillway crest and

$$O \cong 3.9 L E^{1.5} \qquad (11\text{--}8)$$

If a conduit is used, then the hydraulics of culverts become applicable, and E may be measured above either the inlet or outlet center line, depending on whether the conduit will flow part full or full. In either case, the outflow equation is of the form

$$O = KAE^{0.5} \qquad (11\text{--}9)$$

where A is the cross-sectional area and K is determined from the hydraulics of the particular conduit.

Similarly, the storage S can be expressed as a function of the same elevation E, as determined from the topographic characteristics of the reservoir site. A curve of S vs. E can be plotted from topographic data, and

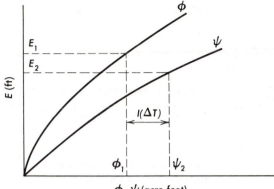

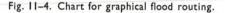

Fig. II-4. Chart for graphical flood routing.

will usually be found to approximate an exponential function of the form

$$S = C_2 E^n \tag{11-10}$$

n often being of the order of magnitude 2.5 and C_2 of course depending directly on the particular topography.

Equation (11-6) can now be rearranged to

$$\phi_1 + I \Delta T = \psi_2 \tag{11-11}$$

where

$$\phi = S - \tfrac{1}{2}(O)(\Delta T) = C_2 E^n - C_1 \frac{\Delta T}{2} E^m \tag{11-12}$$

and

$$\psi = S + \tfrac{1}{2}(O)(\Delta T) = C_2 E^n + C_1 \frac{\Delta T}{2} E^m \tag{11-13}$$

Both ϕ and ψ are computed from Eqs. (11-7) and (11-10) as functions of the elevation E, and can be plotted as in Fig. 11-4.

The routing process is basically a solution of Eq. (11-6) for S_2 for successive time increments. But since O_2 is also unknown, this requires simultaneous solution of Eq. (11-7), which in effect gives O_2 as another function of S_2. Use of the graphs in Fig. 11-4 circumvents what would otherwise be a trial-and-error solution.

The curves are entered with the initial elevation E_1, from which ϕ_1 is read. To this is added $I \Delta T$ for the interval, yielding ψ_2 in accordance with Eq. (11-11). This value on the ψ-curve gives the elevation E_2 at the end of the interval.

It may be noted that ϕ may have negative values at low values of E, specifically when $E < \left(\dfrac{C_1 \Delta T}{2C_2}\right)^{m/n}$. This may be avoided by using small values of ΔT when the reservoir stage is low. However, it is normally sufficiently accurate simply to neglect such negative values in the routing calculations.

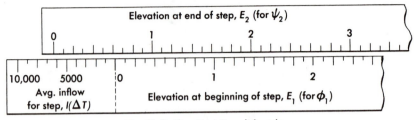

Fig. 11-5. Flood routing slide rule.

An alternative procedure is to prepare a simple cardboard slide rule, as shown in Fig. 11-5. On one scale is plotted the function ψ, in acre-feet, except that the corresponding elevations are ruled on the scale instead of ψ values. On the other scale, the function ϕ is plotted to the right of a zero index and the function $I\,\Delta T$ to the left, both in acre-feet, to the same scale as the ψ-scale. Again, however, elevations corresponding to the ϕ-values are ruled on the scale instead of the ϕ-value themselves.

To make a step computation, the zero on the ψ-scale is set opposite the value of $I\,\Delta T$ for the interval on the ϕ-scale. The final elevation E_2 is then read off the ψ-scale opposite the initial elevation E_1 on the ϕ-scale. In Fig. 11-5, if the initial elevation is 2.00 ft, and $I\,\Delta T$ is 6300 acre-feet, the final elevation is 2.62 ft.

Since discharge O is a function of the elevation E, ϕ and ψ can be computed as functions of O, and the graph of Fig. 11-4, or the slide rule depicted in Fig. 11-5, modified to have values of O, in the desired flow units, in place of E.

The processes described above yield either the stage hydrograph, the outflow hydrograph, or both through Eq. (11-7). It remains to determine whether the stage for the flow thus routed is within the permitted levels of the proposed reservoir, and whether the discharge over the spillway is within prescribed design values downstream of the structure. If not, other designs must be made, either of the spillway, or the downstream works or both.

Thus flood routing through reservoirs describes both outflow discharge and storage as increasing functions of elevation, neglecting the hydrodynamic effects of the flow. Thus it can be readily seen that the hydrographs of inflow to, and outflow from, the reservoir must take the relative positions shown in Fig. 11-6. In that diagram the peak outflow occurs at the time where the outflow graph crosses the inflow graph. This must also be the point of greatest storage, since

$$I - O = \frac{dS}{dt} \tag{11-14}$$

the differential form of the storage equation, Eq. (11-6).

11-11. Equivalent Uniform Flood. A more rapid, though less accurate, means of arriving at the flood storage requirement, is to approximate the design flood hydrograph by an *equivalent uniform flood*, which is a flood

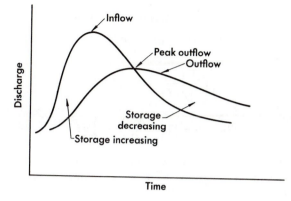

Fig. 11–6. The relation between inflow and outflow hydrographs for a reservoir.

of arbitrary base length T and of uniform intensity I, such that the total flood volume IT is the same as in the actual flood.   (See Fig. 11–7.) If the entire flood were caught and the outflow reduced to zero, the required storage volume would be IT. The actual storage volume needed will of course depend on the permitted maximum outflow O, on the storage-elevation relationship for the reservoir, as expressed by Eq.(11–10), and on the outflow-elevation relationship, as given by Eq.(11–7). The ratio of the permitted outflow to the uniform inflow is the *outflow ratio*, x. Thus,

$$x = \frac{O}{I} \tag{11–15}$$

The ratio of the required storage to the total flood volume is the *detention ratio*, d. Therefore,

$$\text{Required storage volume} = dIT \tag{11–16}$$

The detention ratio depends on x, m, and n. To evaluate d quantitatively, the bookkeeping equation is expressed in differential form:

$$I\,dt = O\,dt + dS \tag{11–17}$$

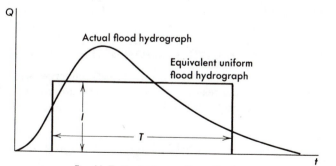

Fig. 11–7. Equivalent uniform flood.

Since $O = C_1 E^m$ and $S = C_2 E^n$, Eq. (11–17) becomes

$$I \, dt = C_1 E^m \, dt + n C_2 E^{n-1} \, dE \tag{11-18}$$

Rearranging gives

$$I \, dt \left(1 - \frac{C_1 E^m}{I} \right) = n C_2 E^{n-1} \, dE$$

Therefore,

$$dt = \left(\frac{n C_2}{I} \right) \left(\frac{E^{n-1}}{1 - C_1 E^m / I} \right) dE \tag{11-19}$$

This can be regarded as the differential equation of the reservoir stage hydrograph. It can be integrated by first expanding into an infinite series by direct division:

$$dt = \frac{n C_2}{I} \left[E^{n-1} + \left(\frac{C_1}{I} \right) E^{m+n-1} + \left(\frac{C_1}{I} \right)^2 E^{2m+n-1} + \left(\frac{C_1}{I} \right)^3 E^{3m+n-1} + \cdots \right] dE$$

By integration, the equation of the stage hydrograph becomes

$$t = \frac{n C_2}{I} \left[\left(\frac{E^n}{n} \right) + \left(\frac{C_1}{I} \right) \left(\frac{E^{m+n}}{m+n} \right) + \left(\frac{C_1}{I} \right)^2 \left(\frac{E^{2m+n}}{2m+n} \right) + \left(\frac{C_1}{I} \right)^3 \left(\frac{E^{3m+n}}{3m+n} \right) + \cdots \right] \tag{11-20}$$

Reintroducing $O = C_1 E^m$ and $S = C_2 E^n$ yields

$$t = \frac{Sn}{I} \left[\left(\frac{1}{n} \right) + \left(\frac{O}{I} \right) \left(\frac{1}{m+n} \right) + \left(\frac{O}{I} \right)^2 \left(\frac{1}{2m+n} \right) + \left(\frac{O}{I} \right)^3 \left(\frac{1}{3m+n} \right) + \cdots \right] \tag{11-21}$$

Now, when t reaches the value T, at the end of the equivalent uniform flood, the stage, storage, and outflow will all attain their maximum values. These values can be inserted in Eq. (11–21) with the outflow ratio, x, replacing the fraction O_{max}/I. Thus,

$$IT = S_{max} n \left[\frac{1}{n} + \frac{x}{m+n} + \frac{x^2}{2m+n} + \frac{x^3}{3m+n} + \cdots \right] \tag{11-22}$$

By Eq. (11–16), the detention ratio, d, is S_{max}/IT, and so

$$d = \frac{1}{n[1/n + x/(m+n) + x^2/(2m+n) + x^3/(3m+n) + \cdots]} \tag{11-23}$$

$$= \frac{1}{1 + x/(1 + m/n) + x^2/(1 + 2m/n) + x^3/(1 + 3m/n) + \cdots} \tag{11-24}$$

For ordinary values of x, m, and n, Eq. (11–24) converges satisfactorily, and thus d can be computed readily. In fact, the ratio m/n can be considered as a single parameter p, and d as a function of only x and p.

$$d = \frac{1}{1 + x/(1 + p) + x^2/(1 + 2p) + x^3/(1 + 3p) + \cdots} \tag{11-25}$$

Thus, with Eqs. (11–15) and (11–25), the storage capacity can be quickly computed for a number of proposed outflow values, and these then subjected to an economic analysis. Values of d for various x and p values have been obtained by J. C. May by using a digital computer and are given in Table 11–2. Equation (11–25) sometimes converges very slowly and a large number of terms must then be used in the expansion for accuracy.

The results of this type of analysis will obviously be influenced to some extent by the selected base length T, which in turn determines I. If Eqs. (11–16) and (11–25) are combined as follows:

$$S_{\text{req}} = (\text{total inflow volume}) \, \frac{1}{1 + (O/I)/(1 + p) + (O/I)^2/(1 + 2p) + \cdots}$$

$$(11\text{–}26)$$

it is immediately evident that the indicated storage requirement increases as I increases, and thus as T decreases, becoming equal to the total volume of the flood for a zero base length.

TABLE 11–2
Detention Ratios

$p = \dfrac{m}{n}$	Outflow Ratio, x							
	0.2	0.3	0.4	0.5	0.6	0.7	0.8	0.9
0.1	0.818	0.726	0.635	0.543	0.450	0.356	0.260	0.160
0.5	0.864	0.794	0.722	0.647	0.569	0.486	0.395	0.289
1.0	0.896	0.841	0.783	0.721	0.655	0.581	0.497	0.391
1.5	0.916	0.870	0.822	0.769	0.711	0.645	0.567	0.465
2.0	0.929	0.890	0.848	0.802	0.751	0.691	0.619	0.522
2.5	0.939	0.905	0.868	0.827	0.781	0.727	0.660	0.567
3.0	0.946	0.916	0.883	0.846	0.804	0.755	0.693	0.605

It will usually be found, however, that if the base length is taken as between $\frac{2}{3}$ and $\frac{7}{8}$ of the duration of the actual inflow hydrograph, the resulting storage requirement will not vary greatly in any case from that indicated by step-routing methods. In view of uncertainties in the design inflow hydrograph, not to mention uncertainties regarding future flood hydrographs, the accuracy may be quite satisfactory. One fairly reasonable basis for selecting T is to assume the equivalent uniform flood ends at the point where the inflow is equal to the specified maximum outflow, and that it begins at such time as the inflow hydrograph slope begins to rise rapidly.

The parameter p depends on the exponents m and n in the outflow and storage equations, respectively. The exponent m depends on the outlet work hydraulics, varying from 0.5 for an orifice to 1.5 or more for a rectangular spillway overfall, or even 2.5 for a V-notch weir. The exponent n depends on the topography and may in some cases vary with elevation. However, it is usually fairly constant in the upper range of elevations, with which most flood-routing problems are concerned. In the absence of exact data for the particular site, the following values of n may be applied:

Site Topography	n
Lake-type topography	1.0 to 1.5
Flood-plain and foothill topography	1.5 to 2.5
Hill-type topography	2.5 to 3.5
Gorge-type topography	3.5 to 4.5

This technique of flood storage determination has the advantage of requiring neither prior design of the outflow works nor complete topographic data. Only such data as needed to estimate m and n are required.

It may be noted that, if the outflow is controlled manually by gates, then the outflow equation, Eq.(11-7), no longer applies. The outflow is no longer controlled by the reservoir stage, but by the downstream conditions permitted. If the rated release O is maintained constant by manipulation of the gates, then the total storage volume required would be simply $(I - O)T$.

There are a number of other flood-routing methods and techniques that have been developed and many are available in the periodical literature. The step-routing procedure is basic and is the most accurate. For preliminary studies, the equivalent uniform flood method as described is typical of the various special approximate methods and is probably as rapid and accurate as any.

11-12. Flow-Routing Through Rivers. Just as flow routing through reservoirs is used to determine sizes and the design of reservoirs and appurtenant hydraulic works, there are problems of design of structures associated with rivers that require flood hydrograph information. Examples are levees, by-pass channels and bridge openings. The process of determining outflow, stage and the timing of the flood is accomplished by flood routing as in the case of reservoirs except that the storage equation is expressed for a reach of river, and thus storage is not a function of outflow alone (or the elevation of the reservoir). A common method of flood-routing in rivers is the Muskingum method, where the relationship for the storage in a reach of river is given as

$$S = KO + Kx(I - O) \qquad (11\text{--}27)$$

The first term on the right side of Eq. (11–27) is called the "prism storage"; it represents the volume that would be stored in the reach if the discharge were uniform throughout. K has derived dimensions of time, and is interpreted as the lag time, or time of travel, through the reach of river. The second term on the right side of Eq. (11–27) is called "wedge storage." It is the storage increment in addition to prism storage that results from the inflow to the reach being different than the outflow at a point in time. The coefficient Kx, where x is less than unity, gives the portion of the potential storage expressed by $(I - O)$ that is wedge storage. The assumption is made that K and x are constant for the reach of river.

If Eq. (11–27) is substituted into Eq. (11–6), with appropriate subscripts, the result is

$$O_2 = C_1 I_1 + C_2 I_2 + C_3 O_1 \tag{11-28}$$

where

$$C_1 = \frac{Kx + \frac{1}{2}\Delta T}{K - Kx + \frac{1}{2}\Delta T}$$

$$C_2 = -\frac{Kx - \frac{1}{2}\Delta T}{K - Kx + \frac{1}{2}\Delta T} \tag{11-29}$$

$$C_3 = \frac{K - Kx - \frac{1}{2}\Delta T}{K - Kx + \frac{1}{2}\Delta T}$$

The so-called coefficients, C_1, C_2, and C_3, can be used to calculate outflow hydrographs from Eq. (11–28) if the inflow hydrograph and an initial value of outflow are known.

To determine K and x for a reach of river, data of inflow and corresponding outflow are needed. Rearranging Eq. (11–27) gives

$$S = K[xI + (1 - x)O]$$

Values of storage, S, determined by

$$S = \sum (I - O) \Delta T \tag{11-30}$$

from the data, can be plotted against the value of $[xI + (1 - x)O]$ for assumed values of x. The value of x selected is that which gives the plotted data the most single-valued line. K is the slope of the line through the data.

Example 11-2

Given below in Table 11–3 are data of the James River at Holcombs Rock, Va., mile 263.2; and of the James River at Bent Creek, Va., mile 222.9 for December 16, 1961 to December 28, 1961. The Bent Creek data were modified to eliminate the flow added between stations by prorating according to area. Find the value of x and K in the Muskingum method for the reach of river.

TABLE 11-3

Computations for Determination of Muskingum K and x for the James River, between Holcombs Rock and Bent Creek, Va.

Date	James River at Holcombs Rock, Va. cfs (I)	James River at Bent Creek, Va. cfs (O)	ΔS cfs-days	S cfs-days	$xI + (1 - x)O$ cfs
Dec. 16	6,980	8,110	—	—	7,880
17	7,640	8,110	—	—	8,020
18	22,800	16,500	6300	6300	17,800
19	32,500	29,300	3200	9500	29,900
20	18,500	21,200	−2700	6800	20,700
21	11,900	13,300	−1400	5400	13,000
22	8,990	9,900	−910	4490	9,720
23	7,200	8,100	−900	3590	7,920
24	6,330	7,100	−770	2820	6,950
25	5,280	6,400	−1120	1700	6,180
26	4,450	5,160	−710	990	5,020
27	4,080	4,400	−320	670	4,340
28	4,000	4,380	−380	290	4,300

The values of $xI + (1 - x)O$ are computed for a trial value of $x = 0.25$. The graph of Fig. 11–8 shows the near linear relationship of S and $xI + (1 - x)O$ through a significant portion of the range. Values of x are selected by a trial-and-error procedure. It is of interest to note that the instantaneous peaks and times of peak at Holcombs Rock and Bent Creek were 34,500 cfs, 0730, December 19 and 30,600 cfs, 1700, December 19 respectively. A line through the data points gives a value of $K = 0.26$ day.

11–13. Flow-Routing by Characteristic Concepts.

The methods of flow routing presented in the previous articles were based on the continuity, or bookkeeping, equation. A more mathematical method exists, based on the continuity equation and the equation of motion, which permits computation of the hydrograph (stage or discharge) at any point in the stream as well as the water surface profile for any time. The solutions are accomplished by finite-difference computation techniques, all related to the method of characteristics. A general derivation follows.

In a differential length of channel, dx, the discharge changes at the rate $\left(\dfrac{\partial Q}{\partial x} \, dx\right) dt$, while the storage changes at the rate (to a first-order approximation) $\left(\dfrac{\partial A}{\partial t} \, dt\right) dx$. With no lateral inflow or outflow to the channel the sum of these two terms must equal zero. When the rate of lateral inflow is q per unit length of channel

$$\frac{\partial Q}{\partial x} + \frac{\partial A}{\partial t} = q$$

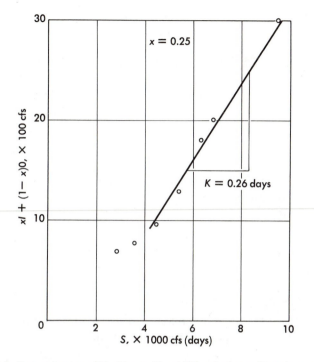

Fig. 11–8. Determination of Muskingum K and X for the James River between Holcombs Rock and Bent Creek, Va.

which becomes

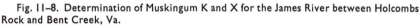

$$uA_x + Au_x + A_t = q \qquad (11\text{–}31)$$

where u is the one-dimensional flow velocity

A is the flow area

x is the distance variable

t is the time variable.

Subscripts denote partial differentiation.

The equation of motion can be derived by considering forces on the differential length of channel and the resulting accelerations. The forces on a unit mass of fluid are

$$-\gamma \frac{\partial h}{\partial x} dx\, A$$

$$+\gamma S_0 A\, dx$$

$$-\gamma S_f A\, dx$$

$$-\rho q u\, dx$$

which are respectively: the unbalanced pressure force in the downstream direction, the component of the weight of the element of fluid in the downstream direction, the friction force expressed as a component of weight, and the force required to give the incoming fluid the momentum of the flow. These forces must balance the acceleration of the fluid, therefore

$$-\gamma \frac{\partial h}{\partial x} A\, dx + \gamma S_0 A\, dx - \gamma S_f A\, dx - \rho q u\, dx = \rho \frac{du}{dt} A\, dx$$

$$= \rho\left(\frac{\partial u}{\partial t} + u\frac{\partial u}{\partial x}\right) A\, dx$$

and then

$$g h_x + g(S_f - S_0) + \frac{qu}{A} + u_t + u u_x = 0 \qquad (11\text{--}32)$$

The assumptions that govern the equations are those of one-dimensional flow and a hydrostatic pressure distribution. The area A is a function of the flow depth, h, so

$$A_x = \frac{dA}{dh} h_x \qquad (11\text{--}33)$$

and

$$A_t = \frac{dA}{dh} h_t \qquad (11\text{--}34)$$

Addition of Eqs. (11–31) and (11–32), and use of the multiplier λ in Eq. (11–31) to ensure homogeneity of terms, results in

$$\lambda \frac{dA}{dh}\left[h_t + \left(u + \frac{g}{\lambda \dfrac{dA}{dh}}\right)h_x\right] + [u_t + (u + \lambda A)u_x]$$

$$+ g(S_f - S_0) + \frac{qu}{A} - q\lambda = 0 \quad (11\text{--}35)$$

Examination of Eq. (11–35) reveals that the terms in brackets are total differentials of u and h, and that

$$\frac{dx}{dt} = u + \frac{g}{\lambda \dfrac{dA}{dh}} = u + \lambda A \qquad (11\text{--}36)$$

which leads to

$$\lambda = \pm \frac{1}{A}\sqrt{\frac{gA}{\left(\dfrac{dA}{dh}\right)}} = \pm \frac{c}{A} \qquad (11\text{--}37)$$

where c is the wave speed, or celerity. We note that if the channel is of rectangular cross-section, or if the channel is very wide, the definition of the wave speed, as

$$c = \sqrt{\frac{gA}{\left(\dfrac{dA}{dh}\right)}} \qquad (11\text{--}38)$$

reduces to

$$c = \sqrt{gh} \tag{11-39}$$

Substitution of Eq. (11–37) into Eqs. (11–35) and (11–36) produces

$$\frac{dx}{dt} = u + c$$
$$\frac{dx}{dt} = u - c \tag{11-40}$$

and

$$\frac{1}{g}\frac{du}{dt} \pm \frac{1}{c}\frac{dh}{dt} + (S_f - S_0) + \frac{g}{gA}(u \mp c) = 0 \tag{11-41}$$

Differentiation of Eq. (11–38) yields

$$\frac{dh}{dt} = \frac{2c}{g}\frac{dc}{dt} \tag{11-42}$$

and substitution into Eq. (11–41) results in

$$\frac{du}{dt} \pm 2\frac{dc}{dt} + g(S_f - S_0) + \frac{q}{A}(u \pm c) = 0 \tag{11-43}$$

and thus

$$\left[\frac{\partial}{\partial t} + (u + c)\frac{\partial}{\partial x}\right](u + 2c) + g(S_f - S_0) + \frac{q}{A}(u - c) = 0$$

and
$$\tag{11-44}$$
$$\left[\frac{\partial}{\partial t} + (u - c)\frac{\partial}{\partial x}\right](u - 2c) + g(S_f - S_0) + \frac{q}{A}(u + c) = 0$$

Eqs. (11–40) and (11–44) form pairs of differential equations in which the first of Eqs. (11–44) is differentiated along paths in the x-t plane described by the first of Eqs. (11–40), and similarly for the second of the equations in the two groups. The partial differential Eqs. (11–31) and (11–32) were transformed into total differential equations linear in the differential terms. Eqs. (11–40) are the characteristic equations.

Solution of the system of equations written above can be accomplished in the x-t plane if appropriate initial and boundary values are prescribed. Initial values of u and c (or alternately h) must be given on the $t = 0$ boundary of the x-t plane. (Several authors referenced at the end of this chapter describe the properties of the solutions.)

Eqs. (11–40) and (11–43) can be written in finite-difference form, viz.,

$$\frac{\Delta x}{\Delta t} = u \pm c \tag{11-45}$$

$$\Delta u \pm 2\Delta c + g(S_f - S_0)\,\Delta t + \frac{q}{A}(u \pm c)\,\Delta t = 0 \tag{11-46}$$

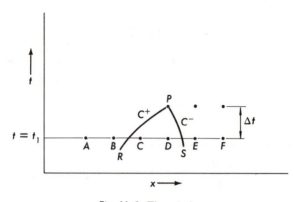

Fig. 11-9. The x-t plane.

In such a form, the solution can be constructed through a grid in the x-t plane, estimating values of u and c in the last term by given values on the boundary or previously calculated values. The two equations permit explicit solution of Δu and Δc. The particular given values of u and c previously determined, or those on the boundary, that are to be used in the estimation are determined by examination of the values of Eqs. (11–40).

The concept of the solution is most easily seen using a fixed grid in the x-t plane, and explicitly solving for u and c. Referring to Fig. 11-9, we assume that the values of u and c are known at the points $A, B, \ldots, F$ at time $t = t_1$, and that we wish to compute the values of u and c at point P. When the first of Eqs. (11–45) and (11–46) are used we approximate $\Delta u \cong u_P - u_R$ and $\Delta c \cong c_P - c_R$. When the second of Eqs. (11–45) and (11–46) are considered the approximations are $\Delta u \cong u_P - u_S$ and $\Delta c \cong c_P - c_S$. In the first case $\Delta x \cong x_P - x_R$ and in the second $\Delta x \cong x_P - x_S$. In both cases $\Delta t = t_P - t_R = t_P - t_S$. Making the substitutions and solving explicitly for u_P and c_S, the unknown values of the dependent variables at point P, we obtain

$$u_P = \frac{(u_R + u_S)}{2} + (c_R - c_S) - g(S_f - S_0)\,\Delta t - \frac{qu}{A}\,\Delta t = 0 \qquad (11\text{–}47)$$

and

$$c_P = \frac{(u_R - u_S)}{2} + (c_R + c_S) + \frac{qc}{A}\,\Delta t = 0 \qquad (11\text{–}48)$$

Values of u_R, u_S, c_R and c_S can be interpolated from values known at A, $B, \ldots, F$. Selection of the appropriate points to be used involves the approximation of the slope of the C^+ and C^- curves by Eq. (11–45), where

$$C^+ : \frac{\Delta x}{\Delta t} = u + c$$

$$C^- : \frac{\Delta x}{\Delta t} = u - c$$

The values of u and c used in evaluation of the friction slope, the last terms in Eqs. (11–47) and (11–48), and in Eqs. (11–45) are estimated by values of those variables at the appropriate points A, B, . . . , F.

In the manner described above, the solution (computation of the variables u and c) can proceed throughout the entire x-t plane, providing initial values and appropriate boundary conditions are given. The result is more than simple flood routing although it is primarily described as such. It gives water-surface profiles for any time in the x-t plane. Other computational schemes have been developed and described in the literature. The significant features of all the methods are presented above, namely, (1) approximating the values of the dependent variables in the region of computation indicated by the C^+ and C^- curves from known or given values at an earlier time step, and (2) determining which known or given values to use by estimation of the slopes of the C^+ and C^- curves.

The difficulty of flow routing by the method and its derivatives described in this article is the gathering of the data which characterizes the stream cross sections and roughness. Values of dA/dh and A as functions of h are needed for the celerity. Depending on the empirical roughness formula used, values of cross section parameters such as hydraulic radius are needed as well as roughness coefficients. Despite the difficulties that determination of these constants present, the methods that have grown out of the characteristic concepts are useful now and promise increasing utility.

PROBLEMS

11–1. Determine the regulation afforded by a reservoir for different rated outflows, for a given stream flow record, as presented in the table below, which shows corrected stream flow record, by months, of the Kettle River, for the eight-year period from 1909 through 1916. Assume a reservoir capacity of 3 billion cu ft.

Kettle River Discharge, in Billions of Cubic Feet

	1909	1910	1911	1912	1913	1914	1915	1916
January	3.11	0.50	0.10	0.50	0.10	0.40	0.20	0.00
February	0.30	1.45	0.30	0.10	0.20	0.10	0.30	0.00
March	0.32	1.00	0.60	0.20	0.20	0.40	0.30	0.50
April	1.74	1.40	1.30	3.90	3.00	2.10	2.10	10.00
May	3.88	0.95	2.30	6.70	3.20	4.00	3.10	6.50
June	1.90	0.50	1.40	1.50	1.20	4.90	2.30	3.00
July	0.50	0.10	1.10	0.50	4.10	3.00	1.30	2.80
August	3.00	0.30	0.70	0.30	0.90	0.70	0.30	0.30
September	0.85	0.20	0.90	0.50	0.80	1.40	0.30	0.50
October	0.80	0.30	1.10	0.20	1.40	0.80	0.90	0.60
November	1.10	0.30	0.50	0.20	1.20	0.40	2.90	0.50
December	0.50	0.20	0.40	0.20	0.60	0.30	1.90	0.80

Required:

a. Mass curve of unregulated stream flow, plotted on large cross-section paper, to scales of 1 in. = 5 months horizontally and 1 in. = 10 billion cu ft vertically.

   b. Determination of dependable outflow that can be maintained 100 per cent of the time, by a graphical study of the drought periods indicated on the mass curve. Assume reservoir full at beginning of period.

   c. Complete regulation diagram for the outflow of (b) above, including mass curve of outflow and quantities of waste water. Use red pencil and superimpose on the mass curve of inflow in (a) above.

   d. Complete regulation diagram for a rated outflow of twice the outflow of (b) above, including mass curve of outflow and quantities of waste and deficiency. Use blue pencil and superimpose on mass curve of inflow in (a) above.

   e. Duration curve of unregulated stream flow, in cfs. versus per cent of time equaled or exceeded.

   f. Duration curves of regulated flow, for both types of regulation, superimposed on unregulated flow curve, as obtained from study of surpluses and deficiencies on the mass curves.

   g. Total waste and deficiency for each condition.

   h. Degree of discharge control and degree of duration control for each condition.

11-2. Prove that the maximum dependable flow that can be provided indefinitely by a reservoir of any arbitrary finite capacity, even if the reservoir is initially full, is less than the average flow in the stream.

11-3. The following annual discharges, in thousands of acre-feet, have been recorded at a certain stream gaging station. The magnitudes and frequencies of these flows (but not necessarily their chronological order) are considered to be truly representative of the watershed. What reservoir capacity should be provided to assure a dependable flow of 400,000 acre-feet per year, neglecting losses?

200	150
250	650
700	500
850	350
300	550

11-4. A 20-year record of annual run-off is available at a proposed dam site as shown below (values in thousands of acre-feet):

Year	Flow	Year	Flow
1943	750	1953	190
1944	580	1954	220
1945	470	1955	210
1946	320	1956	310
1947	300	1957	440
1948	410	1958	830
1949	560	1959	600
1950	800	1960	270
1951	820	1961	320
1952	440	1962	840

a. Construct a "residual mass" (cumulative departure from normal) curve for the above record.

b. Determine the storage necessary to control the outflow to a minimum dependable flow equal to the average inflow, assuming an initially full reservoir and neglecting losses. *Ans.:* 1,094,000 acre-ft.

c. A reservoir of what size would be required to assure a dependable flow of 400,000 acre-ft/yr? *Ans.:* 670,000 acre-ft.

d. Assuming an initially empty reservoir of 500,000 acre-ft capacity, what dependable outflow could be provided? *Ans.:* 358,000 acre-ft/yr.

11-5. The following data represent average monthly flows, in cfs, in a certain river at a proposed reservoir site. Determine, by use of a "residual mass" diagram the dependable flow in cfs that could be provided during this year by a 100,000 acre-ft reservoir. Assume the reservoir to be half full at the beginning of the year, and neglect water losses. Also assume that the same monthly flows are repeated each year.

Month	Average cfs
1	210
2	190
3	80
4	170
5	340
6	620
7	700
8	310
9	280
10	250
11	260
12	220

11-6. The monthly inflows to a reservoir and the monthly demands are shown for a certain year. The reservoir has a total capacity of 300,000 acre-feet, of which the top 50,000 is flood storage and the bottom 50,000 is ration storage. Rationed outflows are one-half the demand outflows and flood release is equivalent to 6000 cfs. Tabulate the month-by-month operation of the reservoir system, assuming storage initially of 50,000 acre-ft.

Inflow (acre-ft)	Demand (acre-ft)
78,000	12,000
85,000	12,000
90,000	15,000
95,000	35,000
38,000	80,000
33,000	105,000
21,000	140,000
15,000	95,000
12,000	35,000
18,000	15,000
26,000	12,000
50,000	12,000

11-7. The following tabulation represents an average year of water supply and demand at a certain reservoir. The reservoir has 80,000 acre-ft of ration storage (release = one-half demand), 200,000 acre-ft of normal storage (release = demand), and 20,000 acre-ft of flood storage (release = 3000 cfs maximum).

Month	Inflow (acre-ft)	Demand (acre-ft)
January	42,400	10,000
February	50,800	15,000
March	65,200	30,000
April	123,400	60,000
May	175,000	80,000
June	114,700	125,000
July	59,200	160,000
August	20,300	175,000
September	17,400	125,000
October	45,000	40,000
November	31,200	25,000
December	37,600	15,000

Assume the reservoir is half-full (140,000 acre-ft) at the beginning of the typical year. Compute:

a. Month-by-month operation of the reservoir.
b. Degree of discharge control. *Ans.:* 83.7%.
c. Degree of duration control. *Ans.:* 63.2%.

11-8. A two-reservoir system is to be constructed on a river for the dual purpose of power and water supply. Each reservoir has a capacity of 250,000 acre-ft, and it is assumed that there is no tributary inflow between the reservoirs. The water demand is taken to be constant at 200,000 acre-ft per month, including losses. The power plant is at the lower reservoir and is to be designed for the maximum power capacity that can be developed from the river, consistent with basic operation for water supply purposes. The effective head on the turbines is related to the reservoir storage in the lower reservoir by the equation, $H = 0.15S^{1/2}$.

The unregulated stream flow is assumed, for design purposes, to have the following monthly pattern (flows in thousands of acre-feet per month):

Month	Flow	Month	Flow	Month	Flow
January	100	May	520	September	360
February	100	June	480	October	200
March	180	July	410	November	140
April	300	August	390	December	100

Assume that both reservoirs are full at the end of October each year. Also assume that contract requirements call for a minimum monthly average power output (neglecting all losses) of 15,000 hp. Determine the following:

a. Monthly operation of system through the year, including storage in each reservoir, outflow from system, head on turbines, monthly average horsepower available, and water demand deficiencies.

b. Total energy developed by the power plant through the year, in horse-power-months.

11-9. A two-reservoir system is proposed for water conservation on a certain river, each reservoir to have a capacity of 400,000 acre-ft, to meet a constant demand below the reservoir of 300,000 acre-ft per month. The unregulated river flow at the lower dam site is shown below for a typical year (flows in thousands of acre-feet):

Month	Flow	Month	Flow	Month	Flow
January	150	May	350	September	370
February	130	June	550	October	300
March	100	July	480	November	310
April	150	August	380	December	300

Assume the reservoir system has 600,000 acre-ft of storage initially. Neglect evaporation losses. For the year shown, tabulate the reservoir operation, showing monthly storage in each reservoir and outflow from bottom reservoir. Calculate the degrees of discharge and duration control for the year. The tributary inflow between reservoirs can be neglected for the analysis; however, operation must be such as to allow for unexpected flood runoff from the intermediate drainage area.

11-10. A two-reservoir system is to be constructed on a river, for the dual purpose of power generation and water supply. Each reservoir has a capacity of 400,000 acre-feet. The unregulated flow in the stream at the two sites is shown in the accompanying table, for a given year. The power plant is at the lower reservoir and is to be designed for the maximum power capacity that can be developed from the river, consistent with basic operation for water supply purposes. The effective head on the turbines is related to the reservoir storage in the lower reservoir by the equation: $H = 0.15(S)^{1/2}$.

Month	Flow at A	Flow at B	Demand
January	100,000	110,000	200,000
February	100,000	120,000	200,000
March	180,000	200,000	200,000
April	300,000	330,000	250,000
May	520,000	570,000	250,000
June	480,000	530,000	250,000
July	410,000	450,000	250,000
August	390,000	430,000	250,000
September	360,000	400,000	250,000
October	200,000	220,000	200,000
November	140,000	150,000	200,000
December	100,000	110,000	200,000

Values in acre-feet; demands include losses.

At the beginning of January, assume reservoir A is empty and Reservoir B is full. Determine the following:

    a. Monthly operation of system through the year, including storage in each reservoir, outflow from system, head on turbines, monthly average horsepower available, and water demand deficiencies.

    b. Total energy developed by the power plant through the year, in horsepower-months.

11–11. Route stream flows through a regulatory system of three reservoirs, with coordination of reservoirs to give optimum control over the stream for purposes of water conservation and flood control. Below are given the monthly unregulated inflows at the sites of the proposed reservoirs, together with monthly demands on the system. Reservoir capacities and operating conditions are the same as assumed in Example 11–1. Data for the two-year period *following* the two-year period shown in the example are as below (in thousands of acre-feet):

Month	Unregulated Inflow at Res. A	Res. B	Res. C	System Demand
January	132	185	217	240
February	107	158	185	285
March	92	133	160	310
April	101	148	165	350
May	91	125	150	380
June	85	95	110	390
July	75	87	105	395
August	81	92	110	395
September	96	131	140	370
October	132	152	180	230
November	161	191	225	175
December	195	237	260	170
January	217	256	310	180
February	304	388	450	190
March	397	441	525	205
April	521	617	780	160
May	355	399	520	150
June	309	383	440	145
July	256	309	385	160
August	219	298	345	180
September	120	162	205	240
October	115	144	180	255
November	111	139	170	250
December	103	135	165	265

Required:

    a. Complete tabulation of river operation analysis for the two years.

    b. Plot of monthly hydrographs for the *4-year period* showing the following, on one sheet:

(1) Unregulated inflow at Reservoir C.

(2) Demand on System.

(3) Regulated outflow from Reservoir C.

c. Duration curves showing flow below Reservoir C versus per cent of time exceeded for:

(1) Unregulated flow.

(2) Regulated flow.

d. Degrees of discharge and duration control. *Ans.:* 92%, 75%.

11–12. Route a design flood through a proposed flood-control reservoir, in order to determine the storage capacity and the maximum flood release required, given the following topographic data at the reservoir site and the hydrograph of the design flood. The flood outlet conduits have hydraulic characteristics such that outflow (acre-ft/hour) $= 500 \sqrt{\text{Head on conduit center line}}$.

Elevation Above Outlet Center Line (feet)	Surface Area (acres)	Time (hours)	Inflow (acre-ft per hour)
10	100	0	100
20	400	4	4,000
30	600	8	8,000
40	1,000	12	13,000
50	1,500	16	18,000
60	2,000	20	19,000
70	2,500	24	17,000
80	3,100	28	14,000
90	3,600	32	9,000
100	4,200	36	4,000
110	4,900	40	2,000
120	5,700	44	700
		48	100
			(constant after 48 hours)

Required:

a. Storage capacity curve, plotted on log-log paper, showing elevation of water surface in feet above outlet versus stored volume in acre-feet.

b. Exponential equation of the above curve.

c. Slide rule, on cardboard, for the reservoir, letting 1 in. equal 30,000 acre-ft. Space intermediate graduations uniformly.

d. Tabulation for routing flood through the reservoir, using the slide rule for computing successive steps.

e. Superimposed hydrographs of reservoir inflow and outflow, for complete duration of flood releases from reservoir.

f. Maximum values of stored volume and outflow.

11–13. Make an approximate study of the relation between flood storage requirements and maximum permitted flood releases, for a reservoir at the site analyzed in the previous problem, with the same flood hydrograph and topographic data as in the previous problem. Compute the following, using the equivalent uniform flood method:

    a. Equivalent uniform flood (values of I and T) for:
        (1) Maximum outflow obtained by step routing process in previous problem.
        (2) Maximum outflow = 3000 acre-ft/hr.
        (3) Maximum outflow = 4000 acre-ft/hr.
        (4) Maximum outflow = 6000 acre-ft/hr.
    b. Flood storage for each of above conditions.
    c. Plot of flood storage requirement versus maximum outflow.

11–14. Releases from flood storage in a reservoir are controlled in accordance with the relation, $Q = 2000H^{3/2}$, where Q = discharge in cfs and H = head on spillway in feet. The reservoir storage, in cubic feet, is defined by the equation, $S = 4(10^6)(H^2 + 100)$. A sudden flood arrives at the reservoir, at a time when all storage except flood storage is already full. The flood has a uniform intensity of 80,000 cfs and a duration of 6 hours. Determine the following:

    a. Maximum total storage in reservoir. *Ans.:* $892(10^6)$ cu ft.
    b. Maximum head on spillway. *Ans.:* 11.1 ft.
    c. Maximum spillway discharge. *Ans.:* 74,000 cfs.
    d. Duration of spillway discharge. *Ans.:* 13.4 hr.

11–15. Releases from flood storage in a reservoir are controlled in accordance with the equation $Q = 4000H^{3/2}$. Reservoir storage is given by $S = 8(10^6)$ $(H^2 + 150)$ where

$$Q = \text{spillway discharge, cfs}$$
$$H = \text{head on spillway, ft}$$
$$S = \text{stored volume, cu ft}$$

The design flood is approximated by a uniform flood of 160,000 cfs intensity and 6-hour duration. Determine:

    (a) Maximum head on spillway, ft.
    (b) Duration of spillway discharge, hrs.

11–16. A flood detention basin has an uncontrolled orifice-type outlet and is located in topography such that the storage volume increases with the $\frac{5}{2}$ power of the elevation. The design flood hydrograph is assumed to be triangular in form, with a duration of 36 hours, and a peak inflow of 25,000 cfs, occurring 12 hours after beginning of flood runoff.

    a. Determine the equivalent uniform flood, assuming it begins one hour later than the actual flood and terminates when the actual flood runoff drops to 5000 cfs.
    b. Determine flood storage required for maximum release of 5000 cfs.
    c. What outlet gate area is required, expressed in terms of the appropriate coefficients of storage and discharge?

11–17. A flood detention reservoir is to be designed to control a design flood with the following hydrograph:

Time (hours)	Inflow (acre-ft/hr)
0	0
4	3,000
8	7,000
12	12,000
16	17,000
20	13,000
24	10,000
28	5,000
32	1,000
36	0

The storage and outflow equations are $S = 1.5 E^{2.5}$ and $O = 500 E^{0.5}$. Compute the storage volume required:

a. By step-routing procedure.
b. By equivalent-uniform-flood method, for maximum outflow obtained in (a).

11-18. The flood storage capacity at a certain reservoir is designed to control flood releases to a maximum of 3000 cfs. The design storm is approximated by an equivalent uniform flood of 12-hour duration and 12,000 cfs average intensity. The outlet works for the flood storage consist of sluicegates and outlet conduits, governed by the equation $Q = K\sqrt{H}$, where Q is the outflow in cfs, H is the head over the conduit entrances, and K is a constant determined by the gate and conduit hydraulics. The flood storage in acre-feet is equal to $400 H^{1.5}$. Compute:
a. Required flood storage. *Ans.* 9750 acre-ft.
b. Required value of K. *Ans.:* 1035.

11-19. Flow values are given below for inflow and outflow of a reach of river. Compute the Muskingum K and X for the reach. Plot the hydrographs.

Day	Inflow (cfs)	Outflow (cfs)
1	7,800	9,400
2	7,800	8,500
3	18,000	14,200
4	43,500*	34,100
5	32,100	36,000†
6	18,200	20,900
7	13,200	15,200
8	10,700	12,200
9	9,000	10,300
10	7,600	9,000
11	6,500	7,700
12	5,900	6,900
13	5,500	6,400

* Peak discharge, 47,600 cfs, day 4, time 1930.
† Peak discharge, 42,000 cfs, day 5, time 0300.

11-20. Using the results of Example 11-2, route the inflow hydrograph of the James River at Holcombs Rock, Va., given in Problem 11-19, and compare to the tabulated outflow.

12

MECHANICS OF SEDIMENTATION

12–1. Introduction. One of the most important and difficult classes of problems encountered by the hydraulic engineer is connected with the erosion, transportation, and deposition of sediment. The control of alluvial rivers for purposes of navigation, channel stability, etc., is perhaps the most important example; the design of stable canals in erodible materials is another; sedimentation in reservoirs and in harbors is still another; and numerous other problems exist, many of first economic magnitude, all related to the phenomena of sedimentation.

Despite the importance of the subject, it is probable that there is a greater differential between the information needed and the information available than in almost any other engineering field. In addition to the multiplicity of variables present in most hydraulic problems, such as the phenomena of non-uniformity and unsteadiness in the flow, both ordinary turbulence and macroturbulence, waves, etc., there are also the peculiar problems associated with sediment transportation—moving bed, constantly changing channel characteristics, flow of mixtures of water and solid particles, etc. Although a great deal of study has been devoted to sedimentation problems, there is still very little, of a rational and quantitative nature, that has been settled and generally adopted in sedimentation methodology.

The gross sediment quantity eroded each year in this country is estimated at about four billion tons, eroded mostly by the processes of sheet erosion, gullying, and stream-channel erosion. About one-fourth of this is carried to the sea and deposited in deltas or on the continental shelf. The remainder is redeposited on flood-plains, in lakes, or other places before reaching the ocean. Of the sediment transported to the ocean, the Mississippi River carries more than three times as much as all other streams in this country put together, about 700 million tons per year.

The sediment deposited in reservoirs in this country has been estimated as enough to correspond to more than 500,000 acre-feet of reservoir capacity each year. Any reservoir contemplated on a silt-carrying stream must be designed with proper allowance for such sediment storage.

12–2. Modes of Sediment Transportation. Sediment in streams may be transported either by rolling or sliding along the bed (*bed load*), by bouncing along the bed (*saltation load*), or in suspension in the turbulently moving

448

water (*suspended load*). In addition, there will usually be a *dissolved load*, consisting of salts and other chemicals in solution, and a *wash load*, consisting of very fine particles carried into and through the channel with no relation to the bed material. For our purposes, the bed load and suspended load are of major importance quantitatively, and the others may be neglected.

Although bed load and suspended load are usually considered separately, there is no sharp line of demarcation between them. Any given particle may be alternately carried in the bed load and in the suspended load. A complete transportation equation should include both in the same function, but attempts to develop such a function have thus far been unsuccessful.

Somewhat more successful have been various attempts to develop empirical equations in terms of parameters selected by the techniques of dimensional analysis. Correlation between laboratory results and field data is still far from satisfactory, however.

Perhaps the most widely used approach in sediment studies at present is that based on a separate analysis of the bed load, in terms of *tractive force* at the bed, and of the suspended load, in terms of turbulent transfer mechanisms, with these two increments added then to give the total load. Since this approach is also helpful in an introductory study of the physical phenomena involved, we shall first consider the bed load and then the suspended load.

12-3. Hydraulic Properties of Sediments. Sediments may be classed either as *cohesive* sediments or *non-cohesive* sediments. Most beds and banks of natural streams are of non-cohesive, or *granular* materials, and the hydraulic analysis can be based on the assumption that there are no cohesive bond forces between adjacent grains. Even in the case of cohesive soils, when the bond is broken between particles, they may function hydraulically as non-cohesive materials. Conversely, sediments may again become cohesive after deposition, by physical or chemical reactions.

In the case of non-cohesive sediments, there are four hydraulic properties of main interest. The sediment *size* is measured in terms of the particle diameter, which may be specified in one of the following ways:

a. Sieve diameter: the minimum-size sieve opening passing the particle.
b. Sedimentation diameter: the diameter of a sphere of the same specific gravity and fall velocity.
c. Nominal diameter: the diameter of a sphere of equal volume.
d. Surface diameter: the diameter of a sphere of equal surface area.

The size variation is usually denoted by a mechanical analysis. A plot showing "per cent finer by weight" versus "size," if plotted on log-probability paper, is usually close to linear in form.

Another hydraulic property of importance is the *shape* of particle. This property is measured either in terms of "sphericity," which is defined as the ratio of the surface area of a sphere of equal volume to that of the particle, or of "roundness," which is the ratio of the average radius of curvature of the individual edges of the particle to the radius of the largest circle that can be inscribed within the cross-section of the particle. Thus a cube is said to have a high sphericity and low roundness; a cylinder has low sphericity and high roundness.

The *specific gravity* of the sediment is the ratio of its solid weight to the weight of an equal volume of water at standard conditions. For most natural sediments, the specific gravity is near 2.65.

The fourth hydraulic property is the *fall velocity*, V_f. This is the equilibrium or terminal velocity which a typical particle will attain in falling through an unlimited extent of fluid, with the fluid usually specified as still water.

If the falling particle is assumed to be measured in terms of its sedimentation diameter, the equilibrium between the weight and the drag resistance is expressed as follows:

$$\tfrac{1}{6}\pi d^3 (\gamma_s - \gamma) = \tfrac{1}{2}\rho C_D V_f{}^2 \frac{\pi}{4} d^2 \tag{12-1}$$

In this equation, d is the sedimentation diameter, γ_s is the specific weight of the sediment, and C_D is a dimensionless drag coefficient which depends on the Reynolds number, particle shape, and particle concentration. Solving for the fall velocity:

$$V_f = \sqrt{\frac{4}{3}\frac{d}{C_D}\, g(S - 1)} \tag{12-2}$$

where S = specific gravity. The factor C_D depends largely on the Reynolds number, which determines whether the relative motion of the water past the falling particle is laminar or turbulent. The Reynolds number in this case is defined as $\dfrac{dV_f}{\nu}$ and the critical value for distinguishing between laminar and turbulent flow is found to be 0.1.

For laminar motion, the Stokes formula is used. That is, $C_D = \dfrac{24}{N_R}$ for spherical particles. Substituting in Eq. (12-2), and simplifying:

$$V_f = \frac{gd^2}{18\nu}(S - 1) \tag{12-3}$$

In this equation, d is the sedimentation diameter in feet; ν is the kinematic viscosity of the water in ft²/sec; g is the acceleration of gravity, and V_f is the fall velocity in ft/sec.

For turbulent motion, the drag coefficient must be determined from experimental curves as a function of N_R, the particle shape, and concentration of particles. Various attempts have been made to define quantitatively a "shape factor," perhaps the most accurate to date being the formula of Alger and Simons:

$$f_s = \frac{c}{\sqrt{ab}} \frac{d_s}{d_n} \tag{12-4}$$

in which a and b are the maximum and minimum dimensions, respectively, as measured in the particle's largest cross-section, which is therefore assumed to be the surface area normal to the direction of fall, as used in the drag force equation. The dimension c is the particle's smallest dimension aligned with the direction of motion. The term d_s/d_n is the ratio of the particle's surface diameter to its nominal diameter, as defined previously.

The shape factor varies from about 0.17 for a flat disk to 1.0 for a sphere. The drag coefficient in turn ranges from about 0.5 for a sphere, at high Reynolds numbers, up to as much as 3.0 for objects of low shape factors. In most cases, sand grains can be assumed approximately spherical.

A heavy concentration of particles may significantly reduce the average fall velocity. For example, if the concentration is 10%, by dry weight, the fall velocity may be decreased by as much as one-third the fall velocity of an isolated particle.

12-4. Movement of Sediment in the Bed.   The classic analysis of bed load, usually taken as a starting point in most bed load studies, was first published by P. duBoys in 1879. This analysis assumes that the bed moves in a sort of laminar flow, each layer sliding over the other, the velocity decreasing linearly with depth. (See Fig. 12-1.) Each layer is assumed to have a thickness δ, the nth layer being the first layer that does *not* move. The velocity differential across a layer is Δv, so that the velocity at the top of the bed is $\Delta v(n-1)$, and the average velocity of movement throughout the bed is $\Delta v(n-1)/2$.

Over the reach L the gravitational component in the direction of flow is $\gamma AL \sin \alpha$, assuming approximately uniform flow in the reach. This is balanced by the frictional resistance at the bed, which can be written as $\tau_0 P \cos \alpha L \sec \alpha$. Thus, since α is assumed to be fairly small, the approximate expression for bed shear stress is obtained:

$$\tau_0 = \gamma \frac{A}{P} S = \gamma R_h S \cong \gamma DS \tag{12-5}$$

In terms of the shear velocity, this is equivalent to

$$v^* = \sqrt{\frac{\tau_0}{\rho}} = \sqrt{gR_h S} \tag{12-6}$$

where R_h is the hydraulic radius and S the bed slope.

The bed shearing force is the cause of bed motion, and is assumed to be distributed through the bed in such a way that the velocity of the elements of the bed decreases linearly with depth, becoming zero at the top of the nth layer. The nth layer does not move, its friction resistance thus balancing

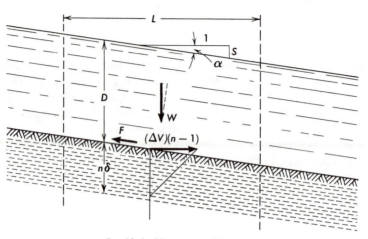

Fig. 12–1. Movement of bed load.

the shear force at that level. The resisting force there is the product of the submerged weight of superposed sediment and its coefficient of friction. Thus, equating these forces gives

$$(\gamma_s - \gamma)n\ \delta LPC_f = \tau_0 LP$$

and therefore the shear stress at the nth layer is

$$\tau_0 = (\gamma_s - \gamma)n\ \delta C_f$$

where γ_s is the specific weight of the bed material and C_f the coefficient of friction. If the bed as a whole is just at the stage of incipient motion, (i.e., if $n = 1$), the *critical tractive stress* is

$$\tau_i = (\gamma_s - \gamma)\ \delta C_f$$

The depth over which motion occurs is thus proportional to the bed shear:

$$\frac{n\ \delta}{\delta} = \frac{\tau_0}{\tau_i} = n$$

The weight of sediment being moved per second past a given point is

$$G_s = \gamma_s A_s V_s = \gamma_s n\ \delta P\left(\frac{1}{2}\right)\Delta v(n - 1)$$

$$= \left(\frac{\gamma_s \delta P\ \Delta v}{2}\right)\left(\frac{\tau_0}{\tau_i}\right)\left(\frac{\tau_0}{\tau_i} - 1\right) \tag{12–7}$$

and, since $\tau_0 = \gamma R_h S$, this can also be written

$$G_s = P\left[\frac{\gamma_s \delta\ \Delta v}{2}\left(\frac{\gamma}{\tau_i}\right)^2\right]R_h S(R_h S - R_{hi} S_i) \tag{12–8}$$

Application of the *duBoys formula*, above, to bed-load calculations is ordinarily restricted to alluvial rivers, for which the typical channel cross-section is wide and relatively shallow. Neglect of side effects renders the function questionable in narrower channels unless special provisions are made for the portion of the shear carried by the channel sides. For the usual assumption of two-dimensional flow, $P = B$, and the bed load per foot of width is

$$g_s = \frac{G_s}{B} = \left[\frac{\gamma_s \delta\, \Delta v}{2} \left(\frac{\gamma}{\tau_i}\right)^2 \right] DS(DS - D_i S_i) \qquad (12\text{-}9)$$

The bracketed factor may be considered a sediment characteristic, ψ:

$$\psi = \left(\frac{\gamma_s \delta\, \Delta v}{2} \frac{\gamma^2}{(\gamma_s - \gamma)^2\, \delta^2 C_f{}^2} \right) = \frac{\gamma_s\, \Delta v}{2(s-1)^2\, \delta C_f{}^2} = \frac{s\gamma(\Delta v/\delta)}{2[C_f(s-1)]^2} \qquad (12\text{-}10)$$

where s is the specific gravity of the bed material, and $\Delta v/\delta$ is the velocity gradient in the sliding bed. It is impracticable to determine ψ by means of

TABLE 12-1
Values of Sediment Characteristic and Critical Tractive Stress

Diameter (mm)	Class	ψ (lb/ft³-sec)	τ_i (lb/ft²)
$\frac{1}{8}$		523,000	0.0162
	Fine sand		
$\frac{1}{4}$		312,000	0.0172
	Medium sand		
$\frac{1}{2}$		187,000	0.0215
	Coarse sand		
1		111,000	0.0316
	Very coarse sand		
2		66,200	0.0513
	Granule gravel		
4		39,900	0.0890

the individual components of Eq.(12-10), primarily because the assumed physical mechanism is not what actually takes place in the bed. Nevertheless, Eq. (12-9) has apparently been adequately verified as qualitatively correct, with ψ determined empirically as a function of grain size. Table 12-1 gives empirical values of ψ and τ_i, as recommended by Straub. The data will be found to satisfy approximately the equation

$$\psi = \frac{111,000}{d_m^{3/4}} \qquad (12\text{-}11)$$

The duBoys formula then becomes

$$g_s = \psi DS \left(DS - \frac{\tau_i}{\gamma} \right) \qquad (12\text{-}12)$$

The depth of flow D is determined, for uniform flow, by the slope S and the unit discharge q. In the Manning formula, for two-dimensional flow,

$$q = \frac{1.5}{n} D^{5/3}S^{1/2} = VD \tag{12-13}$$

Combining Eqs. (12–12) and (12–13), L. G. Straub first suggested the following equation:

$$g_s = \psi\left[\frac{q^{3/5}S^{7/10}}{(1.5/n)^{3/5}}\right]\left[\frac{q^{3/5}S^{7/10}}{(1.5/n)^{3/5}} - \frac{\tau_i}{\gamma}\right]$$

If the τ_i term is dropped as relatively insignificant, then

$$g_s = \psi S^{7/5}\left(\frac{n}{1.5}\right)^{6/5}q^{6/5} \tag{12-14}$$

A similar expression can be derived in terms of the average velocity V:

$$g_s = \psi\left[\frac{V^{3/2}S^{1/4}}{(1.5/n)^{3/2}}\right]\left[\frac{V^{3/2}S^{1/4}}{(1.5/n)^{3/2}} - \frac{\tau_i}{\gamma}\right] \cong \psi S^{1/2}\left(\frac{n}{1.5}\right)^3 V^3 \tag{12-15}$$

Equations (12–9) to (12–15) are limited to two-dimensional flow. If the actual section departs too far from this assumption, the total bed load may be approximated by the following method. In Fig. 12–2, D_i is the depth required to start bed-load movement, assumed to be $\tau_i/\gamma S$ for two-dimensional flow. It is assumed that movement will occur only over that portion of the channel which is below this depth, of width b.

The lower cross-section is then divided into vertical increments dx in width and the assumption made that Eq. (12–9) is applicable in the following approximate form:

$$d(G_s) = \psi S^2 D(D - D_i)dx$$

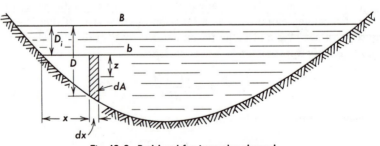

Fig. 12–2. Bed Load for irregular channel.

By integration, the total bed load is then

$$G_s = \psi S^2 \int_0^b D(D - D_i)\, dx = \psi S^2 \int_0^b D\, dA$$

$$= \psi S^2 \int_0^b (2z + D_i)\, dA = \psi S^2 (2\bar{z}A + D_i A) = \psi S^2 A(2\bar{z} + D_i) \quad (12\text{–}16)$$

where $\bar{z}$ is the distance to the centroid of the portion of the area below width b, as shown in Fig. 12–2.

Example 12-1

A given alluvial river has an approximately parabolic cross-section and a bed slope of 1 in 8000. The river width is 1000 feet when the flow depth is 10 feet. The effective bed particle diameter is 2.0 mm. How many tons per day of bed load is being transported when the uniform flow depth is 15.4 feet?

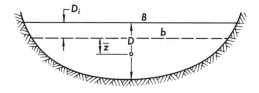

Solution. Depth for initial bed movement

$$= D_i = \frac{\tau_i}{\gamma S} = \frac{0.0513(8000)}{62.4} = 6.58 \text{ ft}$$

Distance to centroid of area below D_i

$$= \tfrac{2}{5}(D - D_i) = 0.4(8.82) = 3.53 \text{ ft}$$

Equation of parabolic cross-section: $B = cD^{1/2}$
  When $B = 1000$ ft, $D = 10$ ft, $\therefore\ B = 10^{5/2} D^{1/2}$
  At $(D - D_i) = (15.4 - 6.58) = 8.82$ ft.
    $b = (10)^{5/2}(8.82)^{1/2}$ ft
Area of cross-section below $D = A$

$$= \tfrac{2}{3}(D - D_i)(b) = \tfrac{2}{3}(D - D_i)(10)^{5/2}(D - D_i)^{1/2}$$

$$= \tfrac{2}{3}(10)^{5/2}(8.82)^{3/2} = 5510 \text{ sq.ft}$$

By duBoys formula: $G = \psi S^2 A(2\bar{z} + D_i)$

$$= \frac{66{,}200(5510)}{(8000)^2}(7.06 + 6.58) = 77.9 \text{ lb/sec}$$

$$= 77.9 \,\frac{3600(24)}{2000} = 3360 \text{ tons/day} \qquad\qquad Ans.$$

12–5. Other Bed-Load Formulas.   In addition to the duBoys formula, Eq. (12–9), and its modification by Straub, Eq. (12–13), numerous other bed-load formulas have been derived by various investigators. Only a few of these will be mentioned here.

Shields' formula was based on the Nikuradse velocity-distribution equations and a qualitative analysis of the drag forces exerted on individual sand grains at the bed. The equation is dimensionally homogeneous and has been successful in correlating numerous laboratory sediment studies, although with considerable scatter of data. It may be written as follows:

$$g_s = \left[\frac{10\gamma^2 V}{d(\gamma_s - \gamma)} \right] DS \left(DS - \frac{\tau_i}{\gamma} \right) \tag{12–17}$$

where d is the sediment diameter. The bracketed factor obviously corresponds to the sediment characteristic ψ in the duBoys formula. The presence of the velocity V, however, indicates that it does not depend solely on the character of the sediment. Combining with the Manning formula, and dropping the initial tractive force term, the Shields formula becomes

$$g_s = \left[\frac{10\gamma^2}{d(\gamma_s - \gamma)} \right] \left(\frac{n}{1.5} \right)^{3/5} S^{17/10} q^{8/5} \tag{12–18}$$

A. A. Kalinske in 1947 suggested a bed-load formula based on the analysis of turbulent fluctuations in velocity at the bed. This investigation was prompted by a suggestion originally made by E. W. Lane and was based also in part on previous studies by C. M. White.

In Fig. 12–3, the force F_i represents the drag force just sufficient to initiate movement of a particle in the bed of diameter d, and W is the submerged weight of the particle. The resultant of these forces passes through the point of support A. Just at the point of movement, the moments of F_i and W about A are in equilibrium. Therefore,

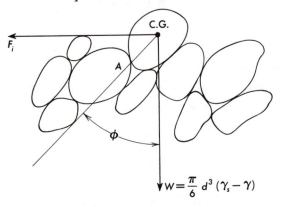

Fig. 12–3. Critical tractive force.

$$F_i = \frac{\pi}{6} d^3 (\gamma_s - \gamma) \tan \phi$$

where ϕ represents approximately the angle of repose of the bed material.

It is then assumed that the critical tractive shear stress at the bed, τ_i, is equal to the above tractive force on an individual particle divided by the *effective bed shear area* (that is the horizontal area of the single particle divided by the ratio of the number of particles actually about to move to the total, a ratio denoted by p). Thus,

$$\tau_i = \frac{(\pi/6)d^3(\gamma_s - \gamma)\tan\phi}{(\pi d^2/4)/p} = \frac{2}{3} pd\,(\gamma_s - \gamma)\tan\phi \qquad (12\text{--}19)$$

The ratio p can be understood to include also the effects due to non-sphericity of the particles. For $\tan \phi = 1$ and $\gamma_s/\gamma = 2.65$, experimental values yield the following approximate equation for critical tractive stress:

$$\tau_i \cong 24d \qquad (12\text{--}20)$$

This corresponds to a value of p equal to about 0.35.

However, Kalinske points out that the force F_i on the particle actually fluctuates because of the turbulent wake behind it. Furthermore, the turbulence of the channel flow itself results in material fluctuations of bed velocities and therefore of bed shear stresses. Assuming velocities fluctuate about the mean in statistically normal fashion, he found that the maximum bed shear stress for given conditions could be as much as three times that computed simply by γDS. Furthermore he found that the actual drag force on an individual particle might be twice the mean force on it. If both these effects are superposed, the critical mean shear stress, for incipient bed movement, need be only

$$\bar{\tau}_i = 4d \qquad (12\text{--}21)$$

The corresponding critical velocity at the bed level is v_i. If the actual velocity is v, then the velocity of motion of the grain itself is approximately $v - v_i$. The rate of bed sediment movement, therefore, is

$$g_s = \gamma_s \left(\frac{\pi}{6} d^3\right)\left(\frac{p}{\pi d^2/4}\right)(v - v_i) = \frac{2}{3}\gamma_s pd(v - v_i) \qquad (12\text{--}22)$$

since $(\pi/6)d^3$ is the volume of one particle, $p/(\pi d^2/4)$ is the number of particles moving per unit area, and $v - v_i$ is the velocity of the bed particles. However v fluctuates normally about a mean value, so that g_s also fluctuates. The temporal mean value of g_s is given in terms of the mean sediment velocity $\bar{v}_s = (\bar{v} - v_i)$ by

$$g_s = \frac{2}{3}\gamma_s pd\bar{v}_s$$

By introducing the shear velocity $v^* = \sqrt{\tau/\rho}$, this becomes

$$g_s = \frac{2}{3}\gamma_s\sqrt{\frac{\tau}{\rho}}\,pd\,\frac{\bar{v}_s}{v^*}$$

By Eq. (3–32), in boundary-layer theory, $v = 11.6v^*$, assuming that the velocity at bed level is the same as at the laminar boundary layer on a smooth wall. Therefore,

$$g_s = 7.7\gamma_s\sqrt{\frac{\tau}{\rho}}\,pd\,\frac{\bar{v}_s}{\bar{v}} \tag{12–23}$$

The ratio $\bar{v}_s/\bar{v}$ is a function of the relative intensity of the turbulence and of the initial velocity ratio $v_i/\bar{v}$. Since the shear varies as the square of the velocity, $v_i/\bar{v} \sim \sqrt{\tau_i/\tau}$. The relation with turbulence intensity is statistical and can be evaluated from the normal error function for any specified turbulence scale. Equation (12–23) can be written in the form

$$g_s = 7.7\gamma_s\sqrt{\frac{\tau}{\rho}}\,pdf\!\left(\frac{\tau_i}{\tau}\right) \tag{12–24}$$

Kalinske found this equation to correlate well with many experimental data, if the function f was chosen to correspond to a turbulence intensity $\sqrt{\Delta(v^2)}/\bar{v}$ of about $\frac{1}{4}$.

However, Rouse has pointed out that Kalinske's equation and data can be expressed in form similar to that of Shields, with the following empirical relation:

$$g_s = \gamma_s\sqrt{\frac{\tau}{\rho}}\,d\left\{10\left[\frac{\tau}{(\gamma_s - \gamma)d}\right]^2\right\} = \gamma_s\sqrt{g}\,\frac{10\gamma^2}{d}\frac{(DS)^{5/2}}{(\gamma_s - \gamma)^2} \tag{12–25}$$

Still another approach is that of Hans Einstein, who employed statistical reasoning to an even greater extent than did Kalinske. Turbulent fluctuations at the bed are such that individual particles move intermittently, in steps of statistically varying lengths and frequencies, with supposedly no such criterion as a limiting tractive condition.

The probability that any one particle is moving at a given time is related to its fall velocity, size, and specific weight, as well as to the hydraulic properties of the flow. The total volume of particles moving per second through an area of unit width is equal to the volume of a single particle per unit horizontal area multiplied by the cross-section area and the probability that any single particle begins to move in that second. Thus,

$$\frac{g_s}{\gamma_s - \gamma} = \frac{(\pi/6)d^3\,dp_s}{(\pi/4)d^2} = \frac{2}{3}d^2p_s \tag{12–26}$$

where p_s is the probability of motion of any single particle in the unit time, thus having the dimension, $\sec^{-1}$.

If the actual time required to move one particle is t, then $p_s t = p$ is the probability of any particle's moving during the time t. Einstein assumed that the time t is linearly proportional to the time required for the given particle to settle in still water through a distance equal to its own diameter; thus

$$t = k \frac{d}{v_s}$$

where v_s is the settling velocity, as given by the Rubey formula,

$$v_s = F \sqrt{gd \left(\frac{\gamma_s - \gamma}{\gamma} \right)} \tag{12-27}$$

and F is a dimensionless function as follows:

$$F = \sqrt{\frac{2}{3} + \frac{36\mu^2 g}{d^3 \gamma (\gamma_s - \gamma)}} - \sqrt{\frac{36\mu^2 g}{d^3 \gamma (\gamma_s - \gamma)}} \tag{12-28}$$

Combining equations gives

$$p = p_s t = \frac{g_s k d}{(2/3)d^2(\gamma_s - \gamma)F\sqrt{gd(\gamma_s - \gamma)/\gamma}}$$

$$= \left(\frac{3g_s k}{2F} \right) \left(\frac{(\gamma/g)^{1/2}}{[(\gamma_s - \gamma)d]^{3/2}} \right) \tag{12-29}$$

Einstein further assumed that the probability p is the probability that the local hydraulic lift force is adequate to overcome the weight of the particle, and thus is a function of the ratio of the two forces:

$$p = f \left(\frac{\text{weight of particle}}{\text{average lift on particle}} \right)$$

$$= f \left[\frac{(\pi/6)d^3(\gamma_s - \gamma)}{C_L(1/2)(\gamma/g)(\pi d^2/4)v^2} \right] \tag{12-30}$$

Since $v = 11.6v^* = 11.6\sqrt{gDS}$, and dropping constant terms in the functional notation, Eq. (12–30) is

$$p = f \left[\frac{(\gamma_s - \gamma)d}{\gamma DS} \right] \tag{12-31}$$

Finally, combining Eqs. (12–29) and (12–31), and including constant terms in the undetermined function, yields

$$g_s = \frac{Fd^{3/2}(\gamma_s - \gamma)^{3/2}}{\rho^{1/2}} f \left(\frac{\gamma_s - \gamma}{\gamma DS/d} \right) \tag{12-32}$$

where the function f must be determined empirically. Einstein found that a logarithmic function correlated the laboratory data, but Rouse has suggested the simpler function:

$$f \left(\frac{\gamma_s - \gamma}{\gamma DS/d} \right) = \frac{40\gamma^4 D^3 S^3}{d^3(\gamma_s - \gamma)^4} \tag{12-33}$$

as fitting the data equally well. Thus, in this form, the Einstein bed-load function becomes

$$g_s = \frac{40\gamma^4 D^3 S^3 F}{\rho^{1/2}\sqrt{(\gamma_s - \gamma)^5 d^3}} \tag{12-34}$$

A formula that has in recent years attained fairly wide use is that of Meyer-Peter and Muller:

$$g_s = 8\sqrt{\frac{g}{\gamma}\left(\frac{S}{S-1}\right)}(\tau - \tau_i)^{3/2} \tag{12-35}$$

This formula is somewhat similar to that of duBoys and has also been shown by Chien to give results comparable to those in the Einstein formula.

Many other bed-load formulas, involving the same type of variables as in the foregoing, have been published. All have been based primarily on laboratory flume studies, and attempts to correlate these with bed-load measurements in actual rivers and canals have not been very successful in general.

A recent formula of a somewhat different type, based on analysis of many data from laboratory flumes, canals, and rivers, is that of Dixon and Westfall:

$$g_s = \left(\frac{1}{2139v_f}\right)V^4 \tag{12-36}$$

where v_f is the fall velocity of the median-sized grain in the bed material and V is the flow velocity, both in ft/sec.

12–6. Suspended Sediment Load. In many streams, perhaps most of them, the suspended load is more important quantitatively than the bed load. This term has particular reference to the sediment particles which are of sizes comparable to those in the bed load and which, indeed, may be in a process of continual interchange with the bed particles. It does not include the wash load, composed of very fine particles which do not tend to settle out of suspension unless in a static pool for a substantial time, nor does it include the dissolved load, composed of molecules of various substances in solution.

The particles in the suspended load tend continually to fall by gravity to the bed; conversely, the turbulence of the water tends to throw bed particles up into the flow area and to maintain them there. For an equilibrium condition, these two tendencies are just balanced at each level.

If the normal fall velocity of the sediment is given by v_s, then the amount of sediment tending to fall through a unit horizontal area is

$$q_s = C_s v_s \tag{12-37}$$

where C_s represents the *sediment concentration* at that level, a dimensionless ratio giving the amount of sediment per unit amount of water.

The transfer of sediment up from the bed and throughout the flow area is essentially a diffusion process, and can be described by the same sort of equation describing diffusion of heat or momentum or other physical quantity. The basic diffusion equation is

$$Q = -\epsilon \frac{dq}{dy} \tag{12-38}$$

where Q is the quantity of material (or energy) being diffused through a given area in a given time, dq/dy is the gradient of concentration of that substance at that area, as measured in the direction of transfer. The minus sign indicates that the transfer process is in the direction of decreasing concentration gradient.

The proportionality constant ϵ is called the diffusion coefficient, its magnitude expressing the effectiveness of the particular transfer mechanism. For given conditions, it will be of at least the same order of magnitude no matter what the substance is which is being diffused. However, it is neither a constant nor is it dimensionless but varies with local turbulence intensities or other factors which may affect the transfer efficiency.

In the case of momentum transfer, the intensity of fluid shear stress at a given point in a turbulent flow is simply a measure of the rate of momentum transfer at the point and can be expressed by means of the diffusion equation, in a form originally attributed to Boussinesq:

$$\tau = \epsilon \frac{d(\rho v)}{dy} = \rho \epsilon \frac{dv}{dy} \tag{12-39}$$

ρv being the momentum per unit volume at the point y.

If the flow is assumed to be characterized by normal turbulence, then the velocity distribution can be expressed in terms of the von Kármán turbulence constant k, as follows:

$$\frac{dv}{dy} = \frac{v^*}{ky} \tag{12-40}$$

which permits ϵ to be expressed as a function of k and y:

$$\epsilon = \frac{\tau}{\rho dv/dy} = \frac{\tau ky}{\rho v^*} = \frac{ky}{\rho v^*} \tau_0 \left(1 - \frac{y}{D}\right)$$
$$= v^* ky \left(1 - \frac{y}{D}\right) \tag{12-41}$$

if it is assumed that the shear stress varies linearly with depth. It may be noted from this equation that the diffusion coefficient has a maximum value of $\frac{1}{4}v^*kD$ at $y = D/2$, and is zero at the boundary and at $y = D$. Its average value is

$$\bar{\epsilon} = \frac{\int_0^D \epsilon \, dy}{D} = \frac{v^* k}{D^2} \int_0^D (Dy - y^2) \, dy = \frac{1}{6} v^* kD \tag{12-42}$$

If k is taken as 0.4, then

$$\bar{\epsilon} = \frac{1}{15} v^* D \qquad (12\text{--}43)$$

However, k may be significantly less than 0.4 for silt-laden flows.

It is usually assumed that the diffusion coefficient for sediment transfer is the same as for momentum transfer, an assumption which has only approximately been verified experimentally. On this basis, the sediment being transfused upward per unit area is given by

$$q_s = -\epsilon \frac{dC_s}{dy} \qquad (12\text{--}44)$$

Equations (12–37) and (12–44) may be combined:

$$C_s v_s = -\epsilon \frac{dC_s}{dy} \qquad (12\text{--}45)$$

assuming equilibrium conditions. This is the basic differential equation for suspended load. Rearranging and integrating gives

$$\int \frac{dC_s}{C_s} = \log_e C_s = -v_s \int \frac{dy}{\epsilon} \qquad (12\text{--}46)$$

The Lane-Kalinske method is based on the assumption that ϵ is constant, at its average value. That is,

$$C_s = e^{-[(v_s/\epsilon)y + K]} = e^{-[15(y/D)(v_s/v^*) + K]}$$

The constant of integration K can be determined only if the concentration is actually known at some location $y = a$. Then,

$$\frac{C_s}{C_a} = e^{-15[(y-a)/D(v_s/v^*)]} = \frac{1}{e^{6[(y-a)/D](v_s/kv^*)}} \qquad (12\text{--}47)$$

On the other hand, a more accurate equation results when the value of ϵ from Eq. (12–41) is inserted in Eq. (12–46):

$$\log_e C_s = -v_s \int \frac{dy}{v^* ky(1 - y/D)} = -\frac{Dv_s}{kv^*} \int \frac{dy}{y(D - y)}$$

$$= -\frac{Dv_s}{kv^*} \left(\frac{1}{D} \log_e \frac{y}{y - D} + C \right)$$

or, by inserting the value C_a at $y = a$,

$$\log_e \frac{C_s}{C_a} = -\frac{v_s}{kv^*} \log_e \frac{(a - D)y}{(y - D)a} = \frac{v_s}{kv^*} \log_e \frac{(D - y)a}{(D - a)y}$$

and finally,

$$\frac{C_s}{C_a} = \left(\frac{D/y - 1}{D/a - 1}\right)^{v_s/kv^*} \tag{12-48}$$

Equation (12–48) is due primarily to Rouse, and has been found to be sufficiently accurate for most purposes. Either equation requires knowledge of the sediment concentration at some point a in order to obtain an absolute measurement of suspended load at the section.

A number of other equations for suspended sediment concentration have been published, but Eq. (12–48) seems to be the most satisfactory one now in use. However, the turbulence constant k, which appears in the exponent, apparently decreases as the sediment concentration increases, to as low as 0.20 for high concentrations.

The total suspended sediment being transported past a given section in a channel is given by

$$G_s = \iint_A vC_s \, dy \, db \tag{12-49}$$

If the flow is two-dimensional, $G_s = Bg_s$, and

$$g_s = \int_0^D vC_s \, dy \tag{12-50}$$

The von Kármán velocity distribution equation, from Eq. (12–40), is

$$\frac{v}{v^*} = \frac{1}{k} \log_e y + \left(\frac{v_{max}}{v^*} - \frac{1}{k} \log_e D\right)$$

$$= \frac{v_{max}}{v^*} - \frac{1}{k} \log_e \frac{D}{y} \tag{12-51}$$

The average velocity in the vertical can be obtained from

$$V = \int_0^D \frac{v \, dy}{D} = \frac{1}{D} \int_0^D \left(v_{max} - \frac{v^*}{k} \log_e D + \frac{v^*}{k} \log_e y\right) dy$$

$$= \frac{1}{D} \left[v_{max} D - \frac{v^*}{k} (\log_e D)D + \frac{v^*}{k} (D \log_e D - D)\right] = v_{max} - \frac{v^*}{k} \tag{12-52}$$

The velocity distribution equation, in terms of the average velocity is therefore

$$v = V + \left(\frac{v^*}{k}\right)\left(1 + \log_e \frac{y}{D}\right) \tag{12-53}$$

Combining Eqs. (12–48), (12–50), and (12–53), the total suspended load per foot of width is:

$$g_s = \frac{v^*C_a}{\left(\frac{D}{a} - 1\right)^{v_s/kv^*}} \int_{y_0}^D \left(\frac{V}{v^*} + \frac{1}{k} - \frac{1}{k} \log_e \frac{D}{y}\right)\left(\frac{D}{y} - 1\right)^{v_s/k^*} (dy) \tag{12-54}$$

The lower limit of integration, y_0, is the smallest distance from the bed at which both the concentration equation and the velocity distribution equation can be expected to apply. This is somewhat arbitrary, but the simplest assumption is to let $y_0 = y$ when $v = 0$ in the velocity distribution equation. That is:

$$y_0 = D(e)^{-(Vk/v^*)-1} \tag{12-55}$$

Equation (12-54) can be readily integrated by a graphical procedure, and charts and tables are available for this purpose if desired. However it is of limited usefulness because it applies only to equilibrium conditions and only for a particular sediment size (corresponding to v_s), and requires measurement of the concentration at some level $y = a$.

For non-equilibrium conditions, Kalinske has derived the following general equation, which corresponds to Eq. (12-45) for the equilibrium case:

$$v \frac{\partial C_s}{\partial x} = v_s \frac{\partial C_s}{\partial y} + \frac{\partial \epsilon_x}{\partial x} \frac{\partial C_s}{\partial x} + \frac{\partial \epsilon_y}{\partial y} \frac{\partial C_s}{\partial y} + \epsilon_x \frac{\partial^2 C_s}{\partial x^2} + \epsilon_y \frac{\partial^2 C_s}{\partial_y{}^2} \tag{12-56}$$

ϵ_x and ϵ_y are the diffusion coefficients in the x and y directions and may be approximately equal. Both are functions of y, as are v and C_s. The equation has not yet been solved, except on the basis of gross simplifying assumptions which render it invalid.

Despite the difficulties inherent in any attempt to obtain an analytical solution for the suspended load of a stream, it appears that field measurements often warrant application of a simple empirical exponential relationship between suspended load and discharge, of the form

$$G_s = cQ^n \tag{12-57}$$

where c and n are constants for a particular gaging station. If such a rating can be established in a given case, it may be quite useful in estimation of long-time average values of silt load.

In actual measurements of the suspended load of a stream, both velocity distribution and sediment concentration measurements are required, to be applied in Eq. (12-50). Since average velocities in the vertical are commonly obtained by averaging the point velocities at 0.2 and 0.8 of the depth, it is convenient to use these locations also for measurements of sediment concentration. Straub found the following formula to be applicable in many rivers:

$$\bar{c} = \frac{3}{8} C_{0.8D} + \frac{5}{8} C_{0.2D} \tag{12-58}$$

where $\bar{c}$ is the average concentration in the vertical. The total suspended load per foot of width is then

$$\begin{aligned}
g_s &= \left(\frac{3}{8} C_{0.8D} + \frac{5}{8} C_{0.2D} \right) \left(\frac{1}{2} V_{0.8D} + \frac{1}{2} V_{0.2D} \right) D \\
&= \frac{D}{16} (3C_{0.8D} + 5C_{0.2D})(V_{0.8D} + V_{0.2D})
\end{aligned} \tag{12-59}$$

However, various types of integrating sediment samplers are now largely replacing point samplers, obviating the need for averaging point values.

12-7. Total Sediment Load. Thus far, it has been impossible to develop any fully rational equation for total sediment load of a stream, although this of course is the most desirable ultimate goal of sedimentation mechanics. As already noted, equations have been developed for both the bed-load and the suspended-load segments of the total load, but even these are far from satisfactorily settled.

Various attempts have been made to determine total sediment load by simply adding the quantities of bed load and suspended load. This is not as easy as it seems, since there is no actual physical division between the two modes of transport and since calculation of suspended load still requires actual knowledge of the concentration at some definite level.

Einstein attempted to extend his bed-load equation to the prediction of suspended-load concentration at a boundary level, with the purpose of inserting it in the suspended-load function as the required reference concentration. However, the results were unsatisfactory quantitatively when compared with actual measurements.

Nevertheless, this approach has been adapted with some success by the U.S. Geological Survey and U.S. Bureau of Reclamation to the field determination of total load. The approach still requires actual measurement of suspended sediment concentrations, with computations of the bed load therefrom by means of the Einstein theory, modified empirically to correlate with U.S.G.S. field data.

A different type of approach is to consider the sediment load as a single entity, without division into bed load and suspended load, and then to develop an empirical formula for sediment transportation on the basis of dimensional analysis and model studies. Laursen has presented a proposed method based on this approach, and Garde and Albertson have also developed a similar procedure.

It is reasonable that the sediment load concentration should be a function of the flow depth and velocity, the channel slope, the density, specific weight, and viscosity of the fluid, and the size and specific weight of the sediment. That is,

$$g_s = f(D, V, S, \rho, \gamma, \mu, d, \gamma_s)\gamma_s q$$
$$= f\left(\frac{DV\rho}{\mu}, S, \frac{d}{D}, \frac{\rho V^2}{\gamma D}, \frac{\rho V^2}{\gamma_s D}\right)\gamma_s q \qquad (12\text{--}60)$$

Provided the flow is ordinarily tranquil, the Froude Number, $\rho V^2/\gamma D$, can probably be neglected. Similarly, if γ_s is considered constant, the Froude-type parameter $\rho V^2/\gamma_s D$ can be neglected.

Finally, the parameter $fV^2/8gD$ can be substituted for the slope S (from the Darcy equation). Then,

$$g_s = \phi\left(\frac{DV}{\nu}, \frac{fV^2}{8gD}, \frac{d}{D}\right)\gamma_s q$$

If the relation is assumed to be exponential in form, then

$$g_s = C\left(\frac{DV}{\nu}\right)^x\left(\frac{fV^2}{8gD}\right)^y\left(\frac{d}{D}\right)^z (\gamma_s DV) \tag{12–61}$$

where x, y, z, and the coefficient C are to be found experimentally. Equation (12–61) can also be regrouped as follows:

$$g_s = \frac{C\gamma_s f^y}{(8g)^y \nu^x} D^{(x+1)-(y+z)} V^{x+2y+1} d^z$$

$$= \psi D^m V^n d^p \tag{12–62}$$

The coefficient ψ depends primarily on the characteristics of the bed material, since the friction factor depends on the bed roughness. If this effect is expressed in terms of the Manning coefficient (since $f = 8gn^2/2.25D^{1/3}$), it being recognized also that the exponents m and p should be negative, then Eq. (12–62) becomes

$$g_s = K \frac{V^a n^b}{D^c d^e} \tag{12–63}$$

Studies of Albertson and Garde yielded an empirical curve that seems approximately to satisfy the following equation:

$$g_s = \frac{1.36}{[\nu(10)^5]^3} \frac{V^4 n^3}{d^{3/2} D} \tag{12–64}$$

where d and D are in feet, V in ft/sec, ν in ft^2/sec, and g_s in (lb/sec)/ft.

However, there is still considerable scatter in the data, quite possibly resulting from the highly uncertain character of the bed roughness effect, whether expressed in terms of the Manning coefficient or in some other way. Equation (12–64) is hardly suitable, therefore, to use as a basis of design, but the same statement is true for any other methods yet devised, short of actual sediment measurements over a full range of design conditions. There still remains urgent need for extensive study leading to a reliable sediment-transport equation.

Example 12–2

In a wide shallow river the discharge is 10 cfs/ft, and the depth is 6.0 ft. The bed slope is 0.0001 and Manning's n is 0.030. Water kinematic viscosity is 0.00001 ft^2/sec. Bed material size is 1 mm, and specific gravity 2.6. Assuming the Albertson-Garde and the Meyer-Peter-Muller formulas to represent sediment transport phenomena in the stream, estimate the suspended sediment load, in lb per sec per foot.

Solution.   Velocity, $V = \dfrac{q}{D} = \dfrac{10}{6} = \dfrac{5}{3}$ ft/sec

$$\text{Grain size, } d = \frac{1}{304.8} \text{ ft}$$

Initial tractive stress, τ_i, from Table 12–1, $= 0.0316$ lb/sq ft
Actual tractive stress, $\tau = \gamma DS = 62.4(6)(0.0001) = 0.0373$ lb/sq ft
Total load, by Eq. (12–64)

$$= \frac{1.36}{[\nu(10)^5]^3} \frac{n^3 V^4}{d^{3/2}D} = \frac{1.36(\frac{5}{3})^4(0.03)^3(304.8)^{3/2}}{(1)^3(6)} = 0.251 \text{ lb/sec/ft}$$

Bed load, by Eq. (12–35)

$$= 8\sqrt{\frac{g}{\gamma}\left(\frac{S}{S-1}\right)}(\tau - \tau_i)^{3/2} = 8\sqrt{\frac{32.2}{62.4}\left(\frac{2.6}{1.6}\right)}(0.0373 - 0.0316)^{3/2}$$

$$= 13(0.719)(0.0057)^{3/2} = 0.004 \text{ lb/sec/ft}$$

Therefore, for the given conditions, and assuming the formulas used are realistically applicable to this stream, almost all the sediment load is in suspension.

Suspended load = Total load − Bed load = 0.251 − 0.004 = 0.247 lb/sec/ft.

Ans.

12–8. Bed Geometry and Flow Resistance in Alluvial Channels.

One of the main obstacles to the development of a reliable and comprehensive sediment transport formula is the complex character of the bed roughness. It is well known that the major cause of energy loss in a flowing fluid is occasioned by roughness elements on the boundary surface. The flow velocity and friction factor or roughness coefficient are directly interrelated. In turn, the sediment-carrying characteristics are directly related to the velocity, as well as to the turbulence structure of the flow. These effects are seen most easily when the transport formulas are expressed in such a form as the Straub-duBoys formula (Eq. 12–15) or the Albertson-Garde formula (Eq. 12–64).

In addition to all the usual difficulties in determining a friction factor or Manning coefficient for a rough boundary surface, the problem is compounded in alluvial channels by the fact that the bed geometry is subject to drastic changes in roughness as the hydraulic parameters change. These changes may be so great as to cause the friction factor in a given stream to change by a factor of as much as ten over the range of discharge carried in a year.

Bed geometry patterns may be roughly categorized as follows:

a. *Undisturbed flat bed.* This condition probably exists only in the initial stages of experimentation in a laboratory flume, and not at all in natural streams. Under very carefully controlled conditions, it is possible in a flume to set such a flat bed in motion, but this is an unstable condition and ripples will soon develop on the surface.

b. *Rippled bed.* At low velocities, a complex ripple pattern will develop on the sand surface. These are small waves with wave lengths less than 1.5 ft and heights less than about 0.1 ft. The ratio of wave length to height ranges between 5 and 25. They may be formed even at velocities too low to initiate sediment motion. They tend to have a "saw-tooth" appearance, with gentle upstream slopes and steep downstream slopes (approximately at the angle of repose). The water flow continuously moves particles up the slope and over the crest, so that the ripples slowly migrate downstream. They occur only rarely in sediments coarser than about 0.6 mm, and their size evidently depends primarily on this grain diameter.

c. *Duned bed.* At larger velocities and sediment transport rates, dunes appear. These features are larger than ripples and, indeed, ripples may continue to occur on the upstream face of the dunes. The length-height ratio is usually between 10 and 40. Dunes usually have the same shape as ripples, with gentle upstream slopes and steeper downstream slopes, and sediment is transported in the same way as they slowly migrate downstream. At higher velocities the ripples tend to disappear off the dune face, and the dunes themselves become more rounded. Normally, neither ripples nor dunes extend over the full channel width, but dunes generally are relatively longer, with the crest length roughly equal to the longitudinal spacing. The size of dunes evidently depends primarily on the flow depth.

d. *Transitional Flat Bed.* At still higher velocities (though still subcritical) the dunes disappear and the bed becomes essentially flat. In this regime, the moving bed load is substantial. The form drag force on the dune is at this point great enough to disintegrate the dune itself and to inhibit the tendency of the flow turbulence to restructure a dune pattern.

e. *Anti-dune Bed.* As the flow velocity approaches critical velocity ($N_F \rightarrow$ 1), the gravity waves on the water surface become large in amplitude and tend to assume the character of a stationary or slowly-moving wave train. The surface waves tend to generate on the sediment bed a corresponding wave train in phase with the surface waves, though of lesser amplitude. These are called anti-dunes, in view of the fact that they often migrate upstream, rather than downstream, presumably because the lowered pressures in the wake behind the sand wave dislodges particles from the downstream face and deposits them on the upstream face of the next wave. When the surface water waves break, the turbulent agitation obliterates the anti-dunes, but they are rapidly re-formed as new surface waves grow. The geometrical cross-section of anti-dunes tends to be sinusoidal, especially at higher velocities.

f. *Chutes and Pools.* At supercritical velocities the cyclic formation and eradication of anti-dunes becomes quite violent and eventually a chute-and-pool structure is developed. The chutes consist of large mounds of sediment over which the water moves at high velocities, followed by deep pools, in which the flow may sometimes be subcritical, with hydraulic jumps at the chute-pool interfaces. In a sense, these mounds are very large anti-dunes, which slowly move upstream.

In addition to the forms noted above, *bars* may be formed on a streambed. These are dune-like forms, except that their length may be equal to or greater than the channel length, and their heights are of the order of magnitude of the depth of flow generating them. Ripples or dunes are normally superposed on the upstream slopes of bars. Their spacing is quite irregular and depends on the general hydraulic geometry of the stream—its meanders, its tributaries, and other characteristics. They occur most regularly on the inside of channel bends and downstream from points of tributary inflow.

It is obvious from the foregoing discussion that the resistance to flow, as measured by the friction factor or roughness coefficient, will be closely related to the prevailing bed geometry. Each ripple or dune constitutes a boundary roughness element of significant size, and therefore is a vorticity generator and drain on the flow energy. In addition, the sand bed itself is analogous to a sand-coated rigid boundary, with each sand grain generating turbulence.

In terms of the turbulent flow regimes defined in Chapter 3, it seems certain that the sand surface of itself generates a low-intensity *wake-interference* flow directly along the boundary, with a friction factor that could be estimated from the *hyper-turbulent* or *normal turbulent* (Nikuradse) relations, depending on Reynolds number. In the flat-bed state, this should constitute essentially the entire friction factor.

This relatively simple approach, however, is complicated even in the flat-bed case by two factors. The first is that additional flow energy is expended in carrying the sediment load; the second is that the presence of the sediment is known to have a damping effect on the turbulence structure. To some extent, these two factors each tend to compensate for the other in their effect on the over-all friction factor, but to how great an extent is not known.

Superimposed on this normal boundary friction of the sand grains are the form drag forces occasioned by the ripples and dunes. A succession of small sand ripples may add, in effect, an *isolated-roughness* component to the friction factor. More likely, however, especially in the case of large ripples, dunes, or anti-dunes, the additional component will be large-scale *hyper-turbulent* flow generated by the *wake-interference* between the large roughness forms.

In principle, if the geometry of these forms could be estimated, the formulas for the appropriate flow regime could be employed as in the methods outlined in Chapter 3 to estimate the friction factor. As yet, however, no reliable means has been devised to determine from the basic hydraulic and sediment parameters just what geometry the bed will assume. To a very limited extent, and with much scatter of data, it can be predicted (on the basis of Froude number, depth–sand diameter ratio, or other parameters) whether the bed will be flat, duned, anti-duned, or what-not, but the actual dimensions of the bed forms are still quite elusive.

Furthermore, the motion of the bed, the non-uniformity and unsteadiness of the flow and many other factors all combine to make the actual prediction of resistance coefficients from the bed geometry in alluvial channels almost a matter of wishful thinking. By the same token, the inductive determination of hydraulic factors from the measurement of bed forms (as in the study of fossil sand ripples in sedimentary rocks, for example) is highly uncertain at best.

Nevertheless, in recent years, a large number of research studies have been focussed upon the determination of friction factors in alluvial channels, and many data have been accumulated from both flumes and natural streams. A number of these are listed for reference in the bibliography at the end of this chapter.

In view of the great number of variables and uncertainties involved in this type of analysis, it seems likely that a similitude approach will be the most useful for some time to come, within the basic constraints of the conservation equations. The bed geometry which controls the friction factor is itself determined in some complex fashion by the independent hydraulic variables (e.g., discharge, channel slope, channel width, fluid properties) and the sediment characteristics (e.g., median grain size, specific gravity, grading of sand mixture). The roughness height, spacing, etc., are not fixed quantities as for a rigid boundary, but adjust to some equilibrium geometry in response to the interaction of the above-listed independent variables. A representative functional equation for the friction factor might be expressed as follows:

$$f = f(V, R, \rho, B, \rho_s, d_m, \gamma, \mu) = 0$$

from which, by dimensional analysis:

$$f = f\left(\frac{V}{\sqrt{gR}}, \frac{RV}{\nu}, \frac{B}{R}, \frac{d_m}{R}, \frac{\rho_s}{\rho}\right) = 0 \qquad (12\text{--}65)$$

or:

$$f = f\left(N_F, N_R, \frac{B}{R}, \frac{d_m}{R}, S\right) \qquad (12\text{--}66)$$

in which N_F is the Froude number $\dfrac{V}{\sqrt{gR}}$, N_R the Reynolds number $\dfrac{RV}{\nu}$, R the hydraulic radius, V the velocity, B the channel width, d_m the median grain diameter, and S the specific gravity of sediment $\dfrac{\rho_s}{\rho}$.

No terms are included for the bed roughness geometry since these would be themselves functions of the independent variables in the function. The hydraulic radius has been specified as the basic length parameter, but it is possible the mean depth or maximum depth might give better correlation of data. In the case of a wide channel, of course, the hydraulic radius and depth are essentially the same.

It seems probable that the form of the function of Eq. (12–66) will be different for each of the different regimes of bed geometry. Not only are the respective friction-factor functions needed, therefore, but also definite criteria for discriminating between the regimes. Although definitive results are not yet available, research is currently very active in this field, and a number of the papers listed at the end of this book will be found helpful.

12–9. Reservoir Sedimentation. If a storage reservoir is constructed on a sediment-carrying stream, it is obvious that a substantial part of the sediment load will be deposited in the reservoir. Gradually the usable storage capacity will thus be eliminated. Therefore adequate provision should be made for sediment storage or removal in the plans for any storage project.

A number of reservoirs in this country have already become virtually useless because of silting. It has been estimated that storage capacity costing $10,000,000 is destroyed in the United States each year by reservoir silting. Over 20 percent of the reservoirs in this country have an estimated useful life of less than 50 years, and 25 per cent more of less than 100 years because of this factor.

The ultimate source of the sediment, of course, is from the lands of the watershed, in the threefold form of sheet erosion, gully erosion, and stream channel erosion. The sediment thus delivered to the stream is then transported down its valley, as far as the hydraulic capacity of the stream permits. When the velocity decreases, through expansion of the cross-sectional area at a lake or reservoir, obviously much of the load will settle out.

Some of the load will be deposited in the valley upstream from the reservoir, in the region of backwater, thus aggrading the valley. More will be laid down in delta-type deposits near the entrance to the reservoir. Some will flow through the lake and outlet conduits, but this will normally be small in proportion to the total.

The deposits in the delta will tend to be gradually shifted downstream. At low stages, the deposits will be re-eroded, with subchannels entrenched in them. The excavated material will then be redeposited in secondary delta deposits downstream.

But probably the bulk of the material deposited in the main storage segment of the reservoir comes from the fine sand, silt, and clay which does not actually settle out in the delta, but moves as a more or less homogeneous suspension, a dense fluid-solid mixture flowing slowly in the form of a *density current*, or *turbidity current*, along the bottom of the reservoir down to the dam itself.

The mechanics of these density currents, or *stratified flows* as they are also called, is yet somewhat obscure. Apparently they occur in two major forms. The usual way is in the form of a slow-moving, perhaps laminar, flow of a fluid stratum of somewhat higher density than clear water. This is the true density current. Occasionally a subsurface landslide occurs in the delta deposit, or elsewhere, and the large mass of sediment suddenly set in motion moves at high velocity down the channel bottom to the dam. The term "turbidity current" is sometimes limited to this type of phenomenon. However, the two terms are often used interchangeably.

In any case, the encroachment on reservoir storage capacity depends quite substantially on this particular mechanism. Considerable research is currently being devoted to the density current phenomenon, but the results do not yet permit effective application in planning and design.

For purposes of reservoir planning, allowance must be made for silt storage on the basis of the following relation:

$$V_s = EQ_s \qquad (12\text{--}67)$$

where V_s is the volume of useful capacity destroyed each year, Q_s is the volume of sediment transported into the reservoir, in acre-ft per year, and E is the reservoir *trap efficiency*.

The trap efficiency must be a function of the flow detention time in the reservoir (which depends on the reservoir capacity, shape, inflow, and outflow operation procedure) and on the sediment characteristics (especially the size distribution and its ability to form density currents). No rational method of predicting the reservoir efficiency is available, but Carl Brown has developed the following empirical formula:

$$E = 1 - \frac{1}{1 + KC/W} \qquad (12\text{--}68)$$

where C/W is the ratio of reservoir capacity in acre-feet to the watershed area in square miles. The coefficient K varies with all the factors mentioned above, ranging from 0.046 to 1.0, with a mean value of about 0.10.

The sediment inflow Q_s, is estimated from sediment transport studies or from actual reservoir sedimentation measurements. The effective volume of reservoir sediment of course depends not only on the weight entering the reservoir, W_s, in pounds per year, but also on its unit weight. But even the latter value changes with time, through the effect of increasing compaction. Preferably the mean specific weight should be used, for the entire mass of deposited sediment at the end of the anticipated time. That is,

$$Q = \frac{W_s}{\gamma_m} \qquad (12\text{--}69)$$

Lane and Koelzer have proposed the following formula for estimating the mean compacted unit weight of reservoir sediment, over the time T:

$$\gamma_m = \gamma_1 X_1 + (\gamma_2 + K_2 \log_{10} T)X_2 + (\gamma_3 + K_3 \log_{10} T)X_3 \qquad (12\text{--}70)$$

where γ_1 = specific weight of sand and coarser sediment after one year

γ_2 = specific weight of silt after one year

γ_3 = specific weight of clay after one year

K_1 = constant for rate of compaction of sand ($K_1 \cong 0$)

K_2 = constant for rate of compaction of silt

K_3 = constant for rate of compaction of clay

X_1 = fractional part of total deposit made up of sand

X_2 = fractional part of total deposit made up of silt

X_3 = fractional part of total deposit made up of clay

T = time in years, for T equal to or greater than one year

γ_m = average specific weight of sediment in reservoir after time T

Measured field values of the γ and K values are given by Lane and Koelzer in Table 12–2.

TABLE 12–2

Values of γ and K for Sand, Silt, and Clay

Reservoir Stage	Sand		Silt		Clay	
	γ_1(pcf)	K_1(pcf)	γ_2(pcf)	K_2(pcf)	γ_3(pcf)	K_3(pcf)
Sediment always submerged or nearly submerged	93	0	65	5.7	30	16.0
Normally a moderate reservoir drawdown	93	0	74	2.7	46	10.7
Normally considerable reservoir drawdown	93	0	79	1.7	60	6.0
Reservoir normally empty	93	0	82	0.0	78	0.0

More recently, Lara and Pemberton have collected extensive quantities of data and have recommended a formula and table (see Table 12–3) for initial unit weight as follows:

$$\gamma_{m_i} = \gamma_1 X_1 + \gamma_2 X_2 + \gamma_3 X_3 \qquad (12\text{–}71)$$

TABLE 12–3

Initial Unit Weights ($T = 0$)

Type of Operation	γ_1(pcf)	γ_2(pcf)	γ_3(pcf)
Normally submerged	97	70	26
Moderate to considerable drawdown	97	71	35
Normally empty	97	72	40
Riverbed sediments	97	73	60

An alternative way of correlating trap efficiency is in terms of the *capacity-inflow ratio* instead of the *capacity-watershed ratio*. The latter ratio does not reflect the influence of general basin humidity as well as the other. Thus, even though a reservoir may have a small capacity (and thus, inferentially, a small trap efficiency), if the inflow is low, then the spillway discharge will be low and the apparent trap efficiency may be quite high.

TABLE 12–4

Median Values of Trap Efficiency E

$\dfrac{C}{I}$ (%)	E (%)	$\dfrac{C}{I}$ (%)	E (%)
0.2	2	2.0	60
0.3	13	3.0	68
0.4	20	4.0	74
0.5	27	6.0	80
0.6	31	10.0	86
0.8	38	20.0	93
1.0	44	100.0	97
1.5	52	1000.0	98

Table 12–4 indicates median values of trap efficiency, E, in terms of the capacity-annual inflow ratio, C/I, expressed in per cent, as determined by Brune. The data will approximately satisfy the following empirical equation (derived by D. M. Crim) except at very low values:

$$E = \frac{C/I}{0.012 + 0.0102 C/I} \tag{12-72}$$

Example 12–3

A storage reservoir is constructed on a stream with an average discharge of 500 cfs and a dominant discharge of 800 cfs. The stream is 200 ft wide, on a bed slope of 0.000324 and has a Manning coefficient of 0.03. The sediment transported is approximately 80% sand, 10% silt, and 10% clay. When deposited in the reservoir, it will always be submerged. The mean sediment diameter can be assumed to be $\frac{1}{4}$ mm. The regulation storage in the reservoir (not including sediment storage) is 90,000 acre ft. For a 100-year anticipated reservoir life, how much total storage capacity should be provided?

Solution. From Table 12–1, $\psi = 312,000 \dfrac{\text{lb}}{\text{ft}^3\text{-sec}}$ and $\tau_i = 0.0172 \text{ lb/ft}^2$

From Manning equation, $D = \left(\dfrac{nQ_{\text{dom}}}{1.5BS^{1/2}}\right)^{3/5} = \left(\dfrac{24}{300(0.018)}\right)^{3/5} = 2.45$ ft.

Assuming sediment load to be given by duBoys' formula,

$$G_s = \psi BDS\left(DS - \frac{\tau_i}{\gamma}\right)$$

$$= 312{,}000(200)(2.45)(0.000324)\left[2.45(0.000324) - \frac{0.0172}{62.4}\right] = 25.6 \text{ lb/sec.}$$

$$W_s = G_s(3600)(24)(365) = 8.05(10)^8 \text{ lb/yr}$$

$$= \text{weight of sediment entering reservoir each year}$$

Mean unit weight $= \gamma_m = \gamma_1 X_1 + (\gamma_2 + K_2 \log_{10} T)X_2 + (\gamma_3 + K_3 \log_{10} T)X_3$

$$= 93(0.8) + [65 + 5.7(2)](0.1) + [30 + 16.0(2)](0.1) = 88.2 \text{ lb/ft}^3$$

Effective volume of sediment inflow in 100 years

$$= Q_s = \frac{G_s}{\gamma_m} = \frac{805(10)^8}{88.2(43{,}560)} = 20{,}900 \text{ acre-feet}$$

Average annual inflow $= \dfrac{500(3600)(24)(365)}{43{,}560} = 362{,}000 \text{ acre-ft/year}$

Required capacity $= 90{,}000 + EQ_s(\frac{1}{100})(E = \text{trap efficiency in } \%)$

$$\frac{C}{I} = \frac{90{,}000 + 209E}{362{,}000}(100) = 24.8 + 0.0577E$$

By Eq. (12–72), $E = \dfrac{\dfrac{C}{I}}{0.012 + 0.0102\dfrac{C}{I}} = \dfrac{24.8 + 0.0577E}{0.265 + 0.000589E}$

Solving, $E = 93$ per cent and $\dfrac{C}{I} = 30$ per cent

$\therefore C = 90{,}000 + 0.93(20{,}900) = 109{,}000 \text{ acre-ft}$ *Ans.*

12–10. Stable Channels in Erodible Material. Another important application of sedimentation mechanics is in the design of stable channels, especially for use as irrigation and drainage canals. Most such channels, for cost reasons, must be unlined and it is necessary therefore that they be so proportioned as to permit neither silting nor scouring in objectionable quantity. That is, the velocities of flow at all points of the cross-section must be sufficient to transport through the canal all sediment that enters it; however, they must at all points be low enough, so that no erosion is caused.

It is obvious that the present state of sedimentation theory is inadequate to permit precise theoretical design of such sections. Much research has been devoted to this problem and substantial improvements in methodology have been effected in recent years, but channel design must still be based in large measure on engineering experience and judgment.

A formula developed by R. C. Kennedy in 1895, on the basis of his study of the irrigation canals in the Punjab in India (now Pakistan), has been widely used for determining the non-silting and non-eroding mean velocity. The formula is as follows:

$$V_0 = CD^x \qquad (12\text{--}73)$$

where V_0 = non-silting, non-eroding velocity, fps

x = 0.5 for clear water

0.64 for moderately silty water as in most canals

C = 0.56 for extremely fine soils

0.84 for fine, light, sandy soils

0.92 for coarse, light, sandy soils

1.01 for sandy loamy silts

1.09 for coarse silt

Fortier and Scobey published in 1926 a table of permissible maximum canal velocities in various materials, based on their studies of canals in the United States. This table, adapted in Table 12–5, has been extensively used in this country as a specification for *maximum* permissible velocities, the Kennedy formula still being used for *minimum* velocities. The Fortier and Scobey list, however, originally took no account of channel depth or sinuosity, and also was based on measurements in "well-seasoned" canals, carrying only moderate amounts of silts or other detritus in suspension.

Lane recommends a reduction of the velocities in the table if the channel is sinuous, as shown in the tabulation below the main table. Similarly, the velocities can be increased somewhat if the depth of flow exceeds 3.0 ft. Thus if the discharge and permitted velocity are known, and therefore also the area, together with the slope and roughness of the channel, the Manning formula can be used to determine the required hydraulic radius of channel. The dimensions of the cross-section can then be calculated on the basis of the geometry of the desired shape of section. The minimum velocity should also be checked for the resulting cross-section.

Thus, for any given channel shape, the required area can be expressed as a function of the width and depth of flow:

$$A = \frac{Q}{V} = f_1(B, D).$$

Similarly the required hydraulic radius, as obtained from the Manning formula, can be expressed as a function of flow width and depth.

$$R_h = \left(\frac{nV}{1.5S^{1/2}}\right)^{3/2} = f_2(B, D)$$

These two equations can then be solved for the required design values of B and D.

TABLE 12–5
Maximum Permissible Velocities in Erodible Channels *

Channel Material	Manning Coefficient, n	Clear Water		Water Transporting Colloidal Silts	
		V (ft/sec)	τ_0 (lb/ft^2)	V (ft/sec)	τ_0 (lb/ft^2)
Fine sand, colloidal	0.020	1.50	0.027	2.50	0.075
Sandy loam, non-colloidal	0.020	1.75	0.037	2.50	0.075
Silt loam, non-colloidal	0.020	2.00	0.048	3.00	0.110
Alluvial silts, non-colloidal	0.020	2.00	0.048	3.50	0.150
Ordinary firm loam	0.020	2.50	0.075	3.50	0.15
Volcanic ash	0.020	2.50	0.075	3.50	0.15
Stiff clay, very colloidal	0.025	3.75	0.260	5.00	0.46
Alluvial silts, colloidal	0.025	3.75	0.260	5.00	0.46
Shales and hardpans	0.025	6.00	0.670	6.00	0.67
Fine gravel	0.020	2.50	0.075	5.00	0.32
Graded loam to cobbles, non-colloidal	0.030	3.75	0.380	5.00	0.66
Graded silts to cobbles, colloidal	0.030	4.00	0.430	5.50	0.80
Coarse gravel, non-colloidal	0.025	4.00	0.300	6.00	0.67
Cobbles and shingles	0.035	5.00	0.910	5.50	1.10

*Adapted from Fortier and Scobey tabulation by U.S. Reclamation Bureau.

These values apply only to well-seasoned, straight channels on mild slopes, with flow depths less than about 3 ft. For flow depths greater than 3.0 ft, increase velocity values by a factor equal to $(\frac{1}{18})(D - 3)$, up to $\frac{1}{4}$ (maximum increase) at $D = 10$ ft. For sinuous channels, decrease values by the following factors:

	Velocity	Shear Stress
Slightly sinuous	5%	10%
Moderately sinuous	13%	25%
Very sinuous	22%	40%

In recent years, a more nearly rational method, based on the concept of tractive force, has been developed by engineers and consultants of the U.S. Bureau of Reclamation. The tractive force, as already noted, is given by the expression $\tau_0 = \gamma R_h S$, where τ_0 is the average bed shear stress, or unit tractive force, R_h is the hydraulic radius, S is the bed slope, and γ is the specific weight of water. However, this is the *average* unit tractive force. Actually the magnitude varies throughout the wetted perimeter, except in the special case of a very wide channel, when τ_0 becomes equal to a constant value of γDS, where D is the depth of flow.

Values of permissible tractive shear stresses corresponding to the permissible velocities, as recommended by U.S. Bureau of Reclamation, have been included in the Fortier-Scobey tabulation (Table 12–5). These may be compared as follows:

$$\tau_0 = \gamma R_h S = \gamma R_h \left(\frac{nV}{1.5 R_h^{2/3}}\right)^2 = \frac{\gamma}{(1.5)^2 R_h^{1/3}} (nV)^2 \cong \frac{28}{R_h^{1/3}} (nV)^2$$

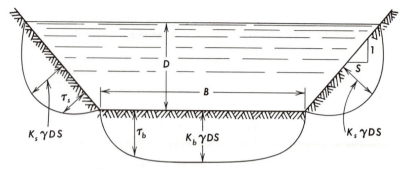

Fig. 12–4. Tractive stress distribution on channel bed.

If R_h is taken as approximately 0.8 ft, then the values of τ_0 in the table will be found to correspond approximately to the corresponding values of n and v.

The tractive force distribution has been determined analytically by U.S.B.R. personnel, by means of rather complex mathematical techniques, to be approximately as indicated in Fig. 12–4 for a trapezoidal channel. The coefficients K_s and K_b, for maximum shear on the sides and bottom, respectively, depend on the width-depth ratio B/D and the side-slope s. However, for channels of ordinary size and shape, the values of K_s and K_b may be taken as $\frac{3}{4}$ and 1, respectively.

Thus, the maximum tractive force on the side slope is only about $\frac{3}{4}$ that on the bottom. However, motion of soil particles on the side slope is also assisted by gravity, so that the resultant effect of tractive force and gravitational force is often such as to cause the critical point for incipient scour to be on the side rather than the bottom. This resultant force is indicated three-dimensionally in Fig. 12–5. W_s is the weight of the soil particle and ϕ is the angle of side slope. $W_s \sin \phi$ is the component of this force down the slope. The tractive force acting horizontally on this particle is $\tau_s a_s$, where a_s

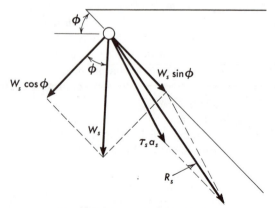

Fig. 12–5. Forces on sediment particle.

is the effective shear area of the particle. The resultant force on the particle, R_s, is then given by

$$R_s = \sqrt{(W_s \sin \phi)^2 + (\tau_s a_s)^2}$$

This force is resisted by the normal component of the weight times the coefficient of friction. Therefore,

$$\sqrt{(W_s \sin \phi)^2 + (\tau_s a_s)^2} = (W_s \cos \phi)C_f = W_s \cos \phi \tan \theta$$

where θ is the angle of repose of the material. Solving for the shear stress, τ_s, yields

$$\tau_s = \frac{W_s}{a_s} \sqrt{\cos^2 \phi \tan^2 \theta - \sin^2 \phi} = \frac{W_s}{a_s} \cos \phi \tan \theta \sqrt{1 - \frac{\tan^2 \phi}{\tan^2 \theta}} \quad (12\text{–}74)$$

Obviously, if the side slope angle ϕ becomes equal to, or greater than, θ, then the permitted value of τ_s becomes zero. On the bottom surface, Eq. (12–74) simplifies to:

$$\tau_b = \frac{W_s}{a_s} \tan \theta \quad (12\text{–}75)$$

The *tractive-force ratio* is defined as

$$K = \frac{\tau_s}{\tau_b}$$

Therefore,

$$K = \cos \phi \sqrt{1 - \frac{\tan^2 \phi}{\tan^2 \theta}} = \sqrt{\cos^2 \phi - \frac{\sin^2 \phi}{\tan^2 \theta}}$$

$$= \sqrt{1 - \sin^2 \phi \left(1 + \frac{1}{\tan^2 \theta}\right)} = \sqrt{1 - \sin^2 \phi \left(\frac{\sin^2 \theta + \cos^2 \theta}{\sin^2 \theta}\right)}$$

$$= \sqrt{1 - \frac{\sin^2 \phi}{\sin^2 \theta}} \quad (12\text{–}76)$$

The required minimum depth of flow to prevent scouring will be either:

$$D = \frac{\tau_s}{K_s \gamma S} \quad \text{or} \quad D = \frac{\tau_b}{K_b \gamma S}$$

depending on whether bed shear or side shear controls. For a balanced design, these would be equal, in which case (assuming $K_s = \frac{3}{4}$ and $K_b = 1$):

$$K = \frac{K_s}{K_b} = \frac{3}{4}$$

and, from Eq. (12–76),

$$\sin \phi = \sqrt{7/16} \, (\sin \theta) \quad (12\text{–}77)$$

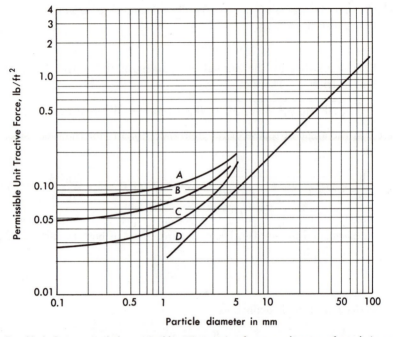

Fig. 12–6. Recommended permissible unit tractive forces on bottom of canals in non-cohesive material (U.S.B.R.): Curve A, for average particle size, with high content of fine sediment in the water, for fine bed material; Curve B, for average particle size, with low content of fine sediment in the water, for fine bed material; Curve C, for average particle size, with clear water, fine bed material; Curve D, for size of particle such that 25 per cent of particles are larger, with coarse bed material.

Therefore, when $\sin \phi < \sqrt{7/16}$ (sin θ) then the bed shear controls and the required depth is:

$$D = \frac{\tau_0}{\gamma S} \qquad (12\text{–}78)$$

When $\sin \phi > \sqrt{7/16}$ (sin ϕ), then side shear controls and the required depth is:

$$D = \frac{\tau_0}{\gamma S}\left(\frac{K}{\frac{3}{4}}\right) \qquad (12\text{–}79)$$

In these formulas, τ_0 is the permitted unit tractive force, as obtained from the Fortier-Scobey table or from the Bureau of Reclamation curve in Fig. 12–6, and K is the tractive force ratio as obtained from Eq. (12–74). Obviously, these equations apply only to non-cohesive soils, since cohesive forces have been neglected in the analysis. In the case of cohesive soils, it can be assumed that the cohesion will counterbalance the down-slope weight component, so that only the tractive force on the bottom need be considered.

Equations (12–74) and (12–75) are impracticable to solve for specific values of the critical tractive forces on the side and bottom because of the unknown value of W_s/a_s in the equations. This factor evidently depends on the properties of the particular soil, especially the particle size, for non-cohesive soils, and the voids ratio and plasticity index for cohesive soils.

The most extensive studies on this problem have been made by the U.S. Bureau of Reclamation. Current practice of this agency (though undoubtedly scheduled for revision with further study) utilizes empirical curves for permissible tractive force on the bottom of an erodible channel, as shown in Figs. 12–6 and 12–7.

The values shown in Table 12–5 calculated from the Fortier-Scobey list of permissible velocities, may also be used as a guide to the selection of a permissible tractive force, particularly for cohesive materials.

The values obtained, either from the curves or tables, apply to straight channels. Lane recommends the reductions shown below the table for sinuous channels.

The problem of designing a stable channel for a given discharge through soil of known properties consists of first determining the permissible bottom tractive force and then, if the material is non-cohesive, the permissible side tractive force, by use of the tractive-force ratio, of Eq. (12–76). The permitted critical tractive force is equated to the actual maximum tractive

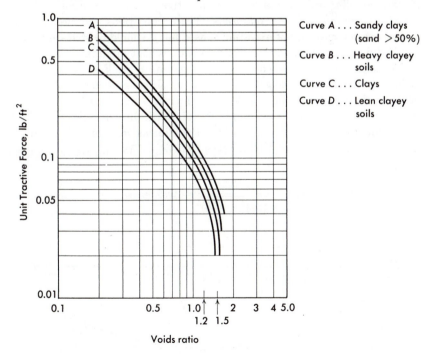

Fig. 12–7. Recommended permissible unit tractive forces on bottom of canals in cohesive materials (U.S.B.R.).

force as in Fig. 12–4 in order to determine the design depth of flow. The width of channel can then be obtained from the Manning formula. Practical modifications may be made to this theoretical section as required.

12–11. Stable Section of Greatest Efficiency.

The method of designing a stable trapezoidal channel, as outlined in the previous article, is intended to assure that the critical tractive force inducing scour at some point of the cross-section is not greater than the permissible tractive force for the given material. However, this force is equaled only at that one point, the rest of the section therefore not being stressed to capacity.

If the section is designed in such a way that every part of the bed reaches the point of incipient motion simultaneously, the depth of flow over each point of the bed will be at its maximum value for stability. Thus, as the depth decreases, the side slope may increase, a cross-section curvilinear in shape thus being established. The final cross-section resulting can be shown to give the minimum water area (therefore minimum excavation and maximum mean velocity) and the minimum top width (therefore maximum mean flow depth) for given discharge in a stable channel of erodible material.

The following analysis of such a stable section of maximum hydraulic efficiency, is based on studies made by the U.S. Bureau of Reclamation. It assumes *clear water* (or at least comparatively clear) flowing in a channel excavated in *non-cohesive* materials. These conditions are probably encountered in a majority of irrigation canals, but it should be stressed that the method does not apply for the other conditions that often prevail.

Consider a curvilinear channel cross-section, as in Fig. 12–8, the angle of inclination being ϕ at any given point (x, y). It is assumed that the tractive force on the bed due to the width of water dx, in a unit length of channel, is simply the component of the weight of water in the direction of flow, as in the usual equation for tractive force in a two-dimensional flow. The area on which this force acts is $\sqrt{(dx)^2 + (dy)^2}$. Therefore, the unit tractive force at this point is

$$\tau_s = \frac{\gamma y S \, dx}{\sqrt{(dx)^2 + (dy)^2}} = \gamma y S \cos \phi$$

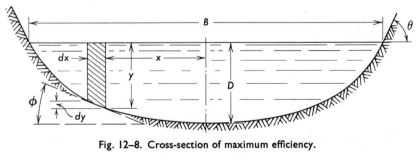

Fig. 12–8. Cross-section of maximum efficiency.

The critical shear stress on a slope is given by Eq. (12–76):

$$\tau_s = K\tau_b = \cos \phi \sqrt{1 - \frac{\tan^2 \phi}{\tan^2 \theta}}\, \gamma DS$$

assuming τ_b is given by γDS. Equating the two expressions for τ_s:

$$y = D\sqrt{1 - \frac{\tan^2 \phi}{\tan^2 \theta}} = \frac{D}{\tan \theta} \sqrt{\tan^2 \theta - \left(\frac{dy}{dx}\right)^2}$$

since $dy/dx = \tan \phi$. Thus the following differential equation of the channel cross-section is obtained:

$$\left(\frac{dy}{dx}\right)^2 + \left(\frac{y}{D}\right)^2 \tan^2 \theta - \tan^2 \theta = 0$$

Separating the variables yields

$$\int \frac{dy}{\sqrt{1 - (y/D)^2}} = \int (\tan \theta)\, dx$$

Integrating gives

$$-D \cos^{-1}\left(\frac{y}{D}\right) = (\tan \theta)x + C$$

At $x = 0$, $y = D$; and therefore $C = -D \cos^{-1} 1 = 0$. The equation of the channel cross-section becomes

$$y = D \cos\left[-\frac{(\tan \theta)x}{D}\right] = D \cos \frac{x \tan \theta}{D} \qquad (12–80)$$

Thus, the equation of a stable section of greatest efficiency is that of a simple cosine curve. If preferred, the equation can be expressed in terms of B rather than $\tan \theta$. When $x = B/2$, $y = 0$, so that

$$\cos \frac{B \tan \theta}{2D} = 0 = \cos \frac{\pi}{2}$$

and therefore

$$D = \frac{B \tan \theta}{\pi} \qquad (12–81)$$

so that, in Eq. (12–80),

$$y = D \cos \frac{\pi x}{B} \qquad (12–82)$$

The cross-section based on Eq. (12–80) is not very efficient hydraulically (the most efficient hydraulic section is that section which gives the maximum hydraulic radius for a given area of flow, and therefore the highest velocity and discharge) because of its substantial departure from a semicircle, the most efficient of all sections. However, it *is* the most efficient *stable* section. This is difficult to prove mathematically, because of the complexity of the expressions for wetted perimeter, etc. However, it can be demonstrated logically by considering two alternate sections of the same area, one more nearly corresponding to a semicircle (therefore more efficient) and one departing still more from the semicircular shape (and therefore less efficient), as shown by the dotted lines in Fig. 12–9. In neither case, of course, can the depth be permitted to be greater than D, for stability.

If the section is altered as in Fig. 12–9a, it may become more efficient, but is unstable, at least at points m and n, where the depth is the same as before but the side slope has been increased. If the section is altered as in Fig. 12–9b, it remains stable (as long as the slope is never increased at points of the same depth) but is wider and shallower. Therefore its wetted perimeter is increased and it is less efficient.

Therefore, the section of Eq. (12–80) gives the smallest width and greatest hydraulic radius for a given area, and therefore has the maximum hydraulic capacity for any stable channel carrying relatively clear water in noncohesive materials.

However, the geometry of the channel makes the calculation of discharges more difficult than in a trapezoidal channel. The area of flow is given by

$$A = 2 \int_0^{B/2} y \, dx = 2D \int_0^{B/2} \cos \frac{x \tan \theta}{D} \, dx$$

$$= \frac{2D^2}{\tan \theta} \left[\sin \frac{x \tan \theta}{D} \right]_0^{B/2} = \frac{2D^2}{\tan \theta} \sin \frac{B \tan \theta}{2D}$$

$$= \frac{2D^2}{\tan \theta} \sin \frac{\pi D}{2D} = \frac{2D^2}{\tan \theta} = \frac{2}{\pi} BD \qquad (12\text{–}83)$$

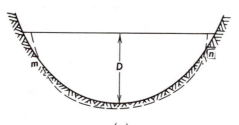

(a)

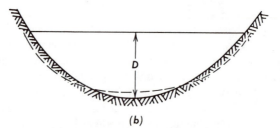

(b)

Fig. 12–9. Comparison of stable sections.

The wetted perimeter is computed as follows:

$$P = 2 \int_0^{B/2} \sqrt{1 + \left(\frac{dy}{dx}\right)^2} \, dx = 2 \int_0^{B/2} \sqrt{1 + \tan^2 \theta \, \sin^2 \frac{x \tan \theta}{D}} \, dx$$

since $dy/dx = -\tan \theta \sin (x \tan \theta / D)$. This expression can be written

$$P = \frac{2D}{\tan \theta} \int_0^{B/2} \sqrt{1 + \tan^2 \theta \, \sin^2 \frac{x \tan \theta}{D}} \, d\left(\frac{x \tan \theta}{D}\right)$$

$$= \frac{2D}{\tan \theta} \int_0^{\pi/2} \sqrt{1 + \tan^2 \theta \, \sin^2 u} \, du$$

where $u = (x \tan \theta)/D$ and is equal to $\pi/2$ when $x = B/2$. This expression can be rewritten as an elliptic integral by introducing the parameter $\psi = \pi/2 - u$, with $d\psi = -du$. Then,

$$P = -\frac{2D}{\tan \theta} \int_{\pi/2}^0 \sqrt{1 + \tan^2 \theta \, \cos^2 \psi} \, d\psi = \frac{2D}{\tan \theta} \int_0^{\pi/2} \sqrt{1 + \tan^2 \theta \, (1 - \sin^2 \psi)} \, d\psi$$

$$= \frac{2D\sqrt{1 + \tan^2 \theta}}{\tan \theta} \int_0^{\pi/2} \sqrt{1 - \frac{\tan^2 \theta}{1 + \tan^2 \theta} \sin^2 \psi} \, d\psi$$

$$= \frac{2D}{\sin \theta} \int_0^{\pi/2} \sqrt{1 - \sin^2 \theta \, \sin^2 \psi} \, d\psi \tag{12-84}$$

This expression is now in the form of a complete elliptic integral of the second kind, written as $E(\sin \theta)$, of "modulus" $\sin \theta$. The function is available most conveniently from tables of elliptic integrals. However, it can also be evaluated by term-by-term integration of the integrand as expanded by the binomial theorem. The resulting expression is

$$E(\sin \theta) = \int_0^{\pi/2} \sqrt{1 - \sin^2 \theta \, \sin^2 \psi} \, d\psi$$

$$= \frac{\pi}{2} \left[1 - \left(\frac{1}{2}\right)^2 \sin^2 \theta - \left(\frac{1 \cdot 3}{2 \cdot 4}\right)^2 \frac{\sin^4 \theta}{3} - \left(\frac{1 \cdot 3 \cdot 5}{2 \cdot 4 \cdot 6}\right)^2 \frac{\sin^6 \theta}{5} \right.$$

$$\left. - \left(\frac{1 \cdot 3 \cdot 5 \cdot 7}{2 \cdot 4 \cdot 6 \cdot 8}\right)^2 \frac{\sin^8 \theta}{7} - \cdots \right] \tag{12-85}$$

Finally, the wetted perimeter of the section is given by

$$P = \frac{2D}{\sin \theta} E(\sin \theta)$$

$$= \frac{2D}{\sin \theta} \left\{ \frac{\pi}{2} \left[1 - \left(\frac{1}{2}\right)^2 \sin^2 \theta - \left(\frac{1 \cdot 3}{2 \cdot 4}\right)^2 \frac{\sin^4 \theta}{3} - \left(\frac{1 \cdot 3 \cdot 5}{2 \cdot 4 \cdot 6}\right)^2 \frac{\sin^6 \theta}{5} - \cdots \right] \right\}$$

$$\tag{12-86}$$

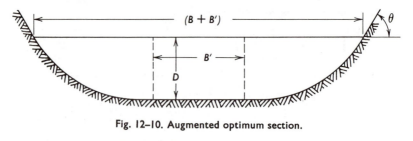

Fig. 12–10. Augmented optimum section.

The hydraulic radius is now given by

$$R_h = \frac{A}{P} = \frac{2D^2/\tan \theta}{(2D/\sin \theta)E(\sin \theta)} = \frac{D \cos \theta}{E(\sin \theta)} \qquad (12\text{–}87)$$

The channel discharge is computed from the Manning formula:

$$Q_o = \frac{1.5}{n} \frac{2D^2}{\tan \theta} \left[\frac{D \cos \theta}{E(\sin \theta)}\right]^{2/3} S^{1/2} = \frac{3D^{8/3}(\cos \theta)^{2/3}S^{1/2}}{n(\tan \theta)[E(\sin \theta)]^{2/3}} \qquad (12\text{–}88)$$

where $E(\sin \theta)$ is evaluated either from a table of elliptic integrals or by Eq. (12–85).

The discharge Q_o of Eq. (12–88) is the discharge that would be obtained in the channel as designed for greatest efficiency for the given material and bed slope. However, the actual design discharge Q may be either smaller or greater than Q_o.

If Q is greater than Q_o, additional channel area must be provided. However, the maximum depth cannot be greater than the stability depth D, as computed from $\tau_b/\gamma S$, τ_b being the permissible tractive force on a level bed. Consequently, a rectangular section is added at the center of the theoretical cosine-curve cross-section, as shown in Fig. 12–10.

The additional width required, B', is determined from the Manning formula, by trial solution:

$$Q = \left(\frac{1.5}{n}\right)\left(\frac{A^{5/3}}{P^{2/3}}\right)S^{1/2}$$

$$= \left(\frac{1.5}{n}\right) \frac{[2D^2/\tan \theta + B'D]^{5/3}}{[(2D/\sin \theta)E(\sin \theta) + B']^{2/3}} S^{1/2} \qquad (12\text{–}89)$$

If, on the other hand, the design discharge Q is less than the theoretical discharge Q_o, then economy would require that a portion of the theoretical channel be eliminated, as sketched in Fig. 12–11, with the actual width and depth, B'' and D'', less than the values B and D for the most efficient stable channel.

An exact solution of this problem would require determination of the hydraulic radius for the partial section in terms of the width B'', with the latter then computed through insertion in the Manning formula. Although

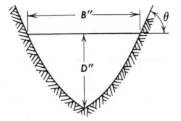

Fig. 12-11. Reduced optimum section.

this is quite possible, it is mathematically tedious. An approximate solution is obtained by assuming that the mean velocity in the partial section is the same as the mean velocity in the theoretical section, and thus that the discharges are proportional to the cross-sectional areas of flow. From Eq. (12-88),

$$Q_o = \frac{2D^2}{\tan \theta} V = \frac{2B^2 \tan \theta}{\pi^2} V \qquad (12\text{-}90)$$

As an approximation it is assumed that the discharge in the channel of Fig. 12-11 is given by

$$Q = \frac{2B''^2 \tan \theta}{\pi^2} V \qquad (12\text{-}91)$$

the same form of equation as Eq. (12-90), keeping the average velocity constant. Then, combining Eqs. (12-90) and (12-91) gives

$$B'' = B\sqrt{\frac{Q}{Q_o}} \qquad (12\text{-}92)$$

Example 12-4

An unlined irrigation canal is to be constructed in non-cohesive material having an angle of repose of $30°$ and particle size of $\frac{1}{4}$ mm. The discharge is 100 cfs, bed slope 0.0001, and Manning coefficient 0.020. Design the most efficient stable channel for these conditions, including center depth, bottom width, surface width, and equation of sides. $E(\sin \theta) = 1.4675$, for $\theta = 30°$.

Solution. From Fig. 12-6, curve C, permitted $\tau_b = 0.029$ lb/ft^2

Max. depth permitted $= D = \dfrac{\tau_b}{\gamma S} = \dfrac{0.029}{62.4(0.0001)} = 4.65$ ft

Surface width for ideal section $= B = \dfrac{\pi D}{\tan \theta} = \dfrac{\pi(4.65)}{\tan 30°} = 25.3$ ft

Equation of sides: $y = D \cos\left(\dfrac{x \tan \theta}{D}\right) = 4.65 \cos\left(\dfrac{x}{8.06}\right)$

Hydraulic radius $= \dfrac{D \cos \theta}{E (\sin \theta)} = \dfrac{4.65(0.866)}{1.4675} = 2.74 \, \text{ft}$

Permitted discharge, $Q_0 = \dfrac{1.5}{n} \dfrac{2D^2}{\tan \theta} R_h^{2/3} S^{1/2}$

$$= (75) \left(\dfrac{2(4.65)^2}{0.577} \right) (2.74)^{2/3}(0.01) = 110 \, \text{cfs}$$

Actual discharge, $Q = 100$ cfs. $\therefore$ reduce B.

Actual surface width, $B'' = B \sqrt{\dfrac{Q}{Q_0}} = \dfrac{25.3}{\sqrt{1.1}} = 24.1 \, \text{ft}$ *Ans.*

(Bottom width $= 0$ ft.)

Center depth $= D'' = D_{\max} \cos \dfrac{(B - B'')}{2(8.06)} = 4.65 \cos \left(\dfrac{1.2}{16.12} \right) = 4.61 \, \text{ft}$ *Ans.*

Corrected equation of sides (origin at channel bottom):

$$y = 4.65 \cos \left(\dfrac{x + 0.6}{8.06} \right) \qquad \qquad \textit{Ans.}$$

12–12. Grassed Channels. Many channels, both natural and artificial, are lined with grasses or other vegetations. This may have the dual effect of increasing bed friction, with resulting velocity reductions, and of stabilizing the soil banks and beds for greater erosion resistance.

The complex nature of the flow turbulence produced by such boundaries, however, makes it very difficult to estimate roughness coefficients. In fact, the latter vary with the nature of the grass and with the season of the year, as well as with the hydraulic radius and the velocity of flow, all of which have an effect on the manner in which the grass contributes to the flow turbulence. Because of its variable nature under these conditions, the Manning roughness coefficient has, in this case, been given a special name, the *retardance coefficient*.

Extensive tests conducted by the U.S. Soil Conservation Service have resulted in a set of empirical design curves and tables for use in the design of stable grassed channels. Examination of these data seem to indicate that the retardance coefficient is mainly dependent on the average depth of the covering grass and the Reynolds number, the latter expressed simply as the product of the average velocity and the hydraulic radius, VR_h. The condition of stand (density of grass, in stems per square foot) and the kind of grass also influence the coefficient.

The S.C.S. has classified the data in terms of five different categories of retardance, as in Table 12–6. The average experimental $n - VR_h$ curves for these five categories of retardance are shown on Fig. 12–12.

Although the grass length is perhaps of paramount importance in determining the degree of retardance, it should be understood that the type of

TABLE 12-6
Degrees of Retardance in Grassed Channels

Stand	Average Grass Length, (in.)	Degree of Retardance
	>30	A. Very high
	11–24	B. High
Good	6–10	C. Moderate
	2–6	D. Low
	<2	E. Very low
	>30	B. High
	11–24	C. Moderate
Fair	6–10	D. Low
	2–6	D. Low
	<2	E. Very low

grass and its stand also are significant. The curves of Fig. 12–12 represent averages only. The S.C.S. data and publications should be consulted for more specific values.

The design of a grassed channel must also include stability determinations, by the method of Art.12–12. However, permissible velocities or tractive forces depend primarily on the type of grass used for the lining. The Soil Conservation Service recommends permissible velocities as shown in Table 12–7, assuming that good covers can be properly maintained.

It should be noted that the bunch grasses listed in the bottom row should not be used on slopes greater than 5 per cent, since they tend to cause flow channeling and high concentrations of velocity.

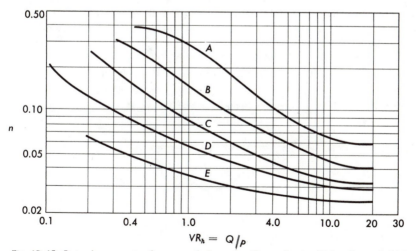

Fig. 12–12. Retardance curves for grassed channels: Curve A, very high; Curve B, high; Curve C, moderate; Curve D, low; Curve E, very low; n = Manning roughness coefficient (or "retardance coefficient"), V = average velocity (ft/sec), R_h = hydraulic radius (ft).

TABLE 12–7
Permissible Velocities in Grassed Channels

Grass	Slope (%)	Permissible Velocity (fps)	
		Erosion-resistant Soils	Easily Eroded Soils
Bermuda	0–5	8	6
	5–10	7	5
	>10	6	4
Buffalo grass, Kentucky bluegrass, smooth brome, blue grama	0–5	7	5
	5–10	6	4
	>10	5	3
Grass mixture	0–5	5	4
	5–10	4	3
Lespedeza sericea, weeping love grass, ischaemum (yellow blue stem) Kudzu, alfalfa, crabgrass	0–5	3.5	2.5
Annuals, common lespedeza, Sudan grass	0–5	3.5	2.5

Example 12–5

A drainage ditch in an erodible soil is to be designed with a stable triangular section, and lined with a good stand of Kentucky bluegrass. The bed slope is 1% and the design discharge 80 cfs. The ditch will be used as soon as the grass starts to grow, but ultimately the grass length will be maintained at about 8 inches. Assuming a 6 in. freeboard, what should be the depth and side slopes for the ditch?

Solution. From Table 12–7, permitted velocity $= 5$ ft/sec

Required area, $A = \dfrac{Q}{V} = \dfrac{80}{5} = 16$ sq ft $= zD^2$

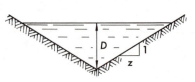

$$R_h = \frac{A}{P} = \frac{16}{2D\sqrt{z^2 + 1}} = \frac{8}{\sqrt{\dfrac{16}{z}(z^2 + 1)}}$$

From Manning equation, $R_h = \left(\dfrac{nV}{1.5S^{1/2}}\right)^{3/2} = 191n^{3/2}$

$\therefore\ VR_h = 955n^{3/2}$. Use curve E, Fig. 12–12.

Assume $n = 0.03$. Then $VR_h = 955(0.03)^{3/2} = 4.86$

From curve E, $n = 0.027$. Then $VR_h = 4.3$

From E, $n = 0.027$. O.K.

$$\therefore\ R_h = \frac{4.3}{5} = 0.86 \text{ ft} = \frac{8}{\sqrt{\dfrac{16}{z}(z^2 + 1)}}$$

Solving, $z = 5.21$. $\therefore\ D = \dfrac{16}{z} = 1.75$ ft required initially.

After full growth, use Curve C (from Table 12–6).

From Manning equation, $Q = \dfrac{1.5}{n} S^{1/2} A R_h^{2/3}$

$$A R_h^{2/3} = \frac{(zD^2)^{5/3}}{(2D\sqrt{z^2 + 1})^{2/3}} = \frac{z^{5/3}}{(2\sqrt{z^2 + 1})^{2/3}} D^{8/3}$$

Final flow depth $= D = \left[\dfrac{80n(2)^{2/3}(5.21^2 + 1)^{2/3}}{5.21(5.21)^{2/3}(0.1)(1.5)}\right]^{3/8} = 6.78n^{3/8}$

$$VR_h = \frac{Q}{P} = \frac{80}{2D\sqrt{5.21^2 + 1}} = \frac{7.54}{D} = \frac{1.11}{n^{3/8}}$$

Assume $n = 0.05$. Then $VR_h = \dfrac{1.11}{(0.05)^{3/8}} = 3.32$

From curve C, $n = 0.047$. $VR_h = 3.48$

From curve C, $n = 0.047$. O.K.

$$\therefore\ D = \frac{7.54}{3.48} = 2.16 \text{ ft. Add 0.5 ft for freeboard.}$$

$\therefore$ Final dimensions are: $D = 2.66$ ft, $z = 5.21$. *Ans.*

PROBLEMS

12–1. A certain reach in a river has its channel split into two arms by a longitudinal bar. The average breadth and depth of the two forks are 800 ft and 8 ft and 400 ft and 5 ft respectively. The length and roughness of the two channels are the same ($n = 0.030$ in each branch). In the upstream channel above the island the breadth is 900 ft, the mean depth 10 ft, and the roughness coefficient 0.030. The mean size of bed particles is 0.25 mm. The total discharge in the stream is 35,000 cfs.

    a. What is the flow in each branch, in cfs?

    b. What is the slope in the main stream, and in each branch, in feet per mile?

    c. Is deposition or erosion likely to take place in each branch? Assume that the suspended load is negligible, and that the bed load in the main stream splits in the branches in proportion to their respective discharges. Use the duBoys formula in estimating bed load.

12–2. The sediment transported by a stream has a mean diameter of $\frac{1}{4}$ mm and specific weight of 162.4 lb/cu ft. The stream has a longitudinal slope of 0.0004 and a Manning coefficient of 0.037. The discharge is 12 cfs/ft and the water temperature is 68°F. Compute the sediment load of the stream, in (lb/sec)/ft, assuming uniform two-dimensional flow, by the following methods:

    a. The bed load by duBoys' formula.

    b. The bed load by the Rouse modification of the Einstein formula.

    c. The total load by the formula based on the Albertson-Garde studies.

12–3. A river bed is composed predominantly of sand averaging 0.25 mm in diameter. The river can be considered approximately rectangular in cross-section, with a width of 200 ft. It has a bed slope of 0.0004 and a Manning roughness coefficient of 0.040. When the discharge in the river is 2000 cfs, what is the rate of bed load transport, in tons per day, as based on the duBoys formula? Assume uniform flow. *Ans.:* 8400 tons/day.

12–4. The sediment transported in uniform two-dimensional flow by a certain river of bed slope 0.0004 has the following characteristics: $d = \frac{1}{8}$ mm, $\gamma_s = 160$ pcf, $v_s = 0.03$ fps, $n = 0.040$. When the discharge is 10 cfs/ft, compute the bed-load transportation, in (lb/sec)/ft, by the following:

    a. DuBoys formula. *Ans.:* 1.6 (lb/sec)/ft.

    b. Shields formula. *Ans.:* 6.3 (lb/sec)/ft.

    c. Rouse-Kalinske formula. *Ans.:* 1.4 (lb/sec)/ft.

    d. Rouse-Einstein formula. *Ans.:* 0.79 (lb/sec)/ft.

12–5. Calculate the following soil properties, for sediment particles $\frac{1}{8}$ mm in diameter, with a specific gravity of 2.65, and angle of repose of 40°.

    a. Fall velocity; assume kinematic viscosity $= 10^{-5}$ ft²/sec.

    b. Critical mean tractive stress, assuming $\frac{1}{3}$ of bed area at point of incipient motion simultaneously.

    c. Ratio of suspended sediment concentrations at the quarter point and midpoint of the flow depth, in terms of the flow depth and bed slope only.

12–6. A river channel has an equilibrium section which approximates a trapezoid, with base width 20 ft and side slopes of 2 horizontal to 1 vertical. The longitudinal bed slope is 0.0001 and the Manning coefficient 0.020. The depth of flow is uniform at 10 ft. The bed material has a critical tractive stress of 0.02 lbs/ft² and a sediment characteristic of 200,000 lb/ft³ sec.

    a. Estimate the total bed-load transportation through this section, in lb per sec.

    b. If the sediment concentration 2 ft below the surface at the channel center-line is 0.04 and the fall velocity of the suspended sediment particles is 0.01 ft/sec, estimate the average suspended sediment in the vertical at this location.

12–7. A storage reservoir is constructed on a stream with an average discharge of 500 cfs and a "bed-building discharge" of 1000 cfs. The stream is 200 ft wide, on a bed slope of 0.000256, with a Manning coefficient of 0.030. The sediment is 80 per cent fine sand, 10 per cent silt, and 10 per cent clay. It will always be submerged after deposition in the reservoir. The effective bed load size is $\frac{1}{8}$ mm. Neglect suspended load. It is desired to provide 80,000 acre-feet of regulation storage. How much additional storage should be provided for sediment storage, for an anticipated 100-year life?

12–8. A storage reservoir is constructed on a stream draining a watershed of 26,000 sq mi. The average annual inflow on the stream is 45 acre-ft/sq mi. A storage reservoir of 2,500,000 acre-ft capacity is to be constructed on the stream. It is desired to estimate the useful capacity that will be destroyed by silting during the estimated 100-year life of the reservoir.

It is estimated, from sediment measurements in the river, that the average annual production of sediment from the watershed is 900 tons/sq mi. This sediment is approximately 60 per cent sand, 25 per cent silt, and 15 per cent clay, and once deposited in the reservoir will remain almost always submerged.

Determine the trap efficiency of the reservoir, and the useful capacity destroyed:

    a. By the capacity-watershed ratio method.

    b. By the capacity-inflow ratio method.

12–9. A storage reservoir on a stream has a capacity of 1,000,000 acre-ft. The stream has an average annual discharge of 500,000 acre-ft and a mean bed slope of 0.0025. The sediment load in the stream is almost entirely bed load, composed of fine sand averaging $\frac{1}{8}$ mm in diameter, transported during the spring floods. These may be considered to last two months and to have an average discharge of 1600 cfs. The river is 200 ft wide and has a Manning coefficient of 0.030. Estimate the loss of capacity due to sedimentation that the reservoir will have suffered after a 50-year period.

12–10. A storage reservoir of 100,000 acre-ft capacity is on a stream draining a watershed of 5000 sq mi above the dam. The stream has a bed slope of 0.0004 and a Manning coefficient of 0.030. The sediment load, consisting predominantly of $\frac{1}{8}$ mm sand, is determined by the duBoys formula. On the basis of a dominant discharge of 1000 cfs, with the stream bed 250 ft wide, determine the approximate reservoir capacity eliminated by sedimentation in 60 years.

12–11. A trapezoidal channel carrying water with some colloidal silts has a longitudinal slope of 0.00008, a roughness coefficient of 0.025, and side slopes of 2 horizontal to 1 vertical. The material in which the canal is excavated is fine sand with an average particle size of 1.0 mm. The angle of repose is 36°.

    a. For a design discharge of 3200 cfs, what should be the base width and flow depth of the channel, determined on the basis of the Fortier-Scobey table?

    b. If the minimum discharge is 2000 cfs, will there be danger of silting in the channel, as based on the Kennedy formula?

12–12. Design the channel of Problem 12–11 on the basis of the U.S.B.R. method of tractive force.

12–13. Design the channel of Problem 12–11 if it is excavated in earth containing non-colloidal coarse gravels and pebbles (25% of which exceeds 1.25 in. in diameter) and is moderately sinuous. The angle of repose is 33.5°. Use the method of tractive force, and assume the slope is 0.0004.

12–14. Design the same channel, by the method of tractive force, if the bed material is compact clay, with a voids ratio of 0.5.

12–15. In an irrigation system there are to be two canals, carrying discharges of 100 cfs and 20 cfs. Both canals have a slope of 0.001 and are to be excavated through material composed of sand and gravel, slightly rounded in shape. The size d_{25} is $\frac{1}{4}$ in. and the angle of repose is 21.5°. Design the most economical stable cross-sections for these canals. Manning n is 0.020.

12–16. A channel is to be excavated in fine sand (average particle diameter is $\frac{1}{4}$ mm, angle of repose is 20°) and will carry clear water. The slope is 0.0001, the design discharge is 200 cfs, and the roughness coefficient is 0.020. Design the channel so that it will have the smallest possible wetted perimeter and still meet the capacity and stability requirements.

12–17. Design the channel of Problem 12–16 as a stable trapezoidal channel and compare the two designs.

12–18. Determine the cross-section of the optimum stable section for a channel in a soil of non-cohesive material, for which the critical tractive stress is 0.1 lb/ft^2, the bed slope is 0.0004, angle of repose is 31°, and roughness coefficient is 0.020. Compute the discharge which can be carried at uniform flow in this channel. *Ans.:* 141 cfs.

12–19. An unlined irrigation canal is to be constructed in non-cohesive material having an angle of repose of 30° and particle size of $\frac{1}{4}$ mm. The discharge is 400 cfs, bed slope is 0.0004 and Manning coefficient 0.020. Design the most efficient stable channel for these conditions, including center depth, bottom width, surface width, and equation of sides.

12–20. An unlined irrigation canal is to be excavated in coarse non-cohesive material of size distribution such that 75% of the grains will pass through a 2.0 mm screen opening. The angle of repose of this material is 30°. The Manning coefficient is 0.030 and the bed slope 0.0001. Determine the maximum depth, surface width, and equation of sides if the channel is designed as an "optimum stable channel," to carry a discharge of 400 cfs.

12–21. Design the best stable *non-curvilinear* channel to carry a design discharge of 150 cfs with a slope of 0.0001 and a Manning coefficient of 0.020. The channel will carry water with some suspended silt and is to be excavated in granular material of average grain size 1.0 mm and angle of repose 30°. Allow a freeboard of 1.0 ft.

12–22. A drainage canal is trapezoidal in cross-section, lined with a good stand of Bermuda grass, on a longitudinal slope of 5 per cent. The design discharge is 100 cfs. Initially the grass averages only about 4 in. in height, but after full development it will be kept about 12 in. high. The canal is in an erosion-resistant cohesive soil, and the side slopes are 1 : 1. Allowing a 6-in. freeboard, what should be the width and depth of the canal? *Ans.:* 20.0 ft, 2.7 ft.

12–23. An irrigation canal is to be designed with a stable trapezoidal section. The design discharge is 80 cfs and the bed slope is 1 per cent. It is to be lined with a good stand of mixed grasses on an erodible sandy soil. Although water will be admitted to the canal as soon as grass begins to grow, it is intended ultimately to keep the grass approximately 15 in. high. Assuming side slopes of 2 horizontal to 1 vertical, and a 6-in. freeboard, what should be the bottom width and depth of the canal?

12-24. A drainage ditch is to be designed with a stable triangular section. It is to be lined with a good stand of Kentucky bluegrass on an erodible soil. The bed slope is 4% and the design discharge 40 cfs. The ditch will be used as soon as the grass starts to grow, but ultimately the grass length will be maintained about 8 inches. Assuming a 6-in. freeboard, what should be the depth and side slopes for the ditch?

13

STREAM CHANNEL MECHANICS

13–1. The Geomorphic Function of Rivers. A river, in addition to its more direct and obvious function as a watershed drainage channel, also performs a very important long-time function. The soil and rock and other detritus eroded on its watershed are emptied into it and gradually transported downstream, eventually to the ocean or to some other base level. In the course of long periods of time, the watershed could presumably be eroded down to almost a plain (a peneplain, in geomorphologic terminology), the sediment so eroded being deposited in deltas or alluvial plains near the river's mouth. However, there is no place known in the present world where such a peneplain has actually been produced.

In the long run, the river can therefore approach true "equilibrium" conditions only asymptotically, and there is no certainty that any stream ever has attained such a condition in the past, and there are none at present. However, for engineering purposes, many streams are sufficiently stable over periods of decades or even centuries to warrant being classed as in equilibrium, with a more or less permanent average discharge, average sediment load, and average channel size and bed slope.

On the other hand, some streams are known to be actively aggrading or degrading their channels and therefore are not in even temporary equilibrium. Furthermore, by virtue of changes in various watershed conditions, streams which have been in practical equilibrium may suddenly become unstable. This is especially true when engineering works have been constructed on the stream or when the watershed characteristics have been modified.

Because of the many uses of a stream and its waters, in the life of man, modifications of its natural regime are being increasingly imposed upon it, in the form of dams, levees, channel dredging, revetments, contractions, etc., not to mention all the ways in which the hydrologic character of the watershed is continually being altered. It is extremely important, therefore, that the engineer who is responsible for these works (or who must deal with the effects of works constructed by others) have a good understanding of the natural tendencies of the individual stream, whether in equilibrium or otherwise, and the hydrologic and geomorphic reactions which it may develop when so disturbed.

13-2. The Concept of Equilibrium. We define, therefore, a stream in equilibrium as one which has developed just the right bed slope and cross-section to transport the discharges and sediment loads which are delivered to it from its watershed. Such a stream is called a *graded stream* by geomorphologists. The *grade* of a stream is considered to be its bed slope; if the grade is too flat for the sediment load, some will be deposited, thus steepening the grade, and if it is steeper than necessary, the stream will erode the bed and thus flatten the grade. Thus, as long as the conditions of discharge and imposed sediment remain unchanged, there is an over-all tendency for the stream to approach a graded condition.

This is an oversimplification, of course, as other hydraulic factors besides the slope determine the capacity of a stream to transmit sediment, and these may react instead of, or along with, the bed in response to a change in conditions. Furthermore, a given stream may be subjected to a wide range of discharges and sediment loads, and thus may be ephemerally unstable even though perennially at grade.

This latter problem requires the concept of a *dominant discharge* for the stream, as that discharge which corresponds to the equilibrium condition of the stream channel and its slope. There are various ways in which this discharge has been defined quantitatively, one fairly common and reasonable definition being that it is that discharge above which half of the total sediment load is carried. This discharge is also called the *bed-building discharge* and the corresponding river stage the *bed-building stage*.

The bed-building stage can be determined quantitatively for a given stream from a duration curve, of sediment transportation versus days equaled or exceeded. The river stage for which the areas under the sediment duration curve on either side are equal is the bed-building stage. The specific relation between sediment load and stage or discharge must also be known in order to convert the mean sediment load into dominant discharge, bed-building stage, or similar parameter. If most of the load is bed load, a formula of the duBoys type, Eq. (12–11), may be used. If both bed load and suspended load are present in appreciable amounts, then one of the total load methods must be used.

13-3. Regime Relations for Stream Variables. A stream in equilibrium, or graded, is also said to be *in regime*. Even though it may not be in the regime condition, if it is being adjusted in the direction of attaining such a condition, it is said to be a regime stream (or *regime canal*).

Exact theoretical formulations of the relationships between the hydraulic geometry of stream channels and their geomorphic behavior are of course not yet developed. However, many field measurements have been made, on both natural and artificial channels, which permit empirical relationships to be derived.

The extensive development of the irrigation systems in the Punjab in

North West British India (now Pakistan) during the latter part of the nine-teenth and early part of the twentieth century, under governmental direction, served as the initial stimulus to the collection and correlation of much of the field data on which the present regime theory is based. R. G. Kennedy published the first such study. Later, important contributions were made by Lindley, Lacey, Inglis, and Blench, each of these authors basing his studies primarily on the Indian canal measurements.

A similar approach has been recently developed in this country by Leopold and other engineers and geologists of the U.S. Geological Survey, based primarily on measurements in regime rivers rather than canals. In Egypt, Leliavsky has developed specialized methods based on his own synthesis of the regime equations and the theoretical formulations of duBoys, Einstein, and others.

Whether or not the regime theory offers the best ultimate approach to problems in stream mechanics is currently a matter of considerable difference of opinion. Nevertheless, it has served as a basis for design of many hydraulic works in different countries and a general understanding of its major im-plications is quite important. The following discussion is taken largely from the work of Blench, who has in essence summarized and codified the work of his predecessors, and from that of Leopold and his colleagues.

Several observational facts are first listed:

1. In alluvial channels, the distribution of bed sand sizes is statistically normal, implying formation by a natural process of weathering and attrition.

2. The sand in the channel bed moves in various ways: at subcritical velocities, it moves in ordinary dunes; at low supercritical velocities it moves as a flat sheet; and at velocities exceeding about $2.5V_c$, it moves in anti-dunes, looking like sine curves and progressing upstream.

3. If the flow has considerable silt or clay in suspension, the material will deposit well-defined banks and even natural levees. Banks are poorly defined in purely sand or gravel rivers.

4. The ratio of width to depth in a regime channel increases as the stream's dominant discharge increases and as the size and abundance of the bed material increases.

5. The slope of a stream apparently decreases as the discharge increases and as the bed-load size and quantity decreases.

6. The slope of a stream is so dependent on its regime conditions of load and discharge that, even if the stream is dammed, it will gradually aggrade up the valley until its regime slope is re-established.

7. The size and spacing of meander loops in an alluvial river increases with increase in the dominant discharge.

8. Straight channels in natural rivers are very rare, there being virtually no reaches that are straight for distances exceeding ten times the channel width.

9. A braided, or anastomosing river (one which flows in two or more anas-

tomosing channels around alluvial islands) usually exhibits steeper slopes and greater width-depth ratios in the divided channels than in the undivided reaches carrying the same flow.

10. The thalweg (line of maximum depth) even in a straight channel follows a meandering path within the channel and apparently under the same laws governing meander patterns in the stream itself.

11. Channel width in a natural channel is largely determined by the discharge at bank-full stage.

12. Sediment load is determined not only by bed shear and grain size (as in the duBoys and similar formulas) but also by the roughness of the bed.

13. The bed roughness is affected both by the grain size and by the bed configuration, which in turn is affected by the discharge, slope, etc.

14. A given stream may, within short distances, have straight, meandering, or braided reaches; similarly, streams of different types may join each other.

15. The hydraulic behavior of ephemeral streams and their channels is essentially the same as that of perennial streams.

16. Even ungraded streams behave so nearly as do graded streams that only flagrant conditions of disequilibrium can be distinguished by quantitative measurements over short periods.

13-4. Regime Equations.

In an open channel of rigid boundaries, uniform flow can be described by a single formula, such as the Manning formula. That is, for a given discharge and channel, one and only one depth can be established uniformly.

If the channel has a movable boundary, however, then the breadth and slope can change, along with the establishment of flow depth. The adjusted slope will largely depend upon the ability of the flow to transport its bed load charge. If the load is too great, some will be deposited, thus increasing the slope, and vice versa. Similarly, the sides may also erode and the width thus increase or, if suspended load is present and velocities small enough, bank deposition may occur. The depth of flow then depends not only on discharge but also on the adjusted width and slope, each of which also depends on discharge. Consequently, three independent equations are necessary to describe uniform (or *in regime*) flow in an erodible channel.

The three equations can be expressed in various forms, but perhaps most conveniently are set up to give the slope, width, and depth for a given discharge under equilibrium conditions. The slope equation primarily expresses equilibrium between the gravitational component along the bed and the frictional resistance to flow, as in the Chezy or Darcy formulas. The depth equation is essentially a statement of equilibrium between tractive forces at the bed and force required to move the equilibrium bed charge. The width equation similarly relates the tractive forces and resistances along the sides.

Observations of Lacey, Blench, and others in the regime canals of India, led them to postulate a slope equation of the same form as the Blasius equation for flow in smooth pipes. The Blasius equation is as follows:

$$S = \left(\frac{f}{4R_h}\right)\left(\frac{V^2}{2g}\right) = \frac{(0.316/(N_R)^{0.25})V^2}{8gR_h} \qquad (13\text{--}1)$$

Lacey's observations yielded the following modification of this equation:

$$S = \frac{V^2}{3.63gD(BV/\nu)^{0.25}} \qquad (13\text{--}2)$$

where V is the velocity, D the depth, B the width, and ν the kinematic viscosity. It is interesting that a smooth-flow function can be used to describe flow in a channel with moving boundaries, and most American hydraulic engineers question its validity. Nevertheless, the data on the regime canals appeared to satisfy it. In terms of discharge, it becomes

$$S = \frac{Q^2}{3.63gDB^{0.25}(Q/A)^{0.25}(1/\nu)^{0.25}A^2}$$

$$= \frac{Q^{1.75}}{3.63(g/\nu^{0.25})A^{1.75}\,DB^{0.25}}$$

If B is assumed to be the mean width, this yields

$$S = \frac{Q^{1.75}}{3.63(g/\nu^{0.25})B^2D^{2.75}} \qquad (13\text{--}3)$$

The depth equation stems from the observation that the Froude number for the flow, V^2/gD, is empirically related to the bed-load quality. Lacey's empirical rule is that

$$\frac{V^2}{gD} = \frac{1.9}{g}\sqrt{d_m} \qquad (13\text{--}4)$$

where d_m is the median diameter of bed particles (sand) in millimeters, and the Froude number is that for which the bed-load becomes very small. Again, in terms of discharge,

$$D = \left(\frac{Q^2}{1.9\,d_m^{1/2}B^2}\right)^{1/3} \qquad (13\text{--}5)$$

However, Eq. (13–4) is considered only approximate and may require substantial modification, particularly for larger sizes or quantities of bed load.

The breadth equation does not have a very obvious dynamic basis but has been empirically correlated with numerous field observations. In the form developed by Lacey, it is

$$B = 2.67Q^{1/2} \qquad (13\text{--}6)$$

Equations (13–3), (13–5), and (13–6) are the three basic regime equations, although they can be and have been written in numerous alternate forms. They give the slope, depth (at thalweg) and breadth (at middepth) toward which a regime channel will tend, in terms of a given discharge and bed-sand size. B and D can be eliminated from Eq. (13–3) and B from Eq. (13–7), which are then simplified as follows:

$$S = \frac{(\nu^{1/4}) \, d_m^{11/24}}{77Q^{1/6}} \tag{13-7}$$

and

$$D = \left(\frac{Q}{13.5\sqrt{d_m}}\right)^{1/3} \tag{13-8}$$

Thus, Eqs. (13–6), (13–7), and (13–8) show the direct dependence of regime width, slope, and depth on the dominant discharge and bed-sand size. Historically, B was used to replace the wetted perimeter and D the hydraulic radius, but it is believed that this approximation is relatively inconsequential in natural regime channels, although more data might yield better values for the constant terms in the equation.

Studies in American rivers indicate that Eq. (13–6) is of correct form for graded streams, but that the constant, 2.67, is actually different for different streams. The same sort of qualification may be placed on the application of Eqs. (13–7) and (13–8) to rivers, the exponents of Q in the equations being fairly accurate but the numerical coefficients uncertain.

The literature on this subject has become quite extensive in recent years, and a great number of modifications in the regime formulas have been proposed by various writers. The references at the end of the chapter may be consulted in this connection. In general, however, the basic formulas as given above are still about as satisfactory as any of the more recent modifications, especially for canals or rivers in non-cohesive materials and containing relatively small amounts of sediment load.

Example 13–1

A stream has a dominant discharge of 50,000 cfs, a median bed particle size of 1.0 mm, and an average kinematic viscosity of 1.0×10^{-5} sq ft/sec. What bed slope, depth, and width would the stream tend to assume over a period of time?

Solution. Using the Lacey regime formulas:

$$S = \frac{\nu^{1/4} d_m^{11/24}}{77Q^{1/6}} = \frac{(1)^{11/24}}{(10^5)^{1/4}(77)(50,000)^{1/6}} = 0.00012 \quad \textit{Ans.}$$

$$D = \left(\frac{O}{13.5\sqrt{d_m}}\right)^{1/3} = \left(\frac{50,000}{13.5\sqrt{1}}\right)^{1/3} = 15.5 \text{ ft} \qquad \textit{Ans.}$$

$$B = 2.67Q^{1/2} = 2.67(50,000)^{1/2} = 596 \text{ ft} \qquad\qquad \textit{Ans.}$$

13–5. River Bends and Meanders. The three regime equations discussed in the previous article were derived for canals which were essentially of straight alignment, and they have been found substantially applicable to natural streams in regime also, as far as the width and depth equations are concerned.

However, the slope equation has not been adequate for natural streams following a strongly sinuous course, because the slope of the stream obviously depends also on the length of its path, and this is determined not by its straight-line course down the flood plain but rather by its actual meandering flow line. Therefore, one or more additional equations are needed to correlate meander dimensions with discharge and the channel geometry.

Regime studies on Indian rivers led Sir Claude Inglis to postulate the simple (in fact, undoubtedly oversimplified) empirical formula,

$$L_m = C_L(Q_{\max})^{1/2} \tag{13–9}$$

in which L_m is the mean length of meander, measured along a straight line between homologous-points on successive meander loops, $Q_{\max}$ is the supposed peak river flood, probably of about 100-year recurrence interval. The constant C_L ranges between 18 and 42, with a rather representative value of 28.

With somewhat less confidence, Inglis proposed a similar equation for width of meander, as measured between outside banks:

$$W_m = C_w(Q_{\max})^{1/2} \tag{13–10}$$

where C_w is of the order of magnitude of $\frac{1}{2}C_L$, but varies considerably, especially as between spilling-type and incised-type channels.

The terms used in Eqs. (13–9) and (13–10) are understood as in the definition sketch shown in Fig. 13–1.

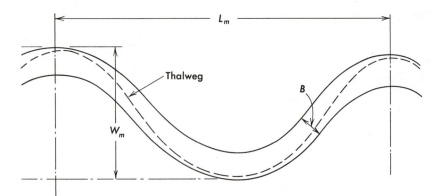

Fig. 13–1. Meander dimensions.

A number of studies have indicated that the meander length, width and radius of curvature are proportional to the square root of the dominant discharge, instead of the peak discharge, and also to the stream width. Fairly good fit is obtained by the following approximate equations.

$$L_m = 30Q^{1/2} = 12B \qquad\qquad (13\text{--}11)$$

$$W_m = 15Q^{1/2} = 6B \qquad\qquad (13\text{--}12)$$

$$R = 6Q^{1/2} = 2.4B \qquad\qquad (13\text{--}13)$$

The constants are typical values only and should be replaced with actual known values whenever possible.

The meander width, W_m, is always less than the *meander belt*, which is that portion of the river's flood plain over which meanders have extended during its geomorphic history. Two other terms are

$$\text{Meander ratio} = \frac{W_m}{L_m} = \frac{C_w}{C_L} \qquad\qquad (13\text{--}14)$$

and

$$\text{Sinuosity (or tortuosity)} = \frac{\text{Length of thalweg}}{\text{Axial length, } L_m} \qquad (13\text{--}15)$$

Common values of sinuosity range from about 1.3 to 4.0.

A *reach* is a relatively long straight stretch of river between bends. A *crossing* is similar to a reach but refers primarily to the relatively short length between adjacent sections of a meander in which the curvature of the flow is being changed from, say, clockwise to counterclockwise.

13-6. Characteristics of Meandering. Whatever the cause, the general phenomena of meandering are fairly well known, from observations on a great many rivers in all parts of the world. The typical patterns are best developed on rivers flowing through alluvial valleys, but are often found even in upstream "youthful" valleys.

Such rivers develop characteristic sinuous patterns which, however, never seem to become completely stabilized but are continually changing. As the currents from a reach or crossing are directed onto the concave bank at a bend, the bank tends to cave and thus cause a lengthening of the bend. Eroded material is deposited in the crossings, so that the channel tends to become deep on the concave sides of the bends and shallow in the crossings. Deposition also takes place on the convex sides of the bends, building the channel farther and farther out towards the middle of the bend. Thus, bends tend to be continually elongated, until a certain limiting position is reached. At this point chutes will tend to develop across the shoals on the convex bank, thus directing the flow toward the downstream end of the concave bank and thereupon moving the bend downstream. Consequently, bends tend not only to elongate in a transverse direction, but also to move downstream.

This typical development, of course, is subject to numerous variations according to local conditions. For example, if particularly resistant banks are encountered on the downstream portions of the concave banks, the downstream migration will be inhibited; also, as the bend reaches a certain maximum sinuosity, a sudden *cutoff* may develop, with the main flow of the stream breaking through the neck of the bend. Thus an entirely new channel is developed, the old channel remaining as an *oxbow* lake.

Another exception to the typical meander development occurs when the banks are very susceptible to erosion. In such a case, the river tends to develop a very wide, shallow channel, maintained in relatively long reaches with few bends. Meanders are best formed when the banks are neither too easily eroded nor too difficult to attack.

Gerard Matthes, from a very wide experience and collection of observations on many streams of all types, says that field data indicate that the most important factors influencing meandering, in order of importance, are the following:

1. General valley slope.
2. Character and amount of bed load.
3. Discharge.
4. Resistance of bed to erosion.
5. Transverse oscillations of water surface, caused chiefly by superelevation at bends, but also by shoals, wind, etc.

Matthes made the further observation that meandering may occur in any type of channel, whether aggrading, graded, or degrading, and that in fact, any stream with at least some bed load flowing in a natural channel, free to adjust its course in erodible materials, will tend to meander.

It has been generally noted that concave banks are scoured most severely during flood stages and that deposition takes place on crossings during such high stages. Conversely, there is a tendency during subnormal stages to scour crossings and deposit in the deep pools at the bends; the latter tendency, however, never compensates for the work done in the opposite direction at high stages, so that the net effect, as described above, is a continual elongating or deepening of bends, with accompanying shoals at the crossings. Matthes has supplemented these general observations by stating that the actual bank-caving occurs, not during the rising or peak stages of floods, but during attack at their falling stages. This is attributed to the fact that, during the rising stages of the flood, the thread of maximum velocity tends to shift away from the low-water channel toward the general axis and slope of the valley. This, together with the increasing discharges and velocities, effects the temporary and partial scouring of bars. As the stage falls, the convex shores receive their greatest deposits, restoring and enlarging the former bars, thus directing the still high velocities sharply against the concave banks, resulting in bank undermining and caving. Consequently, the greatest amount of caving and meander sharpening occurs during the falling stages of floods.

13–7. Basic Mechanics of Meandering. A number of theories of meandering have been devised from time to time, some of which are no longer considered valid and others of which are considered to provide partial explanations. The initiation of meandering has been attributed to the earth's rotation, to excessive slopes, to excessive energy at flood stages, to changes in stage, to changes in load, to irregularities in geologic structure, and to various other factors. Actually, some or many of these may contribute, but it is believed the basic problem can best be approached through consideration of the concept of *grade*, the equilibrium condition toward which all streams continually strive, and the energy exchanges associated therewith.

Slopes in the upper reaches of rivers, where discharges generally are smaller, and the supplied load large, especially in particle size, tend to be steep and velocities correspondingly high. As the large, angular particles are carried downstream, they are abraded and their size gradually reduced. Solution of part of the load aids in this process and, until a stream is perfectly graded, progressive sorting is quite important, so that in general slopes become flatter as the mouth is approached and the load is of a much finer caliber than upstream.

This condition of grade, corresponding to regime conditions, is a longtime-equilibrium condition, about which occur unending temporary shifts and changes due to the perpetual variations in load, discharge, etc., in the river. The graded condition, therefore, even when perfectly attained in the sense of its definition, is not one of permanent channel stability as far as short time periods are concerned. It is the fact of these ever-recurring temporary changes, above and below grade (or, if the stream is not yet graded, about its general trend toward grade) which is the chief explanation for its perpetual tendency toward meandering.

The discharge for which the graded condition in a given stream particularly applies is the *dominant discharge*, which is not, of course, the same as the average discharge, or even the normal discharge, because the sediment characteristics and consequently the grade of the stream are particularly influenced by the higher stages. According to Inglis, the dominant discharge usually corresponds to slightly more than bank-full stage. American hydrologists generally regard the dominant discharge as that corresponding exactly to bank-full stage.

Consider now the energy relations in a moving stream of water. Basically, this energy is derived from that of the sun, through the evaporation, air currents, and precipitation phases of the hydrologic cycle. As rain reaches the ground and begins its journey back to the sea, each pound of water contains potential energy equal in amount to the feet of elevation of the water surface above base level. By the principle of energy conservation, it must have experienced transmutation into other forms of energy by the time it flows to base level. This relation may be expressed by an equation:

Total energy directly after precipitation (= elevation above sea level)
  = energy dissipated as heat through normal turbulent friction of flow
    + energy dissipated by excess turbulence, as at rapids, stream expansions, etc.
    + energy extracted by artificial means, as at turbine plants, etc.
    + energy utilized in transporting sediment load or other debris
    + energy ulilized in erosion of rocks, soils, etc.

In this equation, each energy term is in foot-pounds per pound of water, or corresponding units in other systems. In symbolic form,

$$H = H_b + H_t + H_n + H_s + H_e \qquad (13\text{–}16)$$

A similar equation could be written expressing the relation between the energy contained in the stream at any two points and the energy expended in flowing from the higher point to the lower, except that the energy in the stream at the lower point would not have been completely dissipated, but would still contain potential energy equal to its remaining distance above sea level and also its kinetic energy of flow. Thus,

$$SL = H_f + H_t + H_s + H_e + \frac{V_1^2 - V_2^2}{2g} + (D_2 - D_1) \qquad (13\text{–}17)$$

where S is the average slope of channel bed between points 1 and 2; L is the length of channel, and thus SL the difference in elevation between 1 and 2; H_f is the normal friction loss (depending on velocity of flow, shape of cross-section, sedimentary load, and channel roughness); H_t is the excess turbulence generated and dissipated in friction by whatever abnormal causes may be present (such as falls, rapids, bends, expansions, constrictions, artificial structures, etc.); H_s is the energy, per pound of water, used in transporting the load of sediment, whether as suspended load or bed load or both; V_1 and D_1 are the velocity and depth of flow at point 1, respectively, with a similar notation for these quantities at point 2; and H_e is the energy utilized in eroding the materials of the banks and bottom and placing them in the load being transported. The energy required for erosion of a particle and placing it in motion is greater than that required to keep it moving once it is started. On the other hand, when the available energy drops to the point where the particle can no longer be kept in motion, it simply settles out and ceases to move. However, there is no resulting gain of energy to the stream. Thus, from this point of view, deposition is not the reverse process to erosion. Erosion requires the actual expenditure of stream energy, but there is no corresponding gain of energy when deposition occurs. The ability of a stream to erode, and also to transport, particles is chiefly dependent upon its kinetic energy.

Theoretically, if a stream is graded and is flowing at its dominant discharge, it will transport just those materials which are supplied to it at its upper end, and no more. In other words, the term H_e in Eq. (13–17) vanishes; there is no erosion, nor is there deposition. Actually, such a hypothetical condition would also require that the channel be ideally straight, that its bank and bottom materials be of perfectly uniform homogeneous composition, and that no transverse velocity components develop.

Actually, in a natural river there are always such variations of discharge, entering load, channel characteristics, local disturbances, etc., that there is almost certain to be either erosion or deposition or both in any given stretch of river. The proportion of the available energy that will be used in the various ways described by the equation will depend upon many variables, some of which are interrelated. The normal friction is most affected by velocity, assuming fairly constant channel characteristics. The amount of energy used in transporting load also depends primarily upon velocity since competence increases rapidly with velocity. The amount and caliber of load will also depend on its availability to the stream. The various local special energy losses are also proportional to some power of velocity. Finally, the erosive efficiency of a stream is again particularly dependent upon velocity, as well as upon the erodibility of the contact materials.

The most important of the many variables then, as far as erosion and deposition are concerned, are velocity and contact materials. The latter are fairly well fixed for any given stream by the geology of the region; the velocity, on the other hand, may and usually does vary over a very wide range. Velocity depends upon channel shape, slope, and roughness, which, though somewhat variable, are nevertheless fairly well established for a given stretch of river over rather long periods. The chief factor producing variations in velocity is change in discharge. The discharge in the river is a function of the hydrology and physiography of the watershed and usually fluctuates more or less directly with the precipitation, though with a time lag and usually with considerably flattened peaks and valleys.

All of this leads to the conclusion that phenomena in a stream of an erosional nature are ultimately to be attributed to changes in discharge of the stream. It is perhaps an oversimplification to say that, when the discharge exceeds the dominant discharge for the stream, scouring action is more important than filling, and that the reverse is true when the discharge is smaller than the dominant discharge. Other important local factors have to be considered in any particular case, but this seems to be a perfectly valid generalization in the preliminary approach to a specific problem.

When the discharge is large, then an excess of energy over that required to maintain the graded condition becomes available, since

Total energy/second =
        (Discharge)(Specific weight of water)(Relative elevation)   (13–18)

Some of this is used in increased turbulent friction losses. Some is converted into increased kinetic energy, thereby increasing the competence of the flow. If material of caliber that will yield to the more competent velocities is available, on either the channel bed or the banks or both, some of the increased kinetic energy will be transformed into erosive energy, scouring and introducing additional particles into the tractional and suspended load. Thus, some of the excess energy is absorbed in the kinetic energy of motion of a larger number (and larger sizes) of sedimentary particles. If such an additional load is acquired, a smaller proportion of the total extra energy will be available to maintain higher velocities. On the other hand, if the load is not available, water velocities will be still higher.

Provided the energy for erosion becomes available, work will be done upon those parts of the contact materials that are erodible and, other things being equal, especially on those parts adjacent to high-velocity filaments of the stream. Usually, higher velocities are found in the center of a straight reach, so that the bed tends to be eroded more rapidly than the banks, at least in the "youthful" stretches of the stream, where slopes are steep. This fact is also evident from the fact that bottom deposits in a typical stream are coarser than bank deposits, resulting from the more erosive velocities at the center. If the sides are less resistant than the bottom, a wide, shallow channel, possibly braided, may develop, or meanders may be formed.

Downstream, where slopes become milder, together with greater average discharges and a finer-caliber load, the erosive energy available finds more difficulty in attacking the bed. The bed is usually of considerably greater coarseness than the banks, necessitating a greater competency on the part of bottom filaments than side filaments of velocity in order to acquire additional load. Thus, the sides begin to be attacked at equal or greater rates than the bottom. In the upstream portions of alluvial valleys, where the banks are still fairly coarse, this often results in fairly uniform erosion on both sides, the materials being relatively uncompacted and easily picked up. The channel thus tends to become relatively wide and shallow. Farther downstream, in the alluvial valley proper, the bank materials are finer and have greater compaction and cohesion. Then, any unevenness in the velocity distribution, the load distribution, or the bank characteristics will tend to cause the lateral attacks to concentrate at certain points. Once a non-linear pattern of motion has been established, it tends to perpetuate itself, by principles which will be outlined more fully below. Thus, the characteristic meandering pattern finds its best development almost universally evidenced in the "mature" portions of the streams, the flood plains in the alluvial valleys. Here, the graded condition involves such small slopes that down-cutting of the bed is nearly or completely eliminated, and the available erosive energy of supradominant discharges must be largely derived from the greater velocities induced centrifugally by flow around bends.

Still farther downstream, the sides become even more resistant to erosion, because of finer and more cohesive materials. Also, in the tidal portions of the river, it is possible that the tides prevent the development of velocities sufficiently high to produce much lateral cutting. It is common for the "senile" portions of streams, very near their entrance into the sea, not to have much of sinuosity in their courses, but to flow in a relatively straight, narrow, deep channel or group of distributary channels.

Summarizing the discussion of the foregoing paragraphs, the origin of meandering can be traced directly to the existence in rivers of occasional discharges which are greater than the dominant discharge, which is that flow for which the river is graded or is approaching a graded condition. The excess energy made available to the flow, or at least that portion of it which increases the kinetic energy, can partially be used to erode the bed or banks, until an equilibrium state is reached for the particular discharge. In those stretches where this energy is not sufficient actively to erode the bed or sides unaided, because of lowered slopes and more compact materials, asymmetry in the flow, load, or channel characteristics, will result in more concerted attack on some part or other of the channel. Once the flow becomes non-linear, greater velocities are directed against the concave sides of such incipient bends, because of the centrifugal force effect of the curvilinear flow. This pattern, once inaugurated, is self-perpetuating and immediately begins to produce other bends, soon developing the familiar sinuous type of course.

Assuming then that a meandering course has been established, or at least that an asymmetrical approach flow has resulted in training the high-velocity filaments more strongly against one side of the channel, consider now the flow in the bend itself. As it enters the bend, the inertia of the particles moving curvilinearly serves to increase the elevation of the water surface on the outside of the bend. The net effect is that on a cross-section through the bend, both velocity and surface elevation on the outside of the bend are higher than on the inside.

The high outside velocities act either to scour the bed near the concave shore or to erode or undermine the shore itself. Conversely, the load already carried by the stream as it enters the bend is at least partially dropped on the convex shore, which in turn serves to align the flow more and more sharply against the concave bank.

Assuming the meander arc to be approximately that of a circle, the flow surface becomes equivalent to a forced vortex, the surface being in the ideal case parabolic. Particles in the bottom flow under the outside of the surface are thus subjected to a greater hydrostatic pressure head than are those water particles on the inside bottom, initiating a transverse flow along the bottom. This combines with the longitudinal curvilinear flow, to develop a secondary helicoidal flow in the bend. This spiral flow has been duplicated in models, and is best developed in narrow, deep channels. However, Matthes states that it is rarely observable in natural streams because of their relative shallowness.

If the lateral moving of the channel is inhibited, whether by naturally resistant banks or by artificial revetments, then the bed will scour. It is a matter of common observation, of course, that the deep pools of river channels are found near concave banks, with shallows extending well out from the inner shores of the bends. If the bed and banks are both resistant to erosion (a rare occurrence unless artificially made so, for otherwise, the meander would not have developed at that point), then the excessive energy will tend to be utilized either by a lengthening of the meander or by a downstream shifting.

The erosion of bends is particularly evident during flood stages, especially after the peak has passed and the stage is beginning to drop. The kinetic energy and the centrifugal forces at the bend are greatly augmented at these times of high discharge and, during the postpeak period, the water flows through the relatively narrow channel adjacent to the concave shore. As it then turns into the wider channel, there is a greater cross-sectional area through which the water passes, and also the centrifugal effect no longer acts, so that much of the load is deposited. Crossings thus are usually relatively wide and shallow. Leaving the crossing, the flow is then directed against the next meander and so on.

When the water subsides, and stages become lower than the dominant discharge, somewhat of an opposite effect is produced. Flow is relatively quiescent in the pools, because of the deep sections, and some load is deposited there. Also, because of the high beds at the crossings, velocities must be somewhat accelerated in order to pass the smaller discharges over them, so that the bars in the crossings are scoured to some extent. However, this set of offsetting processes is never of magnitude sufficient to restore the stream position before the flood, assuming all other factors unchanged. Thus there tends to be an over-all continuing process of erosion on concave banks together with deposition on the convex banks and in the crossings.

As this process continues over a period of time, the bend may and usually does lengthen to the point where the increased frictional losses due to the bend itself are such that, together with bank resistance, the lateral movement of the bend is halted. However, it is common that the downstream portion of the concave bank can still be attacked after this point is reached. Also, deposition then continues on the downstream convex bank, so that the meander as a whole gradually moves downstream.

The downstream migration of meanders may be assisted by another type of action, already referred to. As the bend and bank resistance becomes too great for continued stretching of the loop, a point may be reached at which it is easier for the flow to attack the bar deposits on the convex shore than to continue around the bend. Thus, one or more *chutes* may be formed across the inner shoals, whereby the water is trained at high velocities against the downstream portion of the concavity, attacking this bank and thus moving the meander downstream.

On the other hand, it may be that, if the banks are particularly resistant there, the meander width will continue to increase to the point where it actually becomes easier for the water to cut across the neck than to continue to flow around the long, tortuous bend. In this case, the channel is straightened through the formation of a *cutoff*, leaving the old channel containing the quiet water of a so-called *oxbow* lake. These oxbows may continue to carry a part of the river discharge, particularly during floods, but the main portion of the flow is carried through the cutoff, where the gradient is, because of the shorter length, much steeper than that around the bend. The oxbow lake becomes chiefly a sedimentation basin for the suspended load.

After the cutoff has been formed, the same processes will again begin to form a new meander, usually in the same direction as the original but occasionally in the opposite direction.

13–8. Limiting Widths of Meanders.

It is erroneous to suppose, as seems often to have been done, that meanders may increase indefinitely in width. This has been pointed out in the previous article. After a certain limiting width is reached or possibly sooner if something happens to alter the upstream alignment, chutes will form across the convex shoals, thus reducing instead of continuing to increase the curvature. Under certain conditions, cutoffs may be formed. It is even possible for a limiting width of bend to be reached, at which the channel will remain in relative equilibrium for a fairly long period of time. Matthes has emphasized that undue widening of meanders is quite abnormal. Thus, the widely prevalent notion that bends continue indefinitely to move laterally is erroneous, and geological or engineering interpretations that postulate a high degree of such sideward migration and shifting should be scrutinized very carefully.

From a strictly geometrical point of view, the following will indicate approximately the maximum possible development of a meander that can be obtained before a cutoff will eliminate it. The important geometrical elements are indicated in Fig. 13–2.

From the geometry of the figure, the following relations appear:

$$\cos\left(\frac{\theta}{2} - 90°\right) = \frac{R + B/2 + P/2}{2R} = \frac{(1/2)L_m}{2R} \tag{13–19}$$

$$\sin\left(\frac{\theta}{2} - 90°\right) = \frac{W_m - B - 2R}{2R} = \frac{W_m - (B + 2R)}{2R} \tag{13–20}$$

Equations (13–19) and (13–20) may be combined to yield

$$R = \frac{[(1/2)L_m]^2 + (W_m - B)^2}{4(W_m - B)} \tag{13–21}$$

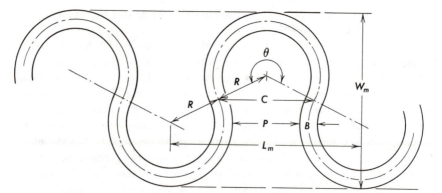

Fig. 13–2. Limiting meander geometry: B = channel width; R = center-line radius; P = neck width; C = chord; θ = angle subtending meander arc; W_m = width, L_m = length.

It may also be noted that

$$\frac{L_m}{2} \gtrless W_m - B, \text{ according as } \frac{\theta}{2} \lessgtr 90° \qquad (13\text{–}22)$$

It is assumed that, as the meander width increases, R and B remain the same. Obviously, therefore, θ will increase and P will decrease. When P becomes zero, a cutoff will have occurred. At this point Eqs. (13–19) and (13–20) may be combined to yield

$$\tan\left(\frac{\theta}{2} - 90°\right) = \frac{W_m - B - 2R}{R + B/2} = 2\,\frac{W_m - (2R + B)}{(2R + B)}$$

from which the meander width is given by

$$W_m = \frac{2R + B}{2}\left[\tan\left(\frac{\theta}{2} - 90°\right) + 2\right] \qquad (13\text{–}23)$$

showing that the width increases with R, B, and θ. The limiting central angle θ is given by

$$\theta = 180° + 2\cos^{-1}\frac{2R + B}{4R} \qquad (13\text{–}24)$$

If the last two equations are combined, then

$$W_m = (2R + B) + \frac{1}{2}\sqrt{(4R)^2 - (2R + B)^2} \qquad (13\text{–}25)$$

For a very small stream $(B \to 0)$, the limiting width reduces to

$$W_m = (2 + \sqrt{3})R \qquad (13\text{–}26)$$

While equations such as the above give an approximate geometrical picture of factors tending to limit meander width, they do not include really fundamental parameters which, as has been pointed out previously, consist of the basic hydrological variables of load and discharge as well as the soil characteristics of the particular region. Particularly fundamental, from the point of view of the energy exchanges involved, is the variation of discharge above and below the dominant value. It would be expected, therefore, that the limiting meander widths in a specific stream are predominantly a function of the dominant discharge of the stream, and the amount and frequency of excessive discharges. These factors also control the radius and breadth simultaneously. A number of empirical equations purporting to relate meander width and discharge, for actually measured natural streams, have been proposed. The same thing has been attempted for meander length and channel width, expressing them as functions of the discharge. The most generally used such equations are those of Inglis, which, however, apply to regime conditions rather than the abnormal limiting conditions discussed above.

Example 13-2

A regime stream has an average discharge of 500 cfs, a dominant discharge of 4000 cfs, and a 100-year discharge of 10,000 cfs. The mean bed particle size is 0.5 mm, and kinematic viscosity is 10^{-5} ft^2/sec. Calculate the following:

(a) Regime slope   (d) Regime meander width

(b) Regime depth   (e) Regime meander length

(c) Regime width   (f) Regime meander radius

(g) Limiting meander width

Solution.

(a) Regime slope by the Lacey equation:

$$= S = \frac{v^{1/4}d_m^{11/24}}{77Q^{1/6}} = \frac{(10)^{-5/4}(\frac{1}{2})^{11/24}}{77(4000)^{1/6}} = 0.000133 \qquad Ans.$$

(b) Regime depth $= D = \left(\dfrac{Q}{13.5(d_m)^{1/2}}\right)^{1/3}$

$$= \left(\frac{4000}{13.5(0.707)}\right)^{1/3} = 7.5 \text{ ft} \qquad Ans.$$

(c) Regime width $= B = 2.67Q^{1/2} = 169$ ft   *Ans.*

(d) Meander width, by Eq. (13–10), $= W_m$

$$= C_w(Q_{max})^{1/2} = 14\sqrt{10,000} = 1400 \text{ ft} \qquad Ans.$$

By Eq. (13–12), $W_m = 6B = 1020$ ft

(e) Meander length, by Eq. (13–9), $= L_m = C_L(Q_{max})^{1/2} = 2W_m = 2800$ ft

                                                        Ans.

By Eq. (13–11), $L_m = 12B > 2040$ ft.

(Note that if C_w is assumed at 10.2, instead of 14, values from the two methods agree.)

(f) Meander radius, $R = \dfrac{\left(\dfrac{L_m}{2}\right)^2 + (W_m - B)^2}{4(W_m - B)}$

$$= \frac{(1400)^2 + (1231)^2}{4(1231)} = 706 \text{ ft} \qquad Ans.$$

If Eq. (13–13) is used, $R = 24B > 406$ ft.

(g) $(W_m)_{\lim} = (2R + B) + \frac{1}{2}\sqrt{(4R)^2 - (2R + B)^2}$

$$= 1581 + \tfrac{1}{2}\sqrt{(2824)^2 - (1581)^2} = 2754 \text{ ft.} \qquad Ans.$$

13–9. Application of the Entropy Principle to Stream Geometry.

Another type of approach to this problem involves the entropy concept, as applied by Langbein, Leopold, and others to the determination of the most probable stream geometry. In all natural processes, the second law of thermodynamics states in effect, any system tends to approach its most probable state, which means its state of greatest disorder, or maximum entropy. With respect to a river, this will have been accomplished when all of the available energy in the water entering the river at higher elevations has been dissipated and the water has returned to its base level with all its particles randomly dispersed throughout the ocean.

Since the total amount of energy to be dissipated is the same regardless of which route the river may follow, this fact in itself is not sufficient to determine the most likely path. However, the entropy principle also implies that the dissipation of energy along the channel will tend to assume the most probable distribution within whatever constraints are imposed by geological and other factors. All such factors permitting, there is no reason why the remaining energy should be dissipated at a higher rate in any one reach of channel than in any other. There should be a tendency, therefore, for the rate of energy loss per unit length of channel to be directly proportional to the remaining elevation as the water flows downstream. That is:

$$\frac{dh}{dx} = -ch \tag{13–27}$$

in which h is the elevation ($=$ energy head) at distance x from the stream source. Integrating:

$$h = He^{-cx} \tag{13–28}$$

with H representing the initial elevation at $x = 0$.

The channel bed, therefore, tends to be exponential in form and to approach the base level asymptotically. The proportionality constant c will depend on the particular stream. Deviations from the basic exponential form will be caused by local variations in topography and geology.

Theoretically it appears that the distance x must become infinite before h actually reduces to zero. However, the constraint of the force of gravity continually re-directs the flow downward to base level. These two tendencies are forced to a compromise in terms of a long meandering path, though of finite length, evidenced especially in the lower reaches of the river.

Even if the channel is initially straight, therefore, the flow will tend to meander, especially as base level is approached. Any slight disturbance in the flow will impart a lateral component to some portion thereof. Once the meandering has been initiated, it will perpetuate itself, as described previously.

The question remains as to the particular meander geometry the flow will carve for itself. Again the entropy principle may be invoked to give at least the most probable answer.

Actual field observations of meanders would perhaps warrant use of any one of several types of curves to approximate them—that is, a sine curve, a series of reversed circular curves, a series of reversed parabolas, etc. All of these will yield reasonable approximations.

It is in accord with the principle of maximum probability, however, that the stream would carve a path tending to minimize the amount of bending required of it in each meander cycle. In order to deflect the stream, a deflecting force must be exerted by its banks. This is equal and opposite to the force exerted by the stream on the banks, and therefore to the erosive potential in the stream at that point.

The "principle of least work" would indicate that these forces would gradually adjust themselves, through carving the channel geometry, in such a way that the total bending required in the given meander cycle is minimal for a given channel length (that is, a given length of thalweg). This situation would imply that the total energy loss attributable to stream curvature is likewise minimal for a given meander cycle. This in turn permits the maximum number of meander cycles in accomplishing the dissipation of the total energy inherent in the stream system. As noted previously, the theoretical exponential curve for the over-all channel length from source to base level requires as great a meandering length as possible.

The mathematical determination of such a curve of minimal total bending involves the calculation of the most probable "random walk" of fixed length between two fixed points. The exact solution of this problem yields an expression involving an elliptic integral and is too complex for practical use.

However, Leopold and Langbein have shown that a close approximation is given by a "sine-generated" curve,—that is a curve for which the angle θ between the curve at any point and the mean down-valley direction is proportional to the sine of the channel distance x from the point of maximum amplitude beginning the cycle.

That is:
$$\theta = \theta_{max} \sin \frac{2\pi x}{L_T} \tag{13-29}$$

In this equation, L_T is the thalweg length, the actual channel distance along the meander, for one complete cycle. Thus, at $x = 0$, $x = \frac{1}{2}L_T$, or $x = L_T$, the deflection angle is 0 and the flow direction is aligned with the main valley floor. When $x = \frac{1}{4}L_T$ or $\frac{3}{4}L_T$, at the inflection points, then θ reaches its maximum and minimum values of $\pm\theta_{max}$. The constant θ_{max} varies with the particular stream.

It can be shown that a curve as defined in Eq. (13–29) has a smaller sum of the squares of the changes of direction along its length, then any other of the familiar geometric curves. It will yield curves of the form shown in Fig. 13–3, which are quite typical of actual meandering streams.

13–10. Results of Model Studies on Stream Meandering. Because of the considerable complexity of the subject, and also because of its economic importance, there have been a number of attempts to evaluate the principles of meandering by means of hydraulic model studies. Inglis attempted primarily to determine the fundamental controlling factors for what he called "modal" meanders, that is the most typical types of meanders for particular streams. He demonstrated fairly definitely that, in alluvial rivers building up their valleys, the magnitude of meanders depends primarily on the dominant discharge, that their shape depends on the material carried and on variations from dominant conditions, and that where the slope of the country exceeds the natural slope for the dominant discharge, the river develops meanders to absorb excess energy. He argued that meandering is mainly a means of utilizing excess energy during a wide range of varying flow conditions.

The Mississippi River Commission has for many years been engaged in extensive study of the meanderings of the Lower Mississippi, obviously an

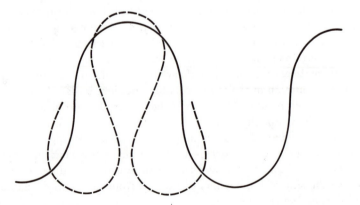

Fig. 13–3. Sine-generated curves—most probable meander geometry.

item of great economic importance. The voluminous report of Fisk attempts to interpret its recent geological history primarily from its past meanders. Also, at the U.S. Waterways Experiment Station in Vicksburg, Mississippi, the Commission for a number of years carried out extensive model studies on those phases of the meander problem pertinent to the Mississippi. A number of reports, published and unpublished, have been prepared on the results of these studies, the first one of importance being that of Tiffany and Nelson.

In 1945, Friedkin published a voluminous and very important report on model studies conducted by him. This report contains extensive photographs (see Fig. 13–4), charts, etc., embodying the quantitative results of all of these studies. A great many variables were isolated, one by one, and their effects on meandering noted. This remains to date as the most extensive quantitative study of this sort. The most important conclusions are enumerated below:

1. *Initiation of meandering.* Even in a straight channel, molded in uniform sand, if the discharge is sufficient to erode the banks, meandering will develop. If, however, lateral erosion is too easily accomplished, the channel cross-sections become wide and shallow, and meandering stops.

2. *Development of meander patterns.* Once the flow has been subjected to a transverse oscillation, it would then develop a series of bends downstream, each being produced as a result of impingement and deflection from the banks and deposition of sand on the inside of the bend.

3. *Development of meandering channel.* As in actual rivers, the cross-sections of the model rivers became deep along the concave banks of bends and shallow in the crossings between bends, the channel profiles thus consisting of a series of alternating deeps and shoals. Also, it was clear that the sand entering a bend was almost all on the convex side, where all or most of it was deposited, thus increasing the confinement of flow against the concave bank

4. *Source, travel, and deposition of sand.* Only the bed load was found to affect the meandering of alluvial rivers, at least directly. The source of load was found to be the caving banks on the concave shores, this load then being carried only a short distance downstream to the next convex bar, where it is then deposited. Thus, there is more or less continuous trading of sand down the river, the rate of trading depending upon the rate of bank caving. This latter varies from bend to bend, so that there is no definite rule as to how much of the sand deposited on a convex shore is from the concave bank next upstream. Some of the eroded sand was deposited in the shoal immediately across the channel because of the slight effects of helicoidal flow.

5. *Effects of changes in stage.* For high stages, there was noted a considerable change in the main directions of flow, it no longer being confined primarily to the low-water channel, but cutting across and eroding the convex deposits, leaving them in the crossings downstream. During lower, but still high stages, most of the erosion was on the concave banks, the

Fig. 13–4. Model study of meandering. (U.S. Army Engineers Waterway Experiment Station.)

point of greatest erosion shifting upstream as stage was lowered. During low water stages, there tended to be deposition in the deep pools and erosion of the crossings. Thus, the over-all picture of flow and sediment erosion and deposition was very markedly affected by changes in discharge and stage.

6. *Effect of sand entering at the head.* If no sand was added at the entrance the effect was at first only local, the upstream channel being degraded and the slope flattened to the graded condition for the smaller load. However the meander pattern downstream, being supported by the continual trading of sand from concave to convex shores, remained unaffected for a long time. However, over a period of time, the degrading would work downstream, until the entire stream would be regraded at a smaller slope, which would thereby reduce and possibly stop the meanderings in the deepened sections. Similarly, if the load at entrance is increased, there would ultimately result a steepening of slope through channel aggradation.

7. *Effect of discharge on size of bends.* The fact observed in natural rivers that large rivers have large bend radii, and small rivers small bend radii, was corroborated in the laboratory, both length and width of meanders being greater for large discharges. Also, it was found that a meandering channel originally shaped by a small discharge could be reshaped to correspond to a higher discharge by periodically raising the discharge to the higher value.

8. *Effect of slope on size of bends.* An increase in slope, for the same discharge and entrance angle, also resulted in an increase in the bend radius, meaning an increase in both length and width of meanders. Since both slope and discharge, when increased, increase the velocity and consequently the available kinetic energy for erosive work, this is to be expected. The higher the velocities in the bends, the more inertia there is to be overcome in manipulating the turn, and therefore the more difficult it is to develop sharp, small-radius meanders.

9. *Effect of entrance angle on bend size.* Up to an entrance angle of 60° (the angle formed with the valley axis where the flow first enters the valley), the size of bends increased with increasing angle, indicating that a change in alignment in one bend tends to affect the bend pattern for a considerable distance downstream. However, when the entrance angle was 90°, or approaching it, then the piling-up effect of the entering water and excessive turbulent dissipation of energy, seemed to remove the ability of the stream to form bends further down.

10. *Degree of sinuosity.* In general, it was found that the degree of sinuosity increased with increasing discharge or with increasing slope or with increasing entrance angle. This, of course, acts to increase friction losses by increasing the channel length, which in turn, for a given amount of total energy available, reduces the amount available for bank erosion until an equilibrium point is reached for the particular discharge. The importance of the entrance angle was indicated in the extreme case of no entrance angle, when no sinuosity at all was developed, the available erosive energy being utilized simply in widening the channel until it was wide and

shallow enough to dissipate the excess energy in normal turbulence without further lateral erosion.

11. *Effect of sand from caving banks on cross-sections.* The tests showed that rapidly eroding banks resulted in wide, shallow cross-sections, and slowly eroding banks in deep narrow cross-sections. They also emphasized the important fact that, over each square foot of the bed, shoaling or deepening takes place, depending upon the sand entering that area and the ability of the flow over that area to carry the sand away. Thus, merely increasing slope or discharge will not necessarily result in deepening due to higher velocities. The higher velocities may also cause bank erosion of a kind to widen and shallow the channel, depending upon bank material and impingement angle.

12. *Effect of bank erosion on slopes.* If the banks erode easily, then the channel becomes wide and shallow, as seen above, which results in a general steepening of over-all slope until a graded condition is reached. If the banks are difficult to erode, the channel becomes generally deeper and the slope thereby is flattened until the velocity is no longer sufficient to scour the bed. This partially accounts for the fact that the Lower Mississippi has a deep channel with flat slopes in its lower reaches, where the bank materials are fine and cohesive, and a relatively shallow channel with steep slopes in its upper sections.

13. *Effect of initial cross-sections.* It was demonstrated that both initially deep and initially shallow sections would develop the same cross-sections if all the other variables were held constant, thus demonstrating that in erodible material a river will shape its own cross-sections in accordance with its flow, slope, bank materials, and alignment, provided only that the initial section is not so wide and shallow that the velocities are not sufficient to move the materials of the bed and banks.

14. *Limiting width and length of meanders.* It was found that, as the width of a meander increased, the rate of concave bank erosion became less, since more of the available energy was being used in friction through the greater length of channel. However, before bank erosion was completely stopped, it was nearly always found that one or more chutes would form across the convex bank shoals, thus diverting the flow from the bank being attacked and shifting the attack downstream, where a new meander would then be initiated. Evidently, chutes would form when the resistance to flow across the shoals became less than the resistance to flow around the complete bend. Of course, the greater the flow the greater was the lateral extension of bends before chutes developed. Chutes may also be formed, perhaps even more commonly, by a change in the upstream alignment of flow, directing the flow across the bars.

15. *Effect of bank material on meander pattern.* Most of the tests were conducted in beds of uniform material, resulting in nearly uniform systems of bends. Tests for comparative purposes, however, conducted in heterogeneous material, though revealing the same general tendencies, resulted in definitely non-uniform meander patterns. This, of course is nearly always the case in nature, and indicates the significance of varying local bank materials. It was very significant that, in none of the many tests

conducted in homogeneous materials, was there a single cutoff developed, demonstrating that the development of a natural cutoff in a meandering stream required the existence of local differences in erodibility, erosion on the lower arm of the bend being slower than on the upper arm. Otherwise, the meanders will all migrate downstream, at a uniform rate, and a cutoff cannot develop.

16. *Reaches and braided rivers.* Reaches between meandering sections of rivers may be developed for one or both of two reasons. Firstly, if the flow does not have a positive angle of attack against a bank, and secondly, if the banks are easily eroded, then a relatively wide, shallow, straight reach will be developed. This type of development reaches its maximum demonstration in the meandering back and forth between bars, all extending over a very wide, shallow channel. Braided rivers are especially developed where banks are easily eroded. They are usually also found in sections of rivers where the slope is steep, as would be expected from the large amount of bed load that is acquired from the easily eroded banks.

13-11. Entrenched Meanders. Meandering stream patterns are generally associated with broad flood plains and it has been long considered that such patterns can only be developed in relatively erodible alluvial materials. The model tests described above seem to demonstrate that, in alluvial valleys at least, the formation of meanders is definitely inhibited when the banks are of tough material, in which case the channel becomes deep and the slope small—the Mississippi in its lowermost reaches being an example.

However, it is also well known now that many rivers cut in solid rock, often in very deep gorges, also exhibit the typical meandering pattern common to most flood-plain rivers, in fact in many cases even more intensively developed. This is true, for example, of the Colorado River and many of its tributaries, both large and small. There has been no little controversy among geologists over the past fifty years as to how these features should be accounted for. One school, probably the most generally followed, has held that such *incised* or *entrenched* meanders represent the meanders of the streams of a former erosion cycle coursing over an ancient peneplain. It is assumed that the peneplain was uplifted, effecting a rejuvenation of its streams, which then began downcutting, all the while adhering to their former meander pattern, which was thereby cut into the underlying strata as the stream formed its youthful valley for the new cycle.

Others hold that there was no need to imagine any ancient peneplains or other uplifted erosion surfaces, but that the valleys could be cut in a meandering pattern in the same way as on an alluvial valley, with lateral corrasion taking place along with the incision of the valley. It has been argued also that streams with entrenched meanders which had symmetrical side slopes on their valleys had been formed strictly by downcutting, presumably beginning from the old meanders on uplifted peneplains, and that similar streams with asymmetrical side slopes (the steeper slope being on the concave side of the bend) had been formed by lateral erosion along with the downcutting, and thus did not represent second-cycle streams at all. Moreover, most such streams actually show evidence of both kinds of effects, making it quite difficult to read their past history merely from the relative asymmetry of their side slopes.

Thus, very little of a definite nature is known about the origin of incised meanders. When one encounters gorges hundreds of feet deep encasing very strongly sinuous meandering rivers, obviously cut down by the rivers, it is very difficult to reconcile the obvious facts, as explained by either of the above theories, with the known principles of meandering.

On the assumption of an initial uplifted surface, on which consequent streams begin to flow, following the topography to begin with and supposedly in fairly straight channels, steep slopes would be developed with strong vertical downcutting. As has been seen, when high velocities are available, particularly if the banks are hard to erode, the high-velocity filaments will be mostly concentrated on the bed of the stream. Also, the heavy bed load carried by the stream will add to its downcutting ability by abrasive action, but will contribute very little to its power to corrade laterally. If any excess energy is available in the flow that can be used for erosion it would certainly be directed at those points adjacent to stream threads of greatest velocity and with the strongest array of "cutting tools" in the form of sediment load, unless of course the channel bed is composed of much more resistant materials than the sides, which would be the case only in a small percentage of such streams.

If the stream, in the course of its downcutting encountered a bed of very resistant rock which thus would come to be a temporary base level, then the portion of the river upstream from this point might possibly become graded to this condition with a relatively low slope. It then might develop a flood plain and system of meanders for a while. However, as the temporary base level is gradually worn down, more and more of this upstream flood plain would be worn away and the channel therein become a part of the steeper downstream slope. Unless this wearing-back process is very slow, it is likely that the accelerated downcutting which would follow would soon obliterate all but the easy, long radius bends. If it were slow, so that the adjustment of slopes would be very gradual, the meander pattern might be preserved, but then there would be only a mature flood plain developed along the entire course; certainly there could not be developed anything like the very narrow, deep, intensely sinuous incised meanders of the Colorado plateau, and many other localities. Even in channels of very low slope, if the banks are tough, the stream will not meander appreciably, but will develop a relatively narrow, deep channel, as found near the mouth of the Mississippi. These considerations make it appear very doubtful that ordinary lateral corrasion, associated with vertical incision, could have produced the *ingrown meander*, as this type of incised meander is sometimes called.

The same objections apply to the theory that meanders on an uplifted peneplain or other former erosion surface were "superposed" onto the under-lying bed rock. As the stream is "rejuvenated," slopes and velocities will be high. Consequently, downcutting will be prevalent. Sharp-radius bends will be obliterated by cutoffs, leaving only bends of easy curvature, even before the old flood-plain deposits are removed. Thus, the meander pattern, if any, superposed on the rocks beneath the alluvium, would not be the one on the flood plain, but would be much less sinuous. Just how much, if any, of the meander pattern would be preserved in the rocks beneath is a matter hardly subject to quantitative test. Laboratory models cannot reproduce the phenomenon, and the only rivers in which the rate of lateral erosion is rapid enough to measure are alluvial rivers, in which the phenomena of meandering and downcutting are not found together.

Furthermore, even if some of the meandering is still retained as the stream reaches and begins cutting into bedrock, the subsequent great preponderance of downcutting over lateral cutting would seem certain to prevent any strongly sinuous meander pattern from developing. Intense meandering, when slopes and velocities are high, would surely require that the bed rock be extremely resistant to erosion, so that excess energy could be dissipated in no other way than by such intense meandering. But if this were so, then the deep meandering gorges could never be cut.

Thus, it appears that neither of the generally held explanations of incised meanders can be valid except in perhaps isolated and exceptional instances. The information now available about the factors influencing meandering, as well as the mechanics of the process itself, makes the formation of incised meanders by any such process as discussed above very questionable. This would appear to be a stimulating and much-needed field for research. In absence of any explanation in terms of usual, normal processes, it seems that the usual geological postulate of uniformity should be set aside here, and some sort of intensely avulsive origin investigated. Great systems of vertical fissures might be imagined, since widened, deepened, and rounded by drainage through them. If erosion processes must account for the complete excavations, however, then it would seem necessary to postulate much greater volumes of water in the streams than now present, together with much less resistant walls than the rocks of which they now consist.

13-12. Engineering Works for River Training and Stabilization.

Practical physical and economic problems associated with river aggradation, degradation, and meandering are both numerous and important. Works constructed to cope with such problems usually are intended to serve directly one or more of the following purposes:

1. Control of floods
2. Maintenance of navigable depths of flow
3. Harbor development
4. Bank stabilization and protection

The accomplishment of these purposes usually involves some combination of the following:

1. Dams
2. Levees
3. Revetments
4. Dikes and groins
5. Cutoffs
6. Dredging
7. Floodways
8. Canals
9. Locks
10. Contraction works

The subject of dams and levees has already been treated briefly. Revetments are materials placed on the river bank slopes to inhibit bank erosion. Dikes and groins are also used for bank stabilization, as well as for contraction works. They are constructed of various materials and patterns, approximately normal to the flow, extending into the channel for the desired width.

Artificial cutoffs are sometimes excavated across the neck of sharp meanders, to shorten lengths and increase slopes and velocities. Similarly, a floodway may be constructed near the river's mouth to carry part of the flood discharge and speed it to the sea.

Navigable channels are maintained by actual removal of bed materials by dredging or by the creation of a succession of deep pools by a series of dams, with lock chambers for the transmission of river traffic past the dams. At some locations, especially at rapids, an artificial canal may be constructed alongside the river.

13–13. Hydraulic Effects of River Modifications. Since the stream is serving to move both water and sediment, any hydraulic study must take both types of factors into consideration. As already discussed, numerous equations describing each of them have been proposed. However, at least for qualitative analyses, it may be assumed that the Manning and duBoys equations are representative of flow in alluvial streams. By assuming two-dimensional flow and neglecting the initial tractive force, these equations become

$$Q = \frac{1.5}{n} BD_m^{5/3} S^{1/2} \tag{13-30}$$

and

$$G_s = \psi BD_m{}^2 S^2 = \frac{111,000}{d_m^{3/4}} BD_m{}^2 S^2 \tag{13-31}$$

The equations involve the seven variables:

Q = water discharge, cfs
G_s = sediment discharge, lb/sec
n = Manning roughness coefficient
B = channel width, ft
D_m = mean depth of flow, ft
S = longitudinal slope, ft/ft
ψ = sediment characteristic, as listed in Table 12–1.

It may be noted also that the data for ψ in Table 12–1 will fit the empirical equation,

$$\psi = \frac{111,000}{d_m^{3/4}} \qquad (13\text{–}32)$$

where d_m is the sand size, in millimeters.

Equations (13–30) and (13–31) apply, of course, to uniform flow conditions. If any one of the seven factors is altered, then one or more of the others must be correspondingly altered to restore uniform conditions.

To illustrate, assume there is an increase in water discharge, with no change in channel width or slope, at least initially. Then the flow depth will increase.

$$D_m = \left(\frac{nQ}{1.5BS^{1/2}}\right)^{3/5} = K_1 Q^{3/5} \qquad (13\text{–}33)$$

and this in turn will lead to an increase in the transport capacity of the flow, in accordance with the following relation:

$$G_s = \psi BS^2(K_1 Q^{3/5})^2 = K_2 Q^{6/5} \qquad (13\text{–}34)$$

There must therefore be an increase in the bed load proportional to the increase in $Q^{6/5}$. Unless this has been supplied by an increased sediment supply due to watershed erosion, it must be supplied by scour of the stream bed itself.

This, of course, represents only a tentative adjustment. If the increased discharge is maintained over a long period, scouring may result in a decrease in bed slope, which in turn would decrease the sediment transport capacity.

As another example, consider the probable results of a dredging project to increase the flow depth D_m. Assuming no change in discharge or width, Eq. (13–30) indicates that slope will decrease as follows:

$$S = \left(\frac{nQ}{1.5BD_m^{5/3}}\right)^2 = \frac{K_3}{D_m^{10/3}} \qquad (13\text{–}35)$$

and then in turn the sediment discharge decreases:

$$G_s = \psi BD_m^{\,2}\left(\frac{K_3}{D_m^{10/3}}\right)^2 = \frac{K^4}{D_m^{14/3}} \qquad (13\text{–}36)$$

Since the transport capacity is now less than the incoming sediment, deposition will take place, gradually restoring the depth to its original value. Thus, dredging must be essentially a continuous operation if all other factors remain constant.

A similar analysis of Eqs. (13–30) and (13–31) will indicate the probable effects of changes in width, roughness, or any of the other factors in the equations. Obviously such calculations are unlikely to give precise quantitative results, since they represent a gross oversimplification of the problem, but the qualitative indications are undoubtedly correct in most cases, and should be used as a preliminary guide to judgment in the planning of any river modification works.

13–14. Dikes, Groins, and Revetments.   Channel contraction works are commonly built of earth fill, dikes, or groins. Dikes and groins may also be used as a means of bank stabilization, but a more common method for the latter utilizes revetments.

Dikes, as employed in river works, may be classified as follows:

1. *Training walls*—used to create a more favorable channel by preventing or inhibiting access to undesirable segments of the existing channel or channels. The walls are parallel to the flow and may be constructed of rubble, piling, or reinforced concrete. The slack water area inshore of the training walls tends to fill up by deposition of suspended sediment.

2. *Spur dikes*, also called *spur jetties*, are set normal or oblique to the flow, at intervals along the shore, and may be constructed of the same materials as the longitudinal dikes or training walls: in fact, both longitudinal and spur dikes are often used conjointly. The purpose of these structures is essentially to confine the main stream flow in a narrower channel away from the shore, either as a means of increasing navigation depths or to prevent bank erosion. This type of structure is also known as a *groin*, especially if relatively short and constructed of masonry.

3. *Permeable dikes*—usually formed of rows of piles or clumps of piles, have the purpose of inducing sedimentation in the area controlled by the dikes. Flow is not precluded, but is retarded sufficiently to cause deposition of the sediment load. The dikes may thus eventually become buried in the sediment so deposited.

The most common type of bank protection consists of revetments, which are mattresses or pavements laid directly on the bank, and extending well below the surface. Some of the materials that are used in revetments are as follows:

1. *Willow fascine mattresses*—bundles of willow trees wired together in mattresses about 1000 ft by 2000 ft in area. Now used only rarely.
2. *Framed mattresses*—layers of willows placed within lumber frameworks.
3. *Tetrahedron block revetment*—tetrahedron-shaped concrete blocks placed on a layer of gravel.
4. *Sacked-concrete revetment*—burlap sacks filled with concrete and placed on the bank immediately after pouring.
5. *Asphalt-mat revetment*—continuous sheets of sand-asphaltic mixture reinforced with wire mesh.
6. *Articulated concrete block mattresses*—assemblages of precast concrete blocks, each 4 ft by 14 in. by 3 in., clipped and wired together and laid on heavy corrosion-resistant reinforcing fabric continuous throughout the mattress.
7. *Stone paving*, either hand-placed or dumped stone, often on a layer of gravel.

13–15. River Straightening and Artificial Cutoffs. It is often expeditious to alleviate tortuous stretches of river by a construction program of straightening the river channel. Bends may be eased by using training walls and other devices to establish more permanent, long-radius channels. Occasionally, it may prove feasible to eliminate a bend altogether by constructing an artificial cutoff channel across the neck of an elongated meander. This serves the purpose of significantly increasing the hydraulic slope and velocity in this section of the river, and also lessens travel distance and necessary maneuvering for water transportation.

Equations (13–30) and (13–31) indicate that such a sudden increase in slope, for a given discharge and channel width, leads to a decrease in depth but to an increase in sediment capacity. Thus,

$$G_s = \psi B S^2 \left[\frac{nQ}{1.5 B S^{1/2}} \right]^{6/5} \propto S^{7/5} \qquad (13\text{–}37)$$

Consequently it would be expected that such a cutoff would lead to scour in the cutoff channel. Furthermore, the smaller depth in the cutoff channel will cause a drawdown in the upstream channel and, therefore, scouring action there as well. On the other hand, there will tend to be backwater and deposition in the river channel below the cutoff.

Thus the advantages accruing from cutoffs tend to be offset by accelerated erosion upstream and augmented flood dangers and silting downstream. However, experience with the Mississippi cutoffs in particular has shown that these dangers can be avoided. This program has apparently been successful as a result of adhering to the following basic principles:

1. An excessive amount of straightening is to be avoided, recognizing that this would tend to induce braiding; some curvature is necessary for the maintenance of a deep channel.

2. The cutoff channel should not be excavated to its full ultimate width; rather a *pilot-cut* channel only is constructed. The pilot channel is gradually excavated by the river itself and thus gradually draws off larger and larger portions of the river discharge. There is thus a gradual adjustment of the entire system to the new conditions, rather than a cataclysmic change as in the sudden development of a natural cutoff.

3. The old channel is left available for flood waters to serve as valley storage in times of peak flows; this minimizes the piling-up effect that might otherwise result downstream from the cutoffs.

4. In so far as practicable, the straightening program should be prosecuted in an upstream direction, in order to maintain maximum sediment capacity in the channel downstream from new cutoffs.

5. Slopes in cutoff channels should not be permitted to exceed those in other stretches of the river where the river has become essentially stable.

6. The tendency of the river to erode its bed as a result of the increased slope must be assisted by supplemental dredging at regions of resistant material.

Effective reduction of flood stages, as well as maintenance of low-water navigable depths, requires a general lowering of bed elevation throughout the entire valley.

7. Revetments must be constructed as necessary to stabilize portions of the new channels and those portions of the upstream and downstream channels subject to attack by the augmented velocities.

13–16. Channel Degradation.

In addition to planned degradation of stream channels, such as occasioned by dredging, dikes, cut-offs, and similar works, certain engineering structures may cause an unwanted and even dangerous channel scouring action. The most important of these are dams, which result in channel degradation for considerable distances downstream, and bridge piers, which may result in deep scour holes adjacent to the piers.

Closure of a dam on a river results in the deposition of part or all of the normal sediment load of the river in the reservoir. The outflow from the reservoir is essentially clear water and therefore competent to initiate bed erosion in the channel downstream. Near the dam, only larger particles will eventually remain in the bed; as the stream gradually picks up its equilibrium load, the median grain size of the permanent bed decreases with distance downstream from the dam. The process will continue until the resultant "armoring" of the channel by winnowing of the fines in the bed material, plus encroachment on the channel by vegetation no longer washed out by occasional floods, plus reduction of bed slope by the degradation process itself, finally reduces the competence of the stream to a new, lower, equilibrium value, just sufficient to transport the sediment contribution from downstream tributaries. In many cases, it is known that the degradation has continued for many years and for great distances downstream.

Theoretical methods are not yet sufficiently developed to permit even approximately reliable forecasts of quantitative rates and depths of degradation. This is not surprising in view of the still uncertain mechanics even of equilibrium sediment transport calculations. However, actual field data collected below many dams indicate that typical degradations of two to four inches per year, extending up to twenty years or more, are quite possible.

Scour at bridge piers similarly is a highly complex non-equilibrium sediment phenomenon. Constriction of the channel by abutments, piers, or other structures results in higher flow velocities, not only because of continuity requirements but also because of curvilinearity of streamlines adjacent to the structures. The correspondingly augmented local shears erode the bed material at the constriction and transport it some distance. The process continues until the flow cross-section has increased to approximately the magnitude of that in the unconstricted portion of the channel. This phenomenon is especially pronounced during flood stages. After the flood crest has passed, the deep scour hole tends to fill up again with incoming sand.

Although a number of attempts have been made to determine quantitative scour relations in laboratory flumes, there does not yet exist a satisfactory theory permitting calculation of scour depths for actual field installations. The scour depends on the geometry of the obstruction, the discharge, the upstream depth and the sediment characteristics. The most reliable way of dealing with this problem is still a combination of practical field experience and a careful movable-bed model study of the actual installation. A very approximate rule-of-thumb, based on studies in India, is

$$D_s = 1.8b\left(\frac{D_0}{b}\right)^{3/4} \tag{13-38}$$

in which D_s is the scour depth, D_0 is the upstream flow depth, and b is the width of pier normal to the flow direction.

13-17. Navigation Locks. There are essentially three basic methods for developing the navigability of a stream. One, the *open-channel method*, consists of improving the channel itself by dredging, straightening, stabilizing, and similar measures, as already discussed. A second is the *canalization method*, constructing a separate navigation canal alongside otherwise non-navigable portions of the stream. The third is the *lock-and-dam method*. This is usually intermediate in cost between open-channel methods and canalization and would therefore be used only when low flows and channel geometry are inadequate for navigation requirements. Furthermore, the stream sediment load must be small and suitable dam sites must be available. Dams are constructed at such available locations, as necessary, in order to raise water levels upstream to navigable depths. Lock chambers are then necessary to move river traffic past the dams. The dams may be either fixed or movable. The latter include such devices as the shutter weir and the bear-trap gate, and may be used when the water depths are great enough at the higher discharges to make lockages unnecessary.

Although the design of a navigation lock is not in itself a problem in stream mechanics, this discussion is included in this chapter because of its primary relation to the field of waterway engineering. Hydraulic analyses in the design of locks include the following:

1. *Selection of site.* This involves determination of upstream depths required, calculation of backwater profiles resulting from dam, evaluation of currents and waves generated by lock operation, determination of spillway width for flood passage, and adjustment with non-navigational factors such as stream sanitation, tributary drainage, fish and wildlife, power development, etc.
2. *Water requirements.* Since pool elevation must be maintained at navigable minimums, the low-water stream flow under these circumstances must equal the algebraic sum of the following: (1) water requirements for lockages; (2) changes in vessel displacements; (3) leakage and seepage;

(4) hydroelectric plant consumption; (5) upstream diversions (or returns); (6) evaporation. In some cases the low-water flow may need to be supplemented by upstream reservoir storage drawdowns in order to meet these requirements.

3. *Geometry of lock chamber and approaches.* The chamber size depends largely upon the size of vessel to be accommodated. Common dimensions are 110 ft × 600 ft, 110 ft × 1200 ft, 60 ft × 360 ft, and 56 ft × 400 ft, although a wide variety of dimensions have been used. The lock wall should have an elevation somewhat above normal pool level, the freeboard depending on wave heights and anticipated operation during flood periods. At least seven feet of freeboard is usually considered necessary. Guide walls are provided along the approaches for directing vessels into the locks and for mooring vessels waiting for reception into the chamber. Orientation of these walls, as well as their character, depends on the hydraulics of the pertinent river currents and discharges from the lock chamber. These must often be determined by model studies. A typical lock is shown in Fig. 13–5.

4. *Filling and emptying system.* This is probably the most important and difficult aspect of the hydraulic design of a navigation lock. Water is admitted to the chamber through culverts, the water being drawn from the headwater pool by gravity flow. Water in the chamber is discharged, also through a culvert system, into the tailwater pool. The time for the filling and emptying cycle should be a minimum, since this is a highly important factor of the cost of river transportation. On the other hand, the trouble with turbulence and waves in the chamber is generally in inverse relation

MC NARY LOCK, COLUMBIA RIVER

Fig. 13–5. McNary Lock and Dam, Columbia River, Washington and Oregon. Note fish ladder adjacent to lock. (U.S. Army Engineers District, Walla Walla.)

to the cycle time. These two conflicting requirements must be compromised as well as possible by efficient design of the filling-and-emptying system.

13-18. Types of Filling-and-Emptying Systems. Most of the hydraulic analyses mentioned above are not problems unique to lock design, and so will not be treated at this point. However, the filling-and-emptying conduits do constitute a special type of problem peculiar to this application, besides being probably the most important single aspect of any particular lock; and therefore they should be discussed briefly in this connection. Usually the system is essentially one of pipes, culverts, and valves; however, the unsteady flows in the system do require special analysis.

Filling-and-emptying systems may be classified as follows:

1. End filling-and-emptying systems
   a. Valves in service gates or gate sills.
   b. Venturi-loop culverts in lock walls adjacent to gates. These are also called stub culverts and have a contracting leg above the gate and an expanding leg below the gate. They are valve-controlled and often have a trapezoidal cross-section.
   c. Submergible end gates. Either vertical-lift or tainter gates may be so controlled as to admit water over or under the gates.
   d. Vertical end gates. Most lock gates are of this type, usually miter gates consisting of two hinged leaves meeting at the center. They may also be vertical tainter gates (or sector gates) which retire into wall recesses when open. Water may be admitted by the gates through the aperture between, when they are partially open.
2. Longitudinal filling-and-emptying systems
   a. Wall-port manifolds. This system consists of a longitudinal culvert running the length of the lock, with manifold connections to the upper pool, lock chamber, and lower pool. Numerous ports open into the lock chamber, and the various segments are controlled by valves.
   b. Bottom lateral manifolds. Here the wall culverts are connected to a series of bottom lateral culverts, with discharge ports in each.
   c. Bottom longitudinal manifolds. These may be used where site conditions preclude side-wall culverts, but where longitudinal distribution of discharge is desired.
3. Combination Systems. Various combinations of the above have been employed, for the purpose of speeding up the operation or of providing a more uniform rate of operation.

In this country, the wall-port manifold system has been most widely used; in Europe the venturi-loop system is fairly common. Other types of end filling-and-emptying systems have usually been limited to low-lift applications, with a few recent exceptions.

13-19. Hydrodynamic Forces in Lock Chamber. Assume for simplicity that water is admitted to a lock chamber through a single port. The valve should be opened gradually rather than instantaneously, since otherwise a strong bore wave would travel back and forth across the chamber, producing excessive hydrodynamic forces on any vessel moored therein. A sloping wave will be set up, as shown in Fig. 13–6, moving at a celerity $C = \sqrt{gD}$, in accordance with the gravity wave equation. Ahead of the wave the velocity V in the chamber is zero. The velocity gradually increases behind the wave, to an average velocity V_p just below the entrance port, at section 1.

In a time dt, the wave front will move a distance $dL = \sqrt{gD}\, dt$. Consider a segment of flow dL, in length, and superimpose a negative velocity $\sqrt{gD}$ on the system, so that the wave boundary is stabilized for purposes of

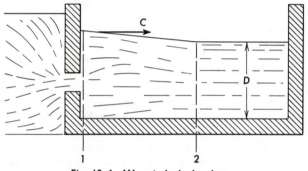

Fig. 13–6. Wave in lock chamber.

analysis, as in Fig. 13–7. The flow velocity changes by dV and the depth by dD in the distance dL. By continuity across the two ends of the segment,

$$\sqrt{gD}D = (\sqrt{gD} - dV)(D + dD)$$

from which

$$dD = \frac{D\, dV}{\sqrt{gD}}$$

If the slope of the water surface is defined as z, then

$$z = \frac{dD}{dL} = \frac{dD}{\sqrt{gD}\, dt} = \frac{1}{g}\frac{dV}{dt}$$

and since the rate of flow in the lock chamber is $Q = BDV$ at this point, where B is the chamber width, it follows that

$$z = \frac{1}{gBD}\frac{dQ}{dt} \qquad (13\text{–}39)$$

where the rate of change of discharge with time, dQ/dt, is controlled mainly by the rate of valve opening. The slope would be vertical for instantaneous opening.

If a vessel of length L and width b is in the chamber, the difference in

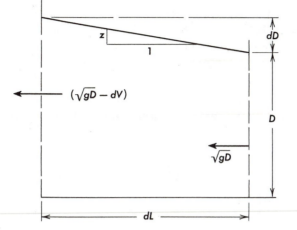

Fig. 13-7. Wave increment.

depths on the two ends results in an unbalanced hydrostatic force on the vessel, which must be resisted by hawser stresses. The net force can be approximated as follows:

$$F = \frac{\gamma b}{2}[(D + zL)^2 - D^2] = \frac{\gamma b}{2}(2zDL + z^2L^2)$$

$$\cong \gamma bzDL \tag{13-40}$$

since z^2L^2 will be small relative to $2zDL$. Substituting from Eq. (13-39) gives

$$F = \gamma bDL\left(\frac{1}{gBD}\right)\left(\frac{dQ}{dt}\right) = \rho L\left(\frac{b}{B}\right)\left(\frac{dQ}{dt}\right) \cong \rho L\frac{dQ}{dt} \tag{13-41}$$

These formulas do not give precise results, because of neglecting the effect the vessel itself has on the wave, but they are quite valid qualitatively. Thus, it is clear that the hawser stresses will depend directly on the size of vessel and the rate of valve opening. The magnitude of the discharge or velocity has no direct effect on the force; therefore large flows can be safely employed, to minimize time requirements, provided only the rates of change are not excessive.

In practice the magnitude of the hawser stresses is usually determined by model testing. The number and arrangement of ports, the size and shape of vessel, the type and manner of operation of the filling-and-emptying system, the arrangement of baffles, and other factors will all have an effect on the turbulence and wave phenomena in the chamber. The U.S. Corps of Engineers recommends that the maximum hawser stress be limited to 1/5000 of the gross weight of the vessel and cargo.

13-20. Lock Chamber Filling Times. Assume that water is admitted to the lock chamber through a single culvert, as shown in Fig. 13-8. At a certain instant t the differential head under which the culvert is operating is H. By neglecting the approach and exit velocity heads, energy considerations yield the following equation:

$$H = (K_i + K_f + K_o + K_v) \frac{Q^2}{2ga^2} \qquad (13\text{–}42)$$

where a is the throat area and K_i, K_f, K_o, and K_v are loss coefficients due to inlet, boundary friction, outlet, and valve constriction, respectively. The instantaneous flow is

$$Q = a\sqrt{2gH} \, \frac{1}{\sqrt{K_i + K_f + K_o + K_v}} \qquad (13\text{–}43)$$

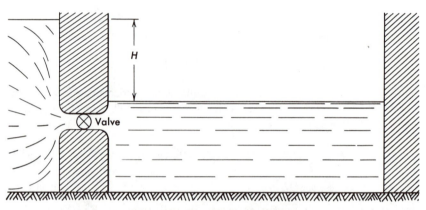

Fig. 13–8. Chamber filling through culvert.

Assuming only a gradual change in discharge, the volume entering the chamber in time dt is $Q\,dt$ and is also equal approximately to $A\,dH$, where A is the horizontal cross-sectional area of the chamber. Thus,

$$dt = -\frac{A\,dH}{Q} = -\frac{A\sqrt{K_i + K_f + K_o + K_v}}{a\sqrt{2gH}}\,dH$$

the negative sign indicating that H decreases as t increases. If the approach and exit velocity terms are not negligible, then,

$$dt = -\frac{A\sqrt{K_i + K_f + K_o + K_v + (a/A_c)^2 - (a/A_a)^2}}{a\sqrt{2gH}}\,dH \quad (13\text{–}44)$$

where A_c and A_a are the flow areas in the lock chamber and approach channel, respectively.

Explicit evaluation of this integral is impossible unless the various co-efficients are known accurately, as well as A_c and A_a. The approach area is presumably a constant, but the lock chamber flow area increases linearly as H decreases.

The inlet coefficient K_i also depends on H; however, with a well-rounded inlet, it is small and can probably be neglected. The friction coefficient, $K_f = f(l/D_e)$, depends on the culvert velocity, which may or may not vary with H, depending upon the valve operation. Similarly K_o may vary somewhat with H, as well as with the flare angle. The most critical term, however, is K_v, which depends upon the valve opening, decreasing as the valve is gradually opened. Presumably the valve will be opened more and more as the head drops, so that K_v also could be regarded as a function of H.

If all these terms are arbitrarily combined into a single coefficient of discharge,

$$C_D = \frac{1}{\sqrt{K_i + K_f + K_o + K_v + (a/A_c)^2 - (a/A_a)^2}} \qquad (13\text{--}45)$$

$$= f(H, \text{ valve type, and operation})$$

then the filling time equation becomes simply

$$dt = -\frac{A\,dH}{aC_D\sqrt{2gH}} \qquad (13\text{--}46)$$

Because of the complex interrelationships between its components, C_D must usually be determined from model studies or estimated from field measurements on similar structures. If more than one culvert is used, then a can be replaced by na, n being the number of similar conduits admitting water.

Since C_D is analogous to a coefficient of velocity for an orifice, it is evident that its limiting value would be unity if all loss terms were eliminated except the exit velocity head. It will be less than unity, therefore, in inverse relation to the hydraulic efficiency of the system, becoming zero when the valve is completely closed.

For preliminary calculations, it is often considered satisfactory to assume C_D constant, at some average value based on experience with similar culvert systems. For this purpose, G. R. Rich recommends the following, for valves fully open:

System	Average Discharge Coefficient
Wall manifold system	0.85–0.95
Short venturi loops	0.75–0.85
Simple rectangular orifices	0.60–0.80

If C_D is taken as a constant, then the filling time, from Eq. (13–46), becomes

$$t = \frac{A}{naC_D\sqrt{2g}} \int_{H_2}^{H_1} \frac{dH}{\sqrt{H}} = \frac{2A}{naC_D\sqrt{2g}} \left(\sqrt{H_1} - \sqrt{H_2}\right) \qquad (13\text{--}47)$$

where H_1 and H_2 are the initial and final head differences, respectively. If equalization of levels is achieved, then of course H_2 is zero and H_1 is the total lift for the lock.

The above calculation assumes the valve opening time is negligibly short with respect to the total filling time. If this is not the case, then the time required for valve opening must be added to the above, allowing for the head reduction occurring in that interval.

Thus, assume for example that C_D increases linearly with time from zero to its full magnitude at the time T_1 when the valve is fully open. For that period,

$$dt = -\frac{A\,dH}{na(t/T_1)C_D\sqrt{2gH}}$$

and

$$\int_0^{T_1} t\,dt = \frac{AT_1}{naC_D\sqrt{2g}} \int_{H_2}^{H_1} \frac{dH}{\sqrt{H}}$$

Integrating yields

$$T_1 = \frac{4A}{naC_D\sqrt{2g}}\,(\sqrt{H_1} - \sqrt{H_2}) \qquad (13\text{--}48)$$

The time required to eliminate the remaining head difference H_2 is

$$\frac{2A}{(na)(C_D)\sqrt{2g}}\sqrt{H_2}$$

from Eq. (13–47), so that the total time for lock filling is in this case

$$T = T_1 + \frac{2A}{naC_D\sqrt{2g}}\sqrt{H_2} = \frac{T_1}{2} + \frac{2A}{naC_D\sqrt{2g}}\,(\sqrt{H_1} - \sqrt{H_2} + \sqrt{H_2})$$

$$= \frac{T_1}{2} + \frac{2A\sqrt{H_1}}{naC_D\sqrt{2g}} \qquad (13\text{--}49)$$

In other words, the time required to fill the chamber under these assumptions is the sum of one-half the valve opening time plus the time that would have been required if the valve had been opened instantaneously. Calculations for the operation of emptying the chamber, for equalization of levels with the tailwater pool, can be made in the same way.

If the filling system is by means of flow over the end gates, then the basic discharge equation relates to flow over a rectangular weir. The effective head producing flow is the headwater elevation minus the elevation of the top of gate. As the gate is lowered, this head increases. However, when the water surface in the lock rises above critical depth level over the gate (about two-thirds of the weir head), then the weir is submerged and the effective head thereafter is the difference in water levels.

If water is admitted through the vertical aperture created by sector or miter gates, the weir equation is again applicable; in this case the effective weir head, however, is the actual difference in water levels. If discharge coefficients can be estimated, it would be possible to make an analysis similar to that for the culvert system to determine filling times.

13-21. Hydraulics of Lock Culverts and Manifolds. For precision in the analysis of flow through the filling system, the portion of the head utilized to accelerate the flow should be included. In a length of conduit dx, the flow is accelerated by a velocity increment dV over a time dt. By the momentum principle,

$$F = \gamma(dh_a)A = \rho Q \, dV = \rho A \frac{dx}{dt} \, dV$$

where A is the cross-sectional area of flow and dx/dt is the instantaneous velocity. Therefore the differential head required to produce the acceleration is

$$dh_a = \left(\frac{dx}{g}\right)\left(\frac{dV}{dt}\right)$$

Over a uniform length of conduit L, the total head required to accelerate the flow is

$$h_a = \left(\frac{L}{g}\right)\left(\frac{dV}{dt}\right) \tag{13-50}$$

If, however, the velocity is changing throughout the length L, by virtue of changing cross-section, this must be modified as follows (assuming A changes linearly with x):

$$h_a = \int_0^L \frac{1}{g}\frac{dx}{dt} \, dV_x = \frac{1}{g}\int_0^L \frac{dx}{dt} \, d\left(\frac{AV}{A_x}\right) = \frac{1}{g}\int_0^L \frac{dx}{dt} \, d\left[\frac{AV}{A + (A_1 - A)x/L}\right]$$

$$= \frac{AL}{g}\int_0^L \frac{dx}{dt} \, d\left[\frac{V}{AL + (A_1 - A)x}\right] = \frac{AL}{g}\frac{dV}{dt}\int_0^L \frac{dx}{AL + (A_1 - A)x}$$

where A is the area at the small end and A_1 the area at the flared end of the changing section. Integrating gives

$$-h_a = \left(\frac{AL}{g}\right)\left(\frac{dV}{dt}\right)\left[\frac{1}{A_1 - A}\log_e\{AL + (A_1 - A)x\}\right]_0^L$$

$$= \left(\frac{AL}{g}\right)\left(\frac{dV}{dt}\right)\left(\frac{1}{A_1 - A}\right)\log_e\frac{A_1}{A} \tag{13-51}$$

If the conduit has a combination of straight and flared reaches, then

$$h_a = \Sigma\left(\frac{L_s}{g}\right)\left(\frac{dV}{dt}\right) + \Sigma\left(\frac{AL_f}{g}\right)\left(\frac{1}{A_1 - A}\right)\left(\log_e\frac{A_1}{A}\right)\left(\frac{dV}{dt}\right)$$

$$= \left(\frac{1}{g}\right)\left(\frac{dV}{dt}\right)\left[\Sigma L_s + \Sigma\frac{A_s L_f}{A_f - A_s}\log_e\frac{A_f}{A_s}\right] \tag{13-52}$$

where the subscripts s and f denote straight and flared reaches, respectively.

Since the bracketed factor is determined solely by the geometry of the conduit, it may be denoted by a single constant C_1. Then, the head expended in accelerating the flow an amount corresponding to an amount dV in the uniform reaches is

$$h_a = \left(\frac{C_1}{g}\right)\left(\frac{dV}{dt}\right) \tag{13-53}$$

At the same time, the head expended in overcoming energy losses is obtained from Eq. (13-42), which becomes

$$h_e = \frac{V^2}{2g}(K_i + K_f + K_o + K_v) \tag{13-54}$$

where V is now understood as the velocity in the uniform reaches. If the approach and exit velocities are not neglected, and if H is the instantaneous difference in pool levels, then

$$\frac{V_a^2}{2g} + H = \frac{V^2}{2g}(K_i + K_f + K_o + K_v) + \left(\frac{C_1}{g}\right)\left(\frac{dV}{dt}\right) + \frac{V_c^2}{2g} \tag{13-55}$$

from which

$$H = \frac{C_1}{g}\frac{dV}{dt} + \frac{V^2}{2g}[K_i + K_f + K_o + K_v + (A_s/A_c)^2 - (A_s/A_a)^2]$$

$$= \frac{1}{g}\left[C_1\frac{dV}{dt} + C_2 V^2\right] \tag{13-56}$$

where C_2 is assumed constant at its average value for the particular system.

By the equation of continuity,

$$A_h\, dH = A_s V\, dt \tag{13-57}$$

Equations (13-56) and (13-57) give the relation between H, V, and t. One of the variables can be chosen arbitrarily, and these two equations then solved simultaneously for the other two variables.

It would, in fact, be possible to eliminate one of the variables, leaving one equation relating the other two.—Thus, from Eq. (13-57),

$$V = \left(\frac{A_h}{A_s}\right)\left(\frac{dH}{dt}\right) \quad \text{and} \quad \frac{dV}{dt} = \left(\frac{A_h}{A_s}\right)\left(\frac{d^2H}{dt^2}\right) \tag{13-58}$$

Inserting these in Eq. (13-56) gives

$$H = \left(\frac{C_1}{g}\right)\left(\frac{A_h}{A_s}\right)\left(\frac{d^2H}{dt^2}\right) + \left(\frac{C_2}{g}\right)\left(\frac{A_h}{A_s}\right)^2\left(\frac{dH}{dt}\right)^2 \tag{13-59}$$

which is of the form

$$\frac{d^2H}{dt^2} + K_1\left(\frac{dH}{dt}\right)^2 - K_2H = 0$$

for which there is no presently known solution. Accordingly the functional relationships can be developed only by the procedure of arithmetical integration.

That is, for a given time interval Δt, Eqs. (13–56) and (13–57) are solved simultaneously, by trial solution, for the corresponding values of ΔH and ΔV, by computing V as $V_1 + \frac{1}{2}\Delta V$, V_1 being the velocity at the beginning of the time increment. In this way, a complete tabulation of corresponding values of t, H, and V can be developed.

The above discussion applies to the case of a filling system consisting of a single culvert only. Even so, a number of important factors have been neglected, such as the effect of valve opening, variation in loss coefficients, and the inertial effect of water entering the lock chamber, which causes an overtravel of equalization levels, with a pendulation of levels above and below the final equilibrium level until the effect is damped out by friction.

If the culvert has several outlets, as in a manifold system, the analysis becomes hopelessly complex. The basic equations are, of course, still of the form of Eqs. (13–56) and (13–57), supplemented by relations expressing continuity requirements at each port. In essence there will be an energy equation and a continuity equation for each port and the continuity equation for the whole system corresponding to Eq. (13–57). For a manifold with n ports, there are thus $2n + 1$ independent equations.

Furthermore, the energy-loss terms associated with the deflection of longitudinal flow in the conduit to the individual ports are difficult to predict accurately. Assuming, however, that the equations could all be written for the given manifold system, it would be possible in principle to solve the $2n + 1$ equations simultaneously, by arithmetic integration and trial solution, in the manner described above for the single culvert. Obviously, this process becomes exceedingly tedious and quite impracticable for a multiport system.

And even when this is done, it still yields only the operation of a given system. If the results are unsatisfactory, then the system must be redesigned and the whole process repeated, and so on, until a trial design is obtained which yields an efficient operation.

Consequently, model testing has been almost universally employed as the practical method by which lock filling-and-emptying systems are designed. Models are operated in accordance with the Froude law, using a model scale of between 15 and 35 to 1. Fairly well-standardized designs are now available for many situations, from extensive model tests by the Corps of Engineers.

For purposes of preliminary design, the required basic conduit size can be determined from the appropriate approximate lock-filling time equation, such as Eq. (13–49). The inlet, bend, and outlet sections should be streamlined if possible. The inlet is usually considerably larger than necessary strictly for streamlining, however, and possibly may take the form of several ports in an intake manifold. One main purpose of this is to minimize trouble with inlet vortices.

Port openings in the lock chamber are spaced as necessary to produce as nearly uniform, turbulence-free filling and emptying as possible. Experience indicates that ports should not be placed in the upper third of the chamber, nor in the lower third if the same manifold is used for emptying, in order to minimize surging in the chamber.

If manifolds are in both walls, opposite ports should be staggered; otherwise, intermixing of opposing jets creates intense turbulence. The water depth in the lock chamber at the beginning of filling should be as great as possible in order to dissipate jet turbulence more effectively.

Various types of culvert valves have been used, the most common in the United States being the tainter type, which has been used with the skin plate facing either upstream or downstream Careful attention must be paid to the possible development of cavitation conditions downstream from the valve, as well as turbulence and air entrainment.

Example 13–3

A navigation lock chamber is filled by two Venturi culverts, for each of which the following data apply:

Initial section:	6 ft × 6 ft	Inlet section loss coefficient:	0.30
Throat section:	4 ft × 4 ft	Throat section friction factor:	0.02
Final section:	6 ft × 6 ft	Outlet section loss coefficient:	0.65
Length of initial section:	10 ft	Lock chamber size. 110 ft × 600 ft	
Length of throat section:	10 ft	Approach pool evaluation:	100 ft
Length of final section:	20 ft	Initial pool elev. in lock:	60 ft

Assume approach velocity and lock chamber velocity are negligible. The average valve loss coefficient during the first 5 seconds of opening is 2.0. At the end of this period (assuming one step computation is sufficient), what is the lock pool elevation and what is the culvert throat velocity?

Solution. Energy equation for culvert:

$$\left(H_1 - \frac{\Delta H}{2}\right) = \left[\sum \frac{L_s}{g}\frac{\Delta V}{\Delta T} + \frac{1}{g}\frac{\Delta V}{\Delta T}\sum \frac{A_s L_f}{A_f - A_s}\log_e\frac{A_f}{A_s}\right]$$

$$+ \left[K_i + f\frac{L}{D} + K_0 + K_v\right]\frac{\left(V_1 + \frac{\Delta V}{2}\right)^2}{2g}$$

$$\therefore \left(40 - \frac{\Delta H}{2}\right) = \frac{1}{g}\frac{\Delta V}{5}\left[10 + \frac{16(10)}{36-16}\log_e\frac{36}{16} + \frac{16(20)}{36-16}\log_e\frac{36}{16}\right]$$

$$+ \left[0.3 + 0.02 \frac{10}{4} + 0.65 + 2.0 \right] \frac{\left(\frac{\Delta V}{2}\right)^2}{2g}$$

$$= \frac{1}{g} \left[5.89(\Delta V) + 0.375(\Delta V)^2 \right] \tag{1}$$

Continuity equation from culvert to chamber:

$$2A_s \left(V_1 + \frac{\Delta V}{2} \right) (\Delta T) = A_h(\Delta H)$$

$$2(16)\left(\frac{\Delta V}{2}\right)(5) = 110(600)(\Delta H)$$

$$\therefore \Delta H = \frac{1}{825} \Delta V \tag{2}$$

Substituting (2) in (1)

$$40 - \frac{1}{1650}(\Delta V) = \frac{1}{5.46}(\Delta V) + \frac{1}{85.9}(\Delta V)^2$$

Solving quadratic, $\Delta V = V_2 = 51.2$ fps.   *Ans.*

Then, $$\Delta H = \frac{51.2}{825} = 0.062 \text{ ft.}$$

Pool elevation $= 60 + 0.06 = 60.06$ ft.   *Ans.*

PROBLEMS

13–1. A given alluvial river has an approximately parabolic cross-section and a bed slope of 1 in 8000. The river width is 1000 ft at a flow depth of 10 ft. The effective bed particle diameter is 2.0 mm and the characteristic ψ (in the duBoys formula) and the critical tractive force are found experimentally to be 66,000 lb/ft^3-sec and 0.05 lb/ft^2, respectively. The Manning coefficient is assumed to be 0.030. The river stages for this reach of river are gaged in feet as follows over a three-month period:[1]

Day	May	June	July	Day	May	June	July
1	4.5	17.0	15.5	16	13.7	19.5	15.1
2	5.0	21.7	14.2	17	12.7	19.6	15.2
3	6.0	24.7	13.3	18	12.1	17.8	14.4
4	6.0	25.1	12.7	19	12.1	17.0	13.2
5	5.2	25.4	15.2	20	11.6	16.7	12.4
6	4.1	24.2	15.6	21	11.5	17.6	11.4
7	7.5	23.3	21.6	22	11.2	18.4	10.4
8	9.8	23.4	22.1	23	10.6	21.8	9.9
9	9.6	23.8	19.0	24	10.4	21.9	9.5
10	9.5	23.5	17.6	25	10.2	19.4	9.1
11	9.5	21.8	17.1	26	10.4	17.5	9.1
12	12.7	20.3	17.7	27	10.4	16.3	9.0
13	15.2	19.5	17.4	28	9.9	16.1	8.7
14	15.1	19.8	16.1	29	9.6	16.5	8.3
15	14.7	18.3	16.4	30	9.3	16.3	7.9
				31	12.4		7.9

[1] A three-month period is too short to determine the bed-building stage with any accuracy, but is used here simply to shorten the calculations.

It is known that this river carries little suspended load, so that it is believed that the duBoys bed load formula will describe the sediment transport function with sufficient accuracy. Determine the following:

  a. At what stage is detritus transportation likely to begin? *Ans.:* 6.4 ft.
  b. What is the total quantity of sediment transported along the bed during the three-month period, in tons? *Ans.:* 382,400 tons.
  c. What is the "bed-building stage" for the stream, as based on the three-month period. *Ans.:* 19.6 ft.

Note: The calculations can be expedited by plotting a stage-duration curve (grouping data in, say, 2-ft depth increments), then a curve of depth versus sediment moved per second, then a sediment-transportation duration curve.

13-2. An alluvial river channel has a width of 980 ft, a roughness coefficient of 0.035, and a slope of 1:10,000. Its sediment load is primarily bed load, with sediment characteristic ψ of 273,000 lb/ft³-sec and τ_i of 0.018 lb/ft². Its "bed-building stage," corresponding to dominant discharge, is believed to be about 18 ft, although the summer low-water stage is only 5 ft. In order to provide a minimum depth of 7 ft for navigation purposes, it is planned to construct a channel contraction, 10 miles long, trapezoidal in cross-section, with side slopes 3 horizontal to 1 vertical.

  a. Before erosion begins in the contraction, it will cause backwater upstream. However, for the low-water discharge, assuming no significant change in water slope, what base width is necessary in the contracted section to assure an immediate low-water depth of 7 ft?
  b. If the contraction works are constructed to a height above high-water stage, with revetments preventing any bank erosion, bed scour will take place until an equilibrium condition is established. At the dominant discharge, what equilibrium depth and slope will be set up in the contracted section?
  c. If the contraction works are constructed only high enough to provide the 7-ft depth for low-water navigation, then scour will occur both along the bed and along the horizontal shoulders. What equilibrium depth and slope would be developed at dominant discharge in this case?

13-3. A stable rectangular river section has a slope S_1, depth D_1, and width B_1. Contraction works are built on the stream, the width of the contracted section being B_2. The roughness and boundary materials remain the same, with initial tractive stress τ_0. Prove that the equilibrium depth in the contracted section is given by the formula,

$$D_2 = D_1 \left[\left(\frac{B_1}{B_2}\right) \frac{\tau_0}{2(\gamma S_1 D_1 - \tau_0)} \left(\sqrt{1 + 4\left(\frac{B_1}{B_2}\right)\left[\left(\frac{\gamma S_1 D_1}{\tau_0}\right)^2 - \frac{\gamma S_1 D_1}{\tau_0}\right]} - 1 \right) \right]^{3/7}$$

13-4. Two forks of a stream come together to form one main channel. It is assumed that the confluence is at a small angle and is accomplished with a minimum of turbulence. The slope of each tributary is 0.0004, the width 100 ft, and the bed-building depth 3 ft. The slope of the main channel is 0.0002 and the width 160 ft. The roughness coefficient for all channels is 0.040. The sediment characteristic in each case (for the duBoys formula) is 100,000 lb/ft³-sec and the initial tractive stress 0.05 lb/ft². Assume two-dimensional flow. Determine whether there will be deposition or erosion in the main channel.

13–5. A stream with a dominant discharge of 200 cfs carves a regime channel in a bed of sand of $\frac{1}{4}$-mm average particle diameter. Assume average water temperature of 75 F, and that the channel remains essentially straight. What channel geometry (width, depth, and slope) would be developed?

13–6. Two tributary streams come together to form one main stream. Their dominant discharges are 400 cfs and 600 cfs, respectively, and the bed load size can be taken as $\frac{1}{4}$ mm. The tributary streams are assumed to have developed as regime channels and the main stream to have developed a regime width. Determine the equilibrium depth and slope of the main channel:

a. By the regime theory.
b. By duBoys' theory.

13–7. A stream with a dominant discharge of 200 cfs develops meanders in material which is predominantly sand and essentially homogeneous. What approximate meander geometry (length, width, radius) should be developed? *Ans.:* 396 ft, 198 ft, 101 ft.

13–8. Assuming one meander loop becomes elongated through encountering resistant material, but that it retains the same radius, what limiting meander width and length may be attained before a cutoff occurs? *Ans.:* 240 ft, 402 ft.

13–9. Repeat Problems 13–7 and 13–8 for a discharge of 400 cfs.

13–10. A regime stream is formed by a dominant discharge of 60,000 cfs and a 100-year discharge of 75,000 cfs. The mean bed particle size is 1.0 mm, and kinematic viscosity $\nu = 10^{-5}$ ft²/sec. Calculate the following:

a. Regime slope
b. Regime depth
c. Regime width
d. Regime meander length
e. Regime meander width
f. Regime meander radius
g. Limiting meander width

13–11. A river channel has a slope of 0.000081, width 300 feet, and bed-building stage of 16 feet. The grain size of sediment load may be taken as 0.5 mm. The Manning coefficient is 0.030. A contracted section 200 feet wide is built along a reach of the stream. What equilibrium depth and slope will be developed in the contracted section at the dominant discharge?

13–12. A certain reach of river has apparently attained a graded condition. However, due to deforestation on an upstream tributary watershed, the character of the bed load material changes to a smaller average size. Determine what the immediate effect will probably be on stream sedimentation characteristics in the vicinity. Also, what would be the long-term effect?

13–13. A stream has a bed slope of 0.000122 and a roughness coefficient of 0.051, and carries a dominant discharge of 1000 cfs. The bed load size is $\frac{1}{4}$ mm. Contraction works of training walls and spur dikes are constructed along a certain reach to improve its navigability.

a. Assuming no initial change in slope, derive formulas showing the relation of flow depth and sediment capacity to channel width. *Ans.:* $D = 124/B^{3/5}$, $G = 72/B^{1/5}$.

b. Describe the probable initial and long-term effects of the contraction works on the stream.

13-14. A lock chamber is 60 ft × 360 ft in plan. A 35 ft × 195 ft barge with 9-ft draft is moored in the chamber. The culvert valves are opened so that the maximum rate of change of rise in the lock chamber is 10 (ft/min)/second. Approximately what total force will the hawsers have to resist? *Ans.:* 795,000 lb.

13-15. If the total lift for the lock of Problem 13–14 is 20 ft, and if a maximum filling time of 10 min is desired, with a valve opening time of 2 min, what size of filling culvert should be used, as a first approximation? Assume two venturi-loop culverts of square cross-section are used. *Ans.:* 5.28 ft × 5.28 ft.

13-16. A lock provides a lift of 8 ft and has a horizontal area 110 ft × 600 ft in size. The filling system consists of two longitudinal manifolds, one in each sidewall, each with an average discharge coefficient of 0.85. The valves are designed to open in 90 sec.

a. What approximate throat area is required for each manifold if the total filling time is limited to 10 min? *Ans.:* 49.3 ft².

b. Approximately how much of the filling will be complete by the time the valve is fully open? *Ans.:* 1.26 ft.

13-17. A navigation lock chamber is 100 ft × 600 ft in plan and is filled and emptied by a wall manifold system in both side walls, with a discharge coefficient of 0.85, 1 when the valve is fully open. Each lateral has a rectangular cross-section 4 feet × 4 feet in size. The total lift is 10 ft.

a. If it takes 60 seconds to open the valve, how much will the water level in the chamber rise by the time the valve is opened?

b. How much additional time will be required before the levels are equalized?

13-18. A navigation lock chamber is filled by two Venturi culverts, for each of which the following data apply

Initial section	6 ft × 6 ft	Inlet-section loss coefficient	0.30
Throat section	4 ft × 4 ft	Throat-section friction factor	0.02
Final section	6 ft × 6 ft	Outlet-section loss coefficient	0.60
Length of contracting section	10 ft	Valve loss coefficient	0.25
Length of expanding section	20 ft	Approach velocity	negligible
Length of throat section	10 ft	Lock-chamber velocity	negligible
Lock chamber size	110 ft × 600 ft	Approach pool elevation:	100 ft

At a certain instant the lock-chamber surface elevation has become 90 ft and the throat velocity 10 fps. Determine the corresponding elevation and velocity 5 sec later.

14

COASTAL HYDRAULICS

14–1. Introduction. Coastal hydraulics is another very broad and complex branch of hydraulic engineering. Especially with the extensive industrial, shipping, recreational, and residential developments along the sea-coasts which are taking place today, it has become a matter of considerable urgency to expand the knowledge of coastal hydraulic processes and to develop design techniques appropriate to them.

The following list of problems associated with coastal hydraulics is indicative of the nature and importance of the subject, although it is by no means a complete listing:

1. Harbor sedimentation and channel maintenance
2. Beach erosion
3. Littoral transposition of sediment
4. Design of sea walls, breakwaters, and other shore structures
5. Hurricane waves and phenomena
6. Tidal hydraulics in estuaries and inlets
7. Saline encroachments in coastal aquifers
8. Design of docks and harbors
9. Design of vessels
10. Design of off-shore structures

Many of these problems, at least in their hydraulic aspects, have long been handled on the basis of judgment, experience, and empiricism. The extreme complexity of the phenomena involved has tended to discourage a more rational approach. Most of them are associated with either wave action or sedimentation or both. The wave patterns are obviously complex and irregular in a high degree, and the sedimentation phenomena are even more difficult to analyze than sedimentation in streams. Nevertheless, within recent years, a large amount of research has been devoted to various aspects of coastal engineering, resulting in significant advances in understanding and methodology.

The sedimentation phenomena in coastal processes, though more complex than in streams and canals, are basically similar, except in so far as influenced by wave action. Consequently, since wave action is likewise the most important aspect of most other problems associated with coastal hydraulics, it is necessary to have some understanding of waves before successful solution of such problems is feasible.

14–2. Types of Waves. There are a number of ways in which water waves may be classified. The following outline is fairly general and comprehensive:

1. *Oscillatory waves*—in which the particles of water do not actually travel with the wave but tend to oscillate about a mean position as the wave passes.
    a. *Deep-water gravity waves*—depth of water greater than one-half the wave length.
    b. *Shallow-water gravity waves*—depth less than one-half the wave length.
    c. *Capillary waves*, or ripples—wave forms controlled by surface tension rather than gravity.
2. *Translatory waves*—in which the water particles associated with the wave are transported with the wave.
    a. *Solitary wave*—a single wave traversing a surface, generated by a single disturbance on, or increment of water added to, the surface.
    b. *Seiches*—long-period oscillations of surface elevation in enclosed basins.
    c. *Surges*—a moving front, either positive or negative, resulting from an increase or decrease in supplied discharge.
3. *Tides*—fluctuations in water level caused by solar and lunar gravitational forces.
4. *Elastic waves*—generated by pressure increment applied at some point in the fluid mass.

The above classification is oversimplified, since each type may have several varieties and since there are gradations between the various types. For present purposes, however, the classification is regarded as adequately representative.

Translatory waves are normally of most importance in river hydraulics, oscillatory waves in coastal hydraulics. Translatory waves in rivers and channels represent a case of unsteady flow, with resulting complications in analysis; however, since all particles move in essentially straight lines, one-dimensional methods of analysis may usually be applied. Oscillatory waves, however, require two-dimensional methods and are usually treated in terms of the theory of potential flow. The subsequent discussions are materially simplified; more accurate mathematical analyses may be found in the listed references.

14–3. Oscillatory Gravity Waves. Consider a simple oscillatory wave motion, as shown in Fig. 14–1. The wave length is λ, the amplitude a and the depth of flow D. If the origin of coordinates is taken at one of the nodes at a time $t = 0$, and the wave is assumed to move in simple harmonic motion, then the equation of the water surface is

$$y = a \sin (mx - nt) = a \sin \left[m\left(x - \frac{n}{m} t \right) \right] \qquad (14\text{–}1)$$

The velocity of propagation, or celerity, of the wave is

$$c = \frac{\lambda}{T} \qquad (14\text{–}2)$$

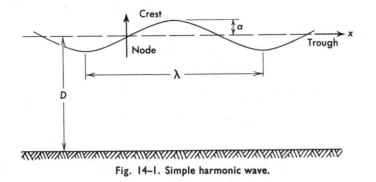

Fig. 14–1. Simple harmonic wave.

where T is the period, or time of travel from one wave crest to the next. In simple harmonic motion, the period is

$$T = \frac{2\pi}{n} \qquad (14\text{--}3)$$

(Note that $y = a \sin mx$ at both $t = 0$ and $t = T$.) The wave frequency is the reciprocal of the period, $n/2\pi$. The wave length is such that the sine function repeats itself. Therefore,

$$\lambda = \frac{2\pi}{m} \qquad (14\text{--}4)$$

Accordingly,

$$c = \frac{\lambda}{T} = \frac{2\pi/m}{2\pi/n} = \frac{n}{m} \qquad (14\text{--}5)$$

The equation of the wave profile can thus also be written

$$y = a \sin \frac{2\pi}{\lambda}(x - ct) \qquad (14\text{--}6)$$

Analysis of the continuity and energy relations for the wave, by the two-dimensional theory of potential flow, will yield the following equation for celerity as a function of flow depth:

$$c = \sqrt{gD\,\frac{\tanh(2\pi D/\lambda)}{2\pi D/\lambda}} \qquad (14\text{--}7)$$

or

$$c = \sqrt{g\,\frac{\lambda}{2\pi}\tanh\frac{2\pi D}{\lambda}} \qquad (14\text{--}8)$$

The profile and celerity equations given above are, mathematically, only first approximations to the theoretical solutions, but are quite satisfactory for most purposes.

For deep-water waves $(D > \lambda/2)$ the hyperbolic tangent[1] of $2\pi D/\lambda$ becomes greater than $\tanh \pi$, or approximately unity—exactly 0.9962[2] at $D = \lambda/2$. The celerity then is

$$c = \sqrt{g \frac{\lambda}{2\pi}} \qquad (14\text{–}9)$$

Equation (14–8) may be solved for the wavelength λ as follows:

$$c^2 = \frac{\lambda^2}{T^2} = g \frac{\lambda}{2\pi} \tanh \frac{2\pi D}{\lambda}$$

from which

$$\lambda = \frac{g}{2\pi} T^2 \tanh \frac{2\pi D}{\lambda} \simeq \frac{g}{2\pi} T^2 = 5.12 T^2 \qquad (14\text{–}10)$$

Inserting this same approximation in the celerity equation gives

$$c = \frac{\lambda}{T} = \frac{g}{2\pi} T \tanh \frac{2\pi D}{\lambda} \simeq \frac{g}{2\pi} T = 5.12 T \qquad (14\text{–}11)$$

Equations (14–8) and (14–10) must be solved by trial to obtain celerity and wave length for a given depth and period. To simplify these determinations, two families of curves have been plotted in Figs. 14–2 and 14–3, giving wave celerity and length, respectively, as functions of period and depth. For shallow-water waves $(D < \lambda/2)$, as $D \to 0$, $\tanh 2\pi D/\lambda \to 2\pi D/\lambda$. The celerity therefore approaches, in Eq. (14–7),

$$c = \sqrt{gD} \qquad (14\text{–}12)$$

These equations for celerity all apply strictly only to waves of small amplitude. However they are sufficiently accurate for nearly all cases except when the wave is almost large and steep enough to form a breaker.

Consider a particle whose initial location, with respect to the described coordinate system, before wave motion begins, is at the point (x, y). In response to the imposed wave action, the particle moves in an elliptical orbit. The horizontal and vertical displacements of the particle from the point (x, y) can be shown to bear the following functional relationships to the time t:

$$\delta_x = \frac{H}{2} \frac{\cosh \left[(2\pi/\lambda)(D + y)\right]}{\sinh (2\pi D/\lambda)} \cos 2\pi \left(\frac{x}{\lambda} - \frac{t}{T}\right) \qquad (14\text{–}13)$$

$$\delta_y = \frac{H}{2} \frac{\sinh \left[(2\pi/\lambda)(D + y)\right]}{\sinh (2\pi D/\lambda)} \sin 2\pi \left(\frac{x}{\lambda} - \frac{t}{T}\right) \qquad (14\text{–}14)$$

Note that y has a negative value, since the positive ordinate direction was assumed upward. H is the wave height, $2a$, at the surface.

These equations describe a sinusoidal motion, with the displacements out

---

[1] $\tanh (x) = \dfrac{\sinh (x)}{\cosh (x)} = \dfrac{\frac{1}{2}(e^x - e^{-x})}{\frac{1}{2}(e^x + e^{-x})}.$

[2] Available from tables of numerical values of hyperbolic functions.

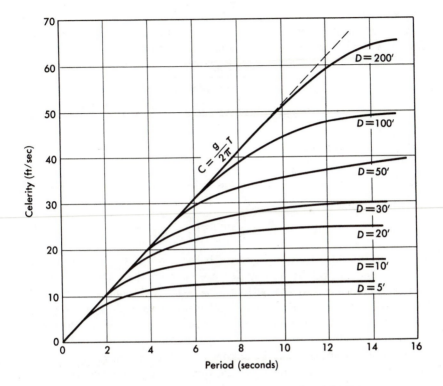

Fig. 14-2. Relation between wave celerity, period, and depth.

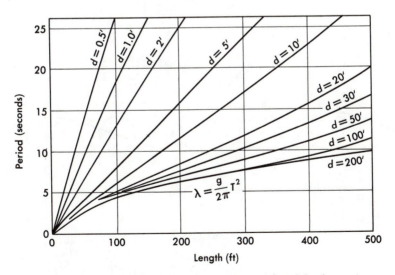

Fig. 14-3. Relation between wave length, period, and depth.

of phase by $\frac{1}{4}T$ in time and $\frac{1}{4}\lambda$ in distance. The full amplitudes of oscillation in the horizontal and vertical directions, α and β, respectively, are obtained from the coefficients in the above equations. Thus,

$$\alpha = H \frac{\cosh\,[(2\pi/\lambda)(D+y)]}{\sinh\,(2\pi D/\lambda)} \tag{14-15}$$

$$\beta = H \frac{\sinh\,[(2\pi/\lambda)(D+y)]}{\sinh\,(2\pi D/\lambda)} \tag{14-16}$$

Thus the displacement equations, (14–13) and (14–14), become:

$$\delta_x = \frac{\alpha}{2} \cos 2\pi \left(\frac{x}{\lambda} - \frac{t}{T}\right) \tag{14-17}$$

$$\delta_y = \frac{\beta}{2} \sin 2\pi \left(\frac{x}{\lambda} - \frac{t}{T}\right) \tag{14-18}$$

At the surface, the amplitudes α_s and β_s become, respectively, $H \coth (2\pi D/\lambda)$ and H. The horizontal amplitudes are greater than the corresponding vertical amplitudes. At the bottom $(y = -D)$, α_b becomes $H/\sinh (2\pi D/\lambda)$ and β_b becomes zero. For large depths, $\sinh\,[(2\pi/\lambda)(D+y)]$ approaches $\cosh\,[(2\pi/\lambda)(D+y)]$, so that β approaches α and therefore the orbits become essentially circular. These characteristics are illustrated in Fig. 14–4.

The orbital velocities may be obtained by differentiating the expressions for displacement. The horizontal and vertical components of velocity are then, respectively,

$$U = \frac{\partial(\delta_x)}{\partial t} = \frac{\pi\alpha}{T} \sin 2\pi \left(\frac{x}{\lambda} - \frac{t}{T}\right) \tag{14-19}$$

and

$$W = \frac{\partial(\delta_y)}{\partial t} = -\frac{\pi\beta}{T} \cos 2\pi \left(\frac{x}{\lambda} - \frac{t}{T}\right) \tag{14-20}$$

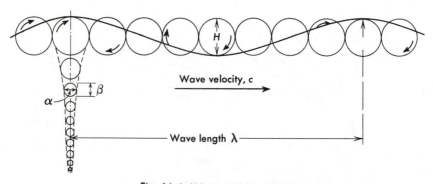

Fig. 14–4. Wave particle orbits.

The kinetic energy represented in all these orbiting fluid particles is obtained by summing the instantaneous kinetic energies of the individual particles over a full wave length, throughout the full depth, and for a unit width of wave. Thus,

$$E_k = \iint \frac{\rho}{2} V^2 \, dx \, dy = \frac{\rho}{2} \int_0^\lambda dx \int_{-D}^0 (U^2 + W^2) \, dy$$

$$= \frac{\rho \pi^2}{2T^2} \int_0^\lambda dx \int_{-D}^0 \left(\alpha^2 \sin^2 \frac{2\pi x}{\lambda} + \beta^2 \cos^2 \frac{2\pi x}{\lambda} \right) dy \qquad (14\text{-}21)$$

Substituting the expressions for α and β from Eqs. (14–15) and (14–16):

$$E_k = \frac{\rho \pi^2 H^2}{2T^2 \sinh^2 \left(\frac{2\pi D}{\lambda} \right)} \int_0^\lambda dx \int_{-D}^0 \left[\cosh^2 \left(\frac{2\pi}{\lambda} \right) (D+y) \sin^2 \left(\frac{2\pi x}{\lambda} \right) \right.$$

$$\left. + \sinh^2 \left(\frac{2\pi}{\lambda} \right) (D+y) \cos^2 \left(\frac{2\pi x}{\lambda} \right) \right] dy$$

$$= \frac{\rho \pi^2 H^2}{2T^2 \sinh^2 \left(\frac{2\pi D}{\lambda} \right)} \int_0^\lambda dx \left[\sin^2 \frac{2\pi x}{\lambda} \left\{ \left(\frac{\sinh \frac{4\pi}{\lambda} (D+y)}{4} \right) + \frac{\pi}{\lambda} (D+y) \right\} \left(\frac{\lambda}{2\pi} \right) \right.$$

$$\left. + \cos^2 \frac{2\pi x}{\lambda} \left\{ \left(\frac{\sinh \frac{4\pi}{\lambda} (D+y)}{4} \right) - \frac{\pi}{\lambda} (D+y) \right\} \left(\frac{\lambda}{2\pi} \right) \right]_{-D}^0$$

$$= \frac{\rho \pi^2 H^2 \lambda}{4\pi T^2 \sinh^2 \left(\frac{2\pi D}{\lambda} \right)} \int_0^\lambda \left[\sin^2 \frac{2\pi x}{\lambda} \left(\frac{\sinh \frac{4\pi D}{\lambda}}{4} + \frac{\pi D}{\lambda} \right) \right.$$

$$\left. + \cos^2 \frac{2\pi x}{\lambda} \left(\frac{\sinh \frac{4\pi D}{\lambda}}{4} - \frac{\pi D}{\lambda} \right) \right] dx$$

$$= \frac{\rho H^2 \lambda^2}{8T^2 \sinh^2 \left(\frac{2\pi D}{\lambda} \right)} \left[\left(\frac{\sinh \frac{4\pi D}{\lambda}}{4} + \frac{\pi D}{\lambda} \right) \left(\frac{\pi x}{\lambda} - \frac{\sin 4\pi x}{4\lambda} \right) \right.$$

$$\left. + \left(\frac{\sinh \frac{4\pi D}{\lambda}}{4} - \frac{\pi D}{\lambda} \right) \left(\frac{\pi x}{\lambda} + \frac{\sin 4\pi x}{4\lambda} \right) \right]_0^\lambda$$

$$= \frac{\rho H^2 \lambda^2 \pi}{8T^2 \sinh^2 \left(\frac{2\pi D}{\lambda}\right)} \left[\sinh \frac{4\pi D}{\lambda} \atop 2 \right] = \frac{\rho H^2 \lambda^2 \pi \sinh \frac{2\pi D}{\lambda} \cosh \frac{2\pi D}{\lambda}}{8T^2 \sinh^2 \frac{2\pi D}{\lambda}}$$

$$= \frac{\rho H^2 \lambda^2 \pi}{8T^2} \coth \left(\frac{2\pi D}{\lambda}\right) \qquad (14\text{--}22)$$

Substituting Eq. (14–10) in Eq. (14–21):

$$E_K = \tfrac{1}{16} \gamma \lambda H^2 \qquad (14\text{--}23)$$

The total energy in a wave motion includes potential energy, too. The potential energy, due to differences in elevation of the water surface, for a given wave, per unit width of wave, is approximately

$$E_p = \frac{\gamma}{2} \int_0^\lambda y^2 \, dx = \frac{\gamma a^2}{2} \int_0^\lambda \sin^2 \frac{2\pi x}{\lambda} \, dx$$

$$= \frac{\gamma a^2 \lambda}{4\pi} \left[\frac{\pi x}{\lambda} - \frac{1}{4} \sin \frac{4\pi x}{\lambda} \right]_0^\lambda$$

$$= \frac{1}{4} \gamma \lambda a^2 = \frac{\lambda}{16} \gamma H^2 \qquad (14\text{--}24)$$

The kinetic energy of the wave is therefore equal to its potential energy, so that the total wave energy is

$$E = \frac{1}{2} \gamma \lambda a^2 = \frac{1}{8} \gamma \lambda H^2 \qquad (14\text{--}25)$$

Thus it is possible to determine, at least approximately, the wave celerity and energy if the wave length and height are known. To determine these latter quantities, however, the mechanics of wave generation must be known. Oscillatory waves are basically generated by wind, the transfer of energy from wind to wave being accomplished by both shear stress and normal stress, that is by both friction drag and form drag. The wave height that can be attained depends upon the velocity, duration, and fetch of the generating wind. A frequently used empirical formula is that of T. Stevenson and D. A. Molitor:

$$H = 0.17\sqrt{V_w F} + 2.5 - \sqrt[4]{F} \qquad (14\text{--}26)$$

where V_w is the wind velocity in miles per hour, and F is the fetch, or the straight-line distance over which the wind can blow without obstruction, in statute miles. The last two terms in the equation are omitted for fetches exceeding 20 miles.

More precise methods have been developed by theoretical and experimental analyses conducted by H. U. Sverdrup and W. H. Munk, for the Hydrographic Office of the U.S. Navy. Their results have been plotted in the form of charts, which permit the determination of wave height and period in terms of wind velocity, wind duration, and fetch. Wave periods are found to average about 10 seconds, with wave lengths about 512 ft and celerities about 51 fps. Maximum values recorded have been about $22\frac{1}{2}$ sec, 2600 ft, and 115 fps, respectively.

C. L. Bretschneider more recently has proposed the equations,

$$H = 0.0555\sqrt{V_w{}^2 F} \tag{14-27}$$

and

$$T = 0.5\sqrt[4]{V_w{}^2 F} \tag{14-28}$$

where H and T are the height, in feet, and the period, in seconds, respectively, of the *significant wave* (in a wave-height frequency distribution associated with a given condition of the sea surface). In these formulas, F is in nautical miles (1 nautical mile = 6080.2 ft) and V_w is in knots (nautical miles per hour). The duration of wind necessary to develop the significant wave is given by

$$t = \frac{F}{1.14T} \tag{14-29}$$

where t is the minimum wind duration in minutes. From Eqs. (14–27) and (14–28) it is noted that $H = 0.222T^2$ for the significant wave. This corresponds to a wave *steepness*, H/λ, of approximately 1/23.

It should be stressed that the various wave formulas and equations in this section apply strictly only to idealized simple harmonic oscillatory waves. Actual waves and wave groups may vary considerably from this theory, and numerous other water wave theories have been developed and compared with empirical wave measurements, both in the laboratory and in open lakes and seas. The foregoing discussion should be considered only as an introduction to the subject. Many of the references in the bibliography at the end of the book will be found to give much more extensive treatments and should be consulted for further study or for design purposes.

Example 14–1

Waves are generated in deep water by a hurricane wind blowing at a speed of 64 knots over a fetch of 100 nautical miles. Determine the following characteristics of the "significant waves" so generated:

(a) Wave height, period, velocity, and wave length.
(b) Equation of wave crest, with respect to an origin through a point on the still water level through which a node passes at time $t = 0$. Include t as a variable.
(c) Total energy in the wave form, per foot of width of wave.

Solution.

(a) By Eq. (14–27), $H = 0.0555V_w\sqrt{F}$

$$= 0.0555(64)\sqrt{100} = 35.6 \text{ ft.} \qquad Ans.$$

By Eq. (14–28), $T = 0.5\sqrt[4]{V_w{}^2F}$

$$= 0.5\sqrt[4]{(64)^2(100)} = 12.6 \text{ secs.} \qquad Ans.$$

By Eq. (14–11), $c = \dfrac{gT}{2\pi} = \dfrac{32.2(12.6)}{6.283} = 65.5 \text{ fps} \qquad Ans.$

By Eq. (14–2) $\lambda = cT = 65.5(12.6) = 825 \text{ ft.} \qquad Ans.$

(b) Equation of wave (see Fig. 14–1):

$$y = \frac{H}{2}\sin\frac{2\pi}{\lambda}(x - ct) = 17.8\sin\frac{2\pi}{825}(x - 65.5t). \qquad Ans.$$

(c) By Eq. (14–25), $E = \frac{1}{8}\gamma\lambda H^2$

$$= \frac{1}{8}(62.4)(825)(35.6)^2 = 8.15(10^6) \text{ ft-lb/ft.} \qquad Ans.$$

14–4. The Solitary Wave. A *solitary wave* is a single disturbance, propagated essentially unaltered in form over long distances at a constant velocity. The main importance of the analysis of this type of wave, in spite of its rare occurrence, is that its behavior has been found to be essentially

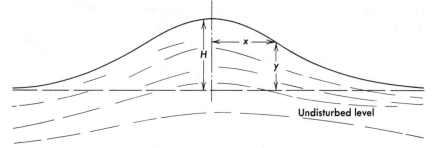

Fig. 14–5. Solitary wave.

the same as that of *each* wave in a train of shoaling oscillatory waves. Tsunamis are also solitary waves and can be studied in these terms.

The solitary wave has a form of the sort sketched in Fig. 14–5, which has been determined both experimentally and analytically to approximate quite closely the following equation:

$$y = H \operatorname{sech}^2 \frac{mx}{2} \qquad (14\text{–}30)$$

where H is the height, (x, y) are coordinates of the profile measured from an origin on the undisturbed surface level directly below the crest, and m is given by the function,

$$m = \sqrt{\frac{3H}{D^2[D + (19/12)H]}} \cong \sqrt{\frac{3H}{D^3}} \qquad (14\text{-}31)$$

D being the depth of the bottom below the undisturbed level.

The theory on which these equations are based is that of irrotational two-dimensional flow, assuming the profile to be transmitted indefinitely in unchanged form. The celerity of the solitary wave is given approximately by the equation

$$c = \sqrt{gD\left(1 + \frac{H}{D}\right)} \qquad (14\text{-}32)$$

In most cases, this is sufficiently close to $c = \sqrt{gD}$.

The volume of water in the wave per foot of width is

$$\text{Volume} = 2 \int_0^\infty y \, dx = 2H \frac{2}{m} \int_0^\infty \text{sech}^2 \frac{mx}{2} \, d\left(\frac{mx}{2}\right)$$

$$= \left[\frac{4H}{m} \tanh \frac{mx}{2}\right]_0^\infty \cong \left(\frac{16}{3} HD^3\right)^{1/2} \qquad (14\text{-}33)$$

The energy in the wave (and therefore the work necessary to generate it) is calculated on the assumption that the kinetic and potential energies are equal. Hence, the energy per foot of width is

$$E = 2E_p$$

$$= 2(2)\frac{\gamma}{2} \int_0^\infty y^2 \, dx = \frac{4\gamma}{m} \int_0^\infty H^2 \, \text{sech}^4 \frac{mx}{2} \, d\left(\frac{mx}{2}\right)$$

$$= \frac{4\gamma(HD)^{3/2}}{\sqrt{3}} \int_0^\infty \text{sech}^4 u \, du = \frac{4\gamma(HD)^{3/2}}{\sqrt{3}}\left[\tanh u - \frac{\tanh^3 u}{3}\right]_0^\infty$$

$$= \frac{8\gamma(HD)^{3/2}}{3\sqrt{3}} \qquad \text{where} \quad u = \frac{mx}{2} \qquad (14\text{-}34)$$

It has been found that the solitary wave reaches its limiting height and will break at a value of H/D of about $\frac{3}{4}$. A theoretical analysis has yielded a value $H/D = \frac{1}{2}\tan 1 = 0.78$. Thus, the maximum energy that can be propagated with a solitary wave is

$$E_{\max} = 8\gamma(\tfrac{3}{4}D^2)^{3/2}/3\sqrt{3} = \gamma D^3 \text{ ft-lb/ft width} \qquad (14\text{-}35)$$

14-5. Transformation of Waves Approaching a Shore.
Waves generated in deep water will be propagated almost indefinitely with little change in form until they approach a shore. As they enter regions of de-

creasing depth, however, the wave characteristics change, the amplitude increasing and the wave length and celerity decreasing, until finally the wave may *break* and terminate in an *uprush* of water on the beach or shore structure. Mass conservation (continuity) requires that the wave period remain essentially constant, and it is found also that energy dissipation is usually negligible until the actual point of breaking and uprush.

Furthermore, the wave may be affected by the phenomena of reflection, refraction, and diffraction. Waves may be wholly or partially reflected as they encounter irregularities in the shallow bottom or coastal structures. The longitudinal form of a wave crest may be refracted or bent if it approaches the shore obliquely. The phenomenon of diffraction (i.e., deflection or fragmentation) occurs typically in the lee of promontories or breakwaters. These phenomena are all analogous to corresponding phenomena experienced with other types of waves (light, acoustic, electromagnetic, etc.) and the wave mechanics data obtained on studies with one type of wave can often be adapted by analogy to other types.

In order to determine design pressures and forces on waterfront structures, as well as the characteristics of beach erosion and deposition, the characteristics of waves must be determined as they approach the shore. This normally requires first the determination of their deep-water characteristics from the nature of the wind in the area of generation, and then a determination of how they are transformed as they near the shore.

Consider first the case of a succession of oscillatory waves approaching a straight shore line, with a uniformly sloping bottom. Assume also that the wave crests are parallel to the shore line, so that the approach is at right angles to the shore and offshore bottom contour lines, as sketched in Fig. 14–6.

It is obvious from Eq. (14–7) that, as the depth decreases, the celerity will decrease. Since the period is assumed constant, the wave length ($= cT$) must

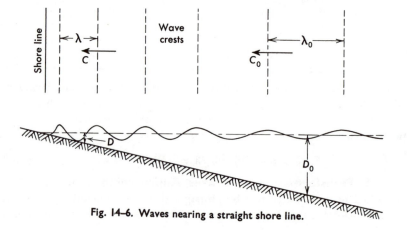

Fig. 14–6. Waves nearing a straight shore line.

also decrease. If the rate of energy transmission (that is, the power transmission) also is constant, then it is reasonable to expect the wave height to increase.

It has been shown that the total energy in a single deep-water oscillatory wave is $\frac{1}{8}\lambda\gamma H^2$ and the wave is moving with a celerity such that its period is $\sqrt{2\pi\lambda/g}$ seconds. However, half of this energy is retained by the orbiting fluid particles and only half is transmitted with the wave form. The total energy represented in a given wave is thus composed half of energy retained by the particles from the previous wave and half of energy carried with the wave form itself. The total energy being *transmitted* by the wave form, therefore, is only $\frac{1}{16}\lambda\gamma H^2$.

The power, or rate of energy, transmission, is thus

$$P = \frac{E}{2T} = \frac{(1/8)\lambda\gamma H^2}{2\lambda/c} = \frac{\gamma}{16}H^2c \qquad (14\text{–}36)$$

in foot-pounds per second per foot of width of wave. On the assumption of constant power transmission, with negligible frictional dissipation, the relation between deep-water and shallow-water wave heights is then

$$H = H_0\left(\frac{c_0}{c}\right)^{1/2} = H_0\left(\frac{\lambda_0}{\lambda}\right)^{1/2} \qquad (14\text{–}37)$$

where H_0 and c_0 represent the height and celerity of wave in deep water, and H and c the corresponding values in shallow water. Actually, the fraction of wave energy which is transmitted with the wave increases somewhat as it approaches the shore. However, Eq. (14–37) is sufficiently accurate for most practical purposes.

By such calculations, the wave characteristics can be approximately determined for any arbitrary location near the shore. However, as noted before, each wave takes on the characteristics of a solitary wave as its steepness becomes great. By Eq. (14–35) the wave presumably will break when

$$E = \tfrac{1}{8}\lambda_0\gamma H_0{}^2 = \gamma D^3$$

that is, when

$$D = \sqrt[3]{\frac{\lambda_0 H_0{}^2}{8}} \quad \text{and} \quad H = \frac{3}{4}D = \frac{3}{4}\sqrt[3]{\frac{\lambda_0 H_0{}^2}{8}} \qquad (14\text{–}38)$$

It may be noted also that waves may break in deep water if the steepness, H/λ, becomes greater than about $\frac{1}{7}$.

Example 14-2

If the waves of Example 14–1 approach a shore, with the direction of wave travel normal to the shore, what will be the wave height when the depth is 60 feet? Approximately at what depth will the wave break?

Solution.

(a) From Eq. (14–10):

$$\lambda = \frac{g}{2\pi} T^2 \tanh \frac{2\pi D}{\lambda} = \frac{g}{2\pi} (12.6)^2 \tanh \frac{2\pi}{\lambda} (60) = 810 \tanh \frac{377}{\lambda}$$

Solving by trial (or from Fig. 14–3), $\lambda = 510$ ft.
In deep water, $\lambda_0 = 825$ ft. (See Example 14–1.)

$$\text{By Eq. (14–37),} \quad H = H_0 \left(\frac{\lambda_0}{\lambda}\right)^{1/2} = 35.6 \left(\frac{825}{510}\right)^{1/2} = 45.2 \text{ ft.} \qquad Ans.$$

(b) Using Eq. (14–38) as an approximation,

$$D = \sqrt[3]{\frac{\lambda_0 H_0{}^2}{8}} = \sqrt[3]{\frac{825(35.6)^2}{8}} = 50.6 \text{ ft.} \qquad Ans.$$

14–6. Refraction of Waves. It is well known that waves approaching a shore obliquely will be refracted so that their crests will finally become almost parallel to the shore before breaking. To evaluate this phenomenon, consider a wave train approaching a straight shore line, with bottom contours all parallel to the shore line, as sketched in Fig. 14–7. The points A and B on a wave crest, separated by a distance ds, advance to A' and B' in a time dt. The direction of ds changes an angular amount $d\alpha$, where α is the angle made with the contour lines.

The wave advance, as measured between A and A', is $c\,dt$ and, as measured between B and B', is $(c + dc)\,dt$, since the celerity decreases as the depth decreases. Therefore, approximately,

$$d\alpha \cong \tan d\alpha = \frac{dc\,dt}{ds} = \left(\frac{dc}{dy}\right)\left(\frac{dy}{ds}\right) dt \qquad (14\text{–}39)$$

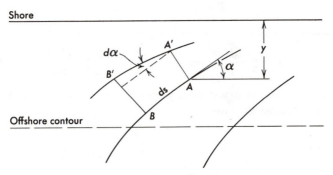

Fig. 14–7. Wave refraction.

where y is the distance of point A from the shore. It is evident from the geometry that

$$\frac{dy}{ds} = \sin \alpha \qquad (14\text{-}40)$$

The component of the celerity measured in the y-direction is, by definition,

$$c \cos \alpha = \frac{dy}{dt} \qquad (14\text{-}41)$$

Combining Eqs. (14–39), (14–40), and (14–41) yields

$$d\alpha = \frac{dc}{c \cos \alpha} \sin \alpha$$

and therefore

$$\frac{dc}{c} = \frac{\cos \alpha \, d\alpha}{\sin \alpha} = \frac{d(\sin \alpha)}{\sin \alpha} \qquad (14\text{-}42)$$

Integrating and combining gives

$$\log \frac{c}{\sin \alpha} = \text{constant}$$

or, more simply,

$$c = k \sin \alpha \qquad (14\text{-}43)$$

where k is a constant determined by the boundary conditions in deep water. If c_0 and α_0 are the wave celerity and direction in deep water, then at the arbitrary point A,

$$\sin \alpha = \sin \alpha_0 \frac{c}{c_0} \qquad (14\text{-}44)$$

Since c can be determined from the bottom depth at A, Eq. (14–44) permits calculation of the direction of the wave crest at that point.

The crest length of course increases as the wave is refracted. The length ds becomes, after being bent through the angle $d\alpha$, a length $ds/\cos d\alpha$. Thus the total crest length, for an initial length of S_0 and angle α_0, becomes, at angle α

$$S = \frac{S_0}{\cos (\alpha_0 - \alpha)} \qquad (14\text{-}45)$$

The height of the refracted wave at any point is controlled by power requirements, it being assumed that no power flows across curves drawn orthogonally to all wave crests and also that energy dissipation is negligible. From Eq. (14–36), by equating power transmitted along a certain crest from deep water to shallow water, if b is the length along a crest corresponding to the length b_0 in deep water, and K is the ratio of the fraction of wave energy transmitted in shallow water to that in deep water, then

$$H^2 b (Kc) = H_0{}^2 b_0 c_0$$

or

$$H = H_0 \sqrt{\frac{c_0}{Kc}} \sqrt{\frac{b_0}{b}} \tag{14-46}$$

The factor $\sqrt{c_0/Kc}$ obviously is dependent on the changing bottom depth, but is usually sufficiently close to unity that it can be neglected in most refraction calculations. The factor $\sqrt{b_0/b}$ is called the *refraction coefficient*, K_D; in the case of a straight shore line, it is the same as $\sqrt{S_0/S}$, from Eq. (14-45). Equation (14-46) is then

$$H = K_D H_0 \tag{14-47}$$

where K_D is determined from a refraction diagram for the given wave direction and period.

A *refraction diagram* is a representation of the form of wave crests as they approach the shore line. Together with another family of lines drawn everywhere orthogonally to the wave crests, these lines representing the directions of wave advance. The steps to be followed in constructing a refraction diagram may be outlined as follows (refer to Fig. 14-8):

1. Sketch the initial wave position in deep water, and subdivide into equal increments ($AB = BC = CD$, etc.)
2. Compute the wave length (or some integral multiple thereof), and sketch the corresponding advance wave.
3. Construct orthogonal lines from the incremental points on the first to their intersection with the next wave. As long as the water is deep the waves will continue to be parallel and equally spaced, and so will the orthogonal lines.

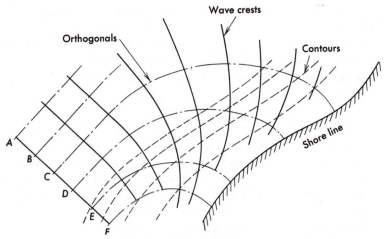

Fig. 14-8. Refraction diagram.

4. As the wave enters shallower water, its celerity and wave length decrease, since $c = \sqrt{(g\lambda/2\pi)\tanh(2\pi D/\lambda)}$ and $\lambda = cT$. Each orthogonal is advanced a distance, perpendicular to the wave crest, equal to the computed wave length (or chosen multiple) for the given depth.

5. This process is continued, wave by wave, until the breaking points are reached.

6. Refraction coefficients are determined from the spacings of the orthogonal lines, and wave heights computed as desired.

The procedure outlined works quite satisfactorily for gradual changes in bottom slope. Model studies are usually required for areas of very steep slope. Various modifications of the above procedure have been developed to speed up the work, and these may be adopted if frequency of use warrants.

It may be noted that convergence of orthogonals indicates increasing wave heights, whereas divergence of orthogonals indicates decreasing heights. Refraction diagrams find their main applications in the hindcasting or forecasting of surf conditions and in the analysis of unusual wave conditions that have been known to occur in the past. This permits development of information for design of waterfront or offshore structures.

Another case of transformation and refraction occurs when waves meet a current, such as an ocean current, tidal current, or flood current. Consider first the case of deep-water waves (of characteristics H_0, T, λ_0, c_0) meeting a current flowing either with or against the waves, in the same direction. The new wave characteristics will become (H, T, λ, c). Since, in deep water, $c = \sqrt{g\lambda/2\pi}$, therefore

$$\frac{\lambda}{\lambda_0} = \left(\frac{c'}{c_0}\right)^2 \tag{14-48}$$

where c' is the velocity relative to the water.

If the current velocity is U and the period remains unchanged, then

$$T = \frac{\lambda_0}{c_0} = \frac{\lambda}{c' \pm U} \tag{14-49}$$

From Eqs. (14–48) and (14–49),

$$\frac{c'}{c_0} = \sqrt{\frac{\lambda}{\lambda_0}} = \sqrt{\frac{c' \pm U}{c_0}} = \frac{1}{2}\left(\sqrt{1 \pm 4\frac{U}{c_0}} + 1\right) \tag{14-50}$$

A following current will thus increase the wave length and velocity, and an opposing current will decrease these quantities. The wave height is determined by the requirement for conservation of power transmission, approximately from Eq. (14–36):

$$\frac{H}{H_0} = \sqrt{\frac{c_0}{c' \pm U}} = \frac{2}{\sqrt{1 \pm 4U/c_0} + 1} \tag{14-51}$$

A more exact analysis of the energy transmission with such a super-imposed current would yield the equation,

$$\frac{H}{H_0} = \sqrt{\frac{2}{(1 \pm 4U/c_0) + \sqrt{1 \pm 4U/c_0}}} \tag{14-52}$$

When the current is at an angle with the wave, as shown in Fig. 14–9, the wave will change not only its characteristics but also its direction, with the deflection angle, δ, measured as $\beta - \alpha$. An extension of the type of analysis outlined above yields the following corresponding equations:

$$\frac{\lambda}{\lambda_0} = \frac{1}{(1 - U/c_0 \sin \alpha)^2} = \frac{\sin \beta}{\sin \alpha} \tag{14-53}$$

$$\frac{H}{H_0} = \frac{1}{(1 - U/c_0 \sin \alpha)^2} \sqrt{\left(\frac{\cos \alpha}{\cos \beta}\right) \frac{(1 - U/c_0 \sin \alpha)^6}{1 + U/c_0 \sin \alpha}} \tag{14-54}$$

14-7. Wave Diffraction. When water waves move past a breakwater or other barrier, the portion of the wave crests passing to the lee side of the barrier are *diffracted*, with their energy spreading out laterally into the protected area behind the breakwater. The wave heights decrease substantially, with corresponding changes in direction and other characteristics.

The analysis of this phenomenon is largely analogous to corresponding analyses for diffraction of light and sound waves. Relatively small amplitude of waves and uniform bottom depth is assumed. A definition sketch for a diffraction analysis for a vertical impermeable breakwater is shown in Fig. 14–10. The y-axis is oriented in the direction of advancing waves and

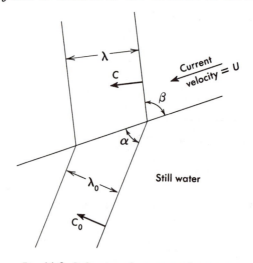

Fig. 14–9. Refraction of waves meeting current.

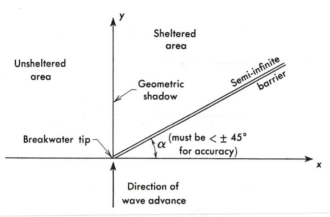

Fig. 14-10. Definition sketch for diffraction.

coincides with the *geometric shadow* line cast by the breakwater bounding the sheltered area. If the wave height in the unsheltered area is H_0, the wave height in the sheltered area is defined by

$$H = K'H_0 \tag{14-55}$$

where K' is the diffraction coefficient, and is a function of the initial wave length λ and the coordinates (x, y) of the position behind the breakwater.

The derivation of the equation for the diffracted surface involves application of light-wave theory to the assumed analogous water waves, using potential theory and complex variables, and will not be given here. In the resulting equation, a new variable, U, must be introduced, where

$$U = \pm\sqrt{\frac{4}{\lambda}\left(\sqrt{x^2 + y^2} - y\right)} \tag{14-56}$$

The positive sign is used for negative values of x, and vice versa. The wave length λ is measured just outside the breakwater tip and may be either a deep water or shallow-water wave length, the only limitation being that the depth is assumed uniform in the affected region.

The diffraction coefficient is a function of U, as given in Table 14-1. It is noted that, along the line of the geometric shadow, U becomes zero, and the diffraction coefficient is 0.5. In the lee region, U is negative, whereas it is positive in the unsheltered region. The angle α is of no significant effect on these values as long as $-45° \leq \alpha \leq 45°$. It may be seen also that the impinging wave height is propagated along a line somewhat external to the geometric shadow (corresponding to $K' = 1.0$), and that there is a slight increase and then a reduction in wave height for a limited region even beyond this. Table 14-1 may be used to compute equations of lines of constant K'. From Eq. (14-56),

TABLE 14–1
Diffraction Coefficient as a Function of U

Diffraction Coefficient K'	U
0.1	—2.25
0.15	—1.44
0.2	—1.02
0.3	—0.528
0.4	—0.225
0.5	0.000
0.6	0.184
0.7	0.341
0.8	0.486
0.9	0.631
1.0	0.779
1.17	1.218
1.0	1.610
0.88	1.878
1.0	2.124

$$\frac{x}{\lambda} = -\frac{U}{\sqrt{2}}\sqrt{\left(\frac{y}{\lambda} + \frac{U^2}{8}\right)} \qquad (14\text{--}57)$$

The negative sign should be used with the radical in order to give the correct sign for x.

Equations can be computed and plotted for as many values of K' as desired. If it is desired to obtain wave patterns also, they may be determined from Table 14–2. The *crest lag* is the distance by which the wave

TABLE 14–2
Crest Lag as a Function of the Diffraction Coefficient

K'	Crest Lag λ	K'	Crest Lag λ
0.1	1.40	0.7	—0.05
0.15	0.60	0.8	—0.06
0.2	0.33	0.9	—0.05
0.3	0.13	1.0	—0.05
0.4	0.06	1.17	0.00
0.5	0.00	1.0	0.00
0.6	—0.03	0.88	—0.03

front lags behind its corresponding position along the line of geometric shadow. The lag is measured along an orthogonal, as shown in Fig. 14–11. Wave crest patterns can thus be sketched for the lee area, and corresponding wave lengths scaled if needed. Assuming no change in period, wave velocities can then be calculated. Heights of waves are determined from the diagram of diffraction coefficients.

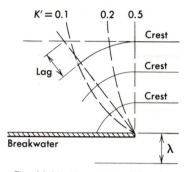

Fig. 14-11. Crest lag in diffraction.

The foregoing discussion has applied specifically to a single breakwater. Often two breakwaters are used, projecting from different points on the shore, leaving a *breakwater gap* through which waves progress shoreward, being diffracted therefore in two directions behind the breakwaters. It is found that, if the gap width is greater than about five wave lengths, the diffraction phenomena on the two sides are independent of each other, and thus the methods already discussed may be used. Mutual interference occurs with smaller gaps.

A definition sketch for this condition is shown in Fig. 14-12. The direction of wave advance is taken to be perpendicular to the breakwaters. At any point (x, y) leeward from the breakwaters, the diffraction coefficient K' is to be determined as a complex function of the parameter U, where

$$U = \sqrt{\frac{4(r \pm y)}{\lambda}} \qquad (14\text{-}58)$$

and r is the radial distance of the point from the breakwater tip. From the figure,

$$r_1 = \sqrt{y^2 + \left(x - \frac{b}{2}\right)^2} \quad \text{and} \quad r_2 = \sqrt{y^2 + \left(x + \frac{b}{2}\right)^2} \qquad (14\text{-}59)$$

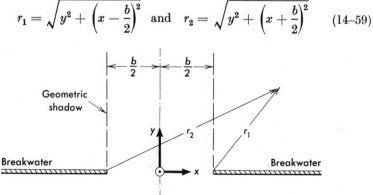

Fig. 14-12. Breakwater gap.

In dimensionless form, these equations are

$$\frac{r_1}{\lambda} = \sqrt{\left(\frac{y}{\lambda}\right)^2 + \left(\frac{x}{\lambda} - \frac{b}{2\lambda}\right)^2} \quad \text{and} \quad \frac{r_2}{\lambda} = \sqrt{\left(\frac{y}{\lambda}\right)^2 + \left(\frac{x}{\lambda} + \frac{b}{2\lambda}\right)^2} \quad (14\text{--}60)$$

The following values of U^2 are calculated:

$$U_1^2 = 4\frac{r_1 - y}{\lambda} \qquad U_3^2 = 4\frac{r_1 + y}{\lambda}$$

$$U_2^2 = 4\frac{r_2 - y}{\lambda} \qquad U_4^2 = 4\frac{r_2 + y}{\lambda} \qquad (14\text{--}61)$$

It is obvious from Eqs. (14–61) that no generalized diffraction diagram applicable to all gaps can be prepared, but rather a separate diagram for each value of the b/λ ratio, as well as for each angle of wave advance. Furthermore the diffraction effects on the surface at (x, y) will not be that of simple addition of the effects from each of the two tips. The diffractive effects on a given wave will be out of phase as they reach the point and must be adjusted correspondingly. The procedure developed requires determination of a specified complex function in each of the four variables (U_1, U_2, U_3, U_4), evaluating their real and imaginary components, adjusting them for phase difference by rotation, and finally adding the resulting components. The resulting complex number has a modulus equal to the value of K' at the point.

When a number of values of K' have been computed, contours of K' can be drawn. Similarly patterns of diffracted wave crests can be computed and

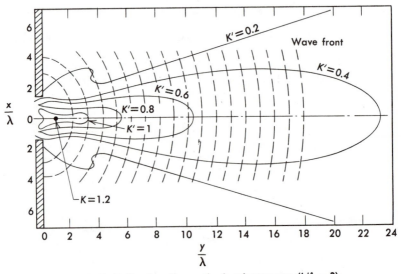

Fig. 14–13. Diffraction diagram for breakwater gap ($b/\lambda = 2$).

drawn, using the *phase difference* caused by diffraction, the latter being the argument of the complex number whose modulus is K'. J. W. Johnson has constructed a number of such diffraction diagrams, one example being as shown in Fig. 14–13, for a gap width ratio, b/λ, of 2.

Many designers believe that, except for very narrow gaps, diffraction coefficients can be determined for each breakwater independently with sufficient accuracy, by the method described for a single breakwater.

Example 14-3

The waves of Example 14–1 are assumed to approach a certain shoreline at a 45° angle.

(a) At a point where the depth is 150 feet and the waves have been refracted to an angle of 30° with the shoreline, what is the wave height?

(b) If, at this point, they encounter a breakwater which is parallel to the shore, what is the wave height on the lee side of the breakwater 100 feet from its tip, as measured directly on the breakwater?

Solution.

(a) For $D = 150$ feet, $T = 12.6$ secs, use Eq. (14–11), or Fig. 14–3.

$$c = \frac{32.2}{2\pi} (12.6) \tanh \frac{2\pi(150)}{(12.6c)} = 64.5 \tanh \frac{74.7}{c}$$

from which $c = 56$ fps.

$$\text{Then } H = H_0 \sqrt{\frac{c_0}{c}} \sqrt{\frac{S_0}{S}} = 35.6 \sqrt{\frac{65.5}{56} \cos (45° - 30°)}$$

(from Eqs. 14–25 and 14–26).

$$\text{Therefore, } H = 35.6 \sqrt{\frac{65.5 \cos 15°}{56}} = 37.8 \text{ ft.} \qquad \qquad Ans.$$

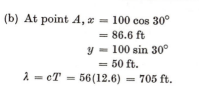

(b) At point A, $x = 100 \cos 30°$

$\qquad\qquad\qquad = 86.6$ ft

$\qquad\qquad y = 100 \sin 30°$

$\qquad\qquad\qquad = 50$ ft.

$\lambda = cT = 56(12.6) = 705$ ft.

$$U = \pm\sqrt{\frac{4}{\lambda}\left(\sqrt{x^2+y^2}-y\right)} = \pm\sqrt{\frac{4}{705}(100-50)} = \pm 0.533$$

Use $U = -0.533$. From Table 14–1, $K' = 0.30$.
∴ $H' = K'H = 0.30(37.8) = 11.3$ ft.   *Ans.*

14-8. Reflection of Waves. When two waves are superposed, their velocities are added vectorially, whereas their depths and pressures are combined by scalar addition. This is of particular importance when a moving wave encounters a solid boundary throughout part or all of the water depth. A negative wave is reflected from the boundary, with the resultant absolute velocity at the boundary reduced to zero.

The height of negative wave is equal to the height of incident wave; at the boundary these heights are superposed, resulting in a standing wave known as *clapotis* at the boundary, its height equal to twice that of the incident wave. A train of oscillatory waves encountering a boundary creates a system of standing waves of amplitude double the amplitude of the approaching waves. These standing waves oscillate from crest to trough positions.

Wave reflection thus may create undesirable surface conditions not only adjacent to the boundary but also for a considerable distance away from the boundary. It may also lead to erosion and undermining of the structure forming the boundary.

If the boundary is permeable or sloping, or if it occupies only a part of the depth, the wave will be only partially reflected. For sufficiently flat slopes, the reflection becomes negligible and the wave becomes a breaker on the slope, with resultant dissipation of its energy. It is also possible to utilize reflection for energy dissipation by installing barriers of varying heights and spacings along the bottom in such a way that the reflected portions of the waves from each barrier tend to cancel each other.

In general, by virtue both of irregularities of wave patterns and of harbor coast and bottom geometry, actual reflection characteristics are exceedingly difficult to analyze in a specific location. Models are very commonly found necessary.

Systematic studies have been made by Caldwell on reflection of solitary waves from various types of structures. By dimensional analysis he derived the following general function:

$$E_a = \frac{E_1 - E_2}{E_1} = K\left(\frac{H}{D}\right)^a\left(\frac{\lambda}{D}\right)^b\left(\frac{W}{D}\right)^c\left(\frac{d}{D}\right)^d\left(\frac{Dc\rho}{\mu}\right)^e (p)^f(\alpha)^g \quad (14\text{–}62)$$

where E_a = ratio of energy absorbed by the structure
             E_1 = energy in incident wave form
             E_2 = energy in reflected wave form
             D = still-water depth

H = incident wave height
λ = incident wave length
c = incident wave celerity
W = thickness of permeable barrier
d = median diameter of rock in barrier
α = angle of slope of barrier with horizontal
p = porosity of permeable barrier = (volume of voids)/volume of structure)
a, b, c, d, e, f, g, K = constants determined by physical analysis or by model study.

Caldwell found that the exponents a, b, and e were each essentially zero. The following equations were then obtained by model tests:

For vertical permeable rock breakwater:

$$E_a = 0.0137 \left(\frac{d}{D}\right)^{0.23} p^{0.87} \tag{14-63}$$

For vertical rock breakwater, with impermeable shoreward face:

$$E_a = 0.0065 \left(\frac{W}{D}\right)^{0.68} \left(\frac{d}{D}\right)^{0.23} p^{0.87} \tag{14-64}$$

For sloping, impermeable seaward face:

$$E_a = 1.5 \, \alpha^{-0.28} \quad \text{for} \quad 6° < \alpha < 37° \tag{14-65}$$

$$E_a = 720 \, \alpha^{-2} \quad \text{for} \quad 37° < \alpha < 75° \tag{14-66}$$

For sloping permeable face, with porosity = 0.45:

$$E_a = 1.42 \left(\frac{d}{D}\right)^{-0.28} (\alpha)^{\left[0.195 \left(\frac{d}{D}\right)^{-0.31}\right]} \tag{14-67}$$

The energy in a solitary wave is determined by Eq. (14–34), which may be written

$$E = \gamma \left(\frac{4}{3} HD\right)^{3/2} \tag{14-68}$$

with E measured in foot-pounds per foot of width of wave. Caldwell concludes that these equations are probably applicable also to progressive oscillatory waves for values of λ/D exceeding about 10.

The Spanish engineer, Iribarren, has developed a very simple formula for demarcation of the critical slope producing reflection, as follows:

$$\tan \alpha = \frac{8}{T} \sqrt{\frac{H}{2g}} \cong \sqrt{\frac{H}{T^2}} \tag{14-69}$$

For flatter slopes, he found that impinging waves will break on the slope; steeper slopes cause surging and reflection. The formula applies only to impermeable slopes. For this slope the reflected wave height is found to be one-half the incident wave height. On the basis of Eqs. (14–62) and (14–68) this would yield an absorbed energy ratio of $1 - 1/2^{3/2}$, or about 0.65.

It may be noted that the parameter H/T^2, in Eq. (14–69) is directly related to the wave steepness H/λ. This follows from Eq. (14–9), for deep-water waves. Thus,

$$\frac{H}{T^2} = \frac{Hc^2}{\lambda^2} = \frac{Hg\lambda}{2\pi\lambda^2} = \left(\frac{H}{\lambda}\right)\left(\frac{g}{2\pi}\right) = 5.12\frac{H}{\lambda} \tag{14–70}$$

14-9. Wave Uprush. *Wave uprush*, or *runup* as it is also called, is another method by which the energy of an incident wave is transformed or dissipated. The clapotis phenomenon mentioned in the preceding article may be considered as a special case of runup, with the kinetic energy transformed momentarily into potential energy.

When a wave breaks on a beach or structure, some of its energy is dissipated in breaking turbulence and the rest is converted into potential energy as it runs up. (See Figs. 14–14 and 14–15.) The height of upwash determines the minimum height to which a structure must be raised to prevent overtopping. The parameter commonly utilized is R/H, the ratio of vertical runup to wave height.

A dimensional analysis of the factors affecting runup will yield the following general equation:

$$\frac{R}{H} = \phi\left(\frac{\lambda}{H}, \frac{D}{H}, \alpha, \frac{E}{\rho c^2 H^2}\right) \tag{14–71}$$

in which E is the incident wave energy and the other terms are as previously defined. The effect of Reynolds number has been assumed to be negligible.

The wave energy is given by Eq. (14–25). The last parameter can then be transformed as follows:

$$\frac{E}{\rho c^2 H^2} = \frac{(1/8)\gamma\lambda H^2}{\rho c^2 H^2} = \frac{g\lambda}{8c^2} = \frac{g\lambda}{8(g\lambda/2\pi)\tanh(2\pi D/\lambda)} = \frac{\pi}{4\tanh(2\pi D/\lambda)} \tag{14–72}$$

Assuming the functional relationships of Eq. (14–71) to be exponential, the following equation is written:

$$\frac{R}{H} = K\left(\frac{\lambda}{H}\right)^a\left(\frac{D}{H}\right)^b(\tan\alpha)^c\left(\tanh\frac{2\pi D}{\lambda}\right)^d \tag{14–73}$$

Model tests at the Waterways Experiment Station and the laboratory of the Beach Erosion Board have given the following equation for breaking wave runup on a continuous, impermeable slope:

$$\frac{R}{H} = 1.02(\tan\alpha)\left(\frac{\lambda}{H}\right)^{1/2}\left(\tanh\frac{2\pi D}{\lambda}\right)^{-1/2} \tag{14–74}$$

Figs. 14-14 and 14-15. Beginning and peak of wave uprush in a test channel, Lake Okeechobee levee studies. (U.S. Army Engineers Beach Erosion Board.).

This equation can be further simplified by noting that

$$\tanh \frac{2\pi D}{\lambda} = \frac{2\pi\lambda}{gT^2} \tag{14–75}$$

so that, with this insertion in Eq. (14–74),

$$\frac{R}{H} = 2.3(\tan\alpha)\left(\frac{T^2}{H}\right)^{1/2} \tag{14–76}$$

In applying this equation, the wave height H is taken as that of the approaching waves before they reach the structure. Also the structure slope must be sufficiently flat to produce breakers, as defined by Eq. (14–69). If this is not the case (that is, if $\tan\alpha > \sqrt{H/T^2}$), then the oncoming waves will not break, but will surge up the structures. On the other hand, more and more of the energy is utilized in reflection as the slope increases. Engineering studies of the U.S. Engineers have led to the recommendation that, for design purposes, for surging waves

$$\frac{R}{H} \simeq 3.0 \tag{14–77}$$

It is not always desirable to use a continuous slope on the seaward side of a coastal structure. Often a composite slope or berm is used, as shown in Fig. 14–16. It is desirable to make the first slope flat enough ($\tan\alpha < \sqrt{H/T^2}$) so that the waves will break on it. At the point of storm water level, a flatter slope may then be extended in order to dissipate some of the remaining energy in friction, thus reducing the final upwash height. It was found that, for the case of a composite of two slopes, Eq. (14–76) could be used satisfactorily with the mean value of $\tan\alpha$. That is,

$$\frac{R}{H} = 2.3\left(\frac{T^2}{H}\right)^{1/2} \frac{\tan\alpha_1 + \tan\alpha_2}{2} \tag{14–78}$$

When the second slope cannot be extended far enough to contain the full runup, then a third slope is used, the intermediate slope then becoming a berm. It has been found that the magnitude of α_3 is not very significant if it is substantially greater than α_2, but it is important that the berm width be

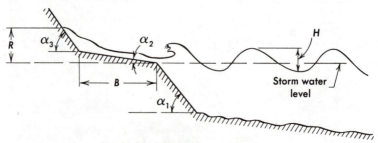

Fig. 14–16. Runup on composite slope.

fairly large, the recommended value being

$$\frac{B}{\lambda} \geq \frac{1}{5} \tag{14–79}$$

Another means of reducing upwash is to use a permeable or artificially roughened slope. It is found that Eq. (14–76) will give satisfactory results in this case if modified as follows:

$$\frac{R}{H} = 2.3(\tan \alpha)\left(\frac{T^2}{H}\right)^{1/2} pr \tag{14–80}$$

where p is a porosity factor and r an empirically determined roughness factor. The porosity factor may be as low as 0.50 for tetrapod blocks or stone. The roughness factor may also become as low as 0.50 for such materials.

It is believed, too, that these factors may be applied to Eq. (14–77) for surging upwash if the slope is rough or permeable. On the other hand, such material is subject to wave attack and uplift pressures caused by receding waves, and should be designed against this possibility.

Example 14–4

A sea wall is to be built to withstand waves of 30-ft height and 10-second period. What wil be the wave run-up for each of the following:

(a) Vertical impermeable wall;
(b) Wall sloping at 30° with horizontal;
(c) Wall at 60° with horizontal;

What per cent of the incident wave energy is reflected in each case?

Solution.

(a) $\tan \alpha = \sqrt{\dfrac{H}{T^2}} = \sqrt{0.30} = 0.548. \quad \therefore \alpha = 28.7°.$

$\quad R \cong H = 30$ ft.   Reflection $= 100\%$ *Ans.*

(b) $\alpha = 30° > 28.7° \therefore$ Waves will surge up slope.

$\quad \therefore R/H \cong 3 \therefore R = 90$ ft.   *Ans.*

Reflection $= (1 - E_a)(100) = [1 - 1.5\alpha^{-0.28}]100$

$$= 100\left(1 - \frac{1.5}{30^{0.28}}\right) = 42.1\% \qquad \textit{Ans.}$$

(c) $\alpha = 60° > 28.7°. \quad \therefore R = 3H = 90$ ft.   *Ans.*

$$\text{Reflection} = 100\left[1 - \frac{720}{(60)^2}\right] = 80\% \qquad \textit{Ans.}$$

14–10. Wave Forces on a Vertical Wall. The design of sea walls, breakwaters, pile-supported shore structures, and other coastal engineering works depends to considerable extent on knowledge of the forces imposed on those structures by waves. A number of theoretical and empirical methods have been published and used, although none as yet are considered completely satisfactory. In general, the approach is first to select a *design wave*. Then, on the basis of knowledge of the wave characteristics, a pressure-intensity relation is developed, representing the effect of the wave on the structure, and this is then used in its design.

Two radically different conditions can be developed. If the ratio H/D (wave height to bottom depth) is less than about 0.75, then a standing wave of height $2H$ is established adjacent to the structure. On the other hand, if the depth is shallow ($H/D > 0.75$), then the waves will break against the structure, with locally very high impact pressures. It is quite possible, because of either sloping bottoms or tidal fluctuations, for both conditions to develop on a given structure at different times. These conditions are illustrated schematically by the pressure diagrams of Fig. 14–17, in which p_u represents the hydrostatic pressure corresponding to the undisturbed water level, p_s the pressure if a standing wave is established, and p_b the pressure resulting from a breaking wave.

Of the numerous theories that have been developed to compute clapotis pressures, the U.S. Engineers recommend the method attributed to M. Sainflou as the most accurate. According to the Sainflou theory, the axis of oscillation of the combined incident and reflected wave motion is raised above the still-water level an amount h_0, as shown in Fig. 14–18.

The orbit center rise is computed by Sainflou to be given approximately by the expression,

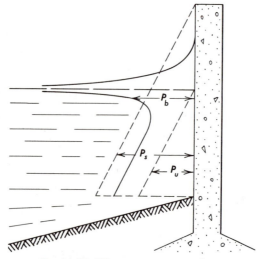

Fig. 14–17. Wave pressures on wall.

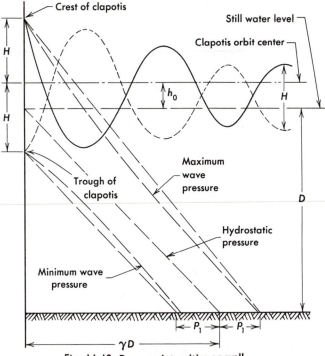

Fig. 14–18. Pressure intensities on wall.

$$h_0 = \frac{\pi H^2}{\lambda} \coth \frac{2\pi D}{\lambda} \tag{14–81}$$

where H is the height of the original free wave and D is the bottom depth at the wall. The clapotis oscillates about this level with an amplitude of $\pm H$. The maximum and minimum pressures at the bottom can be determined from the following relationship:

$$p = \gamma \left[D \pm \frac{H}{\cosh(2\pi D/\lambda)} \right] \tag{14–82}$$

The diagram of pressure intensities is non-linear, as shown in Fig. 14–18. However, it can be approximated, reasonably and conservatively, by a straight-line relationship, in which case the maximum and minimum pressures at any depth d below the still-water level can be computed by

$$p_{max} = \gamma \frac{d + h_0 + H}{D + h_0 + H} \left[D + \frac{H}{\cosh(2\pi D/\lambda)} \right] \tag{14–83}$$

$$p_{min} = \gamma \frac{d + h_0 - H}{D + h_0 - H} \left[D - \frac{H}{\cosh(2\pi D/\lambda)} \right] \tag{14–84}$$

The pressures due solely to the clapotis, when at its crest position, are shown in Fig. 14–19. Here,

$$p_1 = \frac{\gamma H}{\cosh (2\pi D/\lambda)} \tag{14–85}$$

$$p_2 = \gamma \frac{h_0 + H}{D + h_0 + H} \left(D + \frac{p_1}{\gamma} \right) \tag{14–86}$$

With the clapotis in crest position, the total force on the wall due solely to the clapotis is then

$$F_{\max} = \frac{\gamma}{2} \left\{ (D + h_0 + H) \left[D + \frac{H}{\cosh (2\pi D/\lambda)} \right] - D^2 \right\} \tag{14–87}$$

With the clapotis in trough position, a negative thrust is exerted due to the clapotis:

$$F_{\min} = \frac{\gamma}{2} \left\{ (D + h_0 - H) \left[D - \frac{H}{\cosh (2\pi D/\lambda)} \right] - D^2 \right\} \tag{14–88}$$

The corresponding bending moments due solely to the clapotis effect are

$$M_{\max} = \frac{\gamma}{6} \left\{ (D + h_0 + H)^2 \left[D + \frac{H}{\cosh (2\pi D/\lambda)} \right] - D^3 \right\} \tag{14–89}$$

$$M_{\min} = \frac{\gamma}{6} \left\{ (D + h_0 - H)^2 \left[D - \frac{H}{\cosh (2\pi D/\lambda)} \right] - D^3 \right\} \tag{14–90}$$

To the above thrusts and moments should be added the hydrostatic values if applicable. In event that still water exists behind the structure, then the hydrostatic effects on the two sides would cancel each other, and the clapotis thrust and moment would then be the effective values for design.

Pressures due to breaking waves may, under some conditions, reach substantially higher values than for the clapotis condition, apparently because of particles of air trapped between the wall and the wave. From

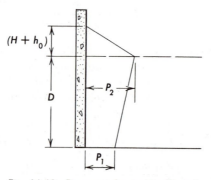

Fig. 14–19. Pressures due to clapotis only.

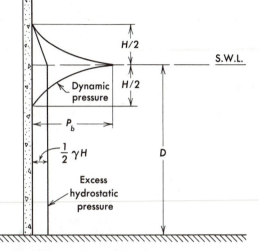

Fig. 14-20. Breaking wave pressures.

many suggested theories, the U.S. Engineers have recommended the method of R. R. Minikin as being the most nearly correct in the light of present information. The pressure distribution postulated by Minikin is shown in two parts in Fig. 14-20, a dynamic pressure centered about the still-water level and an excess hydrostatic pressure resulting from the wave height itself. These are in addition to the still-water hydrostatic pressures, which often may be balanced by equal pressures on the lee side.

The maximum pressure due to the breaking wave is at the still-water level and is given by Minikin as

$$p_{\max} = p_b + p_s = \frac{2\pi g \gamma DH}{\lambda} + \frac{1}{2}\gamma H = \left(\frac{\gamma H}{2}\right)\left(4\pi g \frac{D}{\lambda} + 1\right) \quad (14\text{-}91)$$

The total horizontal force on the wall due to the breaking wave is obtained by integration of the dynamic pressure diagram, assumed to be parabolic, and adding that area to the area of the trapezium representing the excess hydrostatic pressure. Thus,

$$F_H = \frac{1}{3}Hp_b + \frac{1}{2}\gamma H\left(D + \frac{H}{4}\right) = \gamma H\left[\frac{2}{3}\pi g \frac{H}{\lambda}D + \frac{1}{2}\left(D + \frac{H}{4}\right)\right]$$
$$= \frac{\gamma H}{6}\left[4\pi g H \frac{D}{\lambda} + 3\left(D + \frac{H}{4}\right)\right] \quad (14\text{-}92)$$

If the wall is sloping at an angle α with the horizontal, then the dynamic pressure in the above expressions, p_b, should be replaced by $p_b \sin^2 \alpha$.

Example 14-5

A vertical wall is used as a breakwater and must withstand design waves of

10-sec period and 30-ft height. Behind the breakwater, these are diffracted to a height of 10 feet. What is the maximum bending moment exerted on the wall at a point where the bottom depth is: (a) 50 feet; (b) 35 feet?

Solution.

(a) For $D = 50$ feet, $\dfrac{H}{D} = 0.6 < 0.75$. ∴ Waves do not break.

From Fig. 12–3, $\lambda = 360$ feet (assume no change on lee slope).

$$h_0 = \frac{\pi H^2}{\lambda} \coth \frac{2\pi D}{\lambda} = \frac{900\pi}{360} \coth \frac{100\pi}{360} = 11.2 \text{ ft.}$$

On back, $h_0 = \dfrac{100\pi}{360} \coth \dfrac{100\pi}{360} = 1.24$ ft.

The maximum moment on the wall will be caused by a clapotis crest on the front combined with a trough on the back. Thus:

$$M = \frac{\gamma}{6}\left[(D + h_0 + H)^2 \left(D + \frac{H}{\cosh \dfrac{2\pi D}{\lambda}} \right) - D^3 \right]_{\text{front}}$$

$$- \frac{\gamma}{6}\left[(D + h_0 - H)^2 \left(D - \frac{H}{\cosh \dfrac{2\pi D}{\lambda}} \right) - D^3 \right]_{\text{back}}$$

$$= \frac{64}{6}\left[(50 + 11.2 + 30)^2 \left(50 + \frac{30}{\cosh \dfrac{100\pi}{360}} \right) - (50)^3 \right]$$

$$- \frac{64}{6}\left[(50 + 1.24 - 10)^2 \left(50 - \frac{10}{\cosh \dfrac{100\pi}{360}} \right) - (50)^3 \right]$$

$$= 5{,}000{,}000 + 555{,}000 = 5{,}555{,}000 \text{ lb-ft.} \qquad\qquad Ans.$$

(b) For $D = 35$ ft, waves break on front, but clapotis still acts on back.

From Fig. 12–3, $\lambda = 312$ ft; $h_0 = \dfrac{100\pi}{360} \coth \dfrac{70\pi}{312} = 1.4$ ft.

Moment on back due to clapotis:

$$= \frac{64}{6}\left[(35 + 1.4 - 10)^2 \left(35 - \frac{10}{\cosh \dfrac{70\pi}{312}} \right) - (35)^3 \right] = -18{,}900 \text{ lb-ft.}$$

On front, $M = \dfrac{\gamma H}{6}\left[\left(4\pi g H \dfrac{D}{\lambda} \right)(D) + 3D\left(\dfrac{D}{2}\right) + 3\left(\dfrac{H}{4}\right)\left(D + \dfrac{H}{6} \right) \right]$

$$= \frac{64(30)}{6}\left[(4\pi g)(30)\left(\frac{35}{312}\right)(35) + \frac{3}{2}(35)^2 + \frac{3}{4}(30)(40) \right] = 16{,}100{,}000 \text{ lb-ft.}$$

Total moment $= 16,100,000 + 18,900$

$\qquad\qquad = 16,120,000$ lb-ft.   *Ans.*

14-11. Rubble-Mound Breakwaters.   Mounds of stone are frequently used in coastal engineering, either for actual structures or as substructures. The formula developed by the Spanish engineer Iribarren is now widely in use for the design of this type of structure. This formula, slightly modified by Hudson, is as follows:

$$W = KS\gamma \left\{ \frac{H}{(S-1)[\cos \alpha - (\sin \alpha)/C_f]} \right\}^3 \qquad (14\text{-}93)$$

In this equation, W is the minimum weight in pounds of individual capstone required for stability, S is the specific gravity of the rock, measured with respect to the actual specific weight of the impinging water, C_f is the friction coefficient of rock-on-rock (about 1.05), K is an empirical coefficient, and other terms are as defined previously.

The coefficient K is dependent on several variables:

$$K = \phi \left(\alpha, \frac{D}{\lambda}, \text{stone shape} \right) \qquad (14\text{-}94)$$

The functional variation with D/λ is minor and may be neglected. Table 14-3 below gives experimental values of K for the two commonest types of rubble mounds.

TABLE 14-3

Values of Coefficient K in Iribarren-Hudson Equation

tan α	K	
	Quarry Stone	Tetrapod Blocks
0.8	0.003	0.00075
0.67	0.008	0.0022
0.5	0.017	0.0046
0.4	0.028	0.0075
0.33	0.036	0.0090
0.25	0.033	0.0118
0.20	0.030	

Weights of subsurface stones, for depths greater than one wave height, may be reduced in accordance with decrease in orbital velocities with depth. This is accomplished in effect by reducing the wave height used in Eq. (14-93) to correspond with what it would be if the actual orbital velocities at depth d were at the surface. The equivalent wave height, H', is computed as follows:

$$H' = \frac{\pi H^2}{\lambda [\sinh (2\pi d/\lambda)]^2} \qquad (14\text{-}95)$$

More recent tests by Hudson, however, have indicated the following simpler formula to be also more reliable.

$$W = \frac{S\gamma H^3}{K_D(S-1)^3 \operatorname{ctn} \alpha} \tag{14-96}$$

Terms are the same as before, except for K_D, the *damage coefficient*. Based on tests at the Waterways Experiment Station, the Beach Erosion Board recommends the values of K_D shown in Table 14–4 for tentative use, pending field testing.

Of the listed types of specially formed armor units only the tetrapod has been used to any extent in actual structures. Tribars have been used in a few instances. Hollow square concrete blocks with four legs have been used successfully in Japan, and other special shapes have been tried occasionally.

The thickness of the required cover layer can be estimated from the following formula:

$$t = n\left(\frac{W}{S\gamma}\right)^{1/3} \tag{14-97}$$

when n is the number of layers of armor units.

The required number of individual units for a given surface area A is approximately as follows:

$$N = An\left(1 - \frac{p}{100}\right)\left(\frac{S\gamma}{W}\right)^{2/3} \tag{14-98}$$

when p is the porosity, in per cent, as given in Table 14–4.

Equation (14–97) can be used also to determine the required crest width, except that n becomes the number of units side by side on the crest. Usually a minimum of three such units is recommended.

14-12. Wave Forces on Piles. Calculation of forces and bending moments on pile structures has become of increasing interest in recent years, especially with the development of various kinds of offshore platforms. Of the several methods that have been suggested, the most generally accepted at present is one which evaluates the force as composed of two parts, a drag force and an inertial force.

Consider a simple pile subjected to wave attack, as in Fig. 14–21. The drag force results from the horizontal components of the motion of the individual orbiting fluid particles. If it is assumed that the maximum drag force is given by the same type of equation as for steady flow, this force can be written

$$F_D = \int_{-D}^{H/2} \frac{\rho}{2} C_D d U^2 \, dy \tag{14-99}$$

where d is the pile diameter (or breadth), C_D is the drag coefficient and U is the maximum horizontal particle velocity at depth y. From Eqs. (14–15) and (14–19), U is given by

$$U = \frac{\pi\alpha}{T} = \frac{\pi H}{T}\frac{\cosh\left[(2\pi/\lambda)(D+y)\right]}{\sinh\left(2\pi D/\lambda\right)} \tag{14-100}$$

Therefore,

TABLE 14-4
Values of Damage Coefficient in Hudson Formula

Armor Unit	Number of Units in Armor Layer	Damage Coefficient K_D				Porosity (%)
		Structure Trunk		Structure Head		
		Breaking Wave	Non-breaking Wave	Breaking Wave	Non-breaking Wave	
Smooth quarry stone	2	2.1	2.6	2.0	2.4	38
Smooth quarry stone	≥ 3	2.6	3.2	—	—	40
Rough quarry stone	2	2.8	3.5	2.7	3.2	38
Rough quarry stone	≥ 3	3.4	4.3	—	—	40
Modified cube	2	6.0	7.5	—	5.0	47
Tetrapod	2	6.6	8.3	5.0	-6.5	50
Quadripod	2	6.6	8.3	5.0	6.5	50
Hexapod	2	7.2	9.0	5.0	7.0	47
Tribar	2	8.0	10.0	5.0	7.5	54

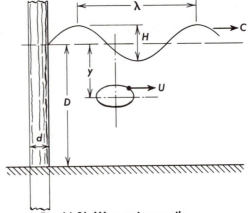

Fig. 14–21. Wave action on pile.

$$F_D = \frac{\rho C_D \, d\pi^2 H^2}{2T^2 \sinh^2 (2\pi D/\lambda)} \int_{-D}^{H/2} \cosh^2 \left[\frac{2\pi}{\lambda} (D + y) \right] dy$$

$$= \frac{\rho C_D \, d\pi^2 H^2}{2T^2 \sinh^2 (2\pi D/\lambda)} \left(\frac{\lambda}{2\pi} \right) \left[\frac{1}{4} \sinh \left\{ 2 \frac{2\pi}{\lambda} (D + y) \right\} + \left(\frac{1}{2} \right) \left(\frac{2\pi}{\lambda} \right) (D + y) \right]_{-D}^{H/2}$$

$$= \frac{\rho}{2} C_D \, dH^2 K_D \tag{14-101}$$

where

$$K_D = \left(\frac{\pi c^2}{8\lambda} \right) \frac{\sinh[(4\pi/\lambda)(D + H/2)] + (4\pi/\lambda)(D + H/2)}{\sinh^2 (2\pi D/\lambda)}$$

$$= \left(\frac{g}{8} \right) \frac{\sinh [(4\pi/\lambda)(D + H/2)] + (4\pi/\lambda)(D + H/2)}{\sinh (4\pi D/\lambda)} \tag{14-102}$$

The calculation of inertial force is based on the assumption that the mass of water displaced by the pile undergoes a maximum acceleration (or deceleration) of $\pm dU/dt$. This assumption yields the equation,

$$F_I = \rho \frac{\pi}{4} d^2 C_m \int_{-D}^{0} \frac{dU}{dt} \, dy \tag{14-103}$$

where C_m is an inertial shape coefficient somewhat analogous to C_D. The inertial force reaches its maximum value as the node passes the pile, since the acceleration is greatest at that time. From Eq. (14–19), this becomes

$$F_I = - \left(\frac{\rho d^2 C_m}{4} \right) \left(\frac{2\pi^2}{T^2} \right) H \int_{-D}^{0} \frac{\cosh [(2\pi/\lambda)(D + y)]}{\sinh (2\pi D/\lambda)} \, dy$$

$$= - \left(\frac{\rho \pi^3 d^2 C_m H}{2T^2 \sinh (2\pi D/\lambda)} \right) \left(\frac{\lambda}{2\pi} \right) \left[\sinh \left\{ \frac{2\pi}{\lambda} (D + y) \right\} \right]_{-D}^{0}$$

$$= \frac{\rho}{2} C_m d^2 H K_i \tag{14-104}$$

where

$$K_i = \frac{\pi^2 c^2}{2\lambda} = \frac{\pi g}{4} \tanh \frac{2\pi D}{\lambda} \tag{14-105}$$

The terms K_D and K_i are only approximate because of the assumptions on which they are based. More precise theoretical analyses are available for higher accuracy if desired, but the equations as given may be satisfactory for preliminary or approximate calculations.

The drag coefficient, C_D, and mass coefficient, C_m, must be determined empirically for given shapes and Reynolds numbers. For approximation purposes, C_D may be taken as 1.2 and C_m as 2.5, for cylindrical piles. These values may be increased about 25 per cent for square piles and about 100 per cent or more for H-piles.

It is noted that the maximum drag force occurs when the wave crest passes (with a maximum of smaller magnitude in the opposite direction when the trough passes). The maximum inertial force, however, occurs when the node passes, approximately 90° out of phase with the peak drag force. Therefore the two forces are not directly added. The true maximum combined force would occur probably somewhere between the crest and node positions. The maximum combined force $F_{\max}$ may be computed approximately as either

$$F_{\max} = F_D \tag{14-106}$$

or

$$F_{\max} = F_I + 0.4 F_D \tag{14-107}$$

whichever is greater.

The lines of action of the inertial and drag forces may be determined by calculating the bending moments on the pile as measured with respect to the bottom. Thus,

$$M_D = \int_{-D}^{H/2} \frac{\rho}{2} C_D dU^2 (D - y)\, dy \tag{14-108}$$

and

$$M_I = \int_{-D}^{0} \left(\frac{\rho \pi d^2 C_m}{4} \right) \left(\frac{dU}{dt} \right) (D - y)\, dy \tag{14-109}$$

Then,

$$y_D = D - \frac{M_D}{F_D} \quad \text{and} \quad y_I = D - \frac{M_I}{F_I} \tag{14-110}$$

where y_D and y_I are the distances below the still-water level to the points of application of F_D and F_I, respectively.

The discussion thus far has reference to forces on a single pile. If a group of piles is so arranged that the gap between adjacent piles is at least 1.5 times the pile diameter, then each pile can be considered independently. If the gap is less than this, the interior piles are subjected to greater forces than as calculated above. The bending moment for the center pile in a group of three, with a gap equal to half the pile diameter, has been measured as up to 2.4 times the corresponding bending moment on a single pile.

Example 14–6

The waves of Example 14–5 attack an off-shore platform supported by piles, each of which is 36 inches in diameter. The bottom depth is 50 ft. Assuming that each pile can be considered independently, what is the approximate maximum force exerted on the pile?

Solution.

$$K_D = \frac{g}{8} \frac{\sinh\left[\frac{4\pi}{\lambda}\left(D + \frac{H}{2}\right)\right] + \frac{4\pi}{\lambda}\left(D + \frac{H}{2}\right)}{\sinh\left(\frac{4\pi D}{\lambda}\right)}$$

$$= \frac{32.2}{8} \frac{\sinh\left[\frac{4\pi}{360}\left(50 + \frac{30}{2}\right)\right] + \frac{4\pi}{360}(65)}{\sinh\frac{4\pi}{360}(50)}$$

$$= \frac{32.2}{8} \frac{\sinh(2.27) + (2.27)}{\sinh(1.75)} = 10.2$$

$$K_I = \frac{\pi g}{4}\tanh\left(\frac{2\pi D}{\lambda}\right) = \frac{32.2\pi}{4}\tanh(0.875) = 17.8$$

$$F_D = \frac{\rho}{2} C_D\, dH^2 K_D = \frac{1.99}{2}(1.2)(3)(30)^2(10.2) = 32{,}800 \text{ lb.}$$

$$F_I = \frac{\rho}{2} C_M\, d^2 H K_I = \frac{1.99}{2}(2.5)(3)^2(30)(17.8) = 12{,}000 \text{ lb.}$$

$$F_I + 0.4 F_D = 25{,}000 \text{ lb.}$$

Since $F_D > (F_I + 0.4F_D)$, maximum force $= F_D = 32{,}800$ lb.   *Ans.*

14–13. Coastal Sedimentation. Many of the most difficult problems encountered in coastal engineering have to do with sediment erosion, transportation, and deposition. Estuarine and harbor sedimentation, beach erosion, and littoral transposition of sediments are among the most important of these problems. Added to all the complexities of river and channel sedimentation, wave action exerts pronounced influences on coastal sediments, so that a truly rational analysis of problems of this kind is as yet not possible.

Coastal sediments come essentially from two sources: (1) sediment brought downstream by rivers to their mouths; (2) sediment produced by wave attack on exposed coastal formations. The sediments so produced are subjected to hydrodynamic sorting, as well as to transportation and final deposition. The transportation may be with both normal (shoreward or seaward) and tangential (littoral) components.

Beaches are usually composed mainly of sand or such other coarser materials as may be available in the source areas and transportable with the available energy. Finer materials, if available, are either kept in suspension by the turbulence of the surf or deposited in harbors or lagoons.

Ordinary wave activity will manifest sufficient energy to move sand as deep as about 30 ft below the surface. Thus, the beach can be defined as the stretch of coast between the 30-ft depth line, below low tide level, up to that portion of the coast which is essentially permanent, such as a cliff, sand dunes, or man-made structures.

Sand particles are lifted by the turbulence accompanying the passage of a wave, tending to settle out again between waves. However they will move a certain distance while in suspension, the direction of motion depending upon the local currents. The turbulence is especially intense under a breaking wave, so that a trough tends to be formed along the plunge line of the breakers, which, of course, moves with the tide.

The currents are of several types: (1) oscillatory currents, moving normal to the shoreline, inward rapidly as the breaker passes, then more slowly seaward until the next breaker; (2) littoral currents, caused by waves breaking at an angle with the beach; (3) rip currents, which flow outward through openings in the sand deposits, fed by the piling-up of waters caused by littoral currents; (4) tidal currents, especially notable in inlets and estuaries; (5) currents fed by stream discharge.

There are important points of similarity between sediment characteristics of streams and beaches. An equilibrium beach may be defined in similar manner to an equilibrium or graded stream, as one which has characteristics (curvature, slope, sand thickness, and roughness) adjusted to each other in such a way that the available energy for sediment transport is just sufficient to transport the sediment supplied. This, of course, has reference to a long-period equilibrium and not to daily or even seasonal permanence. If the beach is not in equilibrium and is either being eroded or extended, it follows that there is a lack of balance between the supply of sediment and the available hydraulic energy and other characteristics.

The division of the sediment load into bed load and suspended load is applicable as well to beaches as to river beds. The basic mechanics of sedimentation as related to turbulence structure must also be similar. The major difference is that the turbulence structure in coastal waters, caused as it is by wave phenomena, is fundamentally different from that in rivers. Furthermore, the transporting currents are usually only the secondary littoral currents in the surf zone instead of the main river flow.

As is true for streams, beaches tend to change character in such a way as to approach an equilibrium condition. The character of a beach will therefore be determined by several variables, among which are the following:

1. Supply of sediment for replenishing sediment removed by littoral or other currents.

2. Characteristics (approach direction, height, steepness, variability, etc.) of waves striking the beach.
3. Coastal geologic formations, which in large measure control the beach curvature, slope, and roughness.
4. Man-made structures or other operations such as dredging.

The underwater beach topography in large measure controls the surf characteristics and the resulting turbulence and sediment transport. Slopes are more or less arbitrarily classified as follows:

$$\text{Steep slope:} \qquad S > \frac{1}{50}$$

$$\text{Intermediate slope:} \qquad \frac{1}{50} > S > \frac{1}{75}$$

$$\text{Flat slope:} \qquad \frac{1}{75} > S$$

As would be expected, the average grain size of beach sand, as well as the beach roughness, tends to increase with increasing slopes.

A typical beach cross-section is sketched in Fig. 14–22 showing both summer and winter profiles. The vertical scale is exaggerated some 25 times. The difference between summer and winter profiles is due to the severe wave and surf conditions following winter storms in the fetch areas affecting the particular coastline.

When violent winter storms occur, the abnormal wave conditions attack the berm and deposit the material in one or more offshore bars. The details of this process are not yet clear, although it is known that bar formation requires some critical wave steepness to be attained. The bars in turn cause

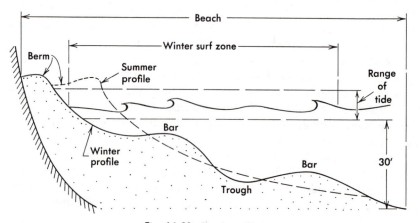

Fig. 14–22. **Beach profiles.**

breakers to form on the larger waves. Smaller waves are transmitted unchanged, to break on a still shallower bar. Breakers also may re-form as smaller waves in the trough between bars and then later rebreak.

Waves of smaller steepness, normally occurring in summer, have the net effect of moving the sand toward the shore, eliminating the bars, and widening the berm. This cycle seems to be repeated seasonally at most beaches. However, occasional very large storms may carry berm sand so far seaward that the subsequent normal surf action cannot reach it, resulting in a permanent beach recession.

Most beach erosion problems, on the other hand, are related to littoral currents. As noted previously, these currents result from waves breaking at an angle with the coast. The resulting continuous movement of suspended sediment with these currents is known as the littoral drift, and is often of very substantial magnitude.

The actual rate of littoral drift is some unknown function of the sediment characteristics and the height, period, and direction of the shoaling waves, as well as of the coastline characteristics. Model studies have indicated that littoral transport reaches a maximum for wave crests which have an angle, in deep water, of about 30° with the shoreline, and which have a steepness of about 0.025. Actual rates in a particular location must usually be estimated on the basis of actual measurements if possible. Quantitative transfer of model data to field conditions is as yet highly uncertain, although qualitative information of considerable value can often be obtained from shoreline models. Littoral drifts as high as 1,000,000 cubic yards per year have actually been observed in the field.

When beach erosion or deposition problems appear to exist and need correction, the following procedure should be followed:

1. Make a thorough study of the entire problem, determining sediment sources, wave characteristics, current magnitudes and directions, magnitudes and rates of drifting, etc.
2. Attempt to determine whether the erosion problem is merely a passing phase in a cycle or is likely to be permanent and continuing.
3. Try to ascertain the original cause of the problem and remedy this if possible, especially if caused by human activities.
4. If new corrective action must be undertaken, attempt to remedy the erosion by an artificial supply of sediment rather than by fixed structures, if economically and technically feasible.
5. Use fixed structures, such as groins and seawalls, only as a last resort, attempting in their design to minimize their potentially harmful effects on other sections of the coast. (See Fig. 14-23.) Thus, groins may retard erosion of the up-current side of the coast but accelerate that on the down-current side, and may even induce harmful rip-currents. Sea walls, intended to inhibit further wave erosion of the beach, may actually hinder the restoration of the beach by waveborne sediment, as well as induce downcoast littoral erosion.

Fig. 14–23. Beach erosion, Michigan. Note protective works near railroad and highway. (U.S. Army Engineers Beach Erosion Board.)

PROBLEMS

14-1. Oscillating surface waves are generated on a deep lake by a wind blowing at 50 mph, with a fetch of 20 mi. Determine the probable wave height, length, period, and celerity.

14-2. At what bottom depth would the celerity of Problem 14-1 begin to be changed by the decreasing depth? What is the celerity at a bottom depth of 50 ft? of 10 ft? Assuming the period does not change, what are the corresponding wave lengths?

14-3. Consider a particle initially 10 ft below the still surface and 40 ft above the bottom. After the wave motion is established, what is the size and character of the orbit of the particle? Repeat for a particle at the surface and one at the bottom.

14-4. Determine the maximum orbital velocities of each of the particles of Problem 14-3, in the horizontal and vertical directions. Determine the average velocities over one-half cycle.

14-5. Determine the total wave energy, for 1 wavelength, in foot-pounds per foot of wave, neglecting friction. *Ans.:* 183,000 ft-lb/ft.

14-6. If a single wave, of the height of each of the oscillatory waves of Problem 14-1, is propagated over the still lake, what are the celerity, volume, and energy of the wave when the bottom depth is 50 ft? At approximately what bottom depth will the wave break?

14-7. A wave train is generated in deep water by a wind blowing in a direction S 45° W over a fetch of 100 nautical miles at a speed of 40 knots. These waves approach a shore which is assumed linear and bearing N 45° W. Offshore contours are parallel to the shore line, with the bottom sloping at a rate of 1 vertical to 5 horizontal.

 a. How far offshore will the waves break? *Ans.:* 158 ft.
 b. Calculate the celerity, height, wavelength, and steepness of the waves when they break.
 c. How far offshore does the bottom begin to exert an appreciable effect on the waves? *Ans.:* 1280 ft.
 d. Calculate the celerity, height, wavelength, and steepness at this point.

14-8. Waves are generated in deep water by wind blowing at a speed of 100 knots over a fetch of 40 nautical miles. Determine the following, for the "significant waves" thus formed:

 a. Equation of wave crest, with respect to an origin through a point on still water level through which a node passes at time $t = 0$. Include t as a variable.
 b. Total energy in the wave form, per foot of wave.
 c. Height of wave at a point near the shore where the depth is 100 ft, assuming shore line parallel to advancing wave crests.
 d. Energy being transmitted with the wave form, at the point where the depth is 100 ft.
 e. Probable depth at which the waves will break.
 f. Maximum orbital velocity experienced by any particle in deep water.

14–9. The same waves as in Problem 14–7 are assumed to approach a shore which runs due north and south and which also has an offshore bottom slope of 1:5.

a. Determine the bearings of wave crests 50 ft, 100 ft, 500 ft, and 1000 ft offshore.

b. Draw a wave refraction diagram for 1000 ft of coastline, and extending 1500 ft offshore, to a scale of $1'' = 200'$, showing crest lines spaced at intervals of $\frac{1}{5}cT$ and orthogonals spaced 100 ft apart in deep water.

c. Determine and locate the breaker line on the refraction diagram.

d. Determine the maximum wave height and steepness attained before breaking.

14–10. A wave train is generated in deep water by a wind blowing over a fetch of 225 nautical miles at a speed of 40 knots. Determine the steepness of the significant wave:

a. After generation in deep water. *Ans.:* 0.0433.

b. At point of breaking, assuming parallel approach to straight shore line. *Ans.:* 0.0561.

c. At point near irregular shore line where bottom depth is 100 ft and where the orthogonals on a refraction diagram have changed from a deep-water spacing of 100 ft to a spacing of 144 ft. *Ans.:* 0.0533.

14–11. Oscillatory waves generated in deep water have a celerity of 50 ft/sec. The waves approach a shoreline which makes an angle of 60° with the direction of wave advance.

a. At a point where the bottom depth has become 50 ft, what angle do the wave crests make with the shore line? *Ans.:* 21.1°.

b. What is the wave height at this point? *Ans.:* 24.6 ft.

c. Assuming that each wave becomes essentially a solitary wave as it nears the shore, at what depth will it break? *Ans.:* 30.0 ft.

14–12. The wave train of Problem 14–10 encounters an offshore current of 5 knots. Determine the absolute velocity, spacing, and height of the waves resulting if:

a. Current is in same direction as waves. *Ans.:* 78.5 fps, 960 ft, 28.2 ft.

b. Current is in opposite direction to waves. *Ans.:* 44.1 fps, 540 ft, 44.1 ft.

c. Current makes angle of 45° with waves, in opposing direction.

14–13. A train of oscillatory significant waves is generated on a lake by a wind blowing at 60 miles per hour over an effective fetch of 25 miles. As the waves move toward the shore, determine the wave height when the bottom depth has decreased to 40 feet for:

a. Straight shore line, perpendicular to direction of wave advance;

b. Irregular shore line, at a point where the orthogonals on a refraction diagram have changed from a deepwater spacing of 100 feet to a spacing of 144 feet.

c. In a region where the advancing waves meet an offshore current of 5 miles per hour, moving opposite to the direction of wave advance.

14-14. A train of oscillatory waves generated in deep water has a period of ten seconds. They approach a shore line with form and off-shore contours as sketched below. Using any appropriate scale, construct an approximate refraction diagram. Estimate the wave height at the apex of the promontory along the shore. Note that the shore actually consists of a vertical cliff and the depth of water at the cliff face is 10 feet.

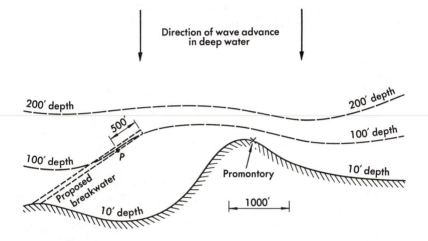

14-15. If a breakwater is placed off the shoreline of Problem 14–14, directly along the 100-ft. depth contour, as shown, estimate the wave height on the lee side of the breakwater, at a point 500 feet back from the tip (point "P" as shown).

14-16. A single semi-infinite breakwater intercepts waves advancing in a direction perpendicular to the breakwater. Compute equations for lines of equal diffraction coefficient, for K' values of 0.1, 0.2, 0.3, 0.5, 0.8, 1.0, 1.17, and 1.0.

14-17. Plot a generalized diffraction diagram on a full-size sheet ($8\frac{1}{2}$ in. $\times$ 11 in.), showing x/λ versus y/λ, for the above curves of equal K'. Use a range of values of $0 < y/\lambda < 300$.

14-18. Superimpose on the diffraction diagram patterns of wave crests, for center line y/λ spacing of 20.

14-19. A single semi-infinite breakwater intercepts waves, advancing in a direction making an angle of 120° with the breakwater. The waves have an incident height of 10 ft and period of 8 sec. The bottom depth is 50 ft. Determine the wave height and velocity at a point A which is 60 ft behind (in direction of wave advance) and 20 ft back behind the breakwater (in direction normal to wave advance). *Ans.:* 4.0 ft, 33.8 fps.

14-20. A wave train is generated in deep water by a wind blowing over a fetch of 225 nautical miles at a speed of 40 knots. It advances parallel to a breakwater, and then through a breakwater gap. The bottom depth at the breakwater is 100 ft and the gap width is designed to be twice the incident wave length. What is the wave height at a point 2000 ft behind and 2000 ft to the right of the gap center line? *Ans.:* 7.6 ft.

14–21. Waves approaching a breakwater have an incident height of 12 ft and period of 6 sec. The bottom depth is 20 ft. Compute the energy reflected by the breakwater, in ft-lb per foot of wave, for the following conditions:

a. Vertical, impermeable wall. *Ans.:* 367,000 ft-lb/ft.
b. Vertical permeable breakwater, 30 ft thick at water line, composed of stone of 30 per cent porosity and 12-in. median diameter.
c. Sloping impermeable breakwater, with $\alpha = 40°$.
d. Sloping impermeable breakwater, with $\alpha = 20°$.
e. Sloping permeable breakwater with $\alpha = 60°$, and stone of 12-in. diameter and 45 per cent porosity. *Ans.:* 206,000 ft-lb/ft.

14–22. A sea wall is to be built to withstand waves of 36-ft height and 9-sec period. The bottom depth is 60 feet. Calculate the following:

a. Energy in the approaching wave, assuming that it has assumed by this point the characteristics of a solitary wave.
b. Wave run-up, assuming the sea-wall is vertical.
c. Wave run-up, assuming the wall slopes at 30° with the horizontal.
d. Wave run-up for a 60° wall.
e. Energy reflected by the wall in each of the three cases above.

14–23. A sea wall is designed to withstand waves of 37-ft height and 12.25-sec period. The bottom depth is 150 ft.

a. What maximum slope can be used to ensure that the waves will break on the wall? *Ans.:* 26.2°.
b. What is the wave runup for this slope? *Ans.:* 85 ft.
c. What is the wave runup if a berm is used, with a slope of 5°? *Ans.:* 50 ft.
d. How wide must the berm be?

14–24. Waves having an incident height of 20 ft and period of 10 sec approach an impermeable sea wall. Compute the wave runup for each of the following:

a. Single wall, of 30° slope. *Ans.:* 60.0 ft.
b. Single wall, of 15° slope. *Ans.:* 27.5 ft.
c. Composite wall, of slopes 15°, 5°, and 45°, respectively. The 5° slope is for the berm, which is adequately long. *Ans.:* 18.2 ft.

14–25. A vertical-wall breakwater is to be designed to withstand design waves of 10-sec period and of 20-ft height. At the wall the bottom depth is 30 ft. Compute the total thrust and bending moment on the wall for the following conditions:

a. Still water behind the breakwater.
b. No water behind the breakwater.
c. Waves of 10-sec period and 5-ft height behind the breakwater.

14–26. For one section of the wall of Problem 14–25, the depth is only 15 ft. Repeat the calculations of Problem 14–25 for this section.

14–27. If the breakwater of Problem 14–25 is designed as a rubble mound of tetrapod blocks, with a slope angle of 30°, what weight of surface blocks should be used? What weight, if quarry stones are used? Plot curves of K values versus α, for both materials, from Table 14–3.

14–28. A vertical-wall breakwater must withstand design waves of 10-sec period and 20-ft height. At the wall, the bottom depth is 20 ft. Compute the maximum total thrust on the wall, per lineal foot for:

a. Still water behind the breakwater. *Ans.:* 157,000 lb/ft.
b. No water behind the breakwater. *Ans.:* 170,000 lb/ft.
c. Waves of 10-sec period and 5-ft height behind the breakwater. *Ans.:* 162,000 lb/ft.

14–29. A vertical-wall breakwater is to be designed to withstand storm waves of 8-sec period and 16-ft height. At the wall, the bottom depth is 15 ft. Behind the breakwater, the wave energy has been so diffracted that the wave height is only 4 ft. Compute the maximum thrust on the wall.

14–30. Derive expressions for bending moment and location of pressure center for drag and inertial wave forces on a circular pile, as based on Eqs. (14–108) through (14–110).

14–31. An offshore structure is supported on concrete piles, each of which is 18 in. in diameter. The bottom depth at the site is 30 ft. The structure is attacked by waves of 12-ft height and 8-sec period. Assuming that each pile can be considered independently, what is the approximate magnitude of the maximum force probably exerted on the pile? *Ans.:* 2280 lb.

REFERENCES

CHAPTER 1

ADDISON, HERBERT. *Land, Water and Food*. London: Chapman and Hall, 1961. 284 pp.

BISWAS, ASIT K. "Hydrologic Engineering Prior to 600 B.C." *Journal of the Hydraulics Division, A.S.C.E.*, Vol. 93 (1967), pp. 115–135.

FINCH, JAMES K. *The Story of Engineering*. Garden City, N.Y.: Doubleday and Co., 1960. 425 pp.

FOX, CYRIL S. *Water*. New York: Philosophical Library, 1952.

KOLUPAILA, STEPONAS. "Early History of Hydrometry in the United States," *Journal of the Hydraulics Division, A.S.C.E.*, Vol. 86 (1960), pp. 1–52.

LANGBEIN, WALTER B., and HOYT, WILLIAM G. *Water Facts for the Nation's Future*. New York: The Ronald Press Co., 1959. 288 pp.

ROUSE, HUNTER, and INCE, SIMON. *History of Hydraulics*. Iowa City, Iowa: Institute

U.S. Department of Agriculture. *Water (Yearbook of Agriculture, 1955)*. Washington: U.S. Govt. Printing Office, 1955. 751 pp.

WITTFOGEL, K. A. *The Hydraulic Civilization: Man's Role in Changing the Earth*. Chicago: University of Chicago Press, 1956.

CHAPTER 2

ALBERTSON, M. L., BARTON, J. R., and SIMONS, D. B. *Fluid Mechanics for Engineers*. Englewood Cliffs, N.J.: Prentice-Hall, Inc., 1960. 567 pp.

DAUGHERTY, R. L. and FRANZINI, J. B. *Fluid Mechanics, with Engineering Applications*. 6th ed. New York: McGraw-Hill Book Co., Inc., 1965. 625 pp.

KING, H. W. and BRATER, E. F. *Handbook of Hydraulics*. 5th ed. New York: McGraw-Hill Book Co., Inc., 1963. 571 pp.

OLSON, REUBEN M. *Essentials of Engineering Fluid Mechanics*. 2nd ed. Scranton: International Textbook Co., 1966. 448 pp.

SABERSKY, R. H. and ACOSTA, A. J. *Fluid Flow*. New York: Macmillan Co. 1964. 393 pp.

SHAMES, IRVING. *Mechanics of Fluids*. New York: McGraw-Hill Book Co., Inc., 1962. 555 pp.

SILBRECH, DONALD A. *Fluid Mechanics*. San Francisco: Wadsworth Publ. Co., 1965. 562 pp.

STREETER, V. L. *Fluid Mechanics*. 4th ed. New York: McGraw-Hill Book Co., Inc., 1966. 707 pp.

SWANSON, W. M. *Fluid Mechanics*. New York: Holt, Rinehart & Winston, 1970. 576 pp.

VENNARD, J. K. *Elementary Fluid Mechanics*. 4th ed. New York: John Wiley & Sons, Inc., 1961. 570 pp.

CHAPTER 3

Basic Equations of Pipe Flow

A.S.C.E. TASK FORCE ON FLOW IN LARGE CONDUITS. "Factors Influencing Flow in Large Conduits," *Journal of the Hydraulics Division, A.S.C.E.* November 1965, pp. 123–153, with discussions. Closure, May 1967, pp. 181–188.

BAKHMETEFF, B. A. *The Mechanics of Turbulent Flow*. Princeton, N.J.: Princeton University Press, 1936.

COLEBROOK, C. F. "Turbulent Flow in Pipes, with Particular Reference to the Transition Region Between the Smooth and Rough Pipe Laws," *Journal of Institution of Civil Enginers*, Vol. 11, 1939.

COLEBROOK, C. F. and WHITE, C. M. "Experiments with Fluid Friction in Roughened Pipes," *Proceedings, Royal Society of London*, Vol. 161 (1937).

MALAIKA, JAMIL. "Flow in Non-Circular Conduits," *Journal of the Hydraulics Division*, *A.S.C.E.*, November 1962, pp. 1–30, with discussions. Closure, November 1964, pp. 233–237.

MOODY, LEWIS F. "Friction Factors for Pipe Flow." *Trans. A.S.M.E.*, Vol. 66 (1944).

NIKURADSE, J. "Gesetzmässigkeiten der turbulenten Strömung in glatten Rohren," *V.D.I. Forschungsheft*, Vol. 356 (1932).

NIKURADSE, J. "Strömungsgesetze in rauhen Rohren," *V.D.I. Forschungsheft*, Vol. 361 (1933).

STREETER, V. L. "Steady Flow in Pipes and Conduits." In *Engineering Hydraulics* (Hunter Rouse, ed.). New York: John Wiley & Sons, Inc., 1950. pp. 387–443.

Turbulent Flow Regimes

CHOW, VEN TE. *Open-Channel Hydraulics*. New York: McGraw-Hill Book Co., Inc., 1959. pp. 194–198.

HASAN, ASGHAR. *Hydraulic Effects of Boundary Roughness in Open Channels at Subcritical Flow*. M.S. Thesis, Blacksburg: Virginia Polytechnic Institute, 1962. 160 pp.

MAY, JAMES C. *Laws of Turbulent Flow in Rough Pipes*. Ph.D. Dissertation, Blacksburg: Virginia Polytechnic Institute, December 1966. 221 pp.

McCONNELL, F. L. *Quasi-Smooth Flow in Pipes*. M.S. Thesis, Blacksburg: Virginia Polytechnic Institute, August 1967. 65 pp.

MORRIS, HENRY M. "A New Concept of Flow in Rough Conduits," *Trans. A.S.C.E.*, Vol. 120 (1955). pp. 373–410.

MORRIS, HENRY M. "Design Methods for Flow in Rough Conduits," *Trans. A.S.C.E.*, Vol. 126 (1961) Part I, pp. 454–490.

NEILL, CHARLES R. "Hydraulic Roughness of Corrugated Pipes," *Journal of the Hydraulics Division, A.S.C.E.*, May 1962. pp. 23–44.

STRAUB, L. G., BOWERS, C. E., and PILCH, M. *Resistance to Flow in Two Types of Concrete Pipe*," *Technical Paper 22, Series B* (December 1960), St. Anthony Falls Hydraulics Laboratory.

WEBSTER, MARVIN J. and METCALF, LAURENCE A. "Friction Factors in Corrugated Metal Pipe," *Journal of the Hydraulics Division, A.S.C.E.*, September 1959, pp. 35–67.

Non-Uniform Flow

ANDERSON, ALVIN G. *Hydraulics of Conduit Bends*. Minneapolis: St. Anthony Falls Hydraulics Laboratory, 1948. 22 pp.

BLAISDELL, F. W. and MANSON, P. W. "Energy Loss at Pipe Junctions," *Journal of the Irrigation and Drainage Division, A.S.C.E.*, September 1967, pp. 59–78.

CAMP, T. R. and LAWLER, J. C., "Water Distribution," in *Handbook of Applied Hydraulics* (Eds. C. V. Davis and K. E. Sorensen). New York: McGraw-Hill Book Co., Inc. 1969, pp. 37:1–58.

CHATURVEDI, M. C. "Flow Characteristics of Axisymmetric Expansions," *Journal of the Hydraulics Division, A.S.C.E.*, May 1963, pp. 61–92.

CROSS, HARDY. *Analysis of Flow in Networks of Conduits or Conductors*. Univ. of Illinois Bulletin 286, November 1936.

KALINSKE, A. A. "Conversion of Kinetic to Potential Energy in Flow Expansions," *Trans. A.S.C.E.*, Vol. 11 (1946), pp. 355–390.

McPHERSON, M. B. "Generalized Distribution Network Head-Loss Characteristics," *Trans. A.S.C.E.*, Vol. 126 (1961), pp. 1190–1234.

PIGGOTT, R. J. S. "Pressure Losses in Tubing, Pipe and Fittings," *Trans., Amer. Soc. Mech. Engrs.*, Vol. 72 (July 1950), p.679.
STREETER, V. L. "Steady Flow in Pipes and Conduits." In *Engineering Hydraulics* (Hunter Rouse, ed.). New York: John Wiley & Sons, Inc., 1950. pp. 412–443.

CHAPTER 4

A.S.C.E. TASK FORCE. "Friction Factors in Open Channels." *Journal of the Hydraulics Division, A.S.C.E.*, March 1963. pp. 97–143.
CHOW, VEN TE. *Open-Channel Hydraulics.* New York: McGraw-Hill Book Co., Inc., 1959. 680 pp.
CHOW, VEN TE. "Open Channel Flow." In *Handbook of Fluid Dynamics* (V. L. Streeter, ed.). New York: McGraw-Hill Book Co., Inc., 1961. pp. 24–1 to 24–59.
DAVIS, CALVIN V. and SORENSEN, KENNETH E. "Canals and Conduits." Section 7 in *Handbook of Applied Hydraulics.* New York: McGraw-Hill Book Co., Inc., 3rd ed., 1969. 34 pp.
HENDERSON, F. M. *Open Channel Flow.* London: Macmillan Co., 1966. 522 pp.
HERBICH, JOHN B. and SHULITS, Sam. "Large-Scale Roughness in Open-Channel Flow." *Journal of the Hydraulics Division, A.S.C.E.*, November 1964. pp. 203–230.
IPPEN, A. T. "Channel Transitions and Controls." In *Engineering Hydraulics* (Hunter Rouse, ed). New York: John Wiley & Sons, Inc., 1950. pp. 496–538.
KING, H. W. and BRATER, E. F. *Handbook of Hydraulics.* 5th ed. New York: McGraw-Hill Book Co., Inc., 1963. 571 pp.
MIROJGAOKER, A. G. and CHARLU, K. L. N. "Natural Roughness Effects in Rigid Open Channels." *Journal of the Hydraulics Division, A.S.C.E.* September 1963. pp. 29–44.
POSEY, C. J. "Gradually Varied Channel Flow." In *Engineering Hydraulics* (Hunter Rouse, ed.). New York: John Wiley & Sons, Inc., 1950. pp. 589–634.
REVELL, RUSSELL W. "Natural Channels." Section 5. In *Handbook of Applied Hydraulics.* 3rd ed. New York: McGraw-Hill Book Co., Inc., 1969. 26 pp.
ROUSE, HUNTER. "Critical Analysis of Open-Channel Resistance." *Journal of the Hydraulics Division, A.S.C.E.* July 1965. pp. 1–25.

CHAPTER 5

General

BAKHMETEFF, B. A. *Hydraulics of Open Channels.* New York: McGraw-Hill Book Co., Inc., 1932.
CHOW, VEN TE. *Open-Channel Hydraulics.* New York: McGraw-Hill Book Co., Inc., 1959. 680 pp.
CHOW, VEN TE. "Open Channel Flow." Section 24 in *Handbook of Fluid Dynamics.* Ed. by V. L. Streeter. New York: McGraw-Hill Book Co., Inc., 1961. 59 pp.
HENDERSON, F. M. *Open Channel Flow.* New York: Macmillan Publ. Co., 1966. 522 pp.
WOODWARD, S. M. and POSEY, C. J. *Hydraulics of Steady Flow in Open Channels.* New York: John Wiley & Sons, Inc., 1941.

Profiles in Gradually Varied Flow

CHEN, C. L. and WANG, C. T. "Non-Dimensional Gradually-Varied Flow Profiles." *Journal of Hydraulics Division, A.S.C.E.*, September 1969, pp. 1671–1686.
MONONOBE, N. "Backwater and Dropdown Curves for Uniform Channels," *Trans. A.S.C.E.*, Vol. 103 (1938), pp. 950–989.
PICKARD, W. F. "Solving the Equations of Uniform Flow," *Journal of Hydraulics Division, A.S.C.E.*, July 1963, pp. 23–38.

POSEY, C. J. "Gradually Varied Channel Flow," Chapter IX in *Engineering Hydraulics*. Ed. by Hunter Rouse. New York: John Wiley & Sons, Inc., 1950. pp. 589–634.

PRASAD, RAMANAND. "Numerical Method of Computing Flow Profiles," *Journal of Hydraulics Division, A.S.C.E.*, January 1970, pp. 75–86.

RAO, N. S. L. and SRIDHARAN, K. "Characteristics of M-1 Backwater Curves." *Journal of Hydraulics Division, A.S.C.E.*, November 1966, pp. 131–139.

U.S. BUREAU OF RECLAMATION: *Guide for Computing Water Surface Profiles*. Denver, Colorado: U.S.B.R., 1957, 164 pp.

VALLENTINE, H. R. "Generalized Profiles of Gradually-Varied Flow," *Journal of Hydraulics Division, A.S.C.E.*, March 1967, pp. 17–24.

VALLENTINE, H. R. "Characteristics of the Backwater Curve," *Journal of Hydraulics Division, A.S.C.E.*, July 1964, pp. 39–49.

Subcritical Transitions

HINDS, JULIAN. "The Hydraulic Design of Flume and Siphon Transitions," *Trans. A.S.C.E.*, Vol. 92 (1928), pp. 1423–1459.

SHUKRY, AHMED. "Flow Around Bends in an Open Flume," *Trans. A.S.C.E.*, Vol. 115 (1950), pp. 751–788.

SIMMONS, W. P. "Transitions for Canals and Culverts," *Journal of Hydraulics Division, A.S.C.E.*, May 1964, pp. 115–154.

SMITH, C. D. "Simplified Design for Flume Inlets," *Journal of Hydraulics Division, A.S.C.E.*, November 1967, pp. 25–34.

SOLIMAN, M. M. and TINNEY, E. ROY. "Flow Around 180° Bends in Open Rectangular Channels," *Journal of Hydraulics Division, A.S.C.E.*, July 1968, pp. 893–908.

Supercritical Transitions

BAGGE, GUNNAR and HERBICH, J. B. "Transitions in Supercritical Open-Channel Flow," *Journal of Hydraulics Division, A.S.C.E.*, September 1967, pp. 23–42.

IPPEN, A. T. "Mechanics of Supercritical Flow," *Trans. A.S.C.E.*, Vol. 116 (1951), pp. 268–295.

IPPEN, A. T. "Transitions for Supercritical Flow." In *Engineering Hydraulics* (Hunter Rouse, ed.), New York: John Wiley & Sons, Inc., 1950. pp. 543–570.

IPPEN, A. T. and DAWSON, J. H. "Design of Channel Contractions," *Trans. A.S.C.E.*, Vol. 116 (1951), pp. 326–346.

IPPEN, A. T. and HARLEMAN, D. R. "Verification of Theory for Oblique Standing Waves," *Trans. A.S.C.E.*, Vol. 121 (1956), pp. 678–694.

KNAPP, ROBT. T. "Design of Channel Curves for Supercritical Flow." *Trans. A.S.C.E.*, Vol. 116 (1951), pp. 296–325.

ROUSE, HUNTER, BHOOTA, B. V., and HSU, E. Y. "Design of Channel Expansions," *Trans. A.S.C.E.*, Vol. 116 (1951), pp. 347–363.

Control Sections

DOERINGSFELD, H. A. and BARKER, C. L. "Pressure-Momentum Theory Applied to the Broad-Crested Weir," *Trans. A.S.C.E.*, Vol. 106 (1941), pp. 934–946.

RAJARATNAM, N. and MURALIDHAR, D. "End Depth for Circular Channels," *Journal of Hydraulics Division, A.S.C.E.*, March 1964, pp. 99–119.

ROUSE, HUNTER. "Discharge Characteristics of the Free Overfall," *Civil Engineering*, Vol. 6 (1936), pp. 257–260.

STRELKOFF, T. S. "Solution of Highly Curvilinear Gravity Flows," *Journal of Engineering Mechanics Division, A.S.C.E.*, June 1964, pp. 195–221.

TRACY, H. J. *Discharge Characteristics of Broad-Crested Weirs*, Washington: U.S. Geological Survey, Circular 397, 1957.

Spatially Varied Flow

CAMP, T. F. "Lateral Spillway Channels." *Trans. A.S.C.E.*, Vol. 105, 1940. pp. 606–617.

CHOW, VEN TE. "Spatially Varied Flow Equations." *Water Resources Research*. October 1969, pp. 1124–1128.

HINDS, JULIAN. "Side-Channel Spillways." *Trans. A.S.C.E.*, Vol. 89, 1926, pp. 881–927.

LI, W. H. "Open Channels with Non-Uniform Discharge." *Trans. A.S.C.E.*, Vol. 120, 1955, pp. 255–274.

SMITH, K. V. H. "Control Point in a Lateral Spillway Channel," *Journal of Hydraulics Division, A.S.C.E.*, May 1967, pp. 27–34.

YEN, B. C. and WENZEL, H. G. "Dynamic Equations for Steady Spatially Varied Flow," *Journal of Hydraulics Division, A.S.C.E.*, March 1970, pp. 801–814.

CHAPTER 6

Articles 6–1 through 6–14 (Dams)

General

CREAGER, W. P., JUSTIN, J. D., and HINDS, JULIAN. *Engineering for Dams.* New York: John Wiley & Sons, Inc., 1945. Vols. I–III. 929 pp.

DAVIS, C. V. and SORENSEN, K. E. (eds.). *Handbook of Applied Hydraulics.* New York: McGraw-Hill Book Co., Inc., 3rd ed. 1969. Chapters 8–19, incl.

HOUK, IVAN E. *Irrigation Engineering: Vol. II, Projects, Conduits and Structures.* New York: John Wiley & Sons, Inc., 1956. 531 pp.

LINSLEY, R. K. and FRANZINI, J. B. *Water Resources Engineering.* New York: McGraw-Hill Book Co., Inc., 1964. pp. 173–214.

MERMEL, T. W. *World Register of Dams.* New York: International Commission on Large Dams, 1970.

SHOKLITSCH, ARMIN. *Hydraulic Structures.* Transl. by L. G. Straub. New York: American Society of Mechanical Engineers, 1937.

SLICHTER, FRANCIS B. "Influences on Selection of the Type of Dam," *Journal of Soil Mechanics and Foundations Division, A.S.C.E.*, May 1967, pp. 1–8.

U.S. BUREAU OF RECLAMATION. *Treatise on Dams.* Denver: The Bureau, 1951. Ch. 1, "Compendium of Dams"; Ch. 4, "Basic Considerations."

U.S. BUREAU OF RECLAMATION. *Design of Small Dams.* Washington: U.S. Govt. Printing Office, 1960.

Masonry Gravity Dams

BUSTAMANTE, JORGE I. "Water Pressure on Dams Subjected to Earthquakes," *Journal of Engineering Mechanics Division, A.S.C.E.* October 1966, pp. 115–127.

CHOPRA, A. K. "Hydrodynamic Pressure on Dams during Earthquakes," *Journal of Engineering Mechanics Division, A.S.C.E.* December 1967, pp. 205–223.

HAMMAD, H. Y. "Seepage Under Dams," *Journal of Soil Mechanics and Foundations Division, A.S.C.E.* July 1963, pp. 25–44.

SANDRU, R. S. "Simplified Procedure for Stress Analysis of Gravity Dams," Paper 2313, *Journal of the Power Division, A.S.C.E.*, December 1959, pp. 173–187.

U.S. BUREAU OF RECLAMATION. *Gravity Dams.* Denver: The Bureau, 1955. 135 pp.

Earth Dams

CASAGRANDE, ARTHUR. "Seepage Through Dams," *Journal New England Water Works Association*, June 1937.

~TERKA, A. J. *Hydraulic Design of Stilling Basins and Energy Dissipators.* Washington, D.C.: U.S. Govt. Printing Office. 1963, 222 pp.

~AJARATNAM, N. and SUBRAMANYA, K. "Profile of the Hydraulic Jump," *Journal of the Hydraulics Division*, A.S.C.E., May 1968, pp. 663–673.

~AND, Walter. "Efficiency and Stability of Forced Hydraulic Jump'" *Journal of the Hydraulics Division*, A.S.C.E., July 1967, pp. 117–127.

~LVESTER, RICHARD. "Hydraulic Jump in All Shapes of Horizontal Channels," *Journal of the Hydraulics Division*, A.S.C.E., January 1964, pp. 23–55.

~ITH, C. D. and YU, J. N. G. "Use of Baffles in Open Channel Expansions," *Journal of the Hydraulics Division*, A.S.C.E., March 1966, pp. 1–17.

Articles 6–23 through 6–26 (Culverts)

~AISDELL, F. W. "Hood Inlet for Closed Conduit Spillways," *Journal of the Hydraulics Division*, A.S.C.E., May 1960, pp. 7–32.

~AISDELL, F. W. "Hydraulic Fundamentals of Closed Conduit Spillways," *Proc. A.S.C.E.*, Separate 354, November 1953.

~AISDELL, F. W. "Flow in Culverts and Related Design Philosophies," *Journal of the Hydraulics Division*, A.S.C.E., March 1966, pp. 19–31.

~AISDELL, F. W. "Hydraulic Efficiency in Culvert Design," *Journal of the Hydraulics Division*, A.S.C.E., March 1966, pp. 11–22.

~OW, VEN TE. "Hydrologic Design of Culverts," *Journal of the Hydraulics Division*, A.S.C.E., March 1962, pp. 39–55.

~RENCH, JOHN L. "Tapered Inlets for Pipe Culverts," *Journal of the Hydraulics Division*, A.S.C.E., March 1964, pp. 255–299.

~ARG, S. P. "Distribution of Head at a Rectangular Concrete Outlet," *Journal of the Hydraulics Division*, A.S.C.E., July 1966, pp. 11–31.

~IGHWAY RESEARCH BOARD. Culvert Hydraulics. Research Report 15-B. Washington: Natl. Research Council, 1953. 71 pp.

~ARSON, C. L. and MORRIS, H. M. *Hydraulics of Flow in Culverts.* Minneapolis: St. Anthony Falls Hydraulic Laboratory, 1948. 162 pp.

~ORRIS, H. M. *Preliminary Flow Tests on a Model Culvert.* Minneapolis: St. Anthony Falls Hydraulics Laboratory, 1949. 26 pp.

~ICE, CHARLES E. "Effect of Pipe Boundary on Head Inlet Performance," *Journal of the Hydraulics Division*, A.S.C.E., July 1967, pp. 149–167.

~RAUB, L. G., ANDERSON, A. G., and BOWERS, C. E. *Effect of Inlet Design on Culvert Capacity.* Minneapolis: St. Anthony Falls Hydraulic Laboratory. Tech. Paper 13, August 1953. 27 pp.

~RAUB, L. G. and MORRIS, H. M. *Hydraulic Data Comparison of Concrete and Corrugated Metal Culvert Pipes.* Minneapolis: St. Anthony Falls Hydraulics Laboratory, 1950. 25 pp.

~.S. BUREAU OF PUBLIC ROADS. *Capacity Charts for the Hydraulic Design of Highway Culverts*, Hydraulic Engineering Circular No. 10 (March 1965). Washington: U.S. Dept. of Commerce, 90 pp.

CHAPTER 7

Pumps and Turbines

~AUMEISTER, THEODORE. "Turbomachinery." In *Handbook of Fluid Dynamics* (V. L. Streeter, ed.). New York: McGraw-Hill Book Co., Inc., 1961. 51 pp.

~AILY, JAMES W. "Hydraulic Machinery." In *Engineering Hydraulics.* New York: John Wiley & Sons, Inc., 1950. pp. 858–992.

~AUGHERTY, R. L. and FRANZINI, J. B. *Fluid Mechanics wih Engineering Applications.* New York: McGraw-Hill Book Co., Inc. 1965. pp. 449–551.

~ISENBERG, PHILIP and TULIN, M. P. "Cavitation." Section 12, in *Handbook of Fluid Dynamics*, (V. L. Streeter, ed.) New York: McGraw-Hill Book Co., Inc., 1961. 46 pp.

FINN, W. D. L. "Finite-Element Analysis of Seepage through Dams," *Journal of Soil Mechanics and Foundation Division*, A.S.C.E., Novermber 1967, pp. 41–48.

U.S. BUREAU OF RECLAMATION. *Earth Dams.* Denver: The Bureau, 1957. 206 pp.

WALKER, F. C. *Development of Earth Dam Design.* Denver: The Bureau, 1958. 28 pp., 12 plates.

Arch Dams

CAIN, WM. "The Circular Arch under Normal Loads," *Trans. A.S.C.E.*, Vol. 85 (1922), pp. 233–283.

COPEN, M. D. and SCRIVNER, L. R. "Arch Dams—State of the Art," *Journal of the Power Division*, A.S.C.E. January 1970, pp. 93–108.

COYNE, ANDRE. "New Dam Techniques," *Proceedings, Institute of Civil Engineers* (London), November 1959, pp. 275 ff.

GLOVER, ROBT. E. "Arch Dams: Review of Experience," Paper 1217, *Journal of the Power Division*, A.S.C.E., April 1957, 48 pp.

HOUK, IVAN E. "Trial Load Analysis of Curved Concrete Dams," *Engineer*, July 5, 1935, pp. 2–5.

HOUK, IVAN E. and KEENER, K. B. "Masonry Dams—Basic Design Assumptions," *Trans. A.S.C.E.*, Vol. 106 (1941), pp. 1115–1130.

LELIAVSKY, SERGE. "Modern Tendencies in Arch Dam Design," *Engineer*, December, 13, 1957 p. 853; December 20, 1957, p. 888; December 27, 1957, p. 930.

PERKINS, W. A. "Analysis of Arch Dams of Variable Thickness," *Trans. A.S.C.E.*, Vol. 118, (1953), pp. 725–770.

POSPISIL, V. and HAYES, M. D. "Design of Boundary Arch Dams." *Journal of the Power Division*, A.S.C.E., January 1970, pp. 73–91.

SERAFIM, J. L. "New Shapes for Arch Dams," *Civil Engineering*, November 1966, pp. 38–43.

SWAMINATHAN, K. V. "Development of Arch Action in Arch Dams," *Journal of Power Division*, A.S.C.E., May 1965, pp. 39–57.

U.S. BUREAU OF RECLAMATION. *Arch Dams.* Denver: The Bureau, 1955. 556 pp.

Buttress Dams

HOLMES, W. H. "Determination of Principal Stresses in Buttresses and Gravity Dams," *Trans., A.S.C.E.*, Vol. 98 (1933), pp. 971 ff.

ROLIN, R. G. *Analysis of the Arches of a Multiple Arch Dam.* Tech. Mem. 539 (1936) and Suppl. (1939). Denver: U.S. Bureau of Reclamation.

SCHORER, HERMAN. "The Buttress Dam of Uniform Strength," *Trans. A.S.C.E.* Vol. 96 (1932), pp. 666 ff.

U.S. BUREAU OF RECLAMATION. *Buttress Dams.* Denver: The Bureau of, 1950. 162 pp.

Rockfill Dams

CURTIS, R. P. and LAWSON, J. D. "Flow over and Through Rockfill Banks," *Journal of Hydraulics Division*, A.S.C.E., September 1967, pp. 1–21.

GROWDON, J. P. "Rockfill Dams: Dams with Sloping Earth Cores," *Journal of the Power Division*, A.S.C.E., Paper 1743 (August 1958), 21 pp.

HUBER, W. G. "Rockfill Dams," *Journal of the Power Division, Proc., A.S.C.E.* Paper 1671 (June 1958).

SNETHLAGE, J. B., SCHEIDENHELM, F. W., and VANDERLIP, A. N. "Rockfill Dams: Review and Statistics," *Journal of the Power Division*, A.S.C.E., Paper 1739, August 1958, 26 pp.

SQUIER, L. R. "Load Transfer in Earth and Rockfill Dams," *Journal of Soil Mechanics and Foundations Division*, A.S.C.E. January 1970, pp. 213–233.

Special Types of Dams

ANWAR, H. O. "Inflatable Dams," *Journal of the Hydraulics Division*, A.S.C.E., May 1967, pp. 99–119.

HOVEY, OTIS E. *Steel Dams.* New York: American Institute of Steel Construction, 1935.

WHITE, L. and PRENTISS, E. A. *Cofferdams.* New York: Columbia University Press, 1950.

Articles 6–15 through 6–21 (Spillways)

General

BAUER, W. J., and BECK, E. J. "Spillways and Stream-bed Protection Work," Section 20 in *Handbook of Applied Hydraulics,* ed. by C. V. Davis and K. E. Sorensen. New York: McGraw-Hill Book Co., Inc., 1969. 54 pp.

CHOW, VEN TE. *Open-Channel Hydraulics.* New York: McGraw-Hill Book Co., Inc., 1959, pp. 327–353, 360–392.

CREAGER, W. P., JUSTIN, J. D., and HINDS, JULIAN. *Engineering for Dams.* New York: John Wiley & Sons, Inc., 1945. pp. 208–245, 357–377, 870–929.

LINSLEY, R. K. and FRANZINI, J. B. *Water Resources Engineering.* New York: McGraw-Hill Book Co., Inc., 1964. pp. 215–250.

U.S. BUREAU OF RECLAMATION. *Treatise on Dams.* Denver: The Bureau, 1950. Ch. 12, "Spillways"; Ch. 13, "Outlet Works."

U.S. BUREAU OF RECLAMATION. *Design of Small Dams.* Washington: U.S. Govt. Printing Office, 1960. Ch. 8, "Spillways."

Overflow Spillways

BLAISDELL, F. W. "Equation of the Free-Falling Nappe," *Proc. A.S.C.E.,* Separate 482.

CAMPBELL, F. B., COX, R. G., and BOYD, M. B. "Boundary Layer Development and Spillway Energy Losses," *Journal of the Hydraulics Division, A.S.C.E.,* May 1965, pp. 149–163.

CASSIDY, JOHN J. "Irrotational Flow over Spillways of Finite Height," *Journal of the Engineering Mechanics Division, A.S.C.E.,* December 1965, pp. 155–173.

JANSEN, ROBT. B. "Flow Characteristics of the Ogee Spillway," *Journal of the Hydraulics Division, A.S.C.E.,* December 1957, 11 pp.

U.S. BUREAU OF RECLAMATION. *Studies of Crests for Overfall Dams.* Denver: U.S. Bur. Recl., 1948, 186 pp.

Side-Channel Spillways

CAMP, THOMAS R. "Lateral Spillway Channels," *Trans., A.S.C.E.,* Vol. 105 (1940), pp. 606–617.

FARNEY, H. S. and ADOLFS, MARKUS. "Side-Channel Spillway Design," *Journal of the Hydraulics Division, A.S.C.E.,* May 1962, pp. 131–154.

HINDS, JULIAN. "Side-Channel Spillways," *Trans. A.S.C.E.,* Vol. 89 (1926), pp. 881–927.

KEULEGAN, G. H. "Spatially Variable Discharge over a Sloping Plane," *Trans. Amer. Geophysical Union,* 1944, pp. 956–959.

SMITH, KENNETH V. H. "Control Point in a Lateral Spillway Channel," *Journal of the Hydraulics Division, A.S.C.E.,* May 1967. pp. 27–49.

Siphon Spillways

GIBSON, A. H. and ASPEY, T. H. "Experiments on Siphon Spillways," *Proceedings Institute of Civil Engrs.,* Vol. 231 (1930–31), p. 203 ff.

NAYLOR, A. H. *Siphon Spillways.* London: E. Arnold and Co., 1935.

ROCK, ELMER. "Design of a High Head Siphon Spillway," *Trans. A.S.C.E.,* Vol. 105 (1940), pp. 1050–1075.

STEVENS, J. C. "On the Behavior of Siphons," *Trans. A.S.C.E.,* Vol. 99 (1934), pp. 986 ff.

Shaft Spillways

BLAISDELL, F. W. *Hydraulics of Closed Conduit Spillways.* Par Parts II–VII. March 1958. Minneapolis: St. Anthony Falls H

BLAISDELL, F. W. and DONNELLY, C. A. "The Box Inlet Drop Spil *Trans. A.S.C.E.,* Vol. 121 (1956), pp. 955–994.

BRADLEY, J. N., WAGNER, W. E., and PETERKA, A. J. "Morning-C A Symposium," *Trans. A.S.C.E.,* Vol. 121 (1956), pp. 311–409.

HEBAUS, GEORGE G. "Crest Losses for Two-Way Drop Inlet," *Jou Division, A.S.C.E.,* May 1969, pp. 919–940.

Chute Spillways

A.S.C.E. TASK COMMITTEE: "Aerated Flow in Open Channels," *Jou Division, A.S.C.E.,* May 1961, pp. 73.

GUMENSKY, D. B. "Design of Side Walls in Chutes and Spillwa Vol. 119 (1954), pp. 355–372.

STRAUB, L. G. and ANDERSON, A. G. "Experiments on Self-Ae Channels," *A.S.C.E. Transactions,* Vol. 125, 1960, pp. 456–486.

Spillway Crest Gates

BRADLEY, J. N. "Rating Curves for Flow over Drum Gates," *T* 119 (1954), pp. 403–433.

BUZZELL, D. A. "Trends in Hydraulic Gate Design," *Trans. A.S.* pp. 27–42.

MAYER, P. R. and BOWMAN, J. R. "Spillway Crest Gates," *Sectio Applied Hydraulics,* ed. by C. V. Davis and K. E. Sorensen. Nev Book Co., Inc., 1969. 16 pp.

TOCH, ARTHUR. "Discharge Characteristics of Tainter Gates," *T* 120 (1955), pp. 290–300.

Articles 6–22 and 6–23 (Stilling Basins)

A.S.C.E. TASK FORCE. "Energy Dissipators for Spillways and Ou *of the Hydraulics Division, A.S.C.E.,* January 1964, pp. 121–14

BLAISDELL, F. W. "Development and Hydraulic Design, St. A Basin," *Trans. A.S.C.E.,* Vol. 113 (1948), pp. 483–520.

BLAISDELL, F. W. and DONNELLY, C. A. "The Box Inlet Drop Spil *Trans. A.S.C.E.,* Vol. 121, pp. 955–986.

CHOW, VEN TE. *Open-Channel Hydraulics.* New York: McGraw- 1959.. pp. 393–438.

ELEVATORSKI, E. A. *Hydraulic Energy Dissipators.* New York: Mo Inc., 1959.

FIALA, GENE R. and ALBERTSON, M. L. "Manifold Stilling Ba *Hydraulics Division, A.S.C.E.,* July 1961, pp. 55–81.

JONES, L. E. "Some Observations on the Undular Jump," *Jour Division, A.S.C.E.,* May 1964, pp. 69–82.

KEIM, S. RUSSELL. "Contra Costa Energy Dissipator," *Journe Division, A.S.C.E.,* March 1962, pp. 109–122.

KINDSVATER, CARL E. "The Hydraulic Jump in Sloping Channel Vol. 109 (1944), pp. 1107–1120.

MORRIS, HENRY M. "Design of Roughness Elements for Energy Di Drainage Chutes," *Highway Research Record,* May 1969, pp. 25–

MORRIS, HENRY M. *Hydraulics of Energy Dissipation in Steep, Rou 19, Research Division, Virginia Polytechnic Institute, Novembe

HAMMOND, ROLT. *Water Power Engineering.* New York: The Macmillan Co., 1958. 302 pp.

HICKS, TYLER. *Pump Selection and Application.* New York: McGraw-Hill Book Co., Inc., 1957. 373 pp.

JOHNSON, VIRGIL E. "Mechanics of Cavitation," *Journal of the Hydraulics Division,* A.S.C.E. May 1963, pp. 252–275.

KARASSIK, I. J. *Engineer's Guide to Centrifugal Pumps.* New York: McGraw-Hill Book Co., Inc., 1964. 29 pp.

KRISTAL, F. A. and ANNETT, F. A. *Pumps.* New York: McGraw-Hill Book Co., Inc, 1953. 373 pp.

KRUEGER, R. E. *Selecting Hydraulic Reaction Turbines.* Denver: U.S. Bureau of Reclamation. 1954. 45 pp.

MOODY, LEWIS F. and ZOWSKI, THADDEUS. "Hydraulic Machinery," Section in 26 *Handbook of Applied Hydraulics* (C. V. Davis and K. E. Sorensen, eds.) New York: McGraw-Hill Book Co., Inc., 1969. 90 pp.

SHEPHERD, D. G. *Principles of Turbomachinery.* New York: The Macmillan Co., 1956. 463 pp.

STEVENS, J. C. and DAVIS, C. V. "Hydroelectric Plants," Section 24 in *Handbook of Applied Hydraulics* (C. V. Davis and K. E. Sorensen, eds.). New York: McGraw-Hill Book Co., Inc., 1969. 42 pp.

Water Hammer and Surge Tanks

DRUML, FRANK U. "An Analysis of a Simple Surge Tank," *Journal of the Hydraulics Division,* A.S.C.E., October 1959. pp. 115–130.

HALLIWELL, A. R. "Velocity of a Water-Hammer Wave in an Elastic Pipe," *Journal of the Hydraulics Division,* A.S.C.E., July 1963, pp. 1–21.

McNOWN, J. S. "Surges and Water Hammer." Chapter 7 in *Engineering Hydraulics,* ed. by Hunter Rouse. New York: John Wiley & Sons, Inc., 1950. pp. 444–495.

PARMAKIAN, JOHN. *Waterhammer Analysis.* Englewood Cliffs, N.J.: Prentice-Hall, 1955. 161 pp.

PARMAKIAN, JOHN. "Waterhammer Design Criteria," *Journal of the Power Division,* A.S.C.E., Paper 1216, April 1957. 8 pp.

PAYNTER, HENRY M. "Fluid Transients in Engineering Systems." Section 20 in *Handbook of Fluid Dynamics* (V. L. Streeter, ed.). New York: McGraw-Hill Book Co., Inc., 1961. 47 pp.

RICH, GEORGE R. *Hydraulic Transients.* 2nd Ed. New York: Dover Publications, Inc., 1963. 409 pp.

RICH, GEORGE R. "Water Hammer." Section 27 in *Handbook of Applied Hydraulics* (C. V. Davis and K. E. Sorensen, eds.). New York: McGraw-Hill Book Co., Inc., 1969. 32 pp.

RICH, GEORGE R. "Surge Tanks." Section 28 in *Handbook of Applied Hydraulics* (C. V. Davis and K. E. Sorensen, eds.). New York: McGraw-Hill Book Co., Inc. 1969. 34 pp.

STREETER, V. L. and WYLIE, E. B. *Hydraulic Transients.* New York: McGraw-Hill Co., Inc., 1967. 329 pp.

STREETER, V. L. "Water Hammer Analysis," *Journal of the Hydraulics Division,* A.S.C.E., November 1969. pp. 1959–1972.

STREETER, VICTOR L. and LAI, CHINTU. "Water-Hammer Analysis Including Fluid Friction," *Journal of the Hydraulics Division,* A.S.C.E., May 1962. pp. 79–112.

TAYLOR, E. H., REISMAN, ARNOLD, and WARD, JACK W. "Unsteady Flow in Conduits with Simple Surge Tanks," *Journal of the Hydraulics Division,* A.S.C.E., February 1959, pp. 1–11.

TULTS, HAROLD. "Simplified Computation of Surge-Tank Action." *Civil Engineering.* February 1955. pp. 63–65.

WOOD, DON J. "Water-Hammer Analysis by Analog Computers," *Journal of the Hydraulics Division,* A.S.C.E., January 1967. pp. 1–11.

WOOD, DON J. "Influence of Line Motion on Water-Hammer Pressures." *Journal of the Hydraulics Division, A.S.C.E.*, May 1969, pp. 941–959.

WOOD, DON J., DORSCH, R. G., and LIGHTNER, CHARLENE. "Wave-Plan Analysis of Unsteady Flow in Closed Conduits." *Journal of the Hydraulics Division, A.S.C.E.*, March 1966, pp. 83–110.

CHAPTER 8

ALLEN, J. *Scale Models in Hydraulic Engineering.* London: Longmans, Green & Co. 1947.

AMERICAN SOCIETY OF CIVIL ENGINEERS. *Hydraulic Models.* ASCE Manual 25. New York, 1942. 110 pp.

BARR, D. I. H. and SMITH, A. A. "Application of Similitude Theory to Correlation of Uniform Flow Data." *Proceedings, Institution of Civil Engineers* Vol. 37, July 1967. pp. 487–509.

CAMPBELL, F. B. and PICKETT, E. B. "Prototype Performance and Model-Prototype Relationship," Section 3 in *Handbook of Applied Hydraulics* (C. V. Davis and K. E. Sorensen, eds.) 3rd ed. New York: McGraw-Hill Book Co., Inc., 1969. 22 pp.

HOLT, MAURICE. "Dimensional Analysis." In *Handbook of Fluid Dynamics* (V. L. Streeter, ed.). New York: McGraw-Hill Book Co., Inc., 1961. pp. 15–1 to 15–25.

MAXWELL, W. H. C. and WEGGEL, J. R. "Surface Tension in Froude Models." *Journal of the Hydraulics Division, A.S.C.E.*, March 1969, pp. 677–701.

MURPHY, GLENN. *Similitude in Engineering.* New York: The Ronald Press Co., 1950. 302 pp.

SIMMONS, W. P. "Models Primarily Dependent on the Reynolds Number." *Journal of the Hydraulics Division, A.S.C.E.*, June 1960, pp. 59–74.

U.S. BUREAU OF RECLAMATION. *Hydraulic Laboratory Practice.* Monograph 18 (1953). Denver: The Bureau. 111 pp.

WARNOCK, J. E. "Hydraulic Similitude." In *Engineering Hydraulics* (Hunter Rouse, ed.). New York: John Wiley & Sons, Inc., 1950. pp. 136–176.

WHITTINGTON, R. B. "A Simple Dimensional Method for Hydraulic Problems." *Journal of the Hydraulics Division, A.S.C.E.*, September 1963, pp. 1–27.

CHAPTER 9

AMERICAN SOCIETY OF CIVIL ENGINEERS. *Hydrology Handbook*, Manual of Engineering Practice, No. 28. New York: A.S.C.E., 1949. 184 pp.

BRUCE, J. P. and CLARK, R. H. *Introduction to Hydrometeorology.* Oxford: Pergamon Press, Ltd., 1966. 319 pp.

CHOW, V. T. (ed.). *Handbook of Applied Hydrology.* New York: McGraw-Hill Book Co., Inc., 1964. 1391 pp.

CHOW, VEN TE (ed.). *The Progress of Hydrology.* Vols. I–III. Proceedings of First International Seminar for Hydrology Professors. Urbana, Illinois: University of Illinois. 1970, 1295 pp.

DeWIEST, R. J. M. *Geohydrology.* New York: John Wiley & Sons, Inc., 1965. pp. 14–128.

EAGLESON, P. S. *Dynamic Hydrology.* New York: McGraw-Hill Book Co., Inc., 1970. 462 pp.

GUMBEL, E. J. *Statistical Theory of Extreme Values and Some Practical Applications.* Applied Mathematics Series, 33, National Bureau of Standards. Washington: U.S. Dept. of Commerce, 1954. 51 pp.

LINSLEY, R. K., KOHLER, M. A., and PAULHUS, J. L. H. *Applied Hydrology.* New York: McGraw-Hill Book Co., Inc., 2nd ed. 1970. 689 pp.

PETTERSSEN, S. *Introduction to Meteorology.* 3rd ed. New York: McGraw-Hill Book Co., Inc., 1969. pp. 92–116, 176–185, 210–219.

WISLER, C. O. and BRATER, E. F. *Hydrology.* 2nd ed. New York: John Wiley & Sons, Inc., 1959. 408 pp.

CHAPTER 10

DeWiest, R. J. M. *Geohydrology.* New York: John Wiley & Sons, Inc., 1965, 366 pp.

Harr, M. E. *Groundwater and Seepage.* New York: McGraw-Hill Book Co., Inc., 1962, 315 pp.

Jacob, C. E. "Flow of Ground Water." In *Engineering Hydraulics* (Hunter Rouse, ed.). New York: John Wiley & Sons, Inc., 1950, pp. 321–386.

Polubarinova-Kochina, P. Ya. *Theory of Ground Water Movement.* Princeton, N.J.: Princeton University Press, 1962. (English translation by R. J. M. DeWiest.)

Richardson, J. G. "Flow Through Porous Media," in *Handbook of Fluid Dynamics* (V. L. Streeter, ed.). New York: McGraw-Hill Book Co., Inc., 1961, pp. 16–1 through 16–112.

Todd, D. K. *Ground Water Hydrology.* New York: John Wiley & Sons, Inc., 1959, 336 pp.

U.S. Bureau of Reclamation: *Studies of Groundwater Movement.* Technical Memorandum 657. Denver: U.S. Bureau of Reclamation, Dept. of Interior. 1960, 180 pp.

Walton, W. C. *Groundwater Resource Evaluation.* New York: McGraw-Hill Book Co., Inc., 1970.

CHAPTER 11

General

Derby, Ray L. "Water Use in Industry," *Journal of the Irrigation and Drainage Division, A.S.C.E.,* Paper 1364, Sept. 1957, 19 pp.

Gulhati, N. D. "Worldwide View of Irrigation Developments," *Journal of the Irrigation and Drainage Division, A.S.C.E.,* Paper 1751, September 1958, 14 pp.

Langbein, Walter B. "Annual Runoff in the United States," *Circular 52* (1949), U.S. Geol. Survey, Washington: U.S. Govt. Printing Office.

Thomas, H. E. "Water Problems," *Water Resources Research,* Vol. 1, No. 3, 1965, pp. 435–445.

Thomas, Robert O. "Water—A Limiting Resource?" *Journal of the Irrigation and Drainage Division, A.S.C.E.,* Paper 1754 (September 1958). 13 pp.

U.S. Geological Survey. Large Rivers of the United States, *Circular 44,* (1949), U.S. Geol. Survey. Washington: U.S. Govt. Printing Office. 5 pp.

U.S. Geological Survey. *Surface Water Supply of the United States.* Washington: U.S. Govt. Printing Office. Annual series of publications of U.S. Geol. Survey for different basins and regions.

Wells, J. V. B. "Surface Water Resources," *Journal of the Sanitary Engineering Division, A.S.C.E.,* Paper 1272 (June 1957). 9 pp.

River Planning and Regulation

Fogarty, Earl R. "Benefits of Water Development Projects," *Journal of the Irrigation and Drainage Division, A.S.C.E.,* Paper 981, May 1956, 8 pp.

Golze, Alfred R. *Reclamation in the United States.* New York: McGraw-Hill Book Co., Inc., 1952. 441 pp.

Hoyt, Wm. G. and Langbein, Walter B. *Floods.* Princeton, N.J.: Princeton University Press, 1955. 480 pp.

Israelsen, Orson W. "The Engineer and Worldwide Conservation of Soil and Water," *Journal of the Irrigation and Drainage Division, A.S.C.E.,* Paper 1775, Sept. 1958, 22 pp.

Koelzer, V. A. "Reservoir Hydraulics" Section 4 in *Handbook of Applied Hydraulics* (ed. by C. V. Davis and K. E. Sorensen). 3rd ed. New York: McGraw-Hill Book Co., Inc., 1969, 24 pp.

Löf, G. O. G., and Hardison, C. H.: "Storage Requirements for Water in the United States." *Water Resources Research.* Vol. 2, No. 3, 1966, pp. 323–354

Rasmussen, Jewell J. "Economic Criteria for Water Development Projects," *Journal of the Irrigation and Drainage Division, A.S.C.E.*, Paper 977, May 1956, 14 pp.

U.S. Chamber of Commerce. *Conservation and Use of Natural Resources.* Washington: Chamber of Commerce of U.S., 1960. 55 pp.

Reservoir Operation Analysis

Butsch, R. J. "Reservoir System Design Optimization," *Journal of the Hydraulics Division, A.S.C.E.* January 1970, pp. 125–130.

Close, E. R., Beard, L. R. and Dawdy, D. R. "Objective Determination of Safety Factor in Reservoir Design," *Journal of the Hydraulics Division, A.S.C.E.*, May 1970, pp. 1167–1177.

Hall, W. A. and Dracup, J. A. *Water Resources Systems Engineering,* New York: McGraw-Hill Book Co., Inc. 1970, 372 pp.

James, L. D. "Economic Derivation of Reservoir Operating Rules," *Journal of the Hydraulics Division, A.S.C.E.* September 1968, pp. 1217–1230.

Langbein, W. B. "Queuing Theory and Water Storage," *Journal of the Hydraulics Division, A.S.C.E.* Paper 1811, October, 1958, 24 pp.

Maas, A. M., *et al. Design of Water Resources Systems,* Cambridge, Mass: Harvard University Press, 1962.

Roefs, T. G. and Bodin, L. D. "Multireservoir Operation Studies," *Water Resources Research.* April 1970, pp. 410–420.

Schweig, Z. and Cole, J. A. "Optimal Control of Linked Reservoirs," *Water Resources Research.* June 1968, pp. 479–497.

Stephenson, David, "Optimum Design of Complex Water Resource Projects," *Journal of the Hydraulics Division, A.S.C.E.*, June 1970, pp. 1229–1246.

Young, G. K. "Finding Reservoir Operating Rules," *Journal of the Hydraulics Division, A.S.C.E.*, November 1967, pp. 297–322.

Young, G. K. and Pisano, M. A. "Nonlinear Programming Applied to Regional Water Resource Planning," *Water Resources Research.* February 1970, pp. 32–42.

Flood Routing

Buil, J. A. "Synthetic Coefficients for Streamflow Routing," *Journal of the Hydraulics Division, A.S.C.E.*, November 1967, pp. 371–386.

Carter, R. W. and Godfrey, R. G. *Storage and Flood Routing.* Washington: U.S. Geological Survey, 1960. pp. 102–104.

Gilcrest, B. R. "Flood Routing." In *Engineering Hydraulics* (H. Rouse, ed). New York: John Wiley & Sons, Inc., 1950. pp. 635–710.

Graves, E. A. "Improved Method of Flood Routing." *Journal of the Hydraulics Division A.S.C.E.*, January 1967, pp. 29–43.

Henderson, F. M. *Open Channel Flow.* New York: Macmillan Co. 1966, pp. 355–404.

Posey, C. J. "Slide Rule for Routing Floods through Storage Reservoirs or Lakes," *Engineering News-Record*, Vol. 114 (1935), pp. 580–581.

Posey, C. J. and Fu-Te, I. "Functional Design of Flood Control Reservoirs," *Trans. A.S.C.E.*, Vol. 105 (1940), pp. 1838–1674.

Method of Characteristics

Amein, Michael and Fang, Ching Seng. *Streamflow Routing (with application to North Carolina rivers),* Chapel Hill: Water Resources Research Institute of the University of North Carolina, Report No. 17 (revised), 1969. 106 pp.

Amein, Michael, "Streamflow Routing on Computer by Characteristics," *Water Resources Research,* Vol. 2, Number 1, First Quarter, 1966, pp. 123–130.

BALLOFFET, A. "One-dimensional Analysis of Floods and Tides in Open Channels," *Journal of the Hydraulics Division, A.S.C.E.*, Vol. 95, No. HY4, Proc. Paper 6695, July 1969, pp. 1429–1451.

CHOW, V. T. *Open-Channel Hydraulics.* New York: McGraw-Hill Book Co., Inc., 1959. pp. 586–600.

GARRISON, JACK M., GRANJU, JEAN-PIERRE P., and PRICE, JAMES T. "Unsteady Flow Simulation in Rivers and Reservoirs," *Journal of the Hydraulics Division, A.S.C.E.*, September, 1969, pp. 1559–1576.

LIGGETT, JAMES A. "Mathematical Flow Determination in Open Channels," *Journal of the Engineering Mechanics Division, A.S.C.E.*, August 1968, pp. 947–963.

STOKER, J. J. *Water Waves.* New York: Interscience Publishers, Inc., 1957. 567 pp.

STOKER, J. J. "Numerical Solution of Flood Prediction and River Regulation Problems," Report I, IMM-NYU 200, New York University, October 1953.

CHAPTER 12

General

ALGER, G. R. and SIMONS, D. B. "Fall Velocity of Irregular Shaped Particles," *Journal of the Hydraulics Division, A.S.C.E.*, May 1968, pp. 721–737.

ANDERSEN, ALVIN G. "Sedimentation." In *Handbook of Fluid Dynamics* (V. L. Streeter, ed.). New York: McGraw-Hill Book Co., Inc., 1961. pp. 18-1–18-35.

A.S.C.E. TASK COMMITTEE ON PREPARATION OF SEDIMENTATION MANUAL. "Sediment Transportation Mechanics" (two parts), *Journal of the Hydraulics Division, A.S.C.E.*, July 1962, pp. 77–127.

A.S.C.E. TASK COMMITTEE ON PREPARATION OF SEDIMENTATION MANUAL. "Sediment Control Methods, *Journal of the Hydraulics Division, A.S.C.E.*, March 1969, pp. 649–675.

A.S.C.E. TASK COMMITTEE ON PREPARATION OF SEDIMENTATION MANUAL. "Sediment Measurement Techniques" (two parts), *Journal of the Hydraulics Division, A.S.C.E.*, September 1969, pp. 1477–1544.

BROWN, CARL B. "Sediment Tranportation." In *Engineering Hydraulics* (H. Rouse, ed.). New York: John Wiley & Sons, Inc., 1950, pp. 769–857.

HENDERSON, F. M. "Sediment Transport." Chapter 10 in *Open Channel Flow.* New York: Macmillan Co., 1966. pp. 504–487.

LELIAVSKY, SERGE. *An Introduction to Fluvial Hydraulics.* New York: Dover Publications, Inc., 1966. 257 pp.

RAUDKIVI, A. J. *Loose Boundary Hydraulics.* Elsmford, N.Y.: Pergamon Press, 1967. 344 pp.

RENARD, K. G. and HICKOK, R. B. "Sedimentation Needs in Arid Regions." *Journal of the Hydraulics Division, A.S.C.E.*, January 1967, pp. 45–60.

U.S. DEPT. OF AGRICULTURE: *Proceedings of the Federal Inter-Agency Sedimentation Conference.* Washington: U.S. Government Printing Office, 1965. 933 pp.

Bed Load

A.S.C.E. TASK COMMITTEE ON PREPARATION OF SEDIMENTATION MANUAL, "Initiation of Motion," *Journal of the Hydraulics Division, A.S.C.E.*, March 1966, pp. 291–314.

BARR, D. I. H. and HERBERTSON, J. G. "Similitude Theory Applied to Correlation of Flume Sediment Data," *Water Resources Research*, Vol. 4, April 1968, pp. 307–316.

BISHOP, A. A., SIMONS, D. B., and RICHARDSON, E. V. "Total Bed-Material Transport," *Journal of the Hydraulics Division, A.S.C.E.*, March 1965, pp. 175–191.

BROOKS, NORMAN H. "Mechanics of Streams with Movable Beds of Fine Sand." *Trans. A.S.C.E.*, Vol. 123 (1958), pp. 526–594.

COLBY, B. R. "Practical Computations of Bed-Material Discharge." *Journal of the Hydraulics Division, A.S.C.E.*, March 1964, pp. 217–246.

EINSTEIN, HANS. "Formulas for the Transportation of Bed Load," *Trans. A.S.C.E.*, Vol. 107 (1942), pp. 561–597.

GRASS, ANTHONY J. "Initial Instability of Fine Bed Sand," *Journal of the Hydraulics Division, A.S.C.E.*, March 1970, pp. 619–632.

KALINSKE, A. A. "Movement of Sediment as Bed Load in Rivers," *Trans. Amer. Geophysical Union*, Vol. 28 (1947), pp. 615–620.

WILSON, K. E. "Bed-Load Transport at High Shear Stress." *Journal of the Hydraulics Division, A.S.C.E.*, November 1966, pp. 49–59.

Suspended Load

A.S.C.E. TASK COMMITTEE ON PREPARATION OF SEDIMENTATION MANUAL: "Suspension of Sediment." *Journal of the Hydraulics Division, A.S.C.E.*, September 1963, pp. 45–75.

APMANN, R. P. and RUMER, R. P. "Diffusion of Sediments in Developing Flow." *Journal of the Hydraulics Division, A.S.C.E.*, January 1970, pp. 109–123.

BROOKS, N. H. "Calculation of Suspended Load Discharge from Velocity and Concentration Parameters." In *Proceedings of Federal Inter-Agency Sedimentation Conference*. Washington: U.S. Govt. Printing Office, 1965, pp. 229–237.

CONOVER, W. J. and MATALAS, N. C. "Statistical Model of Turbulence in Sediment-Laden Streams." *Journal of the Hydraulics Division, A.S.C.E.*, October 1969, pp. 1063–1081.

HINO, MIKIO. "Turbulent Flow with Suspended Particles." *Journal of the Hydraulics Division, A.S.C.E.*, July 1963, pp. 161–185.

HJELMFELT, A. T. and LENAU, C. W. "Non-Equilibrium Transport of Suspended Sediment," *Journal of the Hydraulics Division, A.S.C.E.*, July 1970, pp. 1567–1586.

JOBSON, H. E. and SAYRE, W. W. "Vertical Transfer in Open-Channel Flow," *Journal of the Hydraulics Division, A.S.C.E.*, March 1970, pp. 703–724.

LANE, E. W. and KALINSKE, A. A. "Engineering Calculations of Suspended Sediment," *Trans. Amer. Geophysical Union*, Vol. 22 (1941), pp. 603–607.

SAYRE, WILLIAM W. "Dispersion of Silt Particles in Open-Channel Flow," *Journal of the Hydraulics Division, A.S.C.E.*, May 1969, pp. 1009–1038.

SUTHERLAND, ALEX J. "Proposed Mechanism for Sediment Entrainment by Turbulent Flows," *Journal of Geophysical Research*. December 15, 1967, pp. 6183–6194.

VANONI, V. A. "Transportation of Suspended Sediment by Water," *Trans. A.S.C.E.*, Vol. 111 (1946), pp. 67–133.

WILLIS, J. C. and COLEMAN, N. L. "Unification of Data on Sediment Transport in Flumes by Similitude Principles," *Water Resources Research*, Vol. 5, December 1969, pp. 1330–1336.

Total Load

BAGNOLD, R. A. *Sediment Transport and Stream Power*. Washington: U.S. Geological Survey Circular 421. 1960.

BOGARDI, J. L. "European Concepts of Sediment Transportation," *Journal of the Hydraulics Division, A.S.C.E.*, January 1965, pp. 29–54.

COLBY, B. R. and HUBBELL, D. W. *Simplified Method for Computation of Sediment Discharge with the Modified Einstein's Procedure*. Washington: U.S. Geological Water-Supply Paper 1593. 1961.

EGIAZAROFF, I. V. "Calculation of Non-Uniform Sediment Concentrations," *Journal of the Hydraulics Division, A.S.C.E.*, July 1965, pp. 225–247.

GARDE, R. J. and ALBERTSON, M. L. Discussion of paper by Laursen (see below), *Journal of the Hydraulics Division, A.S.C.E.*, No. 1856 (November 1958), pp. 59–64.

GILBERT, G. K. *The Transportation of Debris by Running Water*. Washington: U.S. Geological Survey Professional Paper 86, 1914.

GUY, H. P., SIMONS, D. B., and RICHARDSON, E. V. *Summary of Alluvial Channel Data from Flume Experiments, 1956–61*. Washington: U.S. Govt. Printing Office. U.S. Geological Survey Professional Paper 462-I. 1966. 96 pp.

LAURSEN, E. M. "The Total Sediment Load of Streams," *Journal of the Hydraulics Division, A.S.C.E.*, No. 1530 (February 1958), pp. 1–36.

MAO, S. W. and RICE, LEONARD. "Sediment-Transport Capability in Erodible Channels," *Journal of the Hydraulics Division, A.S.C.E.*, July 1963. pp. 69–95.

NORDIN, CARL F. "Study of Channel Erosion and Sediment Transport." *Journal of the Hydraulics Division, A.S.C.E.*, July 1964. pp. 173–191.

SHEPPARD, J. R. "Methods and Their Suitability for Determining Total Sediment Quantities." In *Proceedings of Federal Inter-Agency Sedimentation Conference.* Washington: U.S. Govt. Printing Office. 1965. pp. 272–287.

SHROEDER, K. B. and HEMBREE, C. H. "Application of the Modified Einstein Procedure for Computation of Total Sediment Load," *Trans. Amer. Geophysical Union*, Vol. 37 (1956), pp. 197–212.

STEIN, RICHARD A. "Laboratory Studies of Total Load and Apparent Bed Load." *Journal of Geophysical Research*, April 15, 1965, pp. 1831–1842.

TOFFALETI, FRED B. "Definitive Computations of Sand Discharge in Rivers." *Journal of the Hydraulics Division, A.S.C.E.*, January 1969, pp. 225–248.

ZERNIAL, G. A. and LAURSEN, E. M. "Sediment-Transporting Characteristics of Streams." *Journal of the Hydraulics Division, A.S.C.E.*, January 1963, pp. 117–137.

Bed Geometry and Flow Resistance

ALAM, A. M. Z. and KENNEDY, J. F. "Friction Factors for Flow in Sand-Bed Channels." *Journal of the Hydraulics Division, A.S.C.E.*, November 1969, pp. 1973–1992.

A.S.C.E. TASK FORCE ON PREPARATION OF SEDIMENTATION MANUAL. "Nomenclature for Bed Forms in Alluvial Channels." *Journal of the Hydraulics Division, A.S.C.E.*, May 1966, pp. 51–64.

CHANG, F. F. M. "Ripple Concentration and Friction Factor." *Journal of the Hydraulics Division, A.S.C.E.*, February 1970, pp. 417–430.

EINSTEIN, H. A. and BARBAROSSA, N. L. "River Channel Roughness." *Transactions A.S.C.E.*, Vol. 117, 1952, pp. 1112–1132.

ENGELUND, FRANK. "Hydraulic Resistance of Alluvial Streams." *Journal of the Hydraulics Division, A.S.C.E.*, March 1966, pp. 315–326.

GARDE, R. J. and RAJU, K. G. R. "Regime Criteria for Alluvial Streams." *Journal of the Hydraulics Division, A.S.C.E.*, November 1963, pp. 153–164.

GARDE, R. J. and RAJU, K. G. R. "Resistance Relationships for Alluvial Channel Flow." *Journal of the Hydraulics Division, A.S.C.E.*, July 1966, pp. 77–100.

HILL, HARRY M. "Bed Forms Due to a Fluid Stream." *Journal of the Hydraulics Division A.S.C.E.*, March 1966, pp. 127–143.

HILL, H. M., SRINIVASAN, V. S., and UNNY, T. E. "Instability of Flat Bed in Alluvial Channels." *Journal of the Hydraulics Division, A.S.C.E.*, September 1969, pp. 1545–1558.

LIU, H. K. "Mechanics of Sediment Ripple Formation," *Journal of the Hydraulics Division, A.S.C.E.*, April 1957, 23 pp.

LOVERA, FEDERICO and KENNEDY, J. F. "Friction Factors for Flat-Bed Flows in Sand Channels." *Journal of the Hydraulics Division, A.S.C.E.*, July 1969, pp. 1227–1234.

RAUDKIVI, A. J. "Study of Sediment Ripple Formation." *Journal of the Hydraulics Division, A.S.C.E.*, November 1963, pp. 15–33.

RAUDKIVI, A. J. "Analysis of Resistance in Fluvial Channels." *Journal of the Hydraulics Division, A.S.C.E.*, September 1967. pp. 73–84.

SIMONS, D. B. and RICHARDSON, E. V. "Resistance to Flow in Alluvial Channels," *Trans. A.S.C.E.*, Vol. 127-I (1962), pp. 927–1006.

SMITH, K. V. H. "Alluvial Channel Resistance Related to Bed Form," *Journal of the Hydraulics Division, A.S.C.E.*, January 1968, pp. 59–70.

SQUARER, DAVID. "Friction Factors and Bed Forms in Fluvial Channels," *Journal of the Hydraulics Division, A.S.C.E.*, April 1970, pp. 995–1017.

VANONI, V. A. and HWANG, L. S. "Relation between Bed Forms and Friction in Streams." *Journal of the Hydraulics Division, A.S.C.E.*, May 1967, pp. 121–144.

YALIN, M. S. "Geometrical Properties of Sand Waves." *Journal of the Hydraulics Division, A.S.C.E.*, September 1964, pp. 105–119.

Reservoir Sedimentation

A.S.C.E. TASK COMMITTEE ON PREPARATION OF SEDIMENTATION MANUAL. "Density Currents." *Journal of the Hydraulics Division, A.S.C.E.*, September 1963. pp. 77–87.

BROWN, CARL B. "Sedimentation in Reservoirs." In *Engineering Hydraulics* (Hunter Rouse, ed.). New York: John Wiley & Sons, Inc., 1950. pp. 825–834.

BRUNE, G. M. "Trap Efficiency of Reservoirs," *Trans. Amer. Geophysical Union*, Vol. 34, June 1953.

HEINEMANN, HERMAN G. "Volume-Weight of Reservoir Sediment," *Journal of the Hydraulics Division, A.S.C.E.*, September 1962 pp. 181–197.

KOELZER, V. A. and LARA, J. M. "Densities and Compaction Rates of Deposited Sediments." *Journal of the Hydraulics Division, A.S.C.E.*, April 1958, pp. 1–15.

LANE, E. W. and KOELZER, V. A. "Density of Sediments Deposited in Reservoirs." In *A Study of Methods Used in Measurement and Analysis of Sediment Loads in Streams*. Report No. 9. Iowa City: University of Iowa.

LARA, J. M. and PEMBERTON, E. L. "Initial Unit Weight of Deposited Sediments." In *Proceedings of Federal Inter-Agency Sedimentation Conference*, Washington: U.S. Government Printing Office, 1965, pp. 818–845.

MOORE, CHARLES M., WOOD, W. J., and RENFRO, G. W. "Trap Efficiency of Reservoirs, Debris Basins, and Debris Dams," *Journal of the Hydraulics Division, A.S.C.E.*, Vol. 86, February 1960, pp. 69–87.

U.S. DEPT. OF AGRICULTURE. "Sedimentation in Reservoirs." Symposium 4 in *Proceedings of Federal Inter-Agency Sedimentation Conference*. Washington: U.S. Government Printing Office, 1965, pp. 777–933.

Design of Stable Channels

A.S.C.E. TASK COMMITTEE ON PREPARATION OF SEDIMENTATION MANUAL. "Erosion of Cohesive Sediments." *Journal of the Hydraulics Division, A.S.C.E.*, July 1968, pp. 1017–1050.

CHOW, VEN TE. *Open Channel Hydraulics*. New York: McGraw-Hill Book Co., Inc., 1959. pp. 164–188.

DOUBT, P. D. "Design of Stable Channels in Erodible Materials," In *Proceedings of Federal Inter-Agency Sedimentation Conference*. Washington: U.S. Govt. Printing Office, 1965, pp. 373–376.

FORTIER, S. and SCOBEY, F. C. "Permissible Canal Velocities," *Trans. A.S.C.E.*, Vol. 89 (1926), pp. 940–956.

FREDENHAGEN, V. B. and DOLL, E. H. "Grassed Waterways," *Agricultural Engineering*, Vol. 35 (1954), pp. 417–419.

GLOVER, R. E. and FLOREY, Q. L. *Stable Channel Profiles*. Hydr. Lab. Rept. 325 (1951) Denver: U.S. Bureau of Reclamation.

HAYMIE, R. M. and SIMONS, D. B. "Design of Stable Channels in Alluvial Materials," *Journal of the Hydraulics Division, A.S.C.E.*, November 1968, pp. 1399–1420.

KARTHA, V. C. and LEUTHEUSSER, H. J. "Distribution of Tractive Force in Open Channels." *Journal of the Hydraulics Division, A.S.C.E.*, July 1970, pp. 1469–1484.

KELLERHALS, ROLF. "Stable Channels with Gravel-Paved Beds," *Journal of Waterways and Harbors Division, A.S.C.E.*, February 1967, pp. 63–84.

KOUWEN, N., UNNY, T. E., and HILL, H. M. "Flow Retardance in Vegetated Channels," *Journal of the Irrigation and Drainage Division, A.S.C.E.*, June 1969, pp. 329–342.

LANE, E. W. "Design of Stable Channels," *Trans. A.S.C.E.*, Vol. 120 (1955), pp. 1234–1260.

LANE, E. W., LIN, P. N., and LIU, H. K. *The Most Efficient Stable Channel for Comparatively Clear Water in Non-Cohesive Material*. Fort Collins, Colorado: Colorado State University Research Foundation, 1959. 49 pp.

McHENRY, D. and GLOVER, R. E. "Boundary Shear and Velocity Distribution by Membrane Analogy, Analytical and Finite-Difference Methods." In *Sedimentation Studies in Open Channels* (O. J. Olsen and Q. L. Florey, eds.). U.S. Bureau of Reclamation Laboratory Report, No. Sp-34, August 5, 1952.

PARSONS, D. A. "Vegetative Control of Streambank Erosion." In *Proceedings of Federal Inter-Agency Sedimentation Conference*. Washington: U.S. Govt. Printing Office, 1965, pp. 130–136.

PARTHENIADES, EMMANUEL and PAASWELL, R. E. "Erodibility of Channels with Cohesive Boundary." *Journal of the Hydraulics Division, A.S.C.E.*, March 1970, pp. 755–771.

REE, W. O. "Hydraulic Characteristics of Vegetation for Vegetated Waterways," *Agricultural Engineering*, Vol. 30 (1949), pp. 184–189.

REE, W. O. and PALMER, V. J. *Flow of Water in Channels Protected by Vegetative Lining*. U.S. Soil Conservation Service, Technical Bulletin 967 (1949). Washington: U.S. Govt. Printing Office.

SIMONS, D. B., and HAMILTON, J. M. *Stability of Channels in Coarse, Non-Uniform Bed Material*. Report, Colorado State University, Fort Collins. 1969. 31 pp.

STILLWATER OUTDOOR HYDRAULIC LAB. *Handbook of Channel Design for Soil and Water Conservation*. U.S. Soil Conservation Service, SCS-TP-61 (rev. 1954). Washington: U.S. Govt. Printing Office.

TERREL, P. W. and BORLAND, W. M. "Design of Stable Canals and Channels in Erodible Material." *Trans. A.S.C.E.*, Vol. 123 (1958), pp. 101–115.

CHAPTER 13

Regime Theory

ACKERS, PETER. "Experiments on Small Streams in Alluvium," *Journal of the Hydraulics Division, A.S.C.E.*, July 1964, pp. 1–37.

BLENCH, THOMAS. *Mobile-Bed Fluviology*. Univ. of Alberta Press. 2nd ed., 1969.

BLENCH, THOMAS. "Coordination in Mobile-Bed Hydraulics," *Journal of the Hydraulics Division, A.S.C.E.*, November 1969, pp. 1871–1898.

BLENCH, THOMAS. *Regime Behavior of Canals and Rivers*. London: Butterworth Scientific Publications, 1957.

CHIEN, NING. "A Concept of the Regime Theory," *Trans. A.S.C.E.*, Vol. 122 (1957), pp. 785–793.

GILL, M. A. "Rationalization of Lacey's Regime Flow Equations," *Journal of the Hydraulics Division, A.S.C.E.*, July 1968, pp. 983–995.

HORTON, R. E. "Erosional Development of Streams and Their Drainage Basins," *Bulletin, Geological Soc. of America*, Vol. 56 (1945), pp. 275–370.

INGLIS, SIR CLAUDE. "The Behavior and Control of Rivers and Canals," *Res. Pub. Cent. Bd. Irrig., India*, No. 13 Simla, 1949.

KENNEDY, R. G. "The Prevention of Silting in Irrigation Canals," *Min. Proc. Instn. Civ. Engrs.*, Vol. 119, 1895.

LACEY, GERALD. "A General Theory of Flow in Alluvium," *Jour. Institute Civil Engrs.*, Paper 5515, Vol. 27, 1948.

LANGBEIN, W. B. "Geometry of River Channels," *Journal of the Hydraulics Division, A.S.C.E.*, March 1964, pp. 301–312.

LELIAVSKY, SERGE. *An Introduction to Fluvial Hydraulics*. London: Constable & Co., 1955.

LEOPOLD, L. B. and LANGBEIN, W. B. *The Concept of Entropy in Landscape Evolution*. U.S. Geological Survey Professional Paper 500-A, Washington: U.S. Govt. Printing Office, 1962. 20 pp.

LEOPOLD, L. B., WOLMAN, M. G., and MILLER, J. P. *Fluvial Processes in Geomorphology*. San Francisco: W. H. Freeman Co., 1964. 522 pp.

LEOPOLD, L. B. and MADDOCKS, T. "The Hydraulic Geometry of Stream Channels and Some Pyhsiographic Implications," U.S. Geological Survey Professional Paper 252, Washington: U.S. Government Printing Office, 1953.

MACKIN, J. H. "Concept of the Graded River," *Bulletin, Geological Society of America*, Vol. 59 (1948), pp. 463–512.

NEILL, C. R. and Galay, V. J. "Systematic Evaluation of River Regime," *Journal of Waterways and Harbors Division, A.S.C.E.*, February 1967, pp. 25–53.

ROGERS, F. C. and THOMAS, A. R. "Regime Canals," Section 6 in *Handbook of Applied Hydraulics* (C. V. Davis and K. E. Sorensen, eds.), New York: McGraw-Hill Book Co., Inc., 1969, 24 pp.

SIMONS, D. B. and ALBERTSON, M. L. "Uniform Water Conveyance Channels in Alluvial Material," *Journal of the Hydraulics Division*, A.S.C.E., May 1960, pp. 33–71.

WOLMAN, M. G. and BRUSH, L. M. *Factors Controlling the Size and Shape of Stream Channels in Coarse Non-Cohesive Sands.* Washington: U.S. Geological Survey Professional Paper 282-G, 1961. 29 pp.

Stream Meanders

ANDERSON, A. G. "On the Development of Stream Meanders," *Proc. Intl. Assoc. for Hydraulic Research*, June 1967, pp. 370–378.

EINSTEIN, H. A. and SHEN, H. W. "A Study on Meandering in Straight Alluvial Channels," *Journal of Geophysical Research*, December 15, 1964. pp. 5239–5247.

FRIEDKIN, JOSEPH F. "A Laboratory Study of the Meanderings of Alluvial Rivers," Report of Mississippi River Commission, U.S. Waterways Experiment Station, Vicksburg, Mississippi. May 1945.

LEOPOLD, LUNA B. and WOLMAN, M. G. "River Meanders," *Bulletin, Geological Society of America*, Vol. 71 (1960), pp. 769–794.

LEOPOLD, L. B. and LANGBEIN, W. B. "River Meanders," *Scientific American*, June 1966. pp. 60–70.

MAHARD, RICHARD H. "The Origin and Significance of Intrenched Meanders," *Journal of Geomorphology*, Vol. 5, pp. 32–44. 1942.

MORRIS, HENRY M., and WHITCOMB, J. C. *The Genesis Flood.* Nutley, N.J.: Presbyterian and Reformed Publ. Co., 1961. pp. 154–155, 318–324.

SCHEIDEGGER, A. E. "A Thermodynamic Analogy for Meander Systems," *Water Resources Research*, Vol. 3, 1967, pp. 1041–1046.

SCHUMM, S. A. "Meander Wavelength of Alluvial Rivers," *Science.* Sept. 29, 1967, pp. 1549–1550.

SHEN, H. W. and KOMURA, S. "Meandering Tendencies in Straight Alluvial Channels," *Journal of the Hydraulics Division*, A.S.C.E., July 1968. pp. 997–1016.

SHINDALA, A. and PRIEST, M. S. *The Meandering of Natural Streams in Alluvial Materials.* State College, Mississippi: Mississipi Water Resources Research Institute, 1968. 7 pp.

SURKAN, A. J. and VAN KAN, J. "Constrained Random Walk Meander Generation," *Water Resources Research.* December 1969, pp. 1343–1352.

TOEBES, G. H. and SOOKY, A. A. "Hydraulics of Meandering Rivers with Flood Plains," *Journal of Waterways and Harbors Division*, A.S.C.E., May 1967, pp. 213–236.

WERNER, P. E. "On the Origin of River Meanders," *Trans. Amer. Geophysical Union*, Vol. 32, December 1951, pp. 898–902.

YEN, C. L. "Bed Topography Effect on Flow in a Meander," *Journal of the Hydraulics Division*, A.S.C.E., January 1970, pp. 57–73.

River Modification and Stabilization

ANDERSON, A. G. "Hydraulic Design of Bridges for River Crossings," *Highway Research Record*, No. 123, 1966, pp. 1–16.

A.S.C.E. TASK COMMITTEE ON CHANNEL STABILIZATION WORKS "Channel Stabilization of Alluvial Rivers," *Journal of Waterways and Harbors Division*, A.S.C.E., February 1965, pp. 7–37.

BRADLEY, J. N. *Hydraulics of Bridge Waterways.* Washington: U.S. Bureau of Public Roads. 1960. 53 pp.

BROWN, CARL B. "Sediment Transportation," In *Engineering Hydraulics* (Hunter Rouse, ed.). New York: John Wiley & Sons, Inc., 1950, pp. 814–824.

FRANZIUS, OTTO. *Waterway Engineering.* (Transl. by Lorenz G. Straub.) Cambridge, Mass.: Technology Press, 1936.

HAAS, R. B. and WELLER, H. E. "Bank Stabilization by Revetments and Dikes," *Trans. A.S.C.E.*, Vol. 118 (1953), pp. 849–870.

HALES, Z. L., SHINDALA, A. and DENSON, K. E. "Riverbed Degradation Prediction," *Journal of Water Resources Research*, April 1970, pp. 549–556.

KOMURA, SABURO. "Equilibrium Depth of Scour in Long Constrictions," *Journal of the Hydraulics Division, A.S.C.E.*, September 1966, pp. 17–37.

KOMURA, S. and SIMONS, D. B. "River-Bed Degradation Below Dams," *Journal of the Hydraulics Division, A.S.C.E.*, July 1967, pp. 1–14.

LAURSEN, E. M. "Scour at Bridge Crossings," *Transactions Amer. Soc. Civil Engrs.*, Vol. 127, 1962, pp. 166–209.

LAURSEN, E. M. "Some Aspects of the Problem of Scour at Bridge Crossings," In *Proceedings of Federal Inter-Agency Sedimentation Conference*. Washington: Agricultural Research Service, 1965. pp. 304–309.

MATTHES, GERARD H. "Mississippi River Cut-offs," *Trans. A.S.C.E.*, Vol. 113 (1948) pp. 1–39.

SCHUMM, S. A. "River Metamorphosis," *Journal of the Hydraulics Division, A.S.C.E.*, January 1969, pp. 255–273.

SHEN, H. W., SCHNEIDER, V. R., and KARAKI, S. "Local Scour Around Bridge Piers," *Journal of the Hydraulics Division, A.S.C.E.*, November 1969, pp. 1919–1940.

STRATTON, H. J., DOUMA, J. H., and DAVIS, J. P. "Navigation Systems," Section 31 in *Handbook of Applied Hydraulics* (C. V. Davis and K. E. Sorensen, eds.) New York: McGraw-Hill Book Co., Inc., 1969. 38 pp.

TINNEY, E. R. "The Process of Channel Degradation," *Journal of Geophysical Research*, April 1962, pp. 1475–1480.

Navigation Locks

BROWN, F. R. "End Filling and Emptying Systems for Locks," *Journal of Waterways and Harbors Division, A.S.C.E.*, February 1964, pp. 61–77.

CABELKA, J. "Investigation of Various Types of Lock Filling," In *Report on Second Meeting*. Stockholm: International Association for Hydraulic Structure Research, 1948, pp. 461–476.

DAVIS, J. P. and MURPHY, T. E. "Experimental Research on Lock Hydraulic Systems," *Journal of Waterways and Harbors Division, A.S.C.E.*, February 1966, pp. 17–31.

GRIFFIN, A. F., BLEC, C. E., and BLOOR, R. L. "Design Characteristics of Lock Systems in the United States: A Symposium," *Trans. A.S.C.E.*, Vol. 116 (1951), pp. 829–890.

NELSON, M. E. and JOHNSON, H. J., "Navigation Locks: Filling and Emptying Systems for Locks," *Journal of Waterways and Harbors Division, A.S.C.E.*, February 1964, pp. 47–59.

PARISET, E. and GAGNON, A., "High Lift Locks: Some Hydraulic Problems and Solutions," *Journal of Waterways and Harbors Division, A.S.C.E.*, November 1968, pp. 55–75.

RICH, GEORGE R., *Hydraulic Transients*. New York: Dover Publications. 2nd ed., 1963, pp. 311–334.

RICH, GEORGE R., "Navigation Locks," Section 32, in *Handbook of Applied Hydraulics*. Ed. by C. V. Davis and K. E. Sorensen. New York: McGraw-Hill Book Co., Inc., 1969, 20 pp.

RICHARDSON, G. C., "Navigation Lock Gates and Valves," *Journal of Waterways and Harbors Division, A.S.C.E.*, February 1964, pp. 79–102.

U.S. CORPS OF ENGINEERS. "Navigation Locks," Part 116, Ch. 4 in *Engineering Manual, Civil Works Construction*. Washington: U.S. Govt. Printing Office, 1956. 64 pp.

CHAPTER 14

General

AMERICAN SOCIETY OF CIVIL ENGINEERS. *Coastal Engineering*, Vols. I and II. New York: A.S.C.E., 1969. 1585 pp.

IPPEN, A. I. (Ed.). *Estuary and Coastline Hydrodynamics*, New York: McGraw-Hill Book Co., Inc., 1966. 744 pp.

QUINN, A. D. *Design and Construction of Ports and Marine Structures*, New York: McGraw-Hill Book Co., Inc., 1961. 531 pp.

U.S. ARMY CORPS OF ENGINEERS. *Shore Protection, Planning and Design*. Washington, Superintendent of Documents, 1961. 392 pp.

WIEGEL, R. L. *Oceanographical Engineering*. Englewood Cliffs, N.J.: Prentice-Hall, Inc., 1964. 532 pp.

Wave Theory

AMERICAN SOCIETY OF CIVIL ENGINEERS. "Wave Theory and Measurements," Part I in *Coastal Engineering*. New York: A.S.C.E., 1969, pp. 1–272.

BORGMAN, L. E. "Ocean Wave Simulation for Engineering Design," *Journal of Waterways and Harbors Division, A.S.C.E.*, November 1969, pp. 557–583.

BRETSCHNEIDER, CHAS. L. "Hurricane Design-Wave Practices," *Trans. A.S.C.E.*, Vol. 124 (1959), pp. 39–62.

CHAPPELEAR, J. E. "Shallow-Water Waves," *Journal of Geophysical Research*, Vol. 67, November 1962, pp. 4693–4704.

DAILY, J. W. and STEPHAN, JR., S. C. "Characteristics of the Solitary Wave," *Trans. A.S.C.E.*, Vol. 118 (1953), pp. 575–587.

DEAN, R. G. "Relative Validities of Water-Wave Theories," *Journal of Waterways and Harbors Division, A.S.C.E.*, February 1970, pp. 105–118.

DRONKERS, J. J. *Tidal Computations*. New York: John Wiley and Sons, 1964, 518 pp.

KEULEGAN, G. H. "Wave Motion," In *Engineering Hydraulics* (Hunter Rouse, ed.). New York: John Wiley & Sons, Inc., 1950. pp. 711–745.

MASON, MARTIN A. *A Study of Progressive Oscillatory Waves in Water*. Washington: Beach Erosion Board Technical Report No. I, 1942. 39 pp.

MASON, MARTIN A. "Surface Water Wave Theories," *Trans. A.S.C.E.*, Vol. 118 (1953), pp. 546–574.

MILNE-THOMPSON, L. M. *Theoretical Hydrodynamics*. 2nd ed. New York: The Macmillan Co., 1950. pp. 351–401.

O'BRIEN, M. P. et al. *A Summary of the Theory of Oscillatory Waves*. Washington: Beach Erosion Board Technical Report No. 2, 1942. 43 pp.

STOKER, J. J. *Water Waves*. New York: Interscience, 1957. pp. 1–68.

SVERDRUP, H. U., and MUNK, W. H. *Wind, Sea and Swell: Theory of Relations for Forecasting*. Publication 601 (1947). Washington: U.S. Navy Hydrographic Office.

U.S. DEPARTMENT OF COMMERCE. *Gravity Waves*. Washington: U.S. Govt. Printing Office. Natl. Bureau of Standards Circular 521, 1952. 287 pp.

Wave Transformation, Refraction and Diffraction

BEITINJANI, K. I. and BRATER, E. F. "A Study on Refraction of Waves in Prismatic Channels," *Journal of Waterways and Harbors Division, A.S.C.E.*, August 1965, pp. 37–64.

BLUE, F. L., Jr. and JOHNSON, J. W. "Diffraction of Water Waves Passing Through a Breakwater Gap," *Trans. Amer. Geophysical Union*, Vol. 29, 1948, pp. 704–718.

CAMFIELD, F. E., and STREET, R. L. "Shoaling of Solitary Waves on Small Slopes," *Journal of Waterways and Harbors Division, A.S.C.E.*, February 1969, pp. 1–22.

CARR, J. H. and STELZRIEDE, M. E. *Diffraction of Water Waves by Breakwaters*. Washington: Natl. Bur. Standards Circular 521, 1952. pp. 109–125.

DUNHAM, J. W. "Refraction and Diffraction Diagrams," *Proceedings First Conference on Coastal Engineering*, Council on Wave Research, Engineering Foundation, 1951, pp. 33–49.

FREEMAN, J. C. and LeMEHAUTE, B. "Wave Breakers on a Beach and Surges on a Dry Bed," *Journal of the Hydraulics Division, A.S.C.E.*, March 1964, pp. 187–216.

JOHNSON, J. W. "The Refraction of Surface Waves by Currents," *Trans. Amer. Geophysical Union*, Vol. 28 (1947), pp. 867–874.

JOHNSON, J. W. "Generalized Wave Diffraction Diagrams," *Proceedings Second Conference on Coastal Engineering*, Council on Wave Research, Engineering Foundation, 1952, pp. 6–23.

JOHNSON, J. W. "Engineering Aspects of Diffraction and Refraction," *Trans. A.S.C.E.*, Vol. 118 (1953), pp. 617–652.

KARLSSON, THORBJORN. "Refraction of Continuous Ocean Wave Spectra," *Journal of Waterways and Harbors Division, A.S.C.E.*, November 1969, pp. 437–448.

MONKMEYER, P. L. and MURRAY, W. A. "Travel Time for Waves Moving over a Sloping Bottom," *Journal of Waterways and Harbors Division, A.S.C.E.*, August 1968, pp. 389–396.

PUTMAN, J. A. and ARTHUR, R. S. "Diffraction of Water Waves by Breakwaters," *Trans. Amer. Geophysical Union*, Vol. 29 (1948), pp. 481–490.

WIEGEL, R. L. "Diffraction of Waves by Semi-Infinite Breakwater," *Journal of the Hydraulics Division, A.S.C.E.*, January 1962, pp. 27–44.

YU, Y-YUAN. "Breaking of Waves by an Opposing Current," *Trans. Amer. Geophysical Union*, Vol. 33 (1952), pp. 39–41.

Wave Reflection and Run-Up

BRUUN, PER. "Destruction of Wave Energy by Vertical Walls," *Journal of Waterways Division, A.S.C.E.*, March 1956, Paper 912, 13 pp.

CALDWELL, JOSEPH M. *Reflection of Solitary Waves*. U.S. Engineers, Beach Erosion Board Tech. Mem. 11 (1949), 19 pp., 16 pl.

HUNT, IRA A., JR. "Design of Seawalls and Breakwaters," *Trans. A.S.C.E.*, Vol. 126, Part IV, 1961, pp. 542–570.

LeMEHAUTE, B., KOH, R. C. Y., and HWANG, L. S. "A Synthesis on Wave Run-Up," *Journal of Waterways and Harbors Division, A.S.C.E.*, February 1968, pp. 77–92.

MICHE, M. "The Reflecting Power of Maritime Works Exposed to Action of the Swell," *Annales des Ponts et Chaussées*, Paris, May–June 1951.

SAVAGE, RUDOLPH P. *Laboratory Data on Wave Run-up on Roughened and Permeable Slopes*. U.S. Engineers, Beach Erosion Board, Tech. Mem. 109 (1959), 28 pp.

SAVAGE, RUDOLPH P. "Wave Run-Up on Roughened and Permeable Slopes," *Trans. A.S.C.E.*, Vol. 124 (1959), pp. 852–870.

SAVILLE, THORNDIKE, JR. "Wave Run-Up on Shore Structures," *Trans. A.S.C.E.*, Vol. 123 (1958), pp. 139–150.

Wave Forces on Coastal Structures

AMERICAN SOCIETY OF CIVIL ENGINEERS. "Coastal Structures," Part 3 in *Coastal Engineering*. New York: A.S.C.E., 1969, pp. 745–1212.

BORGMAN, L. E. "Spectral Analyses of Ocean Wave Forces on Piling," *Journal of Waterways and Harbors Division, A.S.C.E.*, May 1967, pp. 129–156.

BRETSCHNEIDER, C. L. "Probability Distribution of Wave Force," *Journal of Waterways and Harbors Division, A.S.C.E.*, May 1967, pp. 5–26.

CARR, JOHN H. *Breaking Wave Forces on Plain Barriers*. Navy Department, Bureau of Yards and Docks, Rept. E-11.3 (1954).

CHAPPELEAR, J. E. "Wave Forces on Groups of Vertical Cylinders," *Journal of Geophysical Research*, Vol. 64 (1959), pp. 199–208.

HUDSON, ROBERT Y. "Wave Forces on Breakwaters," *Trans. A.S.C.E.*, Vol. 118 (1953), pp. 653–685.

HUDSON, ROBERT Y. "Laboratory Investigation of Rubble-Mound Breakwaters," *Trans. A.S.C.E.*, Vol. 126 (1961), Part IV, pp. 492–541.

IRIBARREN, C. R. "Generalization of the Formula for Calculation of Rockfill Dikes and Verification of its Coefficients," *Bulletin, U.S. Beach Erosion Board*, Vol. 5 (1951), pp. 4–24.

JEN, YUAN. "Laboratory Study of Inertia Forces on a Pile," *Journal of Waterways and Harbors Division, A.S.C.E.*, February 1968, pp. 59–76.

LEENDERTSE, J. J. *Forces Induced by Breaking Waves on a Vertical Wall.* Pt. Hueneme, California: Tech Report 092, U.S. Naval Civil Engineering Laboratory, 1962.

MINIKIN, R. R. *Winds, Waves and Maritime Structures.* London: Chas. Griffin and Co., Ltd., 1950. 216 pp.

NAGAI, S. "Stable Concrete Blocks on Rubble-Mound Breakwaters," *Journal of the Waterways and Harbors Division, A.S.C.E.*, August 1962, pp. 85–115.

NAGAI. S. "Pressures of Standing Waves on Vertical Wall," *Journal of Waterways and Harbors Division, A.S.C.E.*, February 1969, pp. 53–76.

PIERSON, W. J., and HOLMES, P. "Irregular Wave Forces on a Pile," *Journal of Waterways and Harbors Division, A.S.C.E.*, November 1965, pp. 1–10

SAINFLOU, M. *Essay on Vertical Breakwaters.* (Transl. by C. R. Hatch.) U.S. Engineers 1938, 49 pp.

SIGURDSSON, GUNNAR. "Wave Forces on Breakwater Capstones," *Journal of Waterways and Harbors Division, A.S.C.E.*, August 1962, pp. 27–60.

SVEE, ROALD. "Formulas for Design of Rubble-Mound Breakwaters," *Journal of the Waterways and Harbors Division, A.S.C.E.*, May 1962, pp. 11–21.

U.S. BEACH EROSION BOARD. *Shore Protection Planning and Design.* Tech. Rpt. No. 4, Washington: U.S. Govt. Printing Office, 1961. pp. 116–139.

U.S. ENGINEERS. *Design of Breakwaters and Jetties.* Washington: U.S. Govt. Printing Office, 1957. 30 pp., 42 pl.

WIEGEL, R. L., BEEBE, K. E., and MOON, JAMES. "Ocean Wave Forces on Circular Cylindrical Piles," *Trans. A.S.C.E.*, Vol. 124 (1959), pp. 89–116.

Coastal Sedimentation

AMERICAN SOCIETY OF CIVIL ENGINEERS. "Coastal Sediment Problems," Part 2 in *Coastal Engineering.* New York: A.S.C.E., 1969, pp. 273–744.

BAGNOLD, R. A. "Sand Movement by Waves," *Journal Institute of Civil Engrs.*, Vol. 27, 1947, p. 447.

BASCOM, WILLARD. "Beaches," *Scientific American*, Vol. 203, August 1960, pp. 81–94.

BERG, D. W. and WATTS, G. M. "Variations in Groin Design," *Journal of the Waterways and Harbors Division, A.S.C.E.*, May 1967, pp. 79–100.

BOWEN, A. J. "Rip Currents," *Journal of Geophysical Research*, October 20, 1969, pp. 5467–5490.

BRUUN, PER. *Coast Stability: Forms of Equilibrium of Coasts with a Littoral Drift* Ph.D. thesis, University of Copenhagen, 1954.

BRUUN PER and GERRITSEN, FRANS. "Stability of Coastal Inlets," *Trans. A.S.C.E.*, Vol. 125 (1960), pp. 1228–1265.

CALDWELL, J. M. *Shore Erosion by Storm Waves.* Washington: U.S. Beach Erosion Board, 1959. 17 pp.

EAGLESON, P. S., GLENNE, R., and DRACUP, J. A. "Equilibrium Characteristics of Sand Beaches," *Journal of the Hydraulics Division, A.S.C.E.*, January 1963, pp. 35–57.

KALKANNIS, GEORGE. *Transportation of Bed Material Due to Wave Action.* Washington: Coastal Engineering Research Center, 1964. 68 pp.

U.S. DEPARTMENT OF AGRICULTURE. "Sedimentation in Estuaries, Harbors and Coastal Areas," Symposium 3 in *Proceedings of Federal Inter-Agency Sedimentation Conference.* Washington: Agricultural Research Service, 1965, pp. 593–776.

APPENDIX

Physical Properties of Sea Water

Temperature °F	Specific Weight lbs/ft³	Density slugs/ft³	Dynamic Viscosity lb-sec/ft³ × 10⁵
32	64.18	1.995	
35	64.17	1.994	
40	64.15	1.994	3.416
45	64.13	1.993	3.146
50	64.10	1.992	2.908
55	64.07	1.991	2.700
60	64.04	1.990	2.480
65	64.00	1.989	2.351
70	63.95	1.988	2.205
75	63.90	1.986	2.071
80	63.85	1.984	1.950
85	63.79	1.983	1.842

Largely compiled from the Manual of Engineering Practice No. 25, *Hydraulic Models*, American Society of Civil Engineers, New York, N.Y. 1942.

Physical Properties of Fresh Water

Temperature (°F)	Specific Weight (lbs/ft³)	Density (slugs/ft³)	Dynamic Viscosity (lb-sec/ft²) × 10⁵	Vapor Pressure (ft of water)	Surface Tension (lbs/ft × 10³)
32	62.42	1.940	3.746	0.186	5.18
35	62.42	1.940	3.536	0.230	5.17
40	62.43	1.940	3.229	0.281	5.14
45	62.42	1.940	2.965	0.340	5.12
50	62.40	1.940	2.735	0.411	5.09
55	62.38	1.939	2.534	0.494	5.06
60	62.35	1.938	2.359	0.591	5.04
65	62.34	1.937	2.196	0.705	5.01
70	62.29	1.936	2.050	0.838	4.98
75	62.25	1.935	1.918	0.993	4.95
80	62.20	1.934	1.799	1.17	4.92
85	62.16	1.932	1.692	1.38	4.89
90	62.11	1.931	1.595	1.61	4.86
95	62.05	1.939	1.505	1.88	4.83
100	62.00	1.927	1.424	2.19	4.80
110	61.92	1.923	1.284	2.95	4.74
120	61.73	1.918	1.168	3.91	4.67
130	61.54	1.913	1.069	5.13	4.60
140	61.38	1.908	0.981	6.67	4.54
150	61.20	1.902	0.905	8.58	4.47
160	61.01	1.896	0.838	10.95	4.40
170	60.79	1.890	0.780	13.83	4.34
180	60.57	1.883	0.726	17.33	4.27
190	60.35	1.876	0.678	21.55	4.20
200	60.13	1.868	0.637	26.59	4.13
212	59.83	1.860	0.593	32.58	4.04

Largely compiled from the Manual of Engineering Practice No. 25, *Hydraulic Models*, American Society of Civil Engineers, New York, N.Y. 1942.

INDEX